T0224542

THE CAMBRIDGE COMPANION TO
EINSTEIN

This volume is the first systematic presentation of the work of
Albert Einstein, comprised of fourteen essays by leading historians and philosophers of science that introduce readers to his work.
Following an introduction that places Einstein's work in the context of his life and times, the book opens with essays on the papers
of Einstein's "miracle year," 1905, covering Brownian motion, light
quanta, and special relativity, as well as his contributions to early
quantum theory and the opposition to his light quantum hypothesis. Further essays relate Einstein's path to the general theory of
relativity (1915) and the beginnings of two fields it spawned, relativistic cosmology and gravitational waves. Essays on Einstein's
later years examine his unified field theory program and his critique of quantum mechanics. The closing essays explore the relation between Einstein's work and twentieth-century philosophy, as
well as his political writings.

Michel Janssen is Professor in the Program in the History of Science,
Technology, and Medicine at the University of Minnesota. Before
coming to Minnesota, he was a member of the editorial team of
the Einstein Papers Project, then at Boston University. He has published extensively on the relativity and quantum revolutions of the
early twentieth century.

Christoph Lehner is Research Scholar at the Max Planck Institute
for the History of Science and the coordinator of its project on the
History and Foundations of Quantum Mechanics. He was an editor at the Einstein Papers Project at Boston University and at the
California Institute of Technology, as well as a scientific advisor
for the 2005 exhibit "Einstein, Chief Engineer of the Universe" in
Berlin. The focus of his research is the history and philosophy of
quantum theory.

Courtesy of the Albert Einstein Archives, the Hebrew University of Jerusalem, Israel.

The Cambridge Companion to
EINSTEIN

Edited by

Michel Janssen
University of Minnesota

Christoph Lehner
Max Planck Institute for the History of Science

CAMBRIDGE
UNIVERSITY PRESS

CAMBRIDGE
UNIVERSITY PRESS

University Printing House, Cambridge CB2 8BS, United Kingdom

One Liberty Plaza, 20th Floor, New York, NY 10006, USA

477 Williamstown Road, Port Melbourne, VIC 3207, Australia

4843/24, 2nd Floor, Ansari Road, Daryaganj, Delhi - 110002, India

79 Anson Road, #06-04/06, Singapore 079906

Cambridge University Press is part of the University of Cambridge.

It furthers the University's mission by disseminating knowledge in the pursuit of education, learning and research at the highest international levels of excellence.

www.cambridge.org
Information on this title: www.cambridge.org/9780521535427

First published 2014
Reprinted 2014

A catalogue record for this publication is available from the British Library

Library of Congress Cataloging in Publication data
The Cambridge companion to Einstein / [edited by] Michel Janssen,
University of Minnesota, Christoph Lehner, Max Planck Institute for the
History of Science.
    volumes ; cm. – (Cambridge companions to philosophy)
Includes bibliographical reference and index.
ISBN 978-0-521-82834-5 (hardback : alk. paper : v. 1)
1. Einstein, Albert, 1879–1955.   2. Physicists–Biography.
3. Physics–History–20th century.   I. Janssen, Michel,
1960– editor of compilation.   II. Lehner, Christoph,
1962 editor of compilation.
QC16.E5C36   2014
530.092–dc23   [B]
2013049485

ISBN 978-0-521-82834-5 Hardback
ISBN 978-0-521-53542-7 Paperback

The editors dedicate this volume to the memory of Gregory Swain Nelson (16 May 1964–23 September 2012).

Where the world ceases to be the arena of personal hopes, wishes and wills, where we face it as free beings, admiring, questioning, contemplating, there we enter into the realm of art and science.

Albert Einstein, "The common element in artistic and scientific experience," *Menschen* 4 (1921)

Contents

Contributors

OLIVIER DARRIGOL is a Research Director at CNRS in Paris. He is the author of four books concerning the histories of quantum theory, electrodynamics, hydrodynamics, and optics.

MICHAEL FRIEDMAN is Frederick P. Rehmus Family Professor of Humanities and Director of the Patrick Suppes Center for the History and Philosophy of Science at Stanford University. He has worked on the philosophical foundations of relativity theory and the reception of Einstein's theories in logical empiricism.

DON HOWARD is Professor of Philosophy and Director of the Reilly Center for Science, Technology, and Values at the University of Notre Dame. A former Assistant and Contributing Editor for the *Collected Papers of Albert Einstein*, he has written extensively on Einstein's philosophy of science and on a wide array of topics in the history and philosophy of late-nineteenth- and early-twentieth-century physics as well as the history of the philosophy of science.

MICHEL JANSSEN was an editor at the Einstein Papers Project and is now Professor in the Program in the History of Science, Technology, and Medicine at the University of Minnesota. He is also a regular visitor at the Max Planck Institute for the History of Science in Berlin. His research focuses on the genesis of relativity theory and quantum mechanics.

DANIEL J. KENNEFICK is Assistant Professor of Physics at the University of Arkansas and contributing editor at the Einstein Papers Project. He studies the history of general relativity, in particular gravitational wave theory, and is the author of *Traveling at the Speed of Thought: Einstein and the Quest for Gravitational Waves*.

A. J. KOX is Pieter Zeeman Professor of History of Physics, Emeritus, at the University of Amsterdam and Senior Editor at the Einstein Papers Project (California Institute of Technology). His research is on the

xi

history of nineteenth- and twentieth-century physics, with a special emphasis on The Netherlands and the work of H. A. Lorentz and his contemporaries.

CHRISTOPH LEHNER was an editor at the Einstein Papers Project and is now Research Scholar at the Max Planck Institute for the History of Science in Berlin. His research focuses on the history and philosophy of modern physics, especially quantum mechanics.

JOHN D. NORTON is Professor in the Department of History and Philosophy of Science and Director of the Center for Philosophy of Science, University of Pittsburgh. He works in the history of physics, with a special emphasis on Einstein's discovery of special and general relativity; and also on general topics in philosophy of science. He has been a contributing editor to the *Collected Papers of Albert Einstein* and is a member of its Executive Committee.

JÜRGEN RENN was an editor at the Einstein Papers Project and is now Director of Department I at the Max Planck Institute for the History of Science in Berlin. He is also honorary professor at Humboldt-Universität and at Freie Universität Berlin as well as Adjunct Professor for Philosophy and Physics at Boston University. His research focuses on structural changes in systems of knowledge, particularly in the natural sciences. He is coinitiator of the Berlin Declaration on Open Access to Knowledge in the Sciences and Humanities launched by the Max Planck Society in 2003 and publisher of the innovative book series Edition Open Access.

THOMAS RYCKMAN teaches philosophy of science and philosophy of physics at Stanford University. Among his most recent works are *The Reign of Relativity: Philosophy in Physics 1915–1925* and *Einstein (The Routledge Philosophers)*, coauthored with Arthur Fine.

ROBERT RYNASIEWICZ is Professor of Philosophy at the Johns Hopkins University and works in the history and philosophy of science. Much of his research has focused on Newton and Einstein.

TILMAN SAUER is a lecturer in history and philosophy of science at the University of Bern and an associate editor with the Einstein Papers Project (California Institute of Technology).

ROBERT SCHULMANN is former Director of the Einstein Papers Project. His research focuses on early-twentieth-century history of science from a cultural perspective.

CHRISTOPHER SMEENK is Associate Professor of Philosophy at the University of Western Ontario. His research focuses on history and philosophy of physics, in particular on methodological issues in the development of modern cosmology.

ROGER H. STUEWER is Professor Emeritus of the History of Science and Technology at the University of Minnesota and co–editor-in-chief of the journal *Physics in Perspective*. His research is on the history of twentieth-century physics, especially the history of quantum and nuclear physics. He was awarded the 2013 Abraham Pais Prize for History of Physics.

Preface

MICHEL JANSSEN AND CHRISTOPH LEHNER

Most volumes in the *Cambridge Companion* series deal with philosophers. Following volumes on Galileo, Newton, and Darwin, this is the
fourth *Companion* devoted to a major scientist. The inclusion of figures such as Galileo, Newton, and Einstein in this series reminds us
that natural philosophy traditionally included what we today call physics, and that up to the middle of the twentieth century a clear border
between physics and philosophy did not exist. Few would dispute that
Einstein was the greatest natural philosopher of the twentieth century
in this traditional sense. Not only was he centrally responsible for the
formulation of the two most important fundamental theories of modern
physics, the theory of relativity and quantum theory, he also devoted
considerable effort to explaining and defending his views on the epistemology and methodology of physics. His writings have had an enormous
impact on the development of philosophy of science in the twentieth
century, and beyond that on analytic philosophy more generally. Many
of the philosophers relevant for the rise of analytic philosophy in the
first half of the twentieth century, especially in the German-speaking
countries, such as Moritz Schlick, Hans Reichenbach, Rudolf Carnap, or
Karl Popper, were concerned with interpreting and developing Einstein's
work in a general philosophical context.

This volume is meant to provide an introduction to Einstein's work
that is comprehensive and accessible to the general reader. Most of the
chapters in this volume deal with Einstein's pathbreaking contributions
to physics, in relativity theory, quantum theory, and statistical physics.
However, there are also several chapters on Einstein's reflections on
the foundations of physics (especially quantum mechanics), scientific
methodology, epistemology, and politics. In the introduction to this volume, we provide a more detailed guide to its contents. Here we want to
acknowledge some of the more important debts we accrued in putting
together this *Companion*.

The volume has been a long time in the making. We appreciate the
patience and forbearance of our contributors and the editors at the Press.
Don Howard first suggested that the two of us edit *The Cambridge*

Companion to Einstein. George E. Smith gave us useful advice early on based on his own experience editing *The Cambridge Companion to Newton.* The reader will notice that many of the authors who contributed to this volume, including its two editors, have been involved in one capacity or another with the Einstein Papers Project. The editorial team of this project, now based at the California Institute of Technology, is responsible for the publication of *The Collected Papers of Albert Einstein,* thirteen volumes of which have appeared so far. We are grateful to Diana Kormos Buchwald, the Director of the Einstein Papers Project, for her unwavering support of work on this volume. The Max Planck Institute for the History of Science in Berlin is another major site for Einstein studies. The production of this volume has been part of the activity of Department I of the Institute in this area. We are therefore extremely grateful to Jürgen Renn, its Director, for his generous support and constant encouragement. The cumulative bibliography and the index for this volume were prepared by scholars and staff at the Max Planck Institute. We thank Martin Jähnert and Lindy Divarci for their meticulous work on the bibliography, and Irene Colantoni, Chandhan Srinivasamurthy, and Ross Fletcher for helping us compile the index.

At various points we benefited from the advice of two senior Einstein scholars, Robert Schulmann and John Stachel, though they bear no responsibility for what we did or failed to do with it. One of us (MJ) would like to thank his colleagues in the Program in the History of Science, Technology, and Medicine at the University of Minnesota, especially the program's former director, Alan E. Shapiro. We thank Laurent Taudin for drawing the figures for several chapters and the appendix of this volume. We also thank the staff at Cambridge University Press, especially Beatrice Rehl, Asya Graf, Christine A. T. Dunn, and Emily Spangler, for shepherding our manuscript through the production process.

Finally, we want to express our gratitude to the Hebrew University of Jerusalem for granting us permission to quote from material in the Albert Einstein Archives and for allowing us to reproduce two pictures of Einstein, one on the cover and one as a frontispiece. We thank Chaya Becker of the Albert Einstein Archives for suggesting these two pictures to us. Both were taken in Berlin in 1928. The first somewhat blurry one, used as the frontispiece, shows Einstein squinting as if deep in thought, holding his pen ready to record the solution to whatever riddle he is contemplating in the notebook on his lap. The second one, used on the cover, appears to have been taken moments later, is in sharp focus, and shows Einstein fully relaxed, approvingly looking, it seems, at what he just jotted down in his notebook. This pair of images nicely captures the spirit of the man to whose work this volume is devoted.

Introduction

This volume brings together fourteen essays by historians and philosophers of science on various aspects of the writings of Albert Einstein. Together they are meant to provide a guide to Einstein's work and the extensive literature about it. The essays can be read independently of one another, though most of them gain from being read in conjunction with others. All of them should be accessible to a broad audience. The use of equations, for instance, has been kept to a minimum throughout this volume. The first ten essays deal with Einstein's contributions to physics and with various philosophical implications of these contributions. The next three essays directly address some of Einstein's more philosophical writings and the impact of his work on the twentieth-century philosophy of science. The final essay is on Einstein's political writings. In this introduction we give a brief overview of Einstein's life and career to provide some context for this collection of essays and highlight some themes addressed more fully in the individual contributions to this volume.

Albert Einstein (1879–1955) was born in the Swabian city of Ulm, the first child of upwardly mobile Jewish parents, Hermann and Pauline (née Koch).[1] In 1880, Hermann's featherbed business failed and he moved his family to Munich, where with one of his brothers he started a gas and water installation business. In 1885, they founded an electrotechnical factory. Growing up around dynamos and electromotors, Einstein developed an early interest in electrodynamics, the field in which he would develop his special theory of relativity. In his "Autobiographical Notes,"[2] he recalled two other experiences that drew him to science at an early age: being shown a compass by his father when he was four or five years old and reading a book on Euclidean geometry at the age of twelve (Einstein 1949a, 9).

In 1894, when the family business was faltering, Einstein's father and uncle moved their factory to Pavia in Italy. His parents and his only sibling, his younger sister Maja,[3] moved to Milan and then to Pavia, while Einstein stayed behind in Munich to finish high school at the Luitpold *Gymnasium*. He soon dropped out, however, and joined his family in

1

Pavia. In October 1895, at the age of sixteen, he traveled to Switzerland to take the entrance exam to the Federal Polytechnic, now known as the *Eidgenössische Technische Hochschule* (ETH), in Zurich. Although he did well on the science and mathematics portions of the exam, he failed the exam overall (CPAE 1, 10–12). After completing his secondary education at the Aargau Cantonal School in Aarau,[4] he was admitted to the ETH the following year, still only seventeen, and began his studies to become a high school mathematics and physics teacher. Among his classmates were Marcel Grossmann, on whose notes he relied to pass exams as he frequently skipped class, and Mileva Marić, who would become his first wife.

In 1900, Einstein graduated fourth in a class of five. Initially, he could only find employment as a substitute teacher and a private tutor. With the help of Grossmann he eventually landed a job at the Patent Office in Bern. In June 1902, he took up a position there as patent examiner third class. He had become a Swiss citizen the year before, after having renounced his German citizenship back in 1896. Meanwhile, Marić had given birth to the couple's first child. It appears that this child, a girl they named Lieserl, was given up for adoption and died in childhood, but exactly what happened to her has never been established.[5] Marić had been struggling at the ETH and the pregnancy effectively put an end to her studies. In January 1903, the couple finally married. Einstein's parents had strongly opposed the match. His father had been on his deathbed in October 1902, broken by a string of business failures, when he had finally relented and given his consent.[6] The couple would have two more children, two sons, Hans Albert and Eduard.[7] The scientific partnership they had envisioned when they were both students at the ETH never materialized (Stachel 1996). Marić did not become a member of the mock Olympia Academy that Einstein formed around this time with his friends Maurice Solovine and Conrad Habicht to discuss readings of shared interest, mostly in philosophy. It is not clear whether Marić participated in these discussions even though they were sometimes held at her and Einstein's own apartment.[8]

Thanks to a list by Solovine (Einstein 1956b, viii), we have a record of the readings of the Olympia Academy, which included works by Baruch Spinoza,[9] David Hume, John Stuart Mill, Hermann von Helmholtz, Ernst Mach, and Henri Poincaré. As a teenager, at the recommendation of Max Talmud (later changed to Talmey), a medical student who regularly dined with the Einstein family, he had read Immanuel Kant's *Critique of Pure Reason*. Einstein would make creative use of the ideas of these authors in his own thinking – of Hume and Mach, for instance, in the development of special relativity (Norton 2010) and of Mach in the development of general relativity (see Chapter 6).

Another friend with whom Einstein discussed scientific and philosophical matters during his early years in Berne was Michele Besso. Their lifelong friendship began while they were both students at the ETH. In 1904, on Einstein's recommendation, Besso joined Einstein at the Patent Office. Besso became an important "sounding board" for Einstein's developing ideas. The 1905 paper introducing special relativity famously has no references but acknowledges the help of one person – Besso.

In the years 1902–4, Einstein published three papers on statistical mechanics, now sometimes referred to as the "statistical trilogy" (Einstein 1902b, 1903, 1904; analyzed in Renn 2005c and Uffink 2006). Many of the results presented in these papers had been found earlier by Ludwig Boltzmann and Josiah Willard Gibbs. At the time, Einstein only knew some of Boltzmann's work and none of Gibbs's. Even though Einstein's results were not new, they played an important role in his subsequent work. The interpretation of probabilities as time averages led him to consider fluctuations, which became the central tool in his attempts to understand microphysics. In the final installment of the statistical trilogy, Einstein finally did something highly original. He applied the formalism developed to deal with fluctuations in gases to fluctuations in heat or black-body radiation, setting the stage for some of his signature contributions to quantum theory in the years ahead.

Then, seemingly out of nowhere, came the papers of the *annus mirabilis*, Einstein's year of miracles.[10] Other than the statistical trilogy, there are few sources to document the development of Einstein's thought leading up to them. Largely on the basis of scattered clues in his correspondence with Marić and later recollections of both Einstein and Besso, his main scientific confidant at the time, Robert Rynasiewicz and Jürgen Renn (2006) have speculated that prior to 1905 Einstein was trying to lay a new atomistic foundation for all of physics. By 1905, they argue, Einstein had come to realize that this attempt was premature. The papers of his miracle year, on this view, should be seen as those parts of a much larger effort that Einstein felt were ready to be presented to the scientific community, each establishing some secure "fixed point from which to carry on" (Rynasiewicz and Renn 2006, 6).

In Chapter 1, expanding on their earlier joint paper, Renn and Rynasiewicz discuss the various strands leading to the papers of the *annus mirabilis* as well as the connections between those strands and papers. Using the notion of what Renn (2006) has called a "Copernicus process," they show how Einstein's innovations of 1905 consisted largely of a reconfiguration of existing bodies of knowledge inherited from such acknowledged masters of classical physics as Boltzmann, Hendrik Antoon Lorentz, and Max Planck. The next three essays each focus on one of the three most important papers of the *annus mirabilis* and on Einstein's other contributions to the more specialized field to which each one belongs. John D. Norton (Chapter 2) discusses the relativity paper (Einstein 1905r); A. J. Kox (Chapter 3) the Brownian motion paper (Einstein 1905k); and Olivier Darrigol (Chapter 4) the light quantum paper (Einstein 1905i).

Einstein's strategy of trying to establish fixed points for further development amounted to looking for constraints on future theories covering a particular domain. Einstein adopted this strategy when he recognized that an attempt to formulate a complete theory for the relevant phenomena would be premature. One of us has argued that Einstein's famous distinction between "constructive theories" and "principle theories" (Einstein 1919f),[11] to which we will return toward the end of this introduction, was intended, at least in part, to capture this difference in strategy: trying to find concrete theoretical models for a group of phenomena versus trying to find constraints on such models (Janssen 2009, 40–1). Though Einstein made the principle/constructive distinction in an article on relativity theory, its heuristic use is perhaps best illustrated by Einstein's early work on quantum theory. In this area, Einstein's "fixed points" strategy was especially successful.

A striking example of the strategy is provided by the central argument for the light quantum hypothesis, cautiously called a "heuristic viewpoint" in the title of the only paper of his *annus mirabilis* that Einstein himself, in a letter to Habicht, called "very revolutionary" (CPAE 5, Doc. 72). For this argument, which one commentator has called "Einstein's miraculous argument" (Norton 2006), Einstein considered a box with black-body radiation. The "fixed points" used as premises for the argument are Boltzmann's principle relating probability and entropy and Wien's law for the spectral distribution of the energy of black-body radiation in the high-frequency regime. The latter, in conjunction with some standard results from thermodynamics, gave Einstein a formula for the entropy of this radiation. Using Boltzmann's principle, he turned this into an expression for the (exceedingly small) probability that, due to some random fluctuation, all radiation in the box would at some point be found concentrated in a small subvolume of it (see Norton 2006 for a careful analysis). The expression for this probability has the exact same form as the expression for the probability that, due to a random fluctuation, all molecules of an ideal gas in some container would at some point be found in a small subvolume of that container. Therefore, Einstein concluded, black-body radiation in the high-frequency regime behaves as a collection of ideal gas molecules with energies proportional to the frequency of the radiation. The conclusion is as secure as its premises – the laws of thermodynamics, Boltzmann's principle, and Wien's law.

Einstein provided further support for the light quantum hypothesis by showing that it could readily explain some phenomena that were utterly mysterious from the point of view of the wave theory of light. The most important of these is the photoelectric effect, in which light shining on a metal plate releases electrons from its surface. The energy of these photoelectrons turns out to be proportional to the *frequency* of the light, as predicted by Einstein's hypothesis, and not, as one would have expected on the basis of the wave theory, to its intensity, proportional to the square of the waves' *amplitudes*. Because of this explanatory feat, the light quantum paper is often remembered as the "photoelectric effect paper" but its centerpiece was the fluctuation argument for the particle behavior of light. The photoelectric effect was just icing on the cake.

In 1909, once again deploying his "fixed points" strategy, Einstein (1909b, 1909c) published two further fluctuation arguments (one for energy, one for momentum) showing that it follows from Planck's law for black-body radiation and some general results in statistical mechanics that any satisfactory future theory of light would have to ascribe both particle and wave characteristics to light. Einstein is quite explicit about the strategy he followed to arrive at this result: "We consider Planck's radiation formula as correct and ask ourselves whether some conclusion about the constitution of radiation can be inferred from it" (Einstein 1909c, 823).[12]

Several years later, in a paper that laid the theoretical foundation for the development of the laser decades later, Einstein showed that one can derive Planck's law from the condition for thermal equilibrium in a simple model for the interaction between matter and radiation (Einstein 1916n). In this model, generic Bohr atoms with discrete energy levels emit and absorb corpuscular light quanta. Einstein introduced his famous A and B coefficients for the probability of such emission and absorption. Despite his later dictum that God does not play dice,[13] Einstein thus introduced – reluctantly and, he hoped, temporarily – a stochastic element into the basic laws of physics. Einstein now also explicitly added the assumption that light quanta carry momentum as well as energy. In the second part of the paper, Einstein showed that this assumption is crucial for his model to give the right momentum fluctuations in a situation analogous to the one he had analyzed earlier (Einstein 1909b, 1909c; Einstein and Hopf 1910). While fully recognizing the provisional character of his model (it completely failed to do justice, e.g., to the wave aspects of light), Einstein, as in 1905 and in 1909, had good reason to believe that it was a step in the right direction.

None of his arguments, however, could overcome the resistance of the physics community to the light quantum hypothesis. Interference phenomena clearly showed that light plus light could sometimes give darkness, which seemed to rule out that light could consist of particles. Robert Millikan's confirmation in 1915 of Einstein's formula for the photoelectric effect convinced the physics community of the formula but not of the light quantum hypothesis from which the formula was derived. And Einstein won the 1921 Nobel Prize for his formula for the photoelectric effect, not for the theory behind it.[14] The physics community only came around to Einstein's point of view in 1923, when Arthur Holly Compton published the results of X-ray scattering experiments and showed that they could be analyzed in impressive quantitative detail in terms of relativistic collisions of high-frequency light quanta and electrons (Stuewer 1975). Even then there were some holdouts, notably Niels Bohr. This staunch opposition to the light quantum hypothesis is the central topic of Roger H. Stuewer's essay in this volume (Chapter 5).

In the early 1920s, Einstein considered a dual theory of light in which corpuscular light quanta are guided by waves. He also tried to design experiments to decide between a wave and a particle theory of light (Einstein 1922a; CPAE 13, Doc. 29). These experiments proved to be either inconclusive (as in the case of experiments by Hans Geiger and Walther Bothe in 1921; discussed in Klein 1970a) or fraudulent (as in the case of experiments by Emil Rupp later in the 1920s; discussed in van Dongen 2007a, 2007b). These episodes are touched upon in the essays of Darrigol (Chapter 4) and Christoph Lehner (Chapter 10) in this volume.

A particularly clear-cut example of Einstein's strategy of finding and proceeding from certain fixed points – and one of the examples he had in mind when he made the distinction between principle and constructive theories in 1919 – is the way in which he presented what came to be known as the special theory of relativity in the most famous of his *annus mirabilis* papers, "On the Electrodynamics of Moving Bodies" (Einstein 1905r). In an oft-quoted passage of his "Autobiographical Notes," Einstein recounted why he had opted for a "principle theory"–approach in this case:

By and by I despaired of the possibility of discovering the true laws by means of constructive efforts based on known facts. The longer and more despairingly I tried, the more I came to the conviction that only the discovery of a universal formal principle could lead us to assured results. The example I saw before me was thermodynamics. (Einstein 1949a, 53)

Following the example of thermodynamics, Einstein derived all of his special theory of relativity in the 1905 paper from two postulates, the relativity postulate and the light postulate.

In older popular histories of relativity and in older textbooks, the light postulate used to be presented as a straightforward generalization of the result of the famous Michelson-Morley experiment of 1887.[15] The experiment, in modern terms, showed that the velocity of light is the same for all observers in uniform relative motion with respect to one another. The Michelson-Morley experiment, however, is not mentioned anywhere in the 1905 relativity paper, even though Einstein was well aware of it by 1905 (Stachel 1987, 45). The formulation of the light postulate, moreover, makes it clear that the experiment was not the origin of the postulate. The postulate does not say, as one would expect if it were linked to the Michelson-Morley experiment, that the velocity of light is independent of the motion of the *observer* (although this is a direct consequence of the conjunction of the two postulates), but rather that it is independent of the motion of the *source*. In fact, the result of the Michelson-Morley experiment would readily be explained if the velocity of light *were* dependent on the velocity of its source. The result of the experiment is thus perfectly compatible with the *negation* of the light postulate, which, all by itself, would seem to rule out that the postulate originated in the experiment.

The real origin of the light postulate, as Einstein made clear on many occasions, is to be found in Maxwell's equations for electrodynamics.[16] These equations predict the existence of electromagnetic waves propagating at the speed of light, regardless of the direction in which they travel and regardless of the velocity of the source by which they are emitted. It was thought in the nineteenth century that this prediction could only be true in a frame of reference in which the ether, the medium thought to be the carrier of these electromagnetic waves, is at rest. It followed that Maxwell's equations could also only hold in such privileged frames of reference. Electromagnetic theory thus seemed to violate the relativity principle, familiar from mechanics since the days of Galileo. As there were good reasons to assume that the Earth was moving with respect to the ether, Maxwell's equations were not expected to hold in the Earth's frame of reference. It should thus be possible to detect the Earth's presumed motion through the ether with experiments in optics and electromagnetism. Yet no experiment ever showed any sign of this motion. In the decade before 1905, Lorentz developed a theory that could account for the negative results of most of these so-called ether drift experiments (Janssen 1995, 2002b, 2009). The theory was a combination of a purely mathematical result and a far-reaching physical hypothesis. The mathematical result was that Maxwell's equations have a remarkable symmetry property. They are invariant under what are now called Lorentz transformations. The physical hypothesis, in effect, was that all other physical laws have this same property. This hypothesis fit with the view of several prominent physicists at the time that all laws of physics could ultimately be reduced to those of electrodynamics (McCormmach 1970; Kragh 1999, ch. 6).

Illustrating the claim in the contribution by Renn and Rynasiewicz (Chapter 1) that the breakthroughs of Einstein's *annus mirabilis* consisted to a large extent of a reinterpretation of existing bodies of knowledge, the mathematical formalism used in Einstein's 1905 paper on special relativity is essentially the same as that of Lorentz's theory. Einstein, however, recognized that Lorentz invariance went beyond the laws of electrodynamics and had to do with the structure of space and time.[17] Einstein introduced his new ideas about space and time in the "Kinematical Part" of the paper.[18] In the "Electrodynamic Part," which gave the paper its title, he showed that, with these new ideas, Maxwell's equations do satisfy the relativity principle after all. If they hold in one frame of reference, they hold in all frames in uniform motion with respect to that one. That it had looked as if they do not was because physicists had tacitly used (what are now called) Galilean transformations rather than Lorentz transformations to relate quantities pertaining to two such frames to one another.

One can read Einstein's 1905 relativity paper as an investigation into what has to give for Maxwell's equations to satisfy the relativity principle. Einstein, however, set it up somewhat differently. He investigated what has to give to render *one important consequence of* Maxwell's equations compatible with the relativity principle: the proposition that light propagates with a fixed velocity independently of the velocity of its source. He showed that this requires changes in our commonsense concepts of space and time. Most importantly, he showed that it is a direct consequence of his two postulates that two observers in uniform motion with respect to one another will disagree whether two events taking place at different locations happen simultaneously or not. This result is known as the relativity of simultaneity. The Lorentz transformation equations for the space and time coordinates of two frames of reference in uniform motion with respect to one another incorporate this effect, as well as two further consequences of the postulates, known as length contraction and time dilation (see the Appendix for details).

As mentioned above, the way in which Einstein presented his theory in his relativity paper nicely illustrates his "fixed points" strategy. Einstein could not use Maxwell's equations as part of the foundation of a new physics. Given his conviction that light sometimes displays particle behavior, he had to reckon with the possibility that these equations, which seemed to vindicate the opposing view that light is a pure wave phenomenon, would have to be altered some day. As it happened, only their interpretation had to be changed, but Einstein had no way of knowing this in 1905.[19] Fortunately, all he needed from Maxwell's equations was the prediction that the velocity of light is independent of the velocity of its source. He was willing to bet the house on this one consequence of Maxwell's equations. And the conjunction of this one element and the relativity postulate sufficed to derive the new ideas about space and time that could be used as the foundation for a new physics.

Before he published the light quantum hypothesis and special relativity, Einstein presumably had convinced himself of their compatibility. Two results were especially reassuring on this score. The simple linear relation between the energy and the frequency of his light quanta is preserved under Lorentz transformations (Rynasiewicz 2005, 44). And because of the way velocities add in special relativity, the theory is compatible both with a particle and a wave theory of light.[20]

The new ideas about space and time for which special relativity is best known can be understood independently of the problems in electrodynamics they were introduced to solve (see the Appendix). However, one cannot properly understand the genesis of the theory and its reception by the physics community without taking into account the electrodynamical context.[21] In his contribution to this volume (Chapter 2), Norton thus presents a reconstruction of Einstein's electrodynamically driven path to special relativity. He does so without relying on the mathematical formalism of electrodynamics typically presupposed in such reconstructions in the history of physics literature.[22]

Unlike the light quantum hypothesis, special relativity was accepted by mainstream physicists within a few years. Whereas the light quantum hypothesis called into question one of the watershed achievements of nineteenth-century physics, the displacement of the particle theory of light by the wave theory, special relativity was the natural culmination of work in electrodynamics in the preceding decades. Much of the mathematical formalism of special relativity was already in place by 1905. Einstein mainly deserves credit for recognizing the importance of key elements of this formalism beyond electrodynamics. This is true not just for the new notions of space and time encoded in Lorentz's transformation equations, but also for other general relations in physics, such as, most famously, the equivalence of energy and mass, $E = mc^2$, which Einstein first published in a short paper that brought his *annus mirabilis* to a close (Einstein 1905s). Einstein's signature formula is also discussed in Norton's essay.[23]

In addition to his pathbreaking papers, Einstein wrote his PhD thesis in 1905, a feat less astonishing than it sounds as the dissertation only takes up seventeen pages (Einstein 1905j). Meanwhile, Einstein continued to perform his duties at the Patent Office well enough to earn a promotion. In 1906, the year after his *annus mirabilis*, he was promoted from patent clerk third class to patent clerk second class. It was not until three years later that he quit the Patent Office[24] and began his ascent up the academic ladder. Once begun, the ascent was rapid. In 1909 he went from part-time instructor at the University of Bern to assistant professor at the University of Zurich. That same year he first appeared on the program of the large annual meeting of German natural scientists and physicians, held that year in Salzburg. He used the occasion to present the fluctuation argument for wave-particle duality mentioned above, a result he conceivably arrived at years earlier (Chapter 1). In 1911, he accepted a full professorship in Prague. The next year he was back in Zurich, now as a full professor at his alma mater, the ETH. He did not stay long there either. In 1914, he accepted a salaried position created especially for him at the Prussian Academy of Sciences in Berlin. The position did not carry any teaching obligations and allowed him to spend as much time as he wanted on his research. Moreover, Einstein was promised the directorship of a planned Kaiser Wilhelm Institute for Physics, which was finally founded in 1917 but even then existed only on paper. Through the institute, research grants could be awarded but there was no building designated to it.

Showing just how controversial Einstein's light quantum hypothesis still was in 1914, Planck and Walther Nernst, who aggressively recruited Einstein for the Berlin position, made it clear that they wanted him to come to Berlin *despite* his ideas about light quanta. As Planck wrote in a letter recommending Einstein for membership in the Berlin Academy: "That he may sometimes have missed the target in his speculations, as for example, in his hypothesis of light quanta, cannot really be held against him" (CPAE 5, Doc. 445). Planck and Nernst were interested in other applications of Einstein's quantum ideas. In 1907, Einstein had shown that these ideas can be used to account for the puzzling rapid decrease of the specific heat of solids at low temperatures (Einstein 1907a). The quantization of matter was clearly more palatable to the physics community than the quantization of radiation. Einstein's work on the specific heat of solids had attracted the attention of low-temperature specialist Nernst, who was instrumental in the choice of the fledgling quantum theory as the subject of the first in a series of prestigious and influential conferences held in Brussels and paid for by the Belgian industrialist Ernest Solvay. At this first Solvay conference, held in 1911, Einstein emerged as a leader in the new field (Barkan 1993).

The move to Berlin did not just change Einstein's professional life; it changed his personal life, too. Shortly after Einstein and Marić moved to Berlin with their two young sons, their marriage, which had been deteriorating for years, unraveled. Within a few months, Marić, Hans

Albert, and Eduard went back to Zurich while Einstein stayed behind in Berlin and renewed an affair he had started and broken off there two years earlier with his cousin Elsa Löwenthal.[25] In 1919, after years of bitter dispute,[26] Einstein obtained a divorce from Marić and married Löwenthal, becoming the stepfather to the two daughters of her previous marriage, Ilse and Margot.[27] Löwenthal played an important role in introducing Einstein to politics during World War I, which broke out only months after his move to the German capital. He was one of only four signatories of a manifesto protesting the war drawn up by Löwenthal's personal physician Georg Nicolai, with some input, or so Nicolai claimed, from Einstein (CPAE 6, Doc. 8).[28] This was reflective of Einstein's innate pacifism as he tentatively began to test the political waters (CPAE 8, Introduction, Sec. IV). It was only after the war, however, that he started to take a more serious interest and a more active role in politics. A key factor, as Robert Schulmann argues in Chapter 14, was his strong sympathy with the plight of underprivileged Jews from Eastern Europe, which he increasingly became aware of toward the end of the war. During the war, Einstein had mostly just hunkered down to work on his science.

It must have been disappointing to Planck and Nernst that instead of pursuing his research in quantum physics, Einstein spent much of his time during his first few years in Berlin on his new theory of gravity, which, he hoped, would extend the principle of relativity of his relativity theory of 1905 from uniform motion to arbitrary accelerated motion. The essay by Michel Janssen (Chapter 6) tells the story of Einstein's ultimately unsuccessful quest for a general relativity of motion. As a new theory about gravity and space-time, however, general relativity was a triumphant success, with important implications for both physics and philosophy, as a number of essays in this volume make clear (see, in particular, Chapters 8, 9, and 11–13).

As far as special relativity is concerned, Einstein deserves credit for recognizing the broader significance of the developments in electrodynamics that led up to it and for presenting the theory as a theory of principle based on two simple postulates. However, there is broad consensus among historians of physics that special relativity was the work of several physicists and mathematicians, including, in addition to Einstein, Lorentz, Henri Poincaré, and Hermann Minkowski. General relativity, by contrast, was clearly Einstein's brainchild and his alone. It is widely seen as the crowning achievement of his scientific career. This is not to say that others did not make important contributions to it. In fact, Einstein could not have formulated the theory if it had not been for prior work by a number of physicists and mathematicians. Discussions with friends and colleagues proved crucial at various junctures. Early on, he turned to his former classmate Grossmann to help him with the unfamiliar mathematics he used to formulate the theory. Yet the theory itself was uniquely Einstein's own.

From the point of view of the historian trying to reconstruct how Einstein arrived at his theories, there is another important difference between special and general relativity. In stark contrast to the dearth of documentary evidence for the reconstruction of Einstein's path to special relativity, there is a wealth of source material documenting his path to general relativity. In the case of special relativity, there are no papers or manuscripts on the topic prior to 1905 and the thinking that led Einstein to the theory left only a few scattered remarks in the relatively small number of letters that survive from this period. In the case of general relativity, we can follow the development in a string of publications between 1907 and 1918, in letters to a growing number of correspondents, and in a handful of surviving notebooks with research and lecture notes.[29] The most important of these is the so-called Zurich Notebook (CPAE 4, Doc. 10), which contains the record of Einstein's attempts, with help from Grossmann, to find suitable gravitational field equations shortly after his realization that the gravitational field or rather its potentials should be represented by the components of the so-called metric tensor field.[30]

Einstein began the journey that would take him to general relativity only a couple of years after he completed special relativity. Special relativity required that Newton's theory of gravity be modified, since it involved instantaneous action at a distance. Einstein, however, became intrigued by another tension between special relativity and Newton's theory of gravity. Special relativity, he realized, is hard to reconcile with Galileo's principle that the acceleration of free fall is the same for all bodies (Einstein 1933c; Renn 2007b, 61). Newton had implemented this principle in his theory by setting a body's inertial mass, a measure of its resistance to acceleration, equal to its gravitational mass, a measure of its susceptibility to gravity. In Newton's theory, this equality is ultimately just a cosmic coincidence. What it told Einstein was that any truly satisfactory theory of gravity should make gravity and inertia two sides of the same coin. A striking example of this intimate connection between gravity and inertia, for which Einstein in 1912 coined the term *equivalence principle*, occurred to him when he first started thinking about the problem of gravity in 1907: someone in free fall does not feel his own weight. Einstein later called this "the most fortunate idea of my life" (CPAE 7, Doc. 31, [p. 21]).

It would take Einstein another eight years to develop his initial insights into a fully worked-out theory that makes gravity part of the fabric of what in general will be a curved space-time (Einstein 1915i,

1916e). In this theory, the so-called metric field does double duty, encoding the geometry of space-time and representing the potentials for the gravitational field. As luck would have it, Einstein was reunited with Grossmann at the ETH in 1912, just as he was starting to recognize the importance of the metric field. With Grossmann's help, he mastered the mathematics he needed, the geometry of curved surfaces of Carl Friedrich Gauss and its generalization by Bernhard Riemann and others. In his contribution to this volume, Michael Friedman (Chapter 13) places general relativity in the context of debates over the foundations of geometry, focusing on Einstein's famous 1921 essay "Geometry and Experience" (Einstein 1921c).[31]

Another important resource that Einstein drew on for the formulation of his new theory was Minkowski's four-dimensional geometrical formulation of special relativity (see the Appendix, Section 2) and its further elaboration by Arnold Sommerfeld and Max Laue. Einstein had initially been skeptical about Minkowski's geometrical formulation of special relativity – dismissing it, as he later once put it, as "superfluous learnedness" (Pais 1982, 152) – but came to recognize its value for the formulation of what would become general relativity. The theory of the electromagnetic field of Maxwell and Lorentz, rephrased in Minkowski's four-dimensional language, provided Einstein with a model for constructing a theory for the metric field.

Besides analogies between metric and electromagnetic fields, however, there are also clear disanalogies. Consider, for instance, the sources of the two fields. The sources of the electromagnetic field are electric charges and currents. In the case of the metric field, all forms of energy-momentum act as sources. Since the metric field also carries energy-momentum, it acts as its own source. As a result, the field equations of general relativity, the equations determining the metric field produced by its sources, are notoriously complicated. In his contribution to this volume, Daniel J. Kennefick (Chapter 8) discusses the challenges this posed for the elaboration and application of Einstein's theory, especially in investigations into the possibility and nature of gravitational waves. Kennefick shows how Einstein and later relativists had to navigate carefully through a web of analogies and disanalogies between electromagnetic and gravitational waves.

As the name of his new theory of gravity makes clear, Einstein hoped to accomplish a good deal more with his general theory of relativity than to replace Newton's theory of gravity by a new theory of gravity that does justice to Galileo's principle and the new physics coming out of electromagnetic field theory and special relativity. Einstein's goal was to develop a theory that would generalize the relativity of uniform motion of special relativity to arbitrary accelerated motion. This can be seen clearly in the title of the paper, coauthored with Grossmann, in which the metric field made its first appearance: "Outline (*Entwurf*) of a Generalized Theory of Relativity and of a Theory of Gravitation" (Einstein and Grossmann 1913). This *Entwurf* theory, as it is commonly called in the historical literature, is in many ways close to the final version of general relativity. The main difference is that the field equations of the final theory are *generally covariant* (i.e., they retain their form under arbitrary coordinate transformations), whereas the *Entwurf* field equations are not. In fact, this is why Einstein and Grossmann refer to a *generalized* rather than a general theory of relativity in the title of their 1913 paper.

General covariance, however, should not be conflated with general relativity of motion, as Einstein eventually conceded (Kretschmann 1917, Einstein 1918e). As Janssen argues in Chapter 6, Einstein's search for general covariance was one of a number of failed attempts to establish the relativity of arbitrary motion. Janssen also argues, however, that Einstein's theory does implement a related principle that can be called the relativity of the gravitational field (CPAE 7, Doc. 31, [p. 20]).[32] According to this principle, based on the local equivalence of gravity and acceleration, two observers in arbitrary motion with respect to one another can both maintain to be at rest if they agree to disagree on whether or not there is a gravitational field present. Einstein's criterion for the absence or presence of a gravitational field (the vanishing or nonvanishing of the so-called Christoffel symbols) is different from the criterion used by relativists today (the vanishing or nonvanishing of the so-called Riemann tensor). This modern criterion does not leave room for disagreement between different observers about the presence of a gravitational field. Both as he was developing general relativity and in commenting on it in later years, Einstein often claimed that his theory made all motion relative. One way to make sense of these pronouncements, which when taken literally are simply false, is to take him to be talking about the relativity of the gravitational field instead.

Although general relativity retains absolute *motion*, it can be argued that it does away with absolute *space* (or absolute *space-time*). Support for this claim comes from a pair of arguments, known as the "hole argument" and the "point-coincidence argument," that Einstein put forward in the course of his struggles with general covariance.[33]

Regardless of whether, to what extent, or in what sense Einstein succeeded in banishing absolute motion and/or absolute space from physics, it is clear that his various attempts to do so informed major steps in the development of his theory. His final attempt provides perhaps the best illustration of this observation. In the course of it, he introduced the so-called cosmological constant (Einstein 1917b) hoping it would ensure that his theory satisfy what the following year he officially named Mach's principle (Einstein 1918e), the requirement that the metric field and thereby the space-time structure be fully determined by its material sources. In that case, all reference to motion with respect to space-time can be read as shorthand for motion with respect to the matter generating the metric field. This, in turn, would render all motion relative. With the 1915 version of the field equations, as the Dutch astronomer Willem De Sitter had pointed out in 1916, absolute motion survives because the metric field is determined not just by material sources but also by boundary conditions, the values of the metric field at spatial infinity. Einstein's ingenious response was to propose that the universe is spatially closed so that it would have no boundary (Einstein 1917b). He gave a simple example of such a universe, which is both closed and static. The metric field describing the space-time geometry of this universe, however, was not a solution of the field equations of his 1915 theory. To turn it into one he had to add a term with the cosmological constant to these field equations. This term describes a gravitational push compensating the usual gravitational pull to ensure that his static universe would not collapse. Einstein believed that these amended field equations would not allow any solutions without matter and would therefore satisfy Mach's principle. De Sitter, however, in short order produced a vacuum solution of these new field equations. Einstein tried to find fault with this solution but had to concede in 1918 that it was unobjectionable. This meant that yet another attempt to relativize all motion had failed. However, as Christopher Smeenk shows in his essay in this volume (Chapter 7), the models of Einstein and De Sitter for the global space-time geometry of the universe at large nonetheless did end up launching the field of relativistic cosmology. More recently, with the discovery of the *accelerated* expansion of the universe, even the cosmological constant, which after the rise of expanding models of the universe Einstein allegedly once called the biggest blunder of his career (Gamow 1970, 149–50), has made a spectacular comeback. These developments illustrate how Einstein's failed crusade against absolute motion still resulted in major contributions to physics.

What may have helped Einstein make his peace with the failure of Mach's principle in 1918 is that he came to realize within a year or so that it was based on a view of the relation between matter and space-time that by the early twentieth century had become anachronistic. Both matter and space-time were represented by fields in Einstein's view, the electromagnetic field and the metric field, respectively. Mach's principle thus amounted to the requirement that one field be *reduced* to another. Another option was to search for a theory that would *unify* the two fields, just as special relativity had unified the electric and magnetic field into one electromagnetic field and general relativity had unified gravity and inertia into one inertio-gravitational field, represented by the metric field. Whether or not – and, if so, to what extent – this consideration actually played a role, Einstein in the early 1920s embarked on the search for a unified field theory modeled on general relativity. He never found a theory that he was satisfied with for any length of time but his attempts to find one account for much of his scientific production during the last three decades of his life. This unified field theory program is the topic of Tilman Sauer's essay in this volume (Chapter 9).

Einstein's strategy in his search for a unified field theory is markedly different from the "fixed points" strategy he used in much of his earlier work. For one thing, he was now searching for a "constructive theory." He had preferred such theories all along. When he introduced the principle/constructive distinction, he wrote: "When we say that we have succeeded in understanding a group of natural processes, we invariably mean that a constructive theory has been found which covers the processes in question" (Einstein 1919f). More than a decade earlier he had written in a letter:

A physical theory can be satisfactory only if its structures are composed of elementary foundations. The theory of relativity is ultimately as little satisfactory as ... thermodynamics was before Boltzmann had interpreted the entropy as probability. (Einstein to Arnold Sommerfeld, January 14, 1908 [CPAE 5, Doc. 73])

As he wrote in his "Autobiographical Notes," in the passage we already quoted, Einstein only started looking for a "principle theory" because he was unable to find "the true laws by means of constructive efforts" (Einstein 1949a, 53). When Einstein (1909c, 817) prophesized in his Salzburg lecture that "the next phase of the development of theoretical physics will bring us a theory of light that can be interpreted as a kind of fusion of the wave and emission theories," he made no pretense of knowing what that future theory might look like.[34] In his search for a unification of gravity and electromagnetism, by contrast, he appears to have been confident that the theory he was looking for would be an extension of general relativity in one direction or another.

Another important and related change in Einstein's methodology is that he started to pay more attention to mathematical elegance and less to empirical data. Einstein's attitude toward empirical data was never as cavalier as suggested by the well-known anecdote related by Ilse Rosenthal-Schneider who was with Einstein in 1919 when word reached him that a British eclipse expedition had confirmed the prediction of general relativity of light bending in the gravitational field of the

Sun. What would he have done, Rosenthal-Schneider purportedly asked him, had the outcome been different? Einstein's reply, she reports, was: "Then I would have had to pity our dear God. The theory is correct all the same" (Calaprice 2011, 368). A better example to gauge Einstein's attitude toward empirical results is his reaction when, in 1906, Walter Kaufmann published results of experiments on the velocity-dependence of the electron mass that seemed to favor two other theories over special relativity. Contrary to what one might have expected in view of his alleged response to Rosenthal-Schneider, Einstein did not dismiss the data but admitted that these theories "yield curves that are significantly closer to the observed curve than the curve obtained from the theory of relativity" (Einstein 1907j, 439). That he nonetheless preferred special relativity over these alternative theories was because "their basic assumptions ... are not suggested by theoretical systems that encompass larger complexes of phenomena" (ibid.).

The best-known expression of the hubris one naturally reads into Einstein's ironic response to Rosenthal-Schneider comes from his Herbert Spencer lecture, On the Method of Theoretical Physics, given in Oxford in June 1933: "Our experience hitherto justifies us in believing that nature is the realization of the simplest conceivable mathematical ideas" (Einstein 1933b, 274).[35] "I am convinced," he elaborated, "that we can discover by means of purely mathematical constructions the concepts and the laws ... which furnish the key to the understanding of natural phenomena.... [T]he creative principle resides in mathematics. In a certain sense, therefore, I hold it true that pure thought can grasp reality, as the ancients dreamed" (ibid.). Even on this occasion, however, he reminded his audience that "[e]xperience remains, of course, the sole criterion of the physical utility of a mathematical construction" (ibid.). And he began his lecture warning his audience that "[i]f you want to find out anything from the theoretical physicists about the methods they use, I advise you to stick closely to one principle: do not listen to their words, fix your attention on their deeds" (ibid., 270).

Yet, despite these disclaimers and caveats, there is a noticeable shift from putting empirical data first to putting mathematical elegance first if we compare Einstein's later work on unified field theory to his earlier work. We will examine this shift more carefully toward the end of this introduction. Einstein saw this shift as a lesson he had learned from his struggles to formulate general relativity. As he told one of his correspondents in 1938: "the problem of gravitation has made me into a believing rationalist, i.e., one who looks for the only reliable source of truth in mathematical simplicity."[36] Einstein was presumably referring here to the events of November 1915 (see, however, note 71 for an alternative interpretation). By mid-October of that year, Einstein had come to the conclusion that the *Entwurf* field equations are untenable, and he was trying to find new ones, preferably before someone else would, such as the Göttingen mathematician David Hilbert, who had taken an interest in the theory.[37] Physical considerations – energy-momentum conservation, agreement with Newton's theory of gravity in the appropriate limit, the analogy between metric and electromagnetic field – had guided Einstein both in the original formulation of the *Entwurf* field equations and in their further analysis. However, as documented in the Zurich Notebook, Einstein had only settled on the "physical strategy" leading him to these equations after a different "mathematical strategy" had failed (Renn 2007a, Vol. 1, 10). The latter meant extracting mathematically attractive candidate field equations from the Riemann curvature tensor and then checking whether these candidates were acceptable from a physics point of view. In 1912–13 Einstein had concluded that none of them were. In November 1915, however, he resurrected one of these rejected equations to replace the discredited *Entwurf* equations (Einstein 1915f). It thus looked as if he had suddenly returned to the mathematical strategy; perhaps because he worried that Hilbert might otherwise beat him to the punch. Many of Einstein's statements, made both at the time and with years of hindsight, confirm such a switch in strategy in late 1915 and the episode has been presented accordingly in the literature (see, especially, Norton 2000). This then would seem to be the full story behind Einstein's claim that general relativity had made him "into a believing rationalist."

However, Janssen and Renn (2007) have argued that it was the physical strategy, not the mathematical one, that led Einstein back to the field equations rejected in the Zurich Notebook (see Chapter 12 for discussion). Einstein, on this interpretation, retained the formalism of the *Entwurf* theory and changed only one element – the definition of the gravitational field in terms of gradients of the gravitational potentials, the components of the metric tensor field. There is strong textual evidence for this reconstruction as well. In particular, Einstein identified his old definition of the gravitational field as a "fatal prejudice" and his new one as "the key to the solution."[38] In his presentation of the new field equations, however, Einstein (1915f) strongly suggested that the new theory amounted to a wholesale replacement of the old one and he put great emphasis on the mathematical elegance of his rediscovered field equations. This makes it understandable, Janssen and Renn (2007, 841) argue, that Einstein would come to misremember the completion of general relativity as a triumph for his mathematical strategy. This case of selective amnesia served him well later on when he had to convince himself and others that his strategy for finding a unified field theory could boast of at least one spectacular success.[39] Perhaps we should thus take seriously another disclaimer in Einstein's Spencer lecture. He noted that the way in which a researcher, reflecting on years of experience in his field, "regards its past and present may depend too much on what he hopes for the future and aims at in the present" (Einstein 1933b, 271).

In view of these considerations, one can question whether the experience with general relativity fully explains the shift in Einstein's methodology from a data-driven "principle-theory" approach to a more speculative "constructive-theory" approach relying heavily on considerations of mathematical elegance. In fact, Janssen (2006) has speculated that the turmoil of his personal life during the war years may have played a role in this shift as well.

Consider the following passage from a lecture he gave in April 1918, on the occasion of Planck's sixtieth birthday:[40]

I believe with Schopenhauer[41] that one of the strongest motives that leads men to art and science is escape from every day life with its painful crudity and hopeless dreariness, from the fetters of one's own ever shifting desires. A finely tempered nature longs to escape from personal life into the world of objective perception and thought.... With this negative motive goes a positive one. Man tries to make for himself in the fashion that suits him best a simplified and intelligible picture of the world; he then tries to some extent to substitute this cosmos of his for the world of experience, and thus to overcome it. This is what the painter, the poet, the speculative philosopher, and the natural scientist do, each in his own fashion. Each makes this cosmos and its construction the pivot of his emotional life, in order to find in this way the peace and security which he cannot find in the narrow whirlpool of personal experience. (Einstein 1918j, 29–30)

Admittedly, Einstein is talking here about the motivation rather than the strategy behind Planck's and presumably Einstein's own research, but building mathematical castles in the air would seem to serve the purpose of finding a "simplified and intelligible picture of the world ... to substitute ... for the world of experience" much better than the "fixed points" strategy we saw Einstein use in so much of his early work. As he got older, Einstein, it seems, more self-consciously began to use physics as a means to free himself, as he put it in his "Autobiographical Notes," "from the chains of the 'merely personal'" (Einstein 1949a, 5).

The circumstances of Einstein's life in April 1918 make it easy to understand why, at this particular juncture, he would want to escape from the "merely personal." The armistice that ended hostilities on the Western Front in World War I was still several months away and, although Löwenthal was able to procure scarce food items, she and Einstein were not spared the privations of war. Family pressure was mounting for him to divorce Marić and marry his cousin, as he would the following year.[42] On top of all this, Einstein had fallen seriously ill and was bedridden for several months. At the beginning of 1918, Planck had to present two papers (Einstein 1918a, 1918c) to the Berlin Academy on Einstein's behalf. Physics, it seems, is what during this period allowed Einstein to find a measure of "peace and security" not to be found "in the narrow whirlpool of personal experience."

It was not the first time that Einstein described the role of physics in his life in this way. More than twenty years earlier he had made an eerily similar statement. On that occasion, he was not addressing his colleagues in a special session of the German Physical Society, but was writing a letter to the mother of his first girlfriend, with whom he had just broken up. He wrote:

Strenuous intellectual work and looking at God's Nature are the reconciling, fortifying, yet relentlessly strict angels that shall lead me through all of life's troubles.... And yet, what a peculiar way this is to weather the storms of life – in many a lucid moment I appear to myself as an ostrich.... One creates a small little world for oneself, and as lamentably insignificant it may be in comparison with the perpetually changing size of real existence, one feels miraculously great and important, like a mole in his self-dug hole.[43]

The sanctuary from personal turmoil that Einstein sought in physics may also have colored the extreme rationalist pronouncements in his Herbert Spencer lecture. The lecture, after all, was delivered in June 1933, only months after Adolf Hitler took office as Chancellor of Germany on January 30, 1933.

Einstein had been traveling at the time. He had left for one of his visits to the United States in December 1932. Crossing the Atlantic back to Europe in March 1933, he received word that the Nazis had raided his summer house in Caputh. For the second time, he renounced his German citizenship, which he had reluctantly reacquired in the 1920s when it was determined retroactively and contrary to what he had been told in 1914, that acceptance of his position in Berlin had effectively made him a dual German-Swiss citizen (Grundmann 1998, 265–80, 375–83). He also resigned as a member of the Prussian Academy of Sciences.[44] He would never set foot on German soil again. Right before he went to Oxford, he had visited his son Eduard in an asylum in Zurich – the last time father and son would ever see each other. These events must at least have been at the back of his mind as he was extolling the virtues of pure mathematics in Oxford in June 1933. The conspicuous timing of these pronouncements on the primacy of mathematical speculation may serve as a warning and a reminder that they do not describe the methodology behind all of Einstein's later work. His papers on the quantum theory of ideal gases and Bose-Einstein statistics (Einstein 1925a, 1925b; see Chapter 4) and the famous EPR paper (Einstein, Podolsky, and Rosen 1935; see Chapter 10) have much more in common with his earlier work than his papers on unified field theory.

The "narrow whirlpool of personal experience" from which Einstein periodically felt the urge to escape had greatly expanded in the wake of the British solar eclipse expeditions of 1919, led by Arthur S. Eddington, which confirmed the prediction of light bending of general relativity.[45] Ever since his arrival in Berlin, Einstein had tried in vain to convince the German astronomy establishment to devote resources to testing the theory, especially through supporting the work of his protégé, the young astronomer Erwin Freundlich, who did not help the cause by alienating some of his powerful superiors (Hentschel 1992). In the end, British scientists provided new astronomical evidence for general relativity. The result made headlines on both sides of the Atlantic and made Einstein an overnight sensation and a household name (Rowe 2012). It was this event that prompted the *London Times* to solicit the article best known today for the distinction between principle and constructive theories. Einstein ended the article with a flourish, offering one last "application of the theory of relativity." "[T]oday in Germany," he wrote presciently, "I am called a German man of science, and in England I am represented as a Swiss Jew. If I come to be regarded as a *bête noire*, the descriptions will be reversed and I shall become a Swiss Jew for the Germans and a German man of science for the English!" (Einstein 1919f).

Besides adulation, honorary doctorates, invitations to visit faraway places,[46] and a platform to pronounce on causes dear to his heart such as Zionism and pacifism, Einstein's newfound fame also brought him increased attention from vociferous and often anti-Semitic opponents to relativity.[47] His friend and colleague Paul Ehrenfest, Lorentz's successor in Leyden, tried to convince Einstein to leave Germany and come to the Netherlands. Einstein accepted a visiting professorship in Leyden,[48] but stayed in Berlin out of loyalty to Planck and other colleagues and also, despite the frequent attacks on relativity and on him personally, out of a sense of solidarity with the German people trying to cope with the aftermath of the war and the harsh conditions of the Versailles treaty. Einstein's attitude toward Germany remained ambivalent throughout the Weimar period (see Chapter 14). Yet, even though he was deeply shaken, for instance, by the assassination by ultra-right-wing elements of Jewish foreign minister Walther Rathenau, whom he knew personally,[49] he ended up staying in Germany until the Nazis' rise to power made it impossible for him to stay any longer.

In the face of the Nazi threat, Einstein also abandoned his long-standing pacifism. Most famously, in August 1939, at the urging of his old acquaintance Leo Szilárd, he signed a letter to President Roosevelt, warning him that it might be possible to use the recently discovered process of nuclear fission for a powerful new type of bomb.[50] Because of this letter and because the equivalence of mass and energy expressed in his signature formula $E = mc^2$ underlies the relation between nuclear fission and the release of prodigious amounts of energy, Einstein has sometimes been called "the father of the atomic bomb." He was at pains to explain that he did not deserve this dubious accolade. His letter to Roosevelt was only one of many factors that led to the establishment of the Manhattan Project. In fact, very little was done with Einstein's warning. The United States only got serious about building an atomic bomb after the attack on Pearl Harbor on December 7, 1941, more than two years after Einstein's letter. Once underway, the Manhattan Project, moreover, had no use for Einstein. He was denied security clearance (Jerome 2002, ch. 4).[51] And he would have had little to contribute anyway, as he had not kept up with the rapid development of nuclear physics in the 1930s. Nuclear physics was an extension of quantum mechanics, initially developed to unlock the mysteries of the atom rather than those of the nucleus. Einstein's hope was that some unification of gravity and electromagnetism modeled on general relativity would one day replace quantum mechanics, both in atomic and in nuclear physics, and would then also explain the newly discovered nuclear particles and forces. Yet, although he was not involved in the Manhattan Project, Einstein did contribute to the war effort, as a consultant to the U.S. Navy's Bureau of Ordnance. After the war and the dropping of atomic bombs on Hiroshima and Nagasaki, he came out strongly in support of arms control. He argued for the establishment of a world government, a supranational rather than an international organization such as the United Nations (Isaacson 2007, 489). Between 1946 and 1948, he promoted these causes as chairman of the Emergency Committee of Atomic Scientists, alongside its executive director, Szilárd.

Einstein was an American citizen by then. He had left Europe for good in October 1933 to take up a position in the newly founded Institute for Advanced Study in Princeton. Abraham Flexner, the institute's founding director, had been courting Einstein since early 1932. Originally, Einstein had planned to spend only four or five months a year at the institute and divide the rest of his time between various other institutions in the United States and in Europe. By April 1934, however, he had made up his mind that he would become a full-time member of the institute and live in Princeton permanently. Shortly thereafter, Löwenthal had gone back to Europe one last time to be with her terminally ill daughter Ilse. Ilse died in 1934 in Paris in the apartment of her sister Margot. Leaving her estranged husband behind, Margot emigrated to the United States later that year and moved in with her mother and stepfather. The following year, in the fall of 1935, the three of them plus Helen Dukas, Einstein's secretary and house-keeper since 1928, moved into the house on 112 Mercer Street where Einstein would live for the remaining two decades of his life. Dukas and Margot would likewise live there until their deaths in 1982 and 1986, respectively.[52] Löwenthal died in December 1936, just a year after they moved into their new home. Although Einstein had for many years been keeping the company of younger women, he was saddened by his wife's passing.

Einstein's son Hans Albert and his wife also came to the United States, eventually settling in Berkeley. Marić stayed in Zurich.[53] In 1939, Einstein's sister Maja fled Fascist Italy and joined the household on Mercer Street (and lived there until her death in 1951). On October 1, 1940, Einstein, Margot, and Dukas were sworn in as U.S. citizens, Einstein famously not wearing socks. He was allowed to keep his Swiss citizenship.

Although Einstein was eternally grateful for the refuge the United States afforded him, he remained sharply critical of its society. He was troubled by its militarism, nationalism, racism (by anti-Semitism but even more by segregation), and hysterical anti-Communism. In January 1953, he incurred the wrath of many of his fellow citizens when a letter to President Truman in which he asked that atomic spies Julius and Ethel Rosenberg be spared the death penalty was made public (Isaacson 2007, 525–6).[54] The McCarthyism of the early 1950s strongly reminded Einstein of Germany on the brink of National Socialism in the early 1930s.[55] Part of Einstein's attitude to the United States is nicely illustrated by his reaction in April 1954 when he first heard that J. Robert Oppenheimer, the true "father of the atomic bomb," who had become the director of the Institute for Advanced Study after the war,[56] would be stripped of his security clearance unless he subjected himself to secret hearings in which the deck would be heavily stacked against him. Asked by a friend why he did not accept one of the many offers of faculty positions abroad to avoid this humiliation, Oppenheimer answered: "Damn it, I happen to love this country" (Bird and Sherwin 2005, 5). This kind of patriotism was completely foreign to Einstein. When Abraham Pais informed him of the impending Oppenheimer hearings, Einstein shrugged it off and told Pais that "all Oppenheimer needed to do ... was go to Washington, tell the officials that they were fools, and then go home" (Pais 1982, 11). "The trouble with Oppenheimer," he purportedly added, "is that he loves a woman who doesn't love him – the United States government" (Serber 1998, 183–4).[57] As we mentioned above, Einstein supported the notion of a world government. As he once wrote during World War I, "I consider affiliation with a state to be a business matter, somewhat akin to one's relationship to life insurance."[58] Given his disdain for petty nationalism, Einstein even had mixed feelings about the establishment of a Jewish state. Einstein was deeply concerned about the conflict between Jews and Arabs. Yet, never dogmatic in science or in human affairs, he relented when the State of Israel was declared in 1948. "I have never considered the idea of a state a good one," he wrote to a friend, "but now there is no going back, and one has to fight it out" (Isaacson 2007, 520).[59] In 1952, Einstein was even offered the (largely ceremonial) position of president of Israel, though all involved were relieved when he turned down the offer (Pais 1982, 11; Isaacson 2007, 521–3). And despite his mixed feelings about the United States, Einstein eventually did come to recognize that its aberrations, unlike those in Germany in the 1930s, tended to be transient. Referring to the end of McCarthyism, he wrote to his son a few months before he died: "God's own country becomes stranger and stranger ... but somehow they manage to return to normality. Everything – even lunacy – is mass produced here. But everything goes out of fashion very quickly."[60]

Einstein officially retired from the Institute for Advanced Study in 1945 but kept his office. He could frequently be seen walking between his house on Mercer Street and the institute – alone, deep in thought; in the company of his assistants, with whom he was still pursuing his unified field theory program;[61] or engaged in discussion with another distinguished member of the institute's faculty, Kurt Gödel.[62] His health meanwhile was declining. In 1948, the aneurysm in his abdominal aorta was discovered that would rupture and kill him seven years later. His friend Besso died just one month before him. In his letter of condolence to the family, Einstein reflected: "Now he has departed from this strange world a little ahead of me. That signifies nothing. For those of us who believe in physics, the distinction between past, present, and future is only a stubbornly persistent illusion."[63] Fittingly, one of Einstein's last acts was his signing of the Russell-Einstein manifesto, which called on nations to vow never to use hydrogen bombs (Rowe and Schulmann 2007, 501–5). This manifesto, signed by Max Born, Linus Pauling, and other luminaries in addition to Russell and Einstein, provided the stimulus for the Pugwash conferences on global security issues, the first one of which was held in 1957. On April 18, 1955, a week after signing the manifesto, Einstein died.[64] In accordance with his wishes, he was cremated and his ashes were scattered at an undisclosed location.[65]

Since this volume is part of a series devoted to philosophy, we close this introduction with some remarks specifically on Einstein's contributions to philosophy. In his "Reply to Criticisms" at the end of the Schilpp volume devoted to his work, Einstein (1949b, 684) characterized himself as an "unscrupulous opportunist" in matters of epistemology, blending many different philosophical schools. With this characterization, Einstein implicitly distanced himself from school philosophy. He did not care for philosophy as a separate field of study based on deductive reasoning from established first principles. However, he was deeply concerned with methodological and epistemological questions when they were connected to his search for a new foundation of physics. What makes his work important to philosophy, in turn, is not the development of a new philosophical system. Rather, it lies in its exploration of the implications of the physics of his time for general philosophical questions and the challenge this new physics posed for traditional academic philosophy of nature.

As we saw above, Einstein started reading important books in history and philosophy of physics early in life. Especially during the halcyon days of the Olympia Academy, he discussed many of these with his friends Habicht, Solovine, and Besso. His own later more philosophical writings were typically addressed to a general audience and published accordingly in newspapers and popular journals. In this volume, Einstein's contributions to philosophy are discussed in the essays of Don Howard (Chapter 11), Thomas Ryckman (Chapter 12), and Michael Friedman (Chapter 13). Einstein's philosophical views also play a central role in Christoph Lehner's essay on Einstein's critique of quantum mechanics (Chapter 10).

Howard gives an account of Einstein's complex and ambiguous relationship with logical positivism. Logical positivists saw in the theory of relativity an important pillar of support for their general analysis of science. While Einstein shared their objections against the Kantian doctrine of *a priori* knowledge, he became increasingly critical of positivism in the 1920s and defended a position that Howard sees closely akin to the holism of Pierre Duhem (1906) and W. V. O. Quine (1951).[66] In Chapter 12, Thomas Ryckman traces the rapprochement to Kant of the later Einstein. In Chapter 13, Michael Friedman analyzes Einstein's views about the relation of geometry and physics and shows how Einstein's position vis-à-vis conventionalism and positivism can be understood more fully if one takes into account nineteenth-century discussions of physical geometry by Hermann von Helmholtz, Felix Klein, and Poincaré.

Since these contributions do not parse Einstein's philosophical work chronologically, we conclude this introduction with a short account of Einstein's most important philosophical writings. We have already touched on several key texts, such as the lecture on the occasion of Planck's sixtieth birthday (Einstein 1918e), the article introducing the distinction between principle and constructive theories in the *London Times* (Einstein 1919f), and the Herbert Spencer lecture in Oxford (Einstein 1933b).

The first text that we want to draw attention to is Einstein's inaugural lecture for the Berlin Academy (Einstein 1914k).[67] Einstein used the occasion to present his views on methodology in theoretical physics. As he told his audience,

[t]he theorist's method involves his using as his foundation general postulates or "principles" from which he can deduce conclusions. His work thus falls into two parts. He must first discover his principles and then draw the conclusions which follow from them. (Einstein 1914k, 740)

The hard part, Einstein explained, was finding those principles. Even at this early date, long before the extreme rationalist stance he took in his Spencer lecture of 1933, Einstein clearly recognized that such principles cannot be found through simple induction from experimental data. "The scientist," Einstein wrote, "has to worm these general principles out of nature by perceiving in comprehensive complexes of empirical facts certain general features which permit of precise formulation" (ibid.).[68] He went on to compare the current state of affairs in quantum theory and in relativity theory in view of these methodological pronouncements. In quantum theory, as he saw it, no principles had yet been found on which further deductions could be based; in relativity, by contrast, the principle of the relativity of arbitrary motion did provide, or so he thought, a sensible starting point for deducing a new theory of gravity. Even though there was much more evidence supporting Einstein's work in quantum theory than for his new theory of gravity, this assessment is understandable given his frustration at the time with the former[69] and his audience's skepticism about the latter. Yet, as we saw previously, the "principles" that formed the input for Einstein's "fixed point" strategy in quantum theory turned out to be much more reliable than the generalized relativity principle. Looking at Einstein's characterization of how a theorist is supposed to find his foundational principles, we see that they have to meet two requirements: they must be grounded in "comprehensive complexes of empirical facts" and they must "permit of precise formulation." These two requirements are met both by the postulates on which Einstein based his special theory of relativity and by, say, Boltzmann's principle relating probability and entropy, which Einstein had put to good use in his early work on quantum theory. That he only counted the former as satisfactory principles suggests that such foundational principles had to meet a third requirement: they somehow must characterize the core of the theory to be developed for the relevant phenomena. In quantum theory, such fundamental principles were clearly still missing.

Einstein specified his notion of fundamental principles further when he introduced the famous distinction between "constructive theories" and "theories of principle" in his article in the London Times (Einstein 1919f). Constructive theories describe phenomena in terms of an underlying mechanism, such as in the kinetic theory of gases, which makes them intuitive and adaptable, and allows them to give a complete description of the physical processes. Principle theories are based on generalizations of empirical laws that are presumed to be universally valid, such as the first and second laws of thermodynamics. Such fundamental principles are not hypothetical, like underlying mechanisms, and therefore empirically better confirmed. Furthermore, they allow for logically complete deductions of more specific physical laws, such as the deduction of Wien's displacement law from the laws of thermodynamics.[70] However, they do not lead to a sense of understanding natural processes like constructive theories.

In another article published in 1919, "Induction and Deduction in Physics" (Einstein 1919g), Einstein addressed the epistemological question how empirical knowledge leads to general laws of nature. He criticized the belief in an "inductive method" – central to Mach's philosophy of science – which assumes that fundamental principles result from step-by-step generalization of individual empirical data. Echoing a passage from the inaugural lecture quoted previously, he stressed that there is no general method for the formulation of fundamental principles, but that such principles result from the "intuitive grasp of the essentials of a large complex of facts" (Einstein 1919g). This implies that fundamental principles cannot be proven from the facts; they can only be falsified if an empirical fact is in conflict with a consequence of the theory. Einstein's stressing the holism of theories as opposed to the testability of individual statements comes close to Duhem's view of theories (cf. note 66), as Howard discusses in Chapter 11.

Einstein had sounded the same theme in his lecture for Planck's sixtieth birthday the year before: "The supreme task of the physicist is to arrive at those universal elementary laws from which the cosmos can be built up by pure deduction. There is no logical path to these laws; only intuition, resting on sympathetic understanding of experience can reach them" (Einstein 1918j, 31; 1954, 226). By the time of the Herbert Spencer lecture fifteen years later, Einstein had lost confidence that the "intuitive grasp of the essentials of a large complex of facts" is a reliable guide to the fundamental principles at the heart of a theory. Einstein gives the example of gravitational theory, where the "sympathetic understanding of experience" produced two completely different theoretical systems, the action-at-a-distance forces of Newtonian theory and the space-time curvature of general relativity (Einstein 1933b, 273–4).[71] What Einstein ignored here is that the advent of field theory in the nineteenth century completely changed the assessment of what a satisfactory theory of gravity would look like. For Einstein, however, this comparison between Newtonian theory and general relativity was probably meant only as a lead-in to the point he really wanted to make: "If, then, it is true," he continued, "that the axiomatic basis of theoretical physics cannot be extracted from experience but must be freely invented, can we ever hope to find the right way?" (ibid.). We already quoted his affirmative answer: "Our experience hitherto justifies us in believing that nature is the realization of the simplest conceivable mathematical ideas" (ibid.)

Einstein's belief that theories are free creations of the human mind that are constrained but not determined by the empirically given is close to Poincaré's conventionalism, which also had a major influence on Moritz Schlick. Originally, Schlick had begun a correspondence with Einstein in 1915 and had become Einstein's most important ally in the defense of general relativity on epistemological grounds against neo-Kantian critiques. Einstein had high praise for Schlick's philosophical analysis of the theory of relativity and corresponded with him about foundational and philosophical questions. Schlick had taken Einstein's point-coincidence argument as expressing the conventional nature of geometrical statements and had inferred from general relativity that Kant's doctrine of the synthetic *a priori* was mistaken.

As shown by Friedman in Chapter 13, Einstein followed this interpretation, up to a certain point, in "Geometry and Experience" (Einstein 1921c) and took over Schlick's distinction between a pure geometry, which is axiomatic in form and analytically true, and a physical geometry, which is defined by "coordinating definitions" that connect geometrical concepts to physical entities. Physical geometry thus becomes an empirical science which can be falsified by experiment.[72] Poincaré had postulated that these coordinating definitions are purely conventional and concluded that geometry is immune to empirical refutation: one could always modify the coordinating definitions to preserve the simplicity of Euclidean geometry. Schlick agreed with Poincaré that the choice of geometry is conventional, but he justified the use of non-Euclidean geometry in general relativity by pointing out that what matters is not the maximal simplicity of geometry by itself, but the maximal simplicity of both geometry and physics. However, as Friedman's contribution shows in detail, Einstein distanced himself from the Poincaré-Schlick position stating that, although correct in principle, this position would have never allowed him to find general relativity (Einstein 1949b, 678). Einstein maintains that we are not free to redefine our concepts of space and time at will, at least not at this time in our development. He argues that his thought experiment of the rotating disk[73] depended on a physical concept of rigid rods and clocks, that is, the assumption that what we call space and time measurements is not purely conventional, but physically meaningful. The role that this *caveat* plays in Einstein's thinking can be seen from his critique of Hermann Weyl's proposal for a unified field theory, which Einstein judged as formally beautiful, but unsatisfactory since it did not agree with a physical definition of oscillating atoms as clocks (Einstein 1918g).

In the lecture for Planck's sixtieth birthday, from which we quoted several times already, Einstein distanced himself more explicitly from positivism. When Einstein (1918j) writes about man trying "to make for himself ... a simplified and intelligible picture of the world" which he then "tries ... to substitute ... for the world of experience," he is contrasting the phenomenal world with the scientific picture that theoretical physicists try to make of it. In contrast to the ever-changing and incomprehensible world of experience, the physicist strives for an image of utmost stability and simplicity. This image, for Einstein, is a free creation of the human mind, constrained only by its empirically testable consequences. Therefore, it is always logically possible that different theories will explain the same phenomena. Nevertheless, Einstein maintains that empirically equivalent theories are not equally acceptable. Einstein is very explicit about what makes one theory superior to another. Somewhat surprisingly, he refers to Leibniz's concept of "pre-established harmony" in this context, which gives an important hint as to what he has in his mind. Leibniz's concept was used by the Göttingen mathematicians Klein, Hilbert, and Minkowski to support their claim that new mathematical formalisms frequently turn out to be useful in the formulation of new physical theories.[74]

In his lecture for Planck's birthday, Einstein also referred to the debate between Planck and Mach in 1909–10,[75] in which Planck charged that Mach's positivism does not do justice to the goal of theoretical physics, that is, to go beyond mere description and find absolutes behind the phenomena. While it is not too surprising that Einstein does not oppose Planck in a birthday speech, it is nevertheless significant that Einstein brings up this debate at all and that he presents it as an issue on which Planck and he are in agreement. In Einstein's correspondence, one finds increasingly critical statements about positivism as a theory of science. In a letter to Besso in 1917, for instance, he wrote, referring to the excessive Machian positivism of their mutual acquaintance Friedrich Adler (cf. note 54): "He is riding the Machian horse to exhaustion." In a follow-up letter he added: "It cannot give birth to anything living, it can only stamp out harmful vermin."[76]

As we have seen, Einstein remained an active participant in the development of quantum theory until the mid-1920s. His inability, however, to find a consistent conceptual framework for quantum theory during the

1910s and the 1920s increased his pessimism about the possibilities of inductive theory construction on the basis of generalizations of empirical knowledge. Gradually, Einstein shifted to a program aimed at extending the fundamental equations of general relativity, the unified field theory program discussed in Chapter 9. Such programs had been launched before by theorists like Hilbert, Weyl, Gustav Mie, and Theodor Kaluza. Einstein had initially criticized them because of their speculative character and because of conflicts with empirical evidence. Nevertheless, Einstein had been impressed by the mathematical simplicity and formal beauty of Weyl's unification of general relativity and electromagnetic field theory.[77] In 1923, he began publishing his own research in unified field theory with the explicit hope that such a theory, constructed on the basis of formal criteria, would be able to explain quantum effects and so achieve what he had not been able to do so far: give a consistent theory of the quantum. By the end of the 1920s, Einstein had completely settled on a program of devising new field theories from considerations of mathematical simplicity, only checking their empirical adequacy afterward.

The 1933 Spencer lecture can be seen as Einstein's manifesto for this program. Here, Einstein added to his previous statements about the impossibility of finding fundamental laws through inductive generalization his new "credo" (Holton 1988, 252) that the key creative element in the development of theories was mathematics. This emphatically rationalistic position has often been interpreted as an indication of Einstein's conversion to realism (see, e.g., Holton 1968), which would explain his critical stance toward quantum mechanics.

However, as is argued by Lehner in Chapter 10, Einstein's defense of a heuristics based on mathematics is quite different from epistemological realism, which he explicitly repudiated for the rest of his life. Einstein's critique of quantum mechanics was not based on epistemological realism, but rather on methodological lessons that Einstein believed he had learned from his earlier work in both relativity and statistical mechanics. Guided by his understanding of the theory of relativity as a theory of invariants, Einstein postulated that physics within the theory needs to distinguish between objective facts and their nonobjective descriptions.

Lehner calls Einstein's position methodological realism and identifies this as the basis of Einstein's conviction that quantum mechanics was only a phenomenological statistical description and not a complete theory. In this context, his critique of positivism received new urgency, as the defenders of the emerging Copenhagen orthodoxy explicitly invoked positivism to avoid the questions that Einstein wanted to pose. The EPR paper (Einstein, Podolsky, Rosen 1935), Einstein's most famous attempt to show the incompleteness of quantum mechanics, was formulated with the explicit aim to overcome what Einstein saw as positivistic excuses. Einstein's argument rested on a postulate of separability, which he elaborated in a paper for a special volume, edited by Wolfgang Pauli, of the journal *Dialectica* (Einstein 1948b).[78] A summary of Einstein's positions on both quantum mechanics and the issue of positivism can be found in his "Reply to Criticism" in the Schilpp volume (Einstein 1949b). Here, Einstein also most explicitly expresses his rapprochement to Kant's transcendental idealism (ibid., 674, 678, 680), which Thomas Ryckman discusses in Chapter 12. As Ryckman points out, this reappraisal of Kant has a long prehistory which reaches back into Einstein's work on general relativity and his dissatisfaction with a positivistic theory of science. In the end, Einstein could see Kant as an inspiration for finding a third position beyond the antagonistic positions of realism and positivism.

NOTES

1 Chapter 14 provides more detailed biographical information about Einstein. Information about Einstein and his work can be found in many places.

   Detailed chronologies of Einstein's life through March 1923 can be found in volumes 1, 5, 8, 10, 12, and 13 of *The Collected Papers of Albert Einstein* (these volumes will be cited throughout this volume as CPAE 1, etc.; the companion volumes with English translations of the documents will be cited as CPAE 1E, etc.). These volumes are produced by the *Einstein Papers Project* at the California Institute of Technology (www.einstein.caltech .edu). Many of the scientific and nonscientific manuscripts in the Albert Einstein Archives at the Hebrew University of Jerusalem are available online at www.albert-einstein.org.

   More selective chronologies of Einstein's life can be found in various editions of *The Quotable Einstein* (Calaprice 1996, 2000, 2005, 2011). The third edition contains a helpful section, "Answers to the Most Common Non-Scientific Questions about Einstein" (Calaprice 2005, 335–46).

   The two most recent full-scale biographies of Einstein, written for a general audience, are Neffe (2005) and Isaacson (2007). Starting in the 1990s, popular biographies, such as Fölsing (1993), Highfield and Carter (1993), Brian (1996), and Overbye (2000), began to make use of the materials presented in the CPAE volumes. This led to a tendency to focus on the early part of Einstein's life and career. This is especially true for the biographies of Fölsing and Overbye. The ones by Brian and Isaacson are more balanced, devoting many chapters to the American period. Several older biographies, such as Frank (1947), Seelig (1954), Clark (1971), and Hoffmann (1972), remain worth reading. Though superseded in places by more recent Einstein scholarship, Pais (1982) remains the best scientific biography.

   Several collections of essays on Einstein and his work appeared in connection with the centenary of his birth: French (1979), Holton and Elkana (1982), and Woolf (1980).

   For introductions to Einstein's physics aimed at physics undergraduates and a more general audience, see Cheng (2013), Kennedy (2012), and Stone (2013).

2 Einstein wrote these for the volume of the Library of Living Philosophers devoted to his work (Schilpp 1949). With typical self-deprecation, he referred to these "Autobiographical Notes" as his "own obituary" (Einstein 1949a, 3).

3 In the early 1920s, Maja Einstein (by then Winteler-Einstein; see note 4) started to write a biography of her brother. In 1924, she completed a thirty-six-page draft. See CPAE 1, xlviii–lxvi, for the first half of her biographical sketch.

4 Einstein boarded with Jost Winteler, a teacher at the school, and his wife Pauline. He dated their daughter Marie for awhile. Her brother Paul married Einstein's sister; her sister Anna married Einstein's friend Michele Besso (CPAE 1, 388).

5 Nothing was known about this daughter until the discovery in the 1980s of the "love letters" between Einstein and Marić, fifty-six letters, mostly from Einstein to Marić, exchanged between 1897 and 1903 (CPAE 1; Renn and Schulmann 1992).

6 This story was related to Abraham Pais by Einstein's longtime secretary Helen Dukas (Pais 1982, 47).

7 Hans Albert became a successful hydraulic engineer; Eduard was diagnosed with schizophrenia as a teenager and would end his life in a mental institution.

8 There is no evidence that Marić collaborated with Einstein on the papers of his miracle year or any other papers for that matter. For careful discussion of this issue, see Stachel (2005, liv–lxiii) and Martínez (2005; 2011, ch. 11).

9 In 1929, in response to a telegram from New York's Rabbi Herbert S. Goldstein asking "Do you believe in God," Einstein wrote: "I believe in Spinoza's God, Who reveals himself in the lawful harmony of the world, not in a God Who concerns himself with the fate and the doings of mankind" (New York Times, April 25, 1929; discussed in Jammer 1999, 48–9; Rowe and Schulmann 2007, 17–18; see also Calaprice 2011, 152–3, 321–45).

10 These papers are conveniently collected with introductions based on the editorial headnotes in CPAE 2 in Stachel (1998). Like the "statistical trilogy" (Einstein 1902b, 1903, 1904), the papers of Einstein's annus mirabilis appeared in Annalen der Physik. All papers Einstein published in this journal between 1901 and 1922 are collected (in the original German) in Renn (2005a), with editorial notes on the light quantum (Cassidy 2005), Brownian motion (Renn 2005c), special relativity (Rynasiewicz 2005), and general relativity (Janssen 2005). This material is available online at einstein-annalen.mpiwg-berlin.mpg.de. For introductions to the papers of Einstein's miracle year written for a general audience, see, e.g., Bernstein (2006) and Rigden (2005).

11 The article in the London Times in which Einstein introduced this distinction can also be found, under another title ("What Is the Theory of Relativity?"), in Ideas and Opinions (Einstein1954), an anthology of Einstein's writings still in print today.

12 As Martin Klein (1964, 9) characterized the strategy: "instead of trying to derive the distribution law from some more fundamental starting point … assume its correctness and see what conclusions it implies as to the structure of radiation." For an overview of the experimental background to wave-particle duality, see Wheaton (1983).

13 Einstein to Max Born, December 4, 1926 (Born 1971, 91). See Calaprice (2010) for this and other versions of this remark. See Jammer (1999, 222) for discussion of the remark in the context of Einstein's views on religion (cf. note 9).

14 See Pais (1982, 502–12), Friedman (2001, 126–40), and, especially Elzinga (2006) for accounts of how Einstein won the Nobel Prize.

15 See Chapter 2 for further discussion of "the myth of the Michelson-Morley experiment" with references to a classic paper by Gerald Holton (1969) destroying this myth and papers by John Stachel (1982, 1987) with additional evidence and further considerations on this topic. These papers are reprinted in Holton (1988 [1973]) and Stachel (2002a), respectively, two important collections of papers on Einstein. For a history of the experiments of Michelson and Morley (including facsimiles of their classic papers: Michelson 1881; Michelson and Morley 1886, 1887), see Swenson (1972). For more recent discussion, see Staley (2008) and van Dongen (2009).

16 See, e.g., Einstein to Paul Ehrenfest [before June 20, 1912] (CPAE 5, Doc. 409).

17 For opposing views on the status of Lorentz invariance in special relativity, see Brown (2005), Brown and Pooley (2006), and Janssen (2009).

18 See Martínez (2009) for discussion of some of the roots of Einstein's new kinematics.

19 As he wrote in a letter in 1908: "I even seriously doubt that it will be possible to maintain the general validity of Maxwell's equations for empty space" (Einstein to Arnold Sommerfeld, January 14, 1908 [CPAE 5, Doc. 73]).

20 See the Appendix for a derivation of the addition theorem of velocities in special relativity.

21 See Galison (2003) for an account of the origins of special relativity that downplays the electrodynamical context and emphasizes the technology of clock synchronization instead.

22 For more technical discussion, see, e.g., Miller (1998 [1981]). His book on Einstein and Picasso (Miller 2001) downplays the electrodynamical context and explores affinities between relativity and cubism instead.

23 See the exchange between Ohanian (2009, 2012) and Mermin (2011, 2012) for further discussion of derivations of $E = mc^2$ by Einstein and others.

24 Long after he left the Patent Office, Einstein would periodically act as a consultant in patent disputes (see, e.g., CPAE 7, Docs. 11, 21, 30, 48, 66, 67; CPAE 13, Docs. 18, 28, 226). For further discussion, see Illy (2012).

25 She was a first cousin on his mother's side of the family and a second cousin on his father's side: their mothers were sisters and their fathers were first cousins.

26 Two of Einstein's closest friends, Besso and Heinrich Zangger, mediated between Einstein and Marić in this period (see CPAE 8 and Schulmann 2012).

27 A letter from Ilse Einstein to Georg Nicolai, May 22, 1918 (CPAE 8, Doc. 545) reveals that the year before he married Elsa, Einstein contemplated marrying her daughter (Calaprice 2005, 338–9).

28 On Nicolai, see Zuelzer (1982).

29 Papers and manuscripts are presented in CPAE 3, 4, 6, and 7; letters in CPAE 5 and 8. The main publications are Einstein (1907j, 1911h, 1912c, 1912d);

Einstein and Grossmann (1913, 1914b); and Einstein (1913c, 1914o, 1915f, 1915g, 1915h, 1915i, 1916e). Einstein subsequently extended and consolidated theory (see, e.g., Einstein 1916o, 1917b, 1918a, 1918e, 1918f). He also wrote a popular book on both special and general relativity (Einstein 1917a). The closest he ever came to writing a textbook on relativity is the thoroughly revised published version of his 1921 Princeton lectures, available in English as *The Meaning of Relativity* (Einstein 1956a [1922c]).

30 The Zurich Notebook is presented in facsimile and in transcription and with a detailed commentary in *The Genesis of General Relativity* (Renn 2007a), a set of four volumes collecting both primary and secondary sources on gravitational theory in the period leading up to the birth of general relativity.

31 Reprinted in *Ideas and Opinions* (Einstein 1954).

32 See Janssen (2012) for further discussion.

33 See Chapter 6 for further discussion and references to the extensive literature on this pair of arguments.

34 Einstein did not care for the way in which the wave-particle duality of light was eventually explained in quantum theory, for the first time in a section written by Pascual Jordan of the so-called *Dreimännerarbeit* (Born, Heisenberg, and Jordan 1926; for discussion, see Duncan and Janssen 2008). As Einstein wrote a few years before he died: "All these fifty years of conscious brooding have brought me no nearer to the answer to the question 'What are light quanta?' Nowadays every Tom, Dick, and Harry thinks he knows it, but he is mistaken" (Einstein to Besso, December 12, 1951; quoted, e.g., in Klein 1979, 138).

35 Reprinted in a slightly different version in *Ideas and Opinions* (Einstein 1954). Page references are to this reprint. See Norton (2000), van Dongen (2010), and the essays by Ryckman and Lehner (Chapters 10 and 12) for further discussion of this passage.

36 Einstein to Cornelius Lanczos, January 24, 1938, quoted and discussed in Chapter 12 and (along with similar comments on other occasions) in Holton (1988, 259) and, more comprehensively, in van Dongen (2010, 57).

37 Einstein eventually did find the new field equations before Hilbert (Corry et al. 1997, Sauer 1999, Renn and Stachel 2007).

38 These two characterizations are from Einstein (1915f, 782) and Einstein to Sommerfeld, November 28, 1915 (CPAE 8, Doc. 153), respectively.

39 For further discussion of the development of Einstein's methodology and the role of his work in general relativity in that development, see van Dongen (2010).

40 This lecture is reprinted in *Ideas and Opinions* (Einstein 1954) under the misleading title "Principles of Research." It was originally printed under the title "Motivations for Research."

41 See Howard (1997) for Einstein's reading of Schopenhauer. In addition to the sentiments expressed in this passage, Howard argues, Einstein drew on Schopenhauer's work in his thinking about the foundations of both general relativity and quantum mechanics.

42 This is around the time that the question arose whether Einstein should marry Elsa or her daughter Ilse (see note 27). In late December 1919, Pauline Einstein, dying of stomach cancer, moved in with her son and new

daughter-in-law. When she died less than two months later, Einstein wrote to a close friend: "one feels in one's bones what blood ties mean!" (Einstein to Zangger, February 27, 1920 [CPAE 9, 332]).

43 Einstein to Pauline Winteler, May 1897 (CPAE 1, Doc. 34). Cf. note 4.

44 See *Ideas and Opinions* (Einstein 1954, Pt. IV) for the ensuing declarations and counterdeclarations of Einstein and the leadership of the Prussian Academy (see also Rowe and Schulmann 2007, 271–8).

45 Some authors, notably Earman and Glymour (1980a), have argued that the observations were inconclusive. For an opposing view, see Kennefick (2009, 2012); see also CPAE 9, Introduction, Sec. III. For the reception of general relativity by the astronomy community, especially in the United States, see Crelinsten (2006). Light bending constitutes one of the three "classic tests" of general relativity, the other two being the gravitational redshift (Earman and Glymour 1980b; CPAE 9, Introduction, Sec. IV) and the anomalous advance of the perihelion of Mercury (Earman and Janssen 1993, Janssen 2003a). Einstein did not see these "classic tests" as the best evidence for his theory. In a way that is reminiscent of his reaction to Kaufmann's alleged refutation of special relativity quoted previously, he told one of his correspondents decades later: "Even if the deflection of light, the perihelial movement or line shift were unknown, the gravitation equations would still be convincing because they avoid the inertial system (the phantom which affects everything but is not itself affected). It is really rather strange that human beings are normally deaf to the strongest arguments, while they are always inclined to overestimate measuring accuracies" (Einstein to Born, May 12, 1952; Born 1971, 192).

46 See Eisinger (2013) for an account of Einstein's travels between 1922 and 1933 based on his travel diaries.

47 The antirelativity campaign, spearheaded by Philipp Lenard and Ernst Gehrcke, reached a climax of sorts with a special session on relativity at the annual meeting of German Natural Scientists and Physicians in Bad Nauheim in 1920 (Einstein et al. 1920), which left the German physics community deeply divided (CPAE 7, 111). For discussion of the antirelativity movement, see, e.g., CPAE 7, the headnote, "Einstein's Encounters with German Anti-Relativists" (101–13); Rowe (2006); Rowe and Schulmann (2007, ch. 3); and Wazeck (2009, 2013). An antirelativity event in the Berlin Philharmonic Hall in August 1920, organized by the right-wing journalist Paul Weyland, provoked Einstein to publish a short reaction in a Berlin newspaper, under the title "On the Anti-Relativity, Inc." (Einstein 1920f).

48 His inaugural lecture, "Ether and Relativity" (Einstein 1920j), presented after some delay on October 27, 1920, is reprinted in Einstein (1983). See van Dongen (2012) for an account of how Einstein came to be appointed in Leyden.

49 For his reaction to Rathenau's assassination, see Einstein (1922i), also presented and discussed in Rowe and Schulmann (2007, 122–4).

50 For the text of Einstein's letter to Roosevelt, see, e.g., Rowe and Schulmann (2007, 359–61) or Calaprice (2005, 377–9). For discussion of how Szilárd got Einstein to write this letter, see, e.g., Rhodes (1986, 303–8) or Isaacson (2007, ch. 21).

51 Unbeknownst to the intelligence community, Einstein could actually have posed something of a security risk. In 1998, it was revealed that, during

the war, Einstein had an affair with a Russian spy, Margarita Konenkova (Jerome 2002, postscript).

52 Dukas and Otto Nathan were the executors of Einstein's will. Dukas worked tirelessly until her death in 1982 organizing and adding to the Einstein archive, yet she and Nathan also went to great lengths to block any publication that might tarnish their plaster-saint image of Einstein (Highfield and Carter 1993, ch. 12; Stachel 2002, 98–9).

53 Marić died in 1948, Eduard in 1965, and Hans Albert in 1973.

54 In 1917, Einstein had successfully prevailed on the authorities to spare the life of Friedrich Adler, a physicist he knew, who had assassinated the Austrian prime minister Count Stürgkh (Fölsing 1997, 402–5; Galison 2008). For a facsimile of a draft of Einstein's petition on Adler's behalf, see Renn (2005b, Vol. 3, 317).

55 See, e.g., Einstein to Norman Thomas, March 10, 1954 (EA 61–549; Isaacson 2007, 533; Rowe and Schulmann 2007, 500–1).

56 On Einstein and Oppenheimer, see Schweber (2008).

57 Serber reports that Pais told him this. The additional comment about Oppenheimer's unrequited love for the U.S. government, quoted both by Bird and Sherwin (2005, 503–40) and by Isaacson (2007, 532), is not included in Pais's published versions of the story, which are almost verbatim the same (Pais 1982, 11; 1994, 241; 1997, 326).

58 CPAE 6, Doc. 20; quoted by Calaprice (2011, 289); see also Rowe and Schulmann (2007, 75).

59 Einstein to Hans Mühsam, September 24, 1948 (EA 38–379). Cf. (Rowe and Schulmann 2007, 346).

60 Einstein to Hans Albert Einstein, December 28, 1954, quoted by Isaacson (2007, 537).

61 See Pais (1982, 483–501) for brief descriptions of all of Einstein's collaborators.

62 For the relationship between Einstein and Gödel, see Yourgrau (2005).

63 Einstein to the Besso family, March 21, 1955 (EA 7–245; Calaprice 2011, 113).

64 See Calaprice (2005, 369–73) for Helen Dukas's account of Einstein's last days. Johanna Fantova kept a record of her conversations with Einstein between October 1953 and April 1955 (the last entry is dated April 12). See Calaprice (2005, 349–68) for excerpts of these Fantova diaries.

65 Before Einstein's body was released to the family, however, his brain was removed during an unauthorized autopsy (Paterniti 2000, Abraham 2001).

66 For an introduction to the Duhem-Quine thesis about the underdetermination of theories by evidence, see, e.g., Curd and Cover (1998, Pt. 3, 255–411), which includes a reprint of Quine (1951) and selections of Duhem (1954), the English translation of the second edition of his 1906 book.

67 The lecture, which was originally published without a title, is reprinted under the title "Principles of Theoretical Physics" in *Ideas and Opinions* (Einstein 1954, 220–3).

68 Recall Einstein's characterization of competitors of special relativity when discussing Kaufmann's experiments: "their basic assumptions ... are not suggested by theoretical systems that encompass larger complexes of phenomena" (Einstein 1907j, 439).

69 Einstein's office in Prague was overlooking a park belonging to an insane asylum. Showing his office to his successor Philipp Frank, Einstein told him, pointing to the patients walking outside: "Those are the madmen who do not occupy themselves with the quantum theory" (Frank 1947, 98).

70 The displacement law, derived by Wilhelm Wien in 1893, prescribes how the peak in the distribution of the energy of heat radiation over frequency moves to higher frequencies when the temperature of the emitting black body is increased. The displacement law provides an important constraint on admissible laws for the spectral distribution of black-body radiation. The law Wien proposed in 1896 for this spectral distribution (and which Einstein used in his light quantum paper of 1905) satisfies this constraint, as does the more general law that Planck proposed in 1900 (Kuhn 1978).

71 This provides an alternative interpretation of Einstein's claim that general relativity made him a "into a believing rationalist" (see note 36).

72 See also the brief discussion in "Reply to Criticisms" (Einstein 1949b, 676–8).

73 For discussion of the rotating-disk thought experiment, see Stachel (1989a) and Chapter 6.

74 See Pyenson (1985), Corry (1997, 2004), Rowe (1999), and Chapter 13 for a discussion of the influence of Göttingen mathematics on Einstein. Einstein's own general theory of relativity, based on Riemannian geometry, was used as a prominent example after 1915, but the idea goes back to the late nineteenth century.

75 See Heilbron (1986) for discussion of the Planck-Mach debate.

76 Einstein to Besso, April 29 and May 13, 1917, respectively (CPAE 8, Docs. 331 and 339). For further discussion, see, e.g., Holton (1968).

77 See the correspondence between Einstein and Weyl in CPAE 8.

78 In a letter to Erwin Schrödinger of June 19, 1935 (EA 22–047), Einstein gave a simple example illustrating his separability postulate and his charge that quantum mechanics is incomplete (see Chapter 10 for analysis of this example). In the EPR paper, he told Schrödinger, "the main point has been buried so to speak beneath learnedness."

1   Einstein's Copernican Revolution

1. INTRODUCTION

Copernicus laid the basis for a complete overhaul of the traditional astronomical worldview by simple rearrangement. He removed the Earth from its central position in the universe and moved the Sun from its originally marginal position into the center of his new system. Copernicus made these rearrangements while retaining much of the technical apparatus of traditional astronomy. Thus the Copernican revolution began not by discarding previously accumulated knowledge but by reorganizing it. Copernicus's contribution can serve as a metaphor for other conceptual innovations brought about through the rearrangement of knowledge gathered over generations in research traditions and suggested by the development of new perspectives on that knowledge. One of us, Renn (2006, 10) has introduced the term *Copernicus process* for this kind of innovation. Generally, a Copernicus process can be conceived of as the reorganization of a system of knowledge in which previously marginal elements take on a key role and serve as a starting point for a reinterpretation of the body of knowledge; typically much of the technical apparatus is kept, inference structures are reversed, and the previous conceptual foundation is discarded. Einstein's achievements during his miracle year of 1905 can be described in terms of such Copernicus processes.[1]

The *annus mirabilis* deserves its name for three groundbreaking papers – on the light quantum, Brownian motion, and special relativity, respectively – that Einstein submitted for publication to the *Annalen der Physik* over the short span of three and a half months in the spring of 1905.[2] Between the first and the second of these papers he also produced a doctoral dissertation on a new method for determining atomic dimensions using fluid phenomena. Much has been written about the genesis of these works, but by and large as if they had been produced independently of each other, as outcomes of separate lines of research. As eclectic as Einstein's interests may have been, there are *prima facie* reasons for thinking that the origins of these papers cannot be understood in isolation from his other papers. In his *Autobiographical Notes* Einstein (1949a, 14–53) describes how he probed the foundations of physics in the decade 1895–1905 beginning with critiques of mechanics and electrodynamics, found them inadequate, and then tried to construct an alternative foundation on the basis of known experimental facts. Initially, the endeavor only led to frustration and despair. Yet out of this emerged the *annus mirabilis*. The three papers completed in the spring of 1905 collectively represent a consolidation of lessons learned and insights gained in this failed enterprise. Each paper points to certain limitations of currently accepted physical laws by addressing a borderline problem, that is, a problem that pertains to two distinct domains of classical physics, such as electrodynamics and mechanics. Each overlaps with at least one of the others in the details of the physics. Fluctuations play an essential role in both the light quantum paper and the Brownian-motion paper (as is also emphasized in the contributions by Kox and Darrigol: Chapters 3 and 4). The empirical laws governing radiation play a role both in the relativity paper and in the light quantum paper. Finally, each seeks to establish inductively secured fixed points (Rynasiewicz and Renn 2006, 6) from which to carry on, achieving a conceptual breakthrough by way of a Copernicus process.

The systems of knowledge that formed the input for Einstein's Copernicus processes reached maturity in the hands of such masters of classical theoretical physics as Ludwig Boltzmann (the kinetic theory of gases), Hendrik Antoon Lorentz (the atomistic elaboration of Maxwellian electrodynamics), and Max Planck (the theory of black-body radiation). The young Einstein became familiar with these knowledge systems from a vantage point that differed from that of its originators and took a different attitude to them. While his predecessors approached the problems they encountered as puzzles to be solved within the general frameworks they had helped establish, Einstein looked upon them as revealing fundamental problems with these frameworks.

Indeed early on Einstein ambitiously began to pursue tentative counterproposals to the established theories of classical physics. These were built around speculative atomistic ideas, which formed a hidden connection between his seemingly diverse pursuits. Though hardly providing a viable alternative to classical physics, Einstein's speculations concerning the atomistic structure of matter and radiation served as the starting point for his formulation of statistical mechanics. This formalism, applicable to a wide range of physical systems and phenomena, allowed him to make connections between the macroscopic theory of heat and assumptions about the composition of the microworld. In particular, it made it possible to investigate some of the problematic consequences of the classical mechanistic worldview. Statistical mechanics thus contributed in an essential way to the recognition that the borderline problems that Einstein had identified could not be solved without changes in the foundations of classical physics. Only when this fundamental incompatibility had been demonstrated could Einstein's speculations achieve the status of revolutionary reinterpretations of classical physics. Even so, Einstein had to accept in 1905 that the attempt to replace the theories of classical physics by a unified comprehensive new one was too ambitious and premature. What he offered instead in the papers of his *annus mirabilis* were more piecemeal contributions to a new future synthesis.

In this chapter we elaborate this basic picture, drawing on correspondence and publications documenting Einstein's thinking prior to 1905. We emphasize those strands that connect the papers of his miracle year. For discussions more focused on content and background of these papers separately, we refer to the contributions of Norton, Kox, and Darrigol (Chapters 2, 3, and 4). In Section 2, we sketch how some of the key results of Boltzmann, Lorentz, and Planck were received and reinterpreted by Einstein and such contemporaries as Gibbs, Poincaré, and Ehrenfest, drawing an analogy with Galileo and his followers. In Section 3, we trace the development of the new perspective from which the young Einstein viewed these results. In Section 4, we discuss Einstein's early speculations about alternative theories capable of dealing with some of the borderline problems he had hit upon. In Section 5, we turn to Einstein's development of statistical mechanics as a particularly effective tool both to bring out the challenge of these borderline problems to the foundations of classical physics and to probe the alternative theories that he hoped would get around those difficulties. Finally, in Section 6, we show how these efforts resulted in the celebrated papers of the *annus mirabilis*.

2. EINSTEIN AS A DISCIPLE OF BOLTZMANN, PLANCK, AND LORENTZ

Even the results at the cutting edge of classical physics did not amount to a break with its conceptual framework.[3] Planck's derivation of his law of black-body radiation did not mark the beginning of the quantum theory.[4] Nor should Lorentz's derivation of the Lorentz transformation equations be looked upon as a first result of relativity theory. Instead, these authors obtained these results through stretching and patching up the knowledge system of classical physics. Furthermore, these results were the culminations of research traditions that had their own goals: Boltzmann used kinetic theory to lay a mechanical foundation of thermodynamics; Planck's primary goal was a deeper understanding of the second law of thermodynamics; and Lorentz was trying to develop Maxwell's theory of electromagnetism, carefully separating the roles of ether and matter.

Behind these original objectives were the problems of integrating different conceptual traditions, which at the same time provided the only means to solve these problems. Boltzmann's and Planck's struggles to understand the second law of thermodynamics confronted them with certain problems in connecting mechanics, thermodynamics, and electrodynamics. Lorentz's attempt to develop the consequences of Maxwell's theory for optics suggested an integration of Maxwell's field theory with the atomistic theory of electricity prevalent in continental Europe. Gradual shifts of emphasis within each research program, caused in part by experiments and in part by inner difficulties and contradictions, stretched the conceptual systems that formed the basis of the different theories to their limits.

At this point a comparison with the emergence of classical physics will be helpful.[5] This will illustrate the remarkable mechanisms at work when the structure of a knowledge system changes. The transformation of the preclassical mechanics of Galileo and contemporaries, still based on Aristotelian foundations, to the classical mechanics of the Newtonian era can be understood in terms of a Copernican process (as defined at the beginning of this chapter). Indeed, the role of Galileo's contributions in the emergence of classical physics is comparable to the role of the contributions of Planck, Lorentz, and Boltzmann in the emergence of special relativity, quantum physics, and statistical physics. Like Moses, Galileo did not reach the Promised Land, or, better perhaps, like Columbus, did not recognize it as such. Galileo arrived at the derivation of results such as the laws of free fall and projectile motion by exploring the limits of the system of knowledge of preclassical mechanics.

Consider Galileo's discovery of the parabolic shape of the trajectory of projectile motion. To derive this motion in classical mechanics we need two principles that are in direct contradiction to Aristotelian natural philosophy, the principle of inertia and the principle of superposition of motions. According to Aristotle, all heavy bodies tend in "natural motion" to their natural place, the center of the Earth, unless they are prevented from doing so by some "violent" force. A violent motion such as projectile motion can only continue as long as a violent force keeps it going. In classical physics, the laws of projectile motion readily follow from its fundamental principles: according to the principle of inertia a projectile moves uniformly in the direction in which it was thrown or fired without being propelled by any force. At the same time it accelerates downward according to the law of free fall. The superposition of these two motions gives the projectile's parabolic trajectory. This derivation exhibits such a close and direct connection between the laws of projectile motion and the principles of classical physics that it is hard to avoid the conclusion that Galileo must already have had these principles at his disposal. This, in turn, naturally leads to the claim that Galileo is the founder of classical mechanics. A closer examination of Galileo's works, however, shows that he failed to give a general proof for the parabolic shape of a projectile's trajectory, just as he failed to give a consistent derivation of his law of free fall despite many years of effort. The reason was that his thinking was still deeply rooted in the knowledge system of Aristotelian dynamics.

It was therefore left to Galileo's disciples to formulate the concepts of classical mechanics that are not found explicitly in Galileo's works. Nevertheless, it is easy to read the principles of inertia and of superposition of motions into Galileo's works. In fact, his pupil Bonaventura Cavalieri had already published (in 1632) the first derivation of the trajectory of an obliquely thrown projectile by the time Galileo finished his last work, the *Discorsi* of 1638. Cavalieri uses the principles of inertia and superposition as in the modern proof. He claims only to have stated what was already generally known as Galileo's achievement. Cavalieri, moreover, was not the only disciple of Galileo to produce the derivation of the parabolic trajectory using the new concepts. What seems natural in the works of almost all of his disciples, however, is not found in the works of their master.

So in the transition from preclassical to classical mechanics, the "errors" of preclassical mechanics did not need "correction"; the results of preclassical mechanics could simply be reinterpreted by the next generation as linchpins of a newly constructed system of knowledge, the system of classical mechanics. Within the old system, one had achieved these results through problematic derivations, by introducing ad hoc hypotheses, suppressing ambiguity, extrapolations – in other words, by extending and stretching the old conceptual structure. The results emerging at the edges of the old system, however, could effectively be reinterpreted as determining the concepts of the new system. For example, in his identification of the motion of a projectile as following a parabolic path, inertial motion as a uniform motion along a straight line appeared as a problematic possibility at the edge of Galileo's Aristotelian conceptual framework. Galileo's result was then reinterpreted by his disciples as the starting point of a new conceptual framework in which the concept of inertia would instead play a central role. However, a Copernicus process is more than just a rearrangement of things that are already known into a more coherent form. Rather, it constitutes the beginning of a far-going reinterpretation of the old results in the light of a new conceptual framework, which itself emerges from a reflection on the available knowledge. It therefore makes little sense to try to establish which the more creative act is: the exploration of the old system eventually yielding the insights that trigger its reorganization or the realization of the break in reaction to a reflection on the old system's shortcomings. Sometimes in history it is the exploration which is considered to be the act of genius as in the case of Galileo, and sometimes it is the reflective reinterpretation as is the case with Einstein.

Can Einstein's revolutionary achievements be captured in terms of such Copernicus processes? Did Einstein similarly formulate concepts that were implicitly prepared for him by the masters of classical physics? In this sense Lorentz would then be the one who both completed the ether theory and initiated the theory of relativity; Planck would be the one who put the capstone on classical radiation theory and, at the same time, fathered quantum theory; and Boltzmann would be both the last great master of the mechanical theory of heat and the originator of statistical physics. All three would be in a position similar to Galileo's, who was both a late representative of Aristotelianism and a pioneer of classical mechanics. Einstein, accordingly, would be in a position similar to that of Galileo's disciples.

Another factor that speaks for this interpretation is that, as in the case of Galileo's disciples, the new ideas of modern physics were articulated by different authors at the same time. This is true both for Einstein's early work on statistical mechanics and for his three main contributions of 1905.

Einstein's papers on statistical physics from the years 1902 to 1904 originated independently and roughly simultaneously with the foundational treatise of Gibbs on the principles of statistical mechanics (Renn 1997, Uffink 2006). However, unlike Gibbs, Einstein immediately explored connections between his work on statistical mechanics and the wide array of other interests he was pursuing at the time such as electron theory or black-body radiation.

Paul Ehrenfest (1905, 1906) noted independently of Einstein that Planck's treatment of the energy distribution of heat radiation in equilibrium was effectively based on two hypotheses. One of them states that radiative energy is composed of particles of energy, with the energy proportional to the frequency – a hypothesis corresponding in essence to Einstein's light quantum idea. Whereas Einstein saw his own hypothesis as revolutionary, Ehrenfest saw it merely as a concise and modest interpretation of Planck's theory. Moreover, Einstein, again unlike Ehrenfest, proposed additional applications of the light quantum hypothesis such as the photoelectric effect. Partly for these reasons, Ehrenfest's contribution has received far less attention from historians of science than

Einstein's. Yet, from an epistemological point of view and focusing only on black-body radiation, Ehrenfest's reinterpretation of Planck's seemingly classical results was of similar significance for the emergence of the nonclassical quantum theory as Einstein's.

Henri Poincaré (1905, 1906), before and independently of Einstein, declared the principle of relativity to be one of the central principles of physics. He clearly saw the problematic character of the simultaneity of spatially separated events. And he interpreted the auxiliary time variable that Lorentz had introduced and given the name "local time" as the time registered by moving clocks synchronized with light signals.[6] Einstein does all of this in his relativity theory. Poincaré's insights, however, are dispersed among several of his papers, and at no place are they combined into a coherent theory. What matters in this context, however, is that the central conceptual achievements of relativity are at least indicated in the works of Poincaré, and that Poincaré, like Einstein, reinterpreted results established earlier, particularly those of Lorentz.

Finally, the Polish physicist Marian von Smoluchowski (1906) published an article about the kinetic theory of Brownian motion in the *Annalen der Physik*. The publication of his work was triggered by Einstein's papers, but the results he presented were obtained independently. Smoluchowski's arguments differ from Einstein's in technical details. But, except for a numerical factor, his results and his mathematical description of Brownian motion are equivalent to those of Einstein. There are also parallels to some of Einstein's important intermediate results, particularly in the work of the Australian physicist William Sutherland.

These parallel discoveries support the view that Einstein's breakthrough of 1905 was not the singular achievement of a lone genius but the result of structural changes in systems of knowledge accessible to contemporary scientists as a shared knowledge resource. Where others saw only individual issues, however, Einstein realized that these insights amounted to a break with classical physics, making it necessary to seek an entirely new conceptual basis. What distinguishes Einstein is the sheer exhaustiveness with which he pursued the reinterpretation of the old conceptual frameworks. He thus went beyond a reinterpretation of previous results, implemented the new concepts to yield new insights, and created a network of interrelated theoretical questions, establishing a research program of his own. In that sense he quickly became much more than a disciple of the masters of classical physics.

Both Planck and Lorentz saw Einstein's work as starting from a reinterpretation of the foundations of their own work. In his Nobel Lecture of 1920 Planck suggests that the question whether the derivation of the radiation law was more than "a play with formula, devoid of content" or whether it was "based on a sound physical idea" was not answered by "the proof of the energy distribution law of heat radiation, still less to the special derivation of that law devised by me" (Planck 1920, 127). Rather, the answer came from "the relentless forward-pushing work of those research workers who used the quantum of action to help them in their own investigations and experiments" (ibid.). In particular he cited Einstein's contributions. To Lorentz it was perhaps even more evident than to Planck that Einstein's new concepts gave physical meaning to quantities in his original derivation that he had regarded only as auxiliary variables. In his Columbia Lectures in 1906 Lorentz remarks: "The designations 'effective coordinates', 'effective time' etc. of which we have availed ourselves for the sake of facilitating our mode of expression, have prepared us for a very interesting interpretation of the above results, for which we are indebted to Einstein" (Lorentz 1909, 223). In a note added in the second edition Lorentz reflects: "The chief cause of my failure was my clinging to the idea that the variable t only can be considered as the true time and that my local time t' must be regarded as no more than an auxiliary mathematical quantity" (Lorentz 1915, 321). Galileo's disciples were able to formulate the basic concepts of classical physics because they did not have to follow the difficult route that had led Galileo to his results and did not share his initial conceptual assumptions. They interpreted the work of their master from the shifted perspective of his final results. This enabled them to introduce concepts that were already defined implicitly by these results. Einstein's conceptual innovation can be understood as the outcome of a comparable structural change in the edifice of classical physics. In this sense Einstein can be seen as a disciple of the masters of classical physics, Boltzmann, Lorentz, and Planck. Our first task is to reconstruct the perspectives under which Einstein gained the traditional knowledge of classical physics, which he then reinterpreted and rearranged in a Copernicus process.

3. THE ORIGINS OF EINSTEIN'S NEW PERSPECTIVE

The knowledge on which Einstein could draw when writing his groundbreaking papers of 1905 has to be considered as the result of a complex intellectual experience, which was gained partly through work on problems that have no direct connection with his famous papers. The crucial role of Einstein's perspective as developed through such experience becomes particularly evident when we examine how he appropriated the knowledge of his teachers of classical physics.

This acquisition can be divided into two stages. In the first, Einstein became familiar with the works of the masters and tried to develop speculative alternatives to the dominant theories of classical physics; in the second he studied them again in connection with the work that would lead directly to the papers of 1905. The sources for an exact reconstruction of these two phases are quite limited, although correspondence between the young Einstein and his fiancé Mileva Marić from the time between 1897 and 1903 is available (Renn and Schulmann 1992). Nevertheless, it is possible to identify, in his approach to the central problems of his papers of 1905, essential features of the impact of Einstein's prior knowledge of science.

Einstein's perspective was from early on influenced by his atomistic worldview.[7] This view served to structure the conceptual framework in which he inserted his newly acquired knowledge. It also provided him with a means of interrelating seemingly unrelated scientific subject matter. In the popular science books of Aaron Bernstein (1853–7) that he devoured as a youth, he had already learned how concepts like atoms or ether could help to uncover "mysterious" and "surprising" relations between different areas of knowledge, which were separated by specialization into various scientific disciplines.[8] Furthermore, several of the conceptual tools used by Bernstein and other authors to establish connections between different areas of physics and chemistry left their mark and arguably determined Einstein's reactions to the readings and lectures of his student days.

Besides the atomic structure of matter, Einstein's interests around 1900 included the relation between thermal and electrical properties of matter. He had already dedicated his diploma thesis to this subject, but now he apparently started to regard it from an atomistic point of view and began to develop independent theories about it. In October 1899 he had devised a method to decide whether the specific heat of an electrically charged body is different from that of an uncharged one. Einstein was excited about physical chemistry (Buchwald et al. 2013), ion theory in particular, which had been spectacularly successful in a great variety of areas. He speculated – again on the basis of atoms – about the connection between the optical and thermal behavior of solid bodies and studied Planck's papers on the thermal equilibrium of radiation. He developed the idea that the generation of light might involve the direct conversion of kinetic energy into light energy (see Section 4). He was fascinated by Lenard's production of cathode rays (i.e., a beam of electrons) by ultraviolet light. Einstein's arguments about an atomistic structure of radiation developed from these interests. In addition, he was working at this time on a molecular theory of solutions.

What is interesting about the multitude of Einstein's efforts around the turn of the century is not so much their results – for those remained rather meager – but the character of the problems he worked on and their importance for the further development of his new perspective. Most of these problems belong to the category of borderline problems of classical physics. Einstein's perspective was characterized by the search for a conceptual basis for all of physics, which he hoped to find with the help of his effectively "interdisciplinary" atomism. Many of the arguments known to us from his letters are based on attempts to use atomistic ideas to explain connections between seemingly disparate physical properties. For example, he sought a connection between the electrical and thermal conductivities of metals, which he traced back to mobile charged particles inside metals, or between intermolecular and gravitational forces. In this sense he wrote to Grossmann in the spring of 1901:

I am now convinced that my theory of atomic attraction can also be extended to gases.... That will then also bring the problem of the inner kinship between molecular forces and Newtonian action-at-a-distance forces much nearer to its solution.... It is a glorious feeling to perceive the unity of a complex of phenomena which appear as completely separate entities to direct sensory observation. (Einstein to Grossmann, April 14, 1901 [CPAE 1, Doc. 100])

Einstein's idea of an atom was not defined by a particular theory such as mechanics, but served as a mental model[9] whose exact properties still had to be discovered from experience. Thus he wrote to Mileva about atomism in Boltzmann's theory of gases: "Everything is very nice, but there is too little stress on the comparison with reality" (Einstein to Marić, April 30, 1901 [CPAE 1, Doc. 102]). Of all the themes of the year 1905, it is the interest in the electrodynamics of moving bodies that reaches farthest back in Einstein's life. At sixteen, the young Albert had already written an essay about the ether (CPAE 1, Doc. 5). There he proposed experiments to demonstrate the altered state of the ether, due to the presence of magnetic fields, by its effect on light or other electromagnetic radiation. Apparently his aim was to discover something about the inner composition of the ether. For someone his age the essay shows remarkable familiarity with the concepts and problems of electrodynamics. Young Albert not only knew about Hertz's experiments on the production of electromagnetic waves, but also about the efforts to explain such waves as the result of mechanical deformations in an elastic ether. Around this time he was already asking himself how a light wave would look to an observer moving alongside of it at the speed of light.[10] He would have to see a time-independent wave field, but it seemed that there could be no such thing. The teenage Einstein's thought experiment also raised the question of what the speed of light would be that this observer would measure. The answer obviously depended on the underlying ether model. In a stationary ether, which is not carried along by the moving system, the speed of light relative to the observer would certainly have to change. Thus Einstein hit upon the borderline problems between electromagnetism and mechanics early on.

Einstein's perspective guiding his exploration of the borderline problems of classical physics was not only shaped by his intellectual experience with an emerging interdisciplinary atomism, but also by his voracious reading of a wide range of literature. In addition to the popular scientific literature he read as a youth, he also worked his way through Violle's (1893) comprehensive physics textbook. As a student he intensively studied advanced technical literature such as Föppl's (1894) introduction to Maxwell's theory, and the comprehensive works of Boltzmann, Drude, Helmholz, Kirchhoff, and Lorentz. He also closely followed current scientific journal articles at the forefront of research, such as the important papers by Drude on the electron theory of metals, by Hertz on the electrodynamics of moving bodies, by Lenard on the generation of cathode rays, and by Wien and Planck on heat radiation. In addition, Einstein studied literature involving historical, foundational, and methodological questions, in part together with his friends Maurice Solovine and Conrad Habicht, with whom he founded the "Olympia Academy" in 1902. Works such as those by Mach on the history of mechanics and on heat theory, David Hume's *Treatise on Human Nature*, and Poincaré's *Science and Hypothesis* contributed to his awareness of the constructed character of scientific concepts, the role of conventions in defining concepts such as that of simultaneity, and the tension in scientific theories between formal principles, such as the second law of thermodynamics, and specific models, such as that of atoms. Exploring the limits of classical physics, the young Einstein was thus very well aware of the potential that lay in its heritage, both in terms of concrete achievements and in terms of epistemological reflectivity.

4. SPECULATIVE ALTERNATIVES TO CLASSICAL PHYSICS

Even the first phase of Einstein's reception of the "classics" had far-reaching consequences. From contemporary evidence and later recollections, we know that, as early as 1901, Einstein was searching for alternatives to the established views of classical physics. Sometime between 1900 and 1905 his interdisciplinary atomism led to a radical break with the nineteenth-century tradition of optics and electrodynamics. Einstein decided to work on a corpuscular or "emission" theory of radiation, similar to that proposed by Newton in the seventeenth century. He did this in spite of the overwhelming evidence, obtained from the beginning of the nineteenth century onward, that clearly pointed to a wave theory of light and that had led to an apparently definitive refutation of the corpuscular theory.

As unconventional as it was, such a revitalized corpuscular theory was in line not only with his interdisciplinary atomism, but also with the fact that the question "wave or particle?" was at that time being asked anew – not necessarily for light, but for recently discovered forms of radiation such as X rays. Through his teacher Conrad Wüest, Einstein may have heard of the controversial discussions about the character of these new rays. They had been discovered as a new kind of light by their effect on photographic plates, but they did not exhibit the familiar properties of light and other electromagnetic radiation such as reflection, refraction, interference, and polarization. Thus the question regarding whether X rays consist of particles was being seriously discussed.

But most of all, Einstein's corpuscular theory of radiation seemed to contain the key to a whole set of phenomena that had claimed his attention as a student. These included processes such as the generation and transformation of light, on which new experimental research had been done. We already mentioned this interest in Section 3; let us now look at it in more detail. On April 30, 1901, he wrote to Mileva: "It occurred to me recently that when light is generated, direct conversion of kinetic energy to light may take place because of the parallelism kinetic energy of the molecules – absolute temperature – spectrum (radiating space energy in the state of equilibrium). Who knows when a tunnel will be dug through these hard mountains!" (Einstein to Marić, April 30, 1901 [CPAE 1, Doc. 102]). Soon thereafter he reported to Mileva with enthusiasm about Lenard's work on the generation of cathode rays by ultraviolet light, which convinced him that the inverse process must also exist, that is, that light could be converted directly into kinetic energy.

Einstein's thoughts about the corpuscular nature of radiation also touched on yet another problem, that of heat radiation, mentioned in his letter of April 30. He had read the latest research with critical attention. In a letter to Mileva of early April 1901 he had mentioned his misgivings about Planck's theory. Einstein's objection is directed against Planck's picture of the mechanism for distributing radiation energy over the different frequencies of the radiation. He did not believe that Planck's resonators could bring about this distribution in the way envisaged. In his own analysis of the problem of heat radiation Einstein was in favor of processes that transform light energy directly, as he mentioned in the cited comment from the end of April concerning the parallelism between the kinetic energy of the molecules and the spectrum of heat radiation. But what exactly is the point of this parallelism?

With his remark about the "kinetic energy of the molecules" Einstein refers to the velocity distribution of molecules in a gas of definite temperature, as derived by Maxwell and Boltzmann within the kinetic gas theory. This distribution of molecules over different speeds obeys a probability law represented by a Gaussian bell-shaped curve. In 1896, Wilhelm Wien had proposed an empirically reasonably adequate formula for the distribution of the energy of black-body radiation of a given temperature over different frequencies that is very similar to this Maxwell-Boltzmann formula for the distribution of the energy of a gas of a certain temperature over different velocities. This is the parallel Einstein refers to.[11] The relevant energies enter the respective expressions exponentially, and the analogy becomes even closer if the distribution of radiation energy is converted into a distribution of the number of particles of radiation, whose energy is assumed to be proportional to their frequency. Thus Einstein's parallelism leads to the revolutionary idea that heat radiation can be thought of as a gas of light particles. The problem of equipartition of energy among the different frequencies of radiation, for which, in Einstein's opinion, Planck's resonators were unqualified, can also be dealt with in this new picture, as can be shown by a simple calculation (Büttner et al. 2003).

Einstein must have been struck by the fact that the particle theory of light provides a new perspective on so many problems. It even made the electrodynamics of moving bodies appear in a new light. In a corpuscular theory one would expect there to be agreement between the laws of mechanics and those of electrodynamics, at least if the latter were appropriately changed. In a corpuscular theory the speed of light is constant only with respect to the source of the light – this is the reason why such a theory is also called an emission theory. It is no longer constant with regard to an ether at rest, but depends on the velocity of the light source, just as the velocity of a cannon ball depends on the velocity of the cannon used to fire it. In such a theory the familiar laws of mechanics apply, in particular, the principle of relativity and the ordinary rules for the composition of velocities. Indeed, a corpuscular theory yields a simple explanation of stellar aberration[12] as a consequence of the composition of the speed of light with the speed of the earth, without assuming an ether about whose state of motion one can only speculate. As is clear from his 1905 publication, Einstein was deeply puzzled by the fact that, in considering the interaction between a magnet and a conductor in relative motion to each other, classical electromagnetic theory resorted to two different explanations depending on whether the magnet or the conductor was considered to be at rest. This asymmetry may have even constituted for Einstein the first strong hint at extending the principle of relativity also to electromagnetism (Earman et al. 1983).

5. STATISTICAL MECHANICS AS A BRIDGE BETWEEN CLASSICAL AND MODERN PHYSICS

In the first phase of his encounter with classical physics Einstein developed the speculative alternatives to established theories we have just discussed. These, however, remained conceptually within classical physics or even reverted back to older theories. How did the breakthrough of 1905 emerge from these early speculations? Critical factors were once again the concrete physical problems that moved Einstein in his search for evidence for the reality of atoms and their role in unifying diverse branches of physics. As early as 1901–2 these problems presented him with the challenge of elaborating the kinetic theory of heat, as he had learned it from Boltzmann's (1898) lectures on gas theory, into a general statistical mechanics (Renn 1997, 2008). This approach would enable him to extend the range of applicability of atomistic pictures to new physical processes for which the pertinent microscopic dynamics differed from that of an ideal gas or was even unknown in detail. It was the establishment of statistical mechanics that provided the conditions for turning the speculations of the young Einstein into a revolution of physics.

In view of Einstein's interest in such diverse applications as an atomistic theory of liquids, diffusion processes, the electron theory of metals, and heat radiation, a generalization of the kinetic theory must have appeared to him the most urgent step beyond what Boltzmann had already achieved. At first Einstein saw his development of statistical mechanics merely as "filling a gap" (Einstein 1902b, 417) in Boltzmann's theory and not as creating a conceptual alternative to it. But while Boltzmann had developed kinetic theory to extend the range of the mechanical foundation of physics, Einstein developed statistical mechanics to question and eventually overcome these very foundations.

How does statistical mechanics function as a bridge between classical and modern physics? Whereas kinetic theory typically starts with the interaction between the various atomic constituents of a macroscopic system, such as collisions between molecules of a gas, statistical mechanics does not focus on such processes. It can therefore be applied much more generally than kinetic theory. Statistical mechanics focuses not on time development but on the statistical properties of virtual ensembles. A virtual ensemble denotes a large number of essentially identical copies of one and the same system. All of these copies obey the same dynamics, but they differ in the exact configuration of their atomic constituents: there is a vast number of ways that billions and billions of atoms can zoom about and still make up a gas with the same volume, pressure, and temperature. The practically impossible analysis of the time development of a microscopically described system is now replaced by the evaluation of statistical mean values over the virtual ensemble for the system under consideration.

Individual elements of statistical mechanics can be found in the work of Maxwell and Boltzmann on the kinetic theory of heat, but it was the book by Gibbs (1902), *Elementary Principles of Statistical Mechanics*, that gave the first formulation of statistical mechanics as a complete and independent theory. That same year Einstein published the first of three papers on statistical physics (Einstein 1902b, 1903, 1904).[13] This statistical trilogy not only established statistical mechanics independently of Gibbs, it also provided the basis for Einstein's exploration of heat radiation and for his analysis of Brownian motion and other fluctuation phenomena as evidence for the existence of atoms.

Statistical mechanics constitutes a conceptual innovation that had a great impact on the further development of twentieth-century physics. That so many of its building blocks are found in the work of Maxwell and Boltzmann suggests that the innovation was largely due to a change of perspective, to a reinterpretation of preexisting results in a new light; in other words, that it was the result of a Copernicus process. Einstein was familiar only with Boltzmann's (1898) book, *Lectures on Gas Theory*, and not with his earlier papers. He therefore did not see the book in the light of Boltzmann's earlier achievements and of the goals that had motivated the latter's research. Einstein read the book with his own interests in mind and developed a new interpretation of its results by placing them in a new context.

This new context notably includes the electron theory of metals, but we focus here on the exploration of heat radiation. In spring 1901 he followed, as we have seen, Planck's (1900a, 1901b) publications about black-body radiation with great interest. One of them deals with the determination of atomic dimensions, and therefore was probably of particular interest to Einstein. But he was also interested in Planck's treatment of the interaction between matter and radiation by means of the already-mentioned concept of resonators. Following Planck's model, Einstein developed the idea that on a microscopic scale, matter consists of just such electric resonators as those assumed by Planck. From this speculative hypothesis Einstein inferred that there should be a connection between optical and thermal properties of matter. In particular, Einstein thought that on the basis of this picture one could explain deviations from the law of Dulong and Petit, which establishes a relation between the heat capacity and the atomic structure of a solid. Under his hypothesis, the Dulong-Petit law should be valid only for a body whose internal energy is distributed equally over its constituent elementary resonators. This, in turn, Einstein claimed, is possible only if the body absorbs radiation across the entire spectrum. Hence, transparent bodies, transmitting rather than absorbing visible light, should represent an exception to the Dulong-Petit law as not all of their internal degrees of freedom are "activated."[14]

Einstein's idea of 1901 bears a striking resemblance to his 1907 quantum theory of specific heats (Einstein 1907a).[15] In both cases, deviations from the Dulong-Petit law are explained and relations between radiative and thermal properties are established. Both approaches make use of Planck's studies of radiation in the context of a theory of the molecular constitution of matter. Einstein's later theory is based on the assumption that, contrary to classical physics, the energy of a single resonator can take on only discrete values. Although the speculative idea of the young Einstein did not anticipate such a break with classical physics, it provides a model whose details could later be made to agree with such new insights. What is more important for our story is that this idea also provided the framework for Einstein's critique of Planck's radiation theory. He was convinced that Planck needed a mechanism with which to establish thermal equilibrium between radiation of various wavelengths and matter. In fact, soon after formulating his idea Einstein began to have his doubts, mentioned earlier, about whether resonators with fixed period and damping can produce thermal equilibrium.

This insight undermined not only his own idea but also Planck's theory of heat radiation. As a consequence, Einstein gave up both his early resonator theory of specific heats and his belief that Planck's studies represent a satisfactory answer to the radiation problem. But the first encounter with Planck had placed just that aspect of the problem of radiation at the center of Einstein's attention that was to become one of the germs for the later conceptual innovation: the relationship between heat radiation in equilibrium and the law of equipartition of energy that gave rise to the Dulong-Petit law, which also played a key role for the theory of electrons in metals. Using statistical mechanics, Einstein succeeded in giving this law a basis that made it possible to apply it to non-mechanical systems.

In this way statistical mechanics became a tool for insights into new conceptual foundations of physics, going beyond the possibilities of transferring by analogy properties of the macroworld directly to the microworld. A special role was played by Einstein's interpretation of the connection – first proposed by Boltzmann for mechanical systems like gases – between entropy and probability, which had also been used by Planck when trying to find a basis for the radiation law. In 1904 Einstein derived, within his statistical mechanics, Boltzmann's principle – according to which the entropy of a system is taken as a measure for the probability of its microscopic states – and thus gave it a more general basis. In his light quanta paper of 1905 Einstein used Boltzmann's principle to make inferences about the particle nature of heat radiation, which he believed to be contrary to Planck's ideas, even if this contradiction is not clearly expressed (see Section 6.1).

With the help of Boltzmann's principle, Einstein could in particular extend the parallelism between heat radiation and a gas, which he had already hit upon in 1901. Boltzmann's principle allowed a more concise relation to be established between the Wien radiation law and an ideal gas, a relation that makes a direct connection between Wien's law and the assumption of statistically independent light quanta.[16] Einstein's argument from 1905 is based on a consideration of fluctuations in which Boltzmann's principle is used to make inferences from the entropy of a thermodynamic state to the probability that a certain number of particles are present in a certain volume. In the same way, the probability of finding a certain number of gas atoms in a given volume can be inferred from the entropy of an ideal gas.

From the formula for the spectral distribution of the energy in heat radiation an expression results quite generally for the entropy of the radiation of a definite frequency as a function of the volume. If one takes Wien's law as a basis, one can use Boltzmann's principle to derive the distribution of the density fluctuations of radiation by Einstein's argument. It turns out that if Wien's law is valid, the distribution behaves as if it consisted of independent quanta of light whose energy is proportional to their frequency.[17] Conversely, Wien's law can be obtained, in a similar way as the entropy of an ideal gas, using Boltzmann's connection between entropy and probability, if one assumes the light quantum hypothesis. So this reasoning suggests a direct analogy between an ideal gas and a gas of particles of light, at any rate to the extent that the energy distribution of the latter obeys Wien's law.

In 1904 Einstein must have been familiar with such insights into the microstructure of heat radiation; his third paper on statistical mechanics contains an examination of the fluctuations of heat radiation, which is closely connected to this reasoning. On the surface, the paper only deals with the classical aspects of radiation. But upon closer examination one discovers that even then Einstein was interested in probing the limits of the classical notions. The starting point of his analysis is his formula for the energy fluctuations of a statistical system, which he could write as a function of mean energy and temperature. The observation of such energy fluctuations would allow one to draw conclusions about the microstructure of the statistical system.

At that time Einstein thought that heat radiation was the only system where experience suggested the existence of fluctuations. He assumed that, for radiation in a cavity with linear dimensions comparable to the dominant wavelength in the spectrum of black-body radiation, the energy fluctuations would be of the same order as the mean energy. Now he could calculate from the empirically confirmed laws of heat radiation, namely the Wien displacement law and the Stefan-Boltzmann law, together with his assumption about the size of the radiation cavity, that the fluctuations occurring had to be of the order of the mean energy.

The remarkable feature of this result was that it rested on a macroscopic radiation law instead of specific assumptions about the microstructure of the radiation. For this reason the arguments concerning energy fluctuations in heat radiation could now be used to obtain information about this microstructure. At some point, Einstein must have found out that Maxwell's theory of electrodynamics, which interprets these fluctuations as interference between different components of heat radiation, is not compatible with the conclusion that he was able to draw from the empirically known laws of heat radiation about the nature of these fluctuations. However, the energy fluctuations of heat radiation can be explained with the help of the light quantum hypothesis. More accurately, the fact that the energy spectrum of heat radiation is described by Planck's formula implies that these fluctuations consist of features that have a wave character as well as others that can be explained by the particle nature of the radiation. With this insight, which Einstein published only in 1909 but had probably already achieved before 1905, the wave-particle dualism was born (Rynasiewicz and Renn 2006). It would determine the subsequent development of quantum theory. Einstein advanced it further in 1909 in connection with another model, involving a mirror exposed to heat radiation, which executes Brownian motion due to radiation pressure. We will come back to this model in connection with our discussion of this type of motion.

In his light quantum paper of 1905 Einstein considered a combined model of heat radiation, gas, and resonators that allowed him to infer from the equipartition law a surprising conclusion in conflict with Planck's own analysis: Planck's relationship between resonator and radiation energy[18] leads, in fact, not to Planck's formula for the black-body energy distribution but to a different result, now known as the Rayleigh-Jeans law. Although this result was correctly derived from classical radiation theory, it was entirely unacceptable because it would make a thermal equilibrium of radiation impossible. At the same time, this result calls into question the idea of a continuous ether. Thus it follows from Einstein's reasoning that Planck's radiation formula is incompatible with classical physics. In light of this insight, Einstein's earlier speculations about alternatives to the prevailing views of classical physics now took on a different status; they were no longer unorthodox lines of reasoning in a classical universe of ideas, but the first hints of a totally new cosmos.

Einstein's development of statistical mechanics was also crucial for his work on Brownian motion (Renn 2005c). His perspective on the problem of Brownian motion was set not only by his search for evidence for the atomic hypothesis, but above all by the special set of questions that were the subject of his previous research. In the first of the two articles (Einstein 1901, 1902a) that he later dismissed as "the worthless work of a novice,"[19] he already had the opportunity to familiarize himself with several of the topics that would play a key role in his paper on Brownian motion, for example, with the nature of diffusion and with the application of thermodynamics to the theory of solutions. In 1903 he had already developed an idea of how to calculate the size of ions in a liquid using hydrodynamic arguments and the size of neutral salt molecules using diffusion.[20] This idea developed into the doctoral thesis (Einstein 1905j) that he concluded successfully in 1905, after earlier failed attempts.

The dissertation suggested a new method for measuring the size of atoms; it showed how to find Avogadro's number by considering large sugar molecules dissolved in solution. The procedure consisted in writing two equations for two unknowns,[21] from which Avogadro's number and the size of the molecules can be calculated. One equation described the change in viscosity of the solution when sugar molecules are added, whereas the other used a relation between the diffusion coefficient of sugar molecules and the viscosity of the solution. Here the diffusion coefficient determines the speed with which differences in concentration of a dissolved substance equalize. The first equation was derived from rather complex hydrodynamic calculations. The other equation, which relates diffusion and viscosity, turned out to be crucial also for the analysis of Brownian motion, so that it is useful to consider its origin in more detail.

Einstein considered particles in a liquid and analyzed their behavior with regard to diffusion, on the one hand, and with regard to osmotic pressure, on the other. So he analyzed two thermal processes that determine the behavior of a dissolved substance in a liquid, and which in the final analysis can be reduced to the motion of its individual particles. In either case the point is to find a bridge between the behavior of a substance as a whole and the motion of individual particles. In the case of osmotic pressure the bridge already existed within the theory of solutions, as developed by Jacobus H. van 't Hoff and later by Walter Nernst. This theory provides an exact counterpart to the kinetic theory of gases. It allows the treatment of osmotic pressure, as measured by means of a membrane permeable to the liquid but not to the dissolved particles, in analogy to the gas pressure and its connection to Avogadro's number. In the case of diffusion, the bridge was Stokes's law that gives the relationship between the force acting on a particle in a liquid, the viscosity of the liquid, and the terminal velocity of the particle. If such a force is introduced fictitiously, and assumed to be balanced by osmotic pressure, then the fictitious force can be eliminated to obtain a direct relationship between diffusion, viscosity, and osmotic pressure. But because osmotic pressure is connected to Avogadro's number, Einstein's relation between diffusion, viscosity, and Avogadro's number can be derived, a relation that also involves the radius of the dissolved particles. Thus Einstein ultimately arrived at an expression for the diffusion coefficient that contained the size of atoms.

In his dissertation he used this equation together with the hydrodynamic equation relating atomic size and changes in viscosity to derive values for the size of atoms from experimental data about diffusion and viscosity. In his article on Brownian motion he again derives the viscosity-diffusion equation, this time with the methods of statistical physics, because only in this way could he justify the application of concepts like osmotic pressure to a collection of suspended particles. This concept, after all, was originally introduced only for the thermodynamics of solutions to deal with dissolved molecules, whereas its applicability to a collection of suspended particles was the problematic step.

But how did Einstein manage to predict from such deliberations that fluctuations of particles suspended in a liquid should be observable, when as late as 1904 he was still convinced that such fluctuations could be empirically established only in heat radiation? Apparently it was precisely the combination of his distinct deliberations on the theory of solutions and the theory of radiation that led him to the idea of Brownian motion. Einstein combined the results of his thesis research with those he collected in his study of fluctuations in the context of statistical mechanics and its application to heat radiation. Thus he had all the connections at hand to build a model of observable fluctuations in a material system. Einstein conjectured that this model is realized by the Brownian motion of particles suspended in a liquid.

The decisive insight that led to Einstein's work is that fluctuations of heat radiation can be related directly to a material process if a cavity filled with heat radiation contains a mirror, which should exhibit behavior similar to Brownian motion as a consequence of the incident radiation and the friction force due to radiation pressure. Einstein discussed this thought experiment in detail in later publications (Einstein 1909b, 189–90; 1909c, 496–8), but he apparently already had the idea in 1905, as suggested by later recollections. In January 1952 Einstein wrote to Max von Laue that he was certain in 1905 that Maxwell's theory leads to incorrect fluctuations of radiation pressure and therefore to the wrong Brownian motion of a mirror immersed in heat radiation as described by Planck's law.[22] So the mirror, which has mechanical, electrodynamic, and thermodynamic properties, represents a borderline problem of all these areas, and may therefore be considered as a kind of "missing link" between Einstein's three main concerns at that time. It may have also served as a conduit for transferring insights from one domain of knowledge to another. Thus one could learn from radiation theory that the observability of fluctuations may be enhanced by an appropriate scaling of the dimensions of the system and also that the scale of displacement in the expected *random walk* does not depend on the mass per se but on other dimensions of the problem, the surface area in the case of a mirror and the radius of the suspended particle in the case of a liquid. If that scale were dependent on the mass, as might appear plausible from just considering the energy balance, fluctuations would actually remain invisible, just as a sufficiently large body would have a negligible velocity corresponding to a given mean energy.

In summary, by connecting the dissertation's model of dissolved molecules with the search for observable fluctuation phenomena Einstein was quite naturally led to consider the irregular motions that must be exhibited by suspended particles. The formula Einstein had developed to describe the dependence of the diffusion coefficient on the size of molecules or particles should then have made it possible to derive information about the atomic scale from the irregular motion of suspended particles. This would have been the case had he succeeded in relating the observable motion of individual particles to diffusion. To close this last gap another conceptual step was necessary, which entailed the view of Brownian motion as a process unknown in classical physics until that time.

Before we come to this last step, it is useful to take another look at the problem of Brownian motion as a borderline problem, in the way Einstein presents it at the beginning of his paper on Brownian motion. From the point of view of phenomenological thermodynamics, he argues, small particles suspended in a liquid should reach thermal equilibrium with the surrounding liquid, and should thus certainly no longer execute irregular motion. This is so because, according to classical thermodynamics, the second law is not just a statistical proposition, but lays claim to absolute validity. However, from the perspective of kinetic theory, these particles differ from atoms and molecules merely by their size. Therefore they should always be subject to collisions due to the thermal equilibrium of the fluid, and they should share in this thermal motion.

So the irregular motion of these particles manifests a conflict between two areas of classical physics, similar to the conflict in the electrodynamics of moving bodies between the relativity postulate and the constancy of the speed of light, and also similar to the conflict between the assumption of a continuum of wavelengths and an equipartition of energy in the case of heat radiation. Einstein's reaction to these conflicts was similar in all three cases, and it also differed in a similar way from the reactions of his contemporaries, who tended to regard such conflicts more from a specialist point of view. It was conceivable that such conflicts came from some hidden flaw in the foundations of each of the special areas, but Einstein's survey of the knowledge of classical physics together with his philosophical background made him skeptical of that possibility. Instead he dared to look for new concepts with which he could overcome what he saw as a foundational crisis of classical physics. For the question of Brownian motion, Einstein combined insights of kinetic theory and of thermodynamics, that is, from micro- and macrophysics, without reducing one to the other. Instead he formulated new laws for the area of "mesoscopic" physics, which he thus first established as an autonomous area of physical knowledge.

6. THE REINTERPRETATION OF CLASSICAL PHYSICS

Einstein's papers of the year 1905 are interrelated in multiple ways. For one, they were the result of his pursuit of physical interests ranging widely from the theory of solutions to electrodynamics. Second, they were the result of an attempt to replace classical physics by a comprehensive new scheme that uses atomistic ideas and includes new, non-classical properties to make apparent the connection between hitherto unconnected neighboring areas of knowledge. Because this attempt eventually failed and Einstein's speculative ideas (such as that of an emission theory of light) came to a dead end, there was a strong incentive for him to search on another plane of reflection for solutions, even if only partial ones. The revolutionary papers of 1905 were therefore, third, the result of Copernicus processes in which not the whole of classical physics, but some of its important partial results were reinterpreted and given a new role in the conceptual organization of the knowledge it comprised.

That these papers were ultimately fragments of a larger, more ambitious project, also explains their hidden connections. Thus in the relativity paper Einstein shows that energy and frequency transform in the same way under the transition from one frame of reference to another – as it must be if there is to be compatibility between the new theory of space and time and the light quantum hypothesis, which says that the energy of a particle of light is proportional to its frequency. But Einstein does not make this connection explicit. Similarly, in the papers of 1905 there is no explicit mention of the connection between his treatment of heat radiation and of Brownian motion. Einstein only came back to these connections in publications several years later.

It is not a comprehensive plan for a new physics that emerges from Einstein's three most important papers of 1905. Rather, they are distinguished by addressing, each in its own way, borderline problems of the old physics, in particular the conspicuous conceptual tensions between mechanics, electrodynamics, and thermodynamics (Renn 2007b). By bringing these tensions to a head and showing that problems such as heat radiation, the electrodynamics of moving bodies, and Brownian motion cannot be satisfactorily explained on the basis of classical physics, he opened up a new perspective on the solutions suggested by the masters of classical physics to the problems in the borderline regions of classical physics. His earlier speculative approaches now became heuristic points of view that indicated escape routes from the dilemmas of classical physics. This becomes particularly evident in the title of the paper about light quanta: "About a Heuristic Point of View Concerning the Generation and Conversion of Light." But also the new relativistic kinematics had for Einstein the character of a heuristic framework within which precise solutions to definite problems could be found. If Copernicus processes were, as we have claimed, the essential mechanism that generated the germ of a new physics from the problematical solutions of the borderline problems of classical physics, we should be able to explain in detail how they effected a reorganization of knowledge. In Sections 6.1–6.3, we shall do so for the three main contributions of the *annus mirabilis*.

6.1. The Light Quantum Hypothesis

For the problem of heat radiation, Einstein's speculative thoughts about a particle theory of light were probably the decisive starting point. As we have seen, he apparently established at an early stage a connection between such a particle theory and Wien's formula for the distribution of energy over the different frequencies present in heat radiation. At the same time, Einstein had tested the limits of classical radiation theory, in particular in connection with his deliberations on fluctuation phenomena.

So he knew that the fluctuations of heat radiation cannot be explained solely on the basis of classical radiation theory.[23] By 1904 he had also realized that these fluctuations behaved similarly to those of an ideal gas or of a solution, if heat radiation behaves like a gas of particles of light, as it does in the range of validity of Wien's radiation law. Of course, by then it had been clear to Einstein for some time that not Wien's but Planck's law is empirically correct. So there was no reason to take Wien's formula and its speculative derivation from a particle theory of light too seriously. But now that Einstein had developed arguments for the impossibility of a classical explanation of the thermal equilibrium of radiation, these arguments acquired a new meaning. They provided heuristic clues for a nonclassical radiation theory.

Wien's formula could now be interpreted as the counterpart of the problematic radiation formula required by classical physics, the Rayleigh-Jeans law. Both formulas could be understood from this perspective as limiting cases of Planck's law under opposite circumstances, namely for either high or low values of frequency of the heat radiation. Einstein's deliberations on fluctuations of heat radiation, which amounted to a wave-particle duality, were also in favor of such "bracketing" of Planck's radiation law between two limiting cases.

Einstein's arguments in his 1905 paper about light quanta amount to a reversal of his earlier speculative deliberations about a particle theory of radiation, and this is the essence of the relevant Copernicus process in this case, that is, of the kind of reflection on prior knowledge that effected its reorganization. This is shown particularly clearly in Einstein's use of Boltzmann's principle as a heuristic tool, which was to achieve essential relevance for his later studies of statistical physics in general and radiation theory in particular. In his light quantum paper Einstein did not use Boltzmann's principle in order to draw conclusions from a probability function, as did Planck, to obtain the entropy of the system; such a method would have required that assumptions be made about the microscopic state of the system. Instead he applied the principle in reverse, to gain insight into the hidden microscopic structure from the entropy function that is assumed to be known. Thus he could infer from the entropy function corresponding to Wien's law in the regime where Wien's law is a good approximation to Planck's that heat radiation behaves as a collection of independent particles of light of energy $E = h\nu$ where h is a constant introduced by Planck and ν is the frequency of the quantum of light. Einstein's innovative application of Boltzmann's principle as a heuristic tool was possibly a discovery that he made in connection with his reinterpretation of Wien's law, a reinterpretation that changed a speculative thought into a heuristic point of view showing the way out of the dilemmas of classical physics.

The third part of the light quantum paper deals with phenomena that from his point of view raised problems for the established classical theory, but could be explained with his new concept of light. Among the experimental issues treated with the help of the light quantum hypothesis is photoluminescence, that is, the change of light of a given frequency into light of another frequency; the photoelectric effect, that is, the release of electrons from a surface when exposed to light; and the ionization of gases by ultraviolet light.

Despite the success of the light quantum hypothesis in explaining these phenomena qualitatively and quantitatively, it was received skeptically by his contemporaries.[24] So where did Einstein find the courage and confidence necessary to publish in 1905 his paper on light quanta, and why did he not publish it in 1904 when he already had essential insights into the nonclassical microstructure of radiation? These questions once again direct our attention to the hidden connections between the seemingly diverse results in the explosion of ideas in 1905. Einstein had good reason to be cautious in 1904 and good reason to be cavalier in 1905, reasons closely connected with his other research.

One important element was the relation of the radiation's micro- and macrostructure, a relation that did not become clear to Einstein until he had made significant progress in his work on the electrodynamics of moving bodies. In 1904 he probably already knew the points that argue against the classical theory of radiation, as well as the points in favor of his light quantum hypothesis. All of these arguments had turned the innocent speculation of his student years about a particle structure of radiation into a serious alternative to the now problematic classical theory.

But how viable was the light quantum hypothesis as an alternative, and how was it related to the well-tested radiation theory of Maxwell and Lorentz? In 1904 this was probably still unclear, although probably Einstein already knew at that time that the emission theory of light was untenable. This was also suggested by the wave-particle dualism, as implied by his considerations of fluctuations in heat radiation. By March 1905 at the latest he had recognized that the wave theory of light would as a theory of averages probably never be "replaced by another theory" (Einstein 1905i, 132), as he wrote in his light quantum paper. Against this background Einstein had apparently developed the framework of a new electrodynamics far enough (i.e., he had found the correct field transformations but not the new kinematics that goes with them) to recognize that the light quantum hypothesis is compatible with the Maxwell theory of radiation's macrostructure (Rynasiewicz and Renn 2006). This then was the moment to publish his revolutionary deliberations on the microstructure of radiation if only as a heuristic point of view.

Conversely, Einstein's deliberations on the light quantum hypothesis also had far-reaching consequences for his paper on the electrodynamics of moving bodies, since they transformed his original tentative thoughts about abandoning the ether into an indispensable prerequisite for his subsequent research. Before we turn to these consequences and the last, crucial steps that led Einstein to the special relativity theory of 1905, let us remind ourselves again of the close connection between Einstein's preoccupation with radiation theory and his interest in fluctuation phenomena. It is therefore not surprising that shortly after the publication of the light quantum paper, he finished first his dissertation and then his paper on Brownian motion. What exactly was the origin of the revolutionary concept of Brownian motion as a stochastic process differing in principle from the classical notion of continuous motion of a particle? This question brings us to the second revolutionary breakthrough of the year 1905.

6.2. Brownian Motion

As we have seen, Einstein concluded from statistical mechanics that a suspension of small particles should have an osmotic pressure, as is the case for solutions of molecules. If this pressure is nonuniformly distributed in space it causes a compensating diffusion process, whose properties can be calculated from Stokes's law, which determines the motion of particles in a viscous medium. Thus Einstein arrived, as we have also discussed, at an equation for the diffusion coefficient as a function of viscosity and atomic dimensions. This coefficient occurs in the partial differential equation that establishes a relation between spatial and temporal changes of concentration of a dissolved substance. This equation was first proposed in 1855 by Adolf Fick, who followed the pattern of other transport phenomena as they were treated in the works of Jean Baptiste Joseph Fourier on heat conduction and Georg Simon Ohm on electrical conductivity. Einstein reinterpreted this equation in a fashion analogous to his reinterpretation of Planck's formula for heat radiation and also, as we shall see, analogous to his reinterpretation of the Lorentz transformation equations for the electrodynamics of moving bodies. He essentially maintained the technical framework of the available results, but changed their conceptual significance in the light of newly achieved results about the mechanics and thermodynamics of suspended particles. In the context of this study dealing with fluctuations in position of suspended particles, he gave the traditional diffusion equation a radically new interpretation and thus "invented" Brownian motion as a theoretical concept.[25] That reinterpretation is the essence of the Copernicus process in this case.

Instead of assuming that the diffusion equation describes the overall distribution of a dissolved substance, Einstein interpreted it as a probability distribution of irregular displacements of individual suspended particles. Here he could resort to his earlier experience with probability distributions in his work on statistical mechanics. In this way he succeeded in identifying the irregular motion of the suspended particles as the elementary process on which diffusion is based.

He assumed that there is a time interval that is short compared to the observation time but long enough to treat the movements of a suspended particle during two successive time intervals as independent of each other. The displacement of the suspended particle can then be described by a probability distribution, which specifies the number of particles that have been displaced by a given distance during one time interval. On the basis of this reinterpretation of the diffusion equation and together with Einstein's expression for the diffusion coefficient in terms of atomic dimensions, the solution of this differential equation yields an expression for the mean square displacement of a particle as a function of the time. Einstein suggested using this expression to determine experimentally the true size of atoms and molecules. In addition to the method for determining the size of atoms developed in his recently completed thesis, and to the method resulting from the theory of heat radiation, he had thus obtained yet another way of confirming the atomic hypothesis that, through its experimental verification by Jean Perrin, turned out to be crucial in establishing the reality of atoms.

6.3. Special Relativity

By the beginning of 1905, at the latest, Einstein clearly saw that theoretically there was no principal alternative to the electrodynamics of Maxwell and Lorentz – as long as one was concerned only with time averages. The speculative undertaking of an emission theory of light had been unsuccessful and Einstein saw himself confronted with the task of reconciling his revolutionary insights into the microstructure of radiation with this state of affairs. The concept of radiation pressure presented an important problem in this context since, as we saw, it was crucial for Einstein's considerations of the nonclassical behavior of a mirror exposed to heat radiation. It is therefore hardly surprising that the transformation properties of energy and frequency and the theory of the radiation pressure on a perfectly reflecting mirror are central themes of Einstein's paper on the electrodynamics of moving bodies. All of this suggests that this paper, rather than representing the crowning achievement of a field theory tradition going back to Maxwell and

Lorentz, as is often assumed, was in fact the result of Einstein's failure to find an alternative to this theory. For this reason exactly those applications are at the center of the paper that he had pursued to probe the limits of classical physics.

The problem of reconciling Einstein's insights into the microstructure of radiation with its macrostructure as expressed by classical electrodynamics could be attacked from two sides. One could try to draw inferences from the requirement that the light quantum hypothesis and relativity principle be compatible, or from the requirement that field theory and the relativity principle be compatible. The latter compatibility could be established through an expansion of Lorentz's theory for the electrodynamics of moving bodies. Einstein knew Lorentz's theory in the form in which it was presented in an 1895 monograph (Lorentz 1895). This early version of the theory had some weaknesses in spite of its overall success. Precisely these weaknesses had motivated both Lorentz and Poincaré to continue improving the theory until it finally assumed a form that implied that the Earth's motion relative to the ether will almost always be unobservable. For this purpose, Lorentz introduced transformation equations for the field and a local time that served him as auxiliary devices, short circuiting otherwise laborious calculations. To ensure the unobservability of motion relative to the ether to higher-than-first order, Lorentz had to introduce special assumptions: a length contraction of moving objects, the assumption that material mass behaves as a function of velocity, and the assumption that all forces behave as a function of velocity, as does the electromagnetic force.

On the basis of these achievements, Einstein was able to build a powerful electrodynamics of moving bodies, taking the relativity principle as a given and assuming Lorentz's theory to be correct for time averages.[26] Its main problem was that some of the key elements of this new electrodynamics did not have a clear physical meaning. The best one could say was that in the transition from one inertial frame to another, the space and time coordinates behave as if the Lorentz transformations, and not the Galileo transformations of classical physics, were correct. Still, the plausibility of this "proto-relativity theory" (Rynasiewicz 2005) was apparently sufficient for Einstein to publish the light quantum paper in March 1905.

The success of the light quantum hypothesis presented Einstein with a paradox. Lorentz's theory could be made to agree with the principle of relativity in the way outlined previously, but when considering this theory in the light of his insights into heat radiation, Einstein must have felt as if the rug had been pulled from under his feet. These insights, in particular the impossibility of achieving a thermal equilibrium for a continuum theory of radiation, directly contradicted the basic assumption of that theory: the existence of a stationary ether. The theory of special relativity as created by Einstein in 1905 originated from the combination of this special perspective on the foundational crisis of classical physics with the comprehensive answer to the problem of the electrodynamics of moving bodies given by the work of Lorentz and its continuation in the sense just outlined.

Without an ether, Einstein did not have a basis for the crucial assumption of Lorentz's theory, that the speed of light in the ether is constant. Without an ether he likewise had no basis for the explanation of the contraction of moving bodies, which Lorentz had introduced in order to explain the result of the Michelson-Morley experiment.

So Einstein was now finally in the position of Galileo's followers, who were aware that the Aristotelian foundation of his theory of motion was no longer tenable and hence pressed on to articulate new foundational concepts such as that of inertial motion as implicitly defined by the "ultimate" achievements of the traditional conceptual system. The ultimate achievement in this case is the improvement on Lorentz's theory, what we call the protorelativity theory. The concepts of this protorelativity theory that were in need of physical explanation were above all space, time, and velocity. The meaning of these concepts could no longer be derived from the framework of Lorentz's theory in terms of the behavior of the ether, molecular forces, and so forth. The striking fact that all the critical concepts belong to kinematics may well have suggested to Einstein that it was misleading to search for a meaning of these concepts on the concrete level of the theory of electromagnetism.

Thus Einstein's special perspective made him focus precisely on those elements that held the key for the eventual solution: for one, the strange notion of local time introduced by Lorentz and the auxiliary assumption he had introduced concerning distances in moving systems of reference, and for another, the constancy of the speed of light and the principle of relativity. The idea of building a new theory from such general principles was not unfamiliar to Einstein. In his *Autobiographical Notes* he wrote:

The longer and the more despairingly I tried, the more I came to the conviction that only the discovery of a universal formal principle could lead us to assured results. The example I saw before me was thermodynamics. The general principle was there given in the theorem: the laws of nature are such that it is impossible to construct a *perpetuum mobile* (of the first and second kind). How, then, could such a universal principle be found? After ten years of reflection such a principle resulted from a paradox upon which I had already hit at the age of sixteen. (Einstein 1949a, 53)

Einstein proceeds to describe the paradox of an observer moving at the speed of light, chasing a light ray. He should see a static field, whereas according to the principle of relativity everything indicates that a moving observer should see the light ray in the same way as an observer at rest on the Earth. Against the background of his failure to construct a new electrodynamics this paradox must have seemed to Einstein as a primarily kinematical problem, which articulates the incompatibility between the classical addition of velocities, the relativity principle, and the assumption of a constant speed of light independent of the observer. In contrast to what was the case for Lorentz, the last two elements had emerged for Einstein as being equally important, although they could not immediately be reconciled.

What exactly were the implications of Lorentzian electrodynamics for the kinematic behavior of moving bodies, implications that now had to be explained without the help of an ether? Obviously one had to conclude that bodies and processes in moving systems behave differently than at rest. If one could explain this peculiar behavior on the basis of kinematics alone, rather than at the level of electrodynamics, then maybe the key to the solution was at hand. In order to find this solution it was necessary to correlate the curious notions of space, time, and velocity in the protorelativity theory with the customary use of these notions. What then are the implications of the results from electrodynamics for the meaning of space, time, and velocity on this more general level of understanding? If the space and time coordinates that occur in the Lorentz transformations are interpreted as physically meaningful quantities, then these transformations imply that identically constructed measuring rods and clocks in a system that moves relative to an observer define for this observer different units of length and time than his own, that is, relative to measuring rods and clocks at rest. This behavior, of course, begged for an explanation, but now not at the level of electrodynamics, as in Lorentz's problematical attempt at a physical justification, but at the level of kinematics.

Up to this point practically every step in Einstein's reasoning resulted almost by necessity from his special perspective on Lorentz's theory of electrodynamics. Now a step of reflection was necessary that went substantially beyond this theory or, more accurately, beyond the very foundations of classical physics. For the time being, the first question was whether there was any way to verify the curious behavior of bodies and processes in moving frames of reference. How do measuring rods and clocks behave in such frames? What does it mean to say "this event takes place simultaneously with another one," and how does one check that? It is quite possible that it was Besso who asked Einstein those shrewd, simple questions on a fine day in May 1905, which, according to Einstein's later recollections (Abiko 2000), delivered the conceptual breakthrough.[27]

Such questions resonated with Einstein's philosophical reading, particularly the writings of David Hume, Ernst Mach, and Henri Poincaré that he had studied intensively, mostly together with his friends of the Olympia Academy. With this background reading he came to terms with the idea that the concept of time is not simply given, but represents a complicated construct, and that to ascertain simultaneity at different locations one needs a definition that must be based on a practical method.

It is at this point that Einstein's insight into the nature of simultaneity becomes a plausible consequence of the preceding chain of reasoning, and, at the same time, takes on its significance as the crucial step for the solution of his problems. Einstein's method for ascertaining simultaneity at different locations – the synchronization of spatially distant clocks by light signals that propagate with a finite speed – at first sight has nothing to do with the complicated physical problems he was trying to solve and does not necessarily go beyond classical physics. Instead, this procedure is quite consistent with our everyday ideas about time measurement, and it was even used in the contemporary technical practice, as Einstein knew from the reading of popular scientific texts and from the context of his patent work.[28]

But the recourse to this practical procedure made it apparent that there is a certain arbitrariness in the determination of simultaneity in frames of reference in relative motion with respect to one another. Einstein's procedure works, to start with, only within one frame of reference, whether at rest or in motion, leaving it open how its results are to be coordinated with those in other frames. Against this background it became conceivable for the first time that the behavior of clocks and measuring rods could depend on the motion of a frame of reference, as suggested by Lorentz's theory. This implies that the curious space and time coordinates in the Lorentz transformations, when taken seriously as actually measured quantities, no longer contradicted the foundations of kinematics.

This kinematics followed in essence from Lorentz's results by reversing the direction of inference in his arguments – a typical feature of Copernicus processes. Within the Lorentzian theory, the Lorentz transformations, including auxiliary physical assumptions, ensured that motion with regard to the ether remains unobservable, "as if" the equations of electrodynamics also hold in a uniformly moving frame and measurements of the speed of light lead to the same results in a moving frame as in a frame at rest with respect to the ether. But if one reverses this argument, the Lorentz transformations result from the principle of relativity and from the new nonclassical principle of the constancy of the speed of light. Thus all of physics is reconstructed in such a way that the Lorentz transformations are no longer auxiliary constructs, but take over the role of the classical Galilean transformations, following directly from the fundamental principles. Einstein's introduction of these principles thus corresponds exactly to the formulation of the principles of inertia and superposition of motions by Galileo's disciples on the basis of the physical laws recognized by their master.

Against the background of this reconstruction, Einstein's acknowledgment of Besso's help in his 1905 relativity paper becomes understandable, as their conversation about the concept of time may well have been the crucial moment after all for the origin of the theory of relativity. It apparently led Einstein to the decisive step of reflection that allowed him to interconnect two levels of knowledge – a theoretical and a practical one – in a new way. As we saw, his reflection about the concept of time interlinked the practical knowledge about time measurements at different locations with propositions about the propagation of light, based in the theoretical knowledge about the electrodynamics of moving bodies.

Only by this interlinkage could expert studies on electrodynamics transcend their specific domain of knowledge and affect our ordinary concepts of time and space, turning Einstein's 1905 paper into the starting point of a scientific revolution that did not remain confined to the realm of specialists. The rise of this revolution from an interaction of two levels of knowledge also explains its temporal specificity, that is, it explains why the reflections of a philosopher such as Hume did not yet lead to the insight into the relativity of simultaneity. The postulate of the independence of the speed of light from the motion of the source, which finally defined Einstein's concept of time, only resulted from the long-term development of the system of knowledge of classical physics, and represents the quintessence of nineteenth-century electrodynamics and its problems at the borderline with mechanics. Whoever investigates scientific revolutions primarily from the points of view of their immediate temporal and local conditions runs the danger of underestimating such long-term processes.

That a lowly patent clerk could, with three papers written over the short span of three and a half months, uproot classical physics only becomes understandable once Einstein's revolution – more accurately his multiple revolutions – are explained in terms of Copernicus processes, the reflection on previously accumulated knowledge leading to its reorganization. Just as Galileo's followers could exploit his end results to establish the conceptual foundation of classical physics by turning some of his arguments upside down, the young Einstein was able to formulate revolutionary new concepts by preserving some of the deepest insights of classical physics, albeit with a new interpretation. The crucial role of new fixed points for Copernicus processes is reflected in Einstein's (1919f) famous distinction between constructive theories and theories of principle. This distinction actually emerged from the experience of his *annus mirabilis* that the failure of his earlier attempts at a constructive foundation of physics could be turned into a success by searching instead, at another level of reflection, for theories of principle, even if that meant that he could not present a grand unified theory and that he had to settle for a more piecemeal approach toward a new physics.

NOTES

1 This chapter is largely based on Renn (2006, ch. 4) and also draws on Rynasiewicz and Renn (2006). We are grateful to Lindy Divarci and to the editors of this volume for comments and suggestions.

2 See Einstein (1905i, 1905k, 1905r) and also Einstein to Conrad Habicht, May 18 or 25, 1905 (CPAE5, Doc. 27).

3 For the following, see Renn (2006, ch. 3).

4 For further discussion of Planck's work on black-body radiation and references to the extensive literature on this topic, see Chapter 4.

5 For the following, see Renn (1993) and Damerow et al. (2004).

6 For a discussion of local time, see Chapter 2.

7 This is also emphasized in Chapter 3.

8 See Bernstein (1853-7). For the following discussion, see also Büttner et al. (2003), Renn (1997), and the introduction to Renn and Schulmann (1992).

9 For the concept of mental model, see Gentner and Stevens (1983). For its use in the context of Einstein's work, see Renn and Sauer (2007, 127).

10 This thought experiment is also discussed in Section 4.3 of Chapter 2.

11 For extensive discussions, see Rynasiewicz and Renn (2006) and Büttner et al. (2003). Einstein also drew attention to this parallel in the opening sentence of his famous paper (Einstein 1917c) in which he introduced his quantum theory of radiation (discussed in Sec. 9 of Chapter 4).

12 For a discussion of this phenomenon and its role in Einstein's thinking, see Section 4.5 of Chapter 2.

13 These papers are also discussed in Section 4 of Chapter 3.

14 Einstein to Marić, March 23 and 27, 1901 (CPAE 1, Docs. 93 and 94).

15 This theory is discussed at the end of Section 5 of Chapter 4.

16 See Section 2 of Chapter 4 for discussion of black-body radiation and the various laws – Wien, Planck, and Rayleigh-Jeans – proposed for the spectral distribution of its energy.

17 See Section 4 of Chapter 4 (esp. Equations 8 and 9) for discussion of this argument, which plays a central role in Einstein's (1905i) light quantum paper (see Sec. 6.1).

18 See Equation 2 in Chapter 4.

19 See Einstein to Johannes Stark, December 7, 1907 (CPAE 5, Doc. 66). These two papers are also discussed in Section 2 of Chapter 3.

20 See Einstein to Michele Besso, March 17, 1903 (CPAE 5, Doc. 7).

21 See the two equations in Section 3 of Chapter 3.

22 See Einstein to Max von Laue, January 17, 1952, quoted and discussed in Rynasiewicz and Renn (2006, 25).

23 For extensive discussion, see Rynasiewicz and Renn (2006).

24 See Chapter 5.

25 For extensive discussion, see Renn (2005c).

26 For an extensive discussion, see Rynasiewicz (2005).

27 See also the acknowledgment in Einstein (1905r).

28 For the historical context, see Galison (2003).

2 Einstein's Special Theory of Relativity and the Problems in the Electrodynamics of Moving Bodies That Led Him to It

1. INTRODUCTION

Modern readers turning to Einstein's famous 1905 paper on special relativity may not find what they expect. Its title, "On the Electrodynamics of Moving Bodies," gives no inkling that it will develop an account of space and time that will topple Newton's system. Even its first paragraph just calls to mind an elementary experimental result due to Faraday concerning the interaction of a magnet and conductor. Only then does Einstein get down to the business of space and time and lay out a new theory in which rapidly moving rods shrink and clocks slow and the speed of light becomes an impassable barrier. This special theory of relativity has a central place in modern physics. As the first of the modern theories, it provides the foundation for particle physics and for Einstein's general theory of relativity; and it is the last point of agreement between them. It has also received considerable attention outside physics. It is the first port of call for philosophers and other thinkers, seeking to understand what Einstein did and why it changed everything. It is often also their last port. The theory is arresting enough to demand serious reflection and, unlike quantum theory and general relativity, its essential content can be grasped fully by someone merely with a command of simple algebra. It contains Einstein's analysis of simultaneity, probably the most celebrated conceptual analysis of the century.

Many have tried to emulate Einstein and do in their fields just what Einstein did for simultaneity, space, and time. For these reasons, many have sought to understand how Einstein worked his magic and came to special relativity. These efforts were long misled by an exaggeration of the importance of one experiment, the Michelson-Morley experiment, even though Einstein later had trouble recalling if he even knew of the experiment prior to his 1905 paper.[1] This one experiment, in isolation, has little force. Its null result happened to be fully compatible with Newton's own emission theory of light. Located in the context of late-nineteenth-century electrodynamics when ether-based, wave theories of light predominated, however, it presented a serious problem that exercised the greatest theoretician of the day.

Another oversimplification pays too much attention to the one part of Einstein's paper that especially fascinates us now: his ingenious use of light signals and clocks to mount his conceptual analysis of simultaneity. This approach gives far too much importance to notions that entered briefly only at the end of years of investigation. It leaves us with the curious idea that special relativity arrived because Einstein took the trouble to think hard enough about what it means to be simultaneous. Are we to believe that the generations who missed Einstein's discovery were simply guilty of an oversight of analysis?[2] Without the curious behavior of light, as gleaned by Einstein from nineteenth-century electrodynamics, no responsible analysis of clocks and light signals would give anything other than Newtonian results.

Why did special relativity emerge when it did? The answer is already given in Einstein's 1905 paper. It is the fruit of nineteenth-century electrodynamics. It is as much the theory that perfects nineteenth-century electrodynamics as it is the first theory of modern physics.[3] Until this electrodynamics emerged, special relativity could not arise; once it had emerged, special relativity could not be stopped. Its basic equations and notions were already emerging in the writings of H. A. Lorentz and Henri Poincaré on electrodynamics. The reason is not hard to understand. The observational consequences of special relativity differ significantly from Newtonian theory only in the realm of speeds close to that of light. Newton's theory was adapted to the fall of apples and the slow orbits of planets. It knew nothing of the realm of high speeds. Nineteenth-century electrodynamics was also a theory of light and the first to probe extremely fast motions. The unexpected differences between processes at high speeds and those at ordinary speeds were fully captured by electrodynamics. But their simple form was obscured by elaborate electrodynamical ornamentations. Einstein's achievement was to strip them of these ornamentations and to see that the odd behavior of rapidly moving electrodynamical systems was not a peculiarity of electricity and magnetism, but imposed by the nature of space and time on all rapidly moving systems.

This chapter will present a simple statement of the essential content of Einstein's special theory of relativity, including the inertia of energy, $E = mc^2$. It will seek to explain how Einstein extracted the theory from electrodynamics, indicating the subsidiary roles played by both experiments and Einstein's conceptual analysis of simultaneity.

All efforts to recount Einstein's path face one profound obstacle, the near complete lack of primary source materials. This stands in strong contrast to the case of general relativity, where we can call on a seven-year record of publication, private calculations, and an extensive correspondence, all prior to the completion of the theory.[4] For special relativity, we have a few fleeting remarks in Einstein's correspondence prior to the 1905 paper and brief, fragmented recollections in later correspondence and autobiographical statements. The result has been an unstable literature, pulled in two directions. The paucity of sources encourages accounts that are so lean as to be uninformative. Yet our preoccupation with the episode engenders fanciful speculation that survives only because of the lack of source materials to refute it. My goal will be an account that uses the minimum of responsible conjecture to map paths between the milestones supplied by the primary source materials.

2. BASIC NOTIONS

2.1. Einstein's Postulates

Einstein's special theory of relativity is based on two postulates, stated by Einstein in the opening section of his 1905 paper (Einstein 1905r). The first is the *principle of relativity*. It just asserts that the laws of physics hold equally in every inertial frame of reference.[5] That means that any process that can occur in one frame of reference according to these laws can also occur in any other. This gives the important outcome that no experiment in one inertial frame of reference can distinguish it intrinsically from any other. For that same experiment could have been carried out in any other inertial frame with the same outcome. The best such an experiment can reveal is motion with respect to some other frame, but it cannot license the assertion that one is absolutely at rest and the other is in true motion.

While not present by name, the principle of relativity has always been an essential part of Newtonian physics. According to Copernican cosmology, the Earth spins on its axis and orbits the Sun. Somehow Newtonian physics must answer the ancient objection that such motions should be revealed in ordinary experience if they are real. Yet, absent astronomical observations, there is no evidence of this motion. All processes on Earth proceed just as if the Earth were at rest. That lack of evidence, the Newtonian answers, is just what is expected. The Earth's motions are inertial to very good approximation; the curvature of the trajectory of a spot on the Earth's surface is small, requiring twelve hours to reverse its direction. So, by the conformity of Newtonian mechanics to the principle of relativity, we know that all mechanical processes on the moving Earth will proceed just as if the Earth were at rest. The principle of relativity is a commonplace of modern life as well. All processes within an airplane cabin, cruising rapidly but inertially, proceed exactly as they would at the hangar. We do not need to adjust our technique in pouring coffee for the speed of the airplane. The coffee is not left behind by the plane's motion when it is poured from the pot.

Einstein's second postulate, the *light postulate*, asserts that "light is always propagated in empty space with a definite velocity c which is independent of the state of motion of the emitting body" (Einstein 1905r, 892). Einstein gave no justification for this postulate in the introduction to his paper. Its strongest justification came from Maxwell's electrodynamics. That theory had identified light with waves propagating in an electromagnetic field and concluded that just one speed was possible for them in empty space, $c = 300,000$ km/sec, no matter what the motion of the emitter.

2.2. Relativity of Simultaneity

Einstein pointed out immediately that the two postulates were "apparently irreconcilable" (Einstein 1905r, 891). His point was obvious. If one inertially moving observer measures c for the speed of some light beam, what must be measured by another inertially moving observer who chases after the light beam at high speed – say 50 percent of c or even 99 percent of c? That second observer must surely measure the light beam slowed. But if the light postulate respects the principle of relativity, then the light postulate must also hold for this second, inertially moving observer, who must still measure the same speed, c for the light beam.

How could these conflicting considerations be reconciled? Einstein's solution to this puzzle became the central conceptual innovation of special relativity. Einstein urged that we only think the two postulates are incompatible because of a false assumption we make tacitly about the simultaneity of events separated in space. If one inertially moving observer judges two events, separated in space, to be simultaneous, then we routinely assume that any other observer would agree. That is the false assumption. According to Einstein's result of the *relativity of simultaneity*, observers in relative motion do not agree on the simultaneity of events spatially separated in the direction of their relative motion.

To demonstrate this result, Einstein imagined two places A and B, each equipped with identically constructed clocks, and a simple protocol to synchronize them using light signals. In simplified form, an observer located at the midpoint of the platform holding A and B waits for light signals emitted with each clock tick. The observer would judge the clocks properly synchronized if the signals for the same tick number

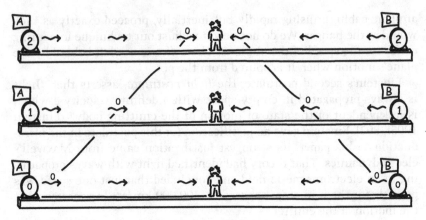

Figure 2.1. *Checking the synchrony of two clocks.*

arrive at the observer at the same time, for the signals propagate at the same speed c in both directions. The check of synchrony is shown in Figure 2.1, where the platform at successive times is displayed as we proceed up the page.

Now imagine how this check of synchrony would appear to another observer who is moving inertially to the left and therefore sees the platform move to the right, as shown in Figure 2.2. To this observer, the fact that the two zero-tick signals arrive at the same time is proof that the two clocks are *not* properly synchronized. For the moving observer would judge the platform observer to be rushing away from clock A's signal and rushing toward that of clock B. So signals emitted by clock A must travel farther to reach the platform observer O than signals emitted by clock B. The moving observer would judge the zero tick of clock A to occur before the zero tick of clock B; and so on for all other ticks. The light postulate is essential for this last step, which depends upon the moving observer *also* judging light signals in both directions to propagate at c; without this postulate, the relativity of simultaneity cannot be derived.

Since observers can use clocks to judge which events are simultaneous, it now follows that they disagree on which pairs of events are simultaneous. The platform observer would judge the events of the zero tick on each of clocks A and B to be simultaneous. The moving observer would judge the zero tick on clock A to have happened earlier.

This simple thought experiment allows us to see immediately how it is possible for Einstein's two postulates to be compatible. We saw that the constancy of the speed of light led to the relativity of simultaneity. We merely need to run the inference in reverse. Let us make the

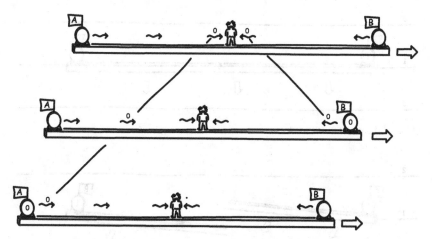

Figure 2.2. *Check of clock synchrony as seen by a moving observer.*

physical assumption that space and time are such that clocks are in true synchrony when set by this procedure. Then, using properly synchronized clocks in our frame of reference, whichever it may be, we will always judge the speed of light to be c. Suppose we chase after a light signal, no matter how rapidly? Since we will have changed frames of reference, we will need to resynchronize our clocks. Once we have done that, we will once again measure a speed c for the light signal.

2.3. Kinematics of Special Relativity

Much of the kinematics of special relativity can be read from the relativity of simultaneity. One effect can be seen in the preceding figures. Figure 2.1 shows that the platform observer will judge there to be as many light signals moving from left to right over the platform as from right to left. A direct expression of the relativity of simultaneity is that the moving observer will judge there to be more signals traversing from A to B, laboriously seeking to catch the fleeing end of the platform; while there will be fewer traversing from B to A, since they approach an end that moves to meet them.

To see another effect, imagine that the horizontal platform moves vertically and that it passes horizontal lines, aligning momentarily with each as it passes, as shown in Figure 2.3.

That alignment depends on judgments of simultaneity: that the event "A passes line 1" is simultaneous with the event "B passes line 1," for example. Another observer who also judges the platform to move to the right would *not* judge these two events to be simultaneous. That

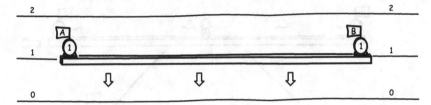

Figure 2.3. *Vertical motion.*

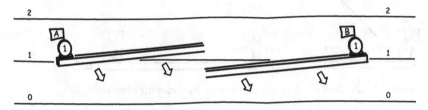

Figure 2.4. *Vertical motion seen by a horizontally moving observer.*

observer would judge the A event to occur before the B event. The outcome, as shown in Figure 2.4, is that the horizontal motion would tilt the platform so that it is no longer horizontal. That rotation is a direct expression of the relativity of simultaneity. A manifestation of this rotation arises in stellar aberration, discussed in Section 4.5.

The more familiar kinematical effects of special relativity also follow from the relativity of simultaneity simply because the measurement of any property of a moving process requires a judgment of simultaneity. For example, we may measure the length of a rapidly moving car by placing two marks *simultaneously* on the roadway as the car passes, one aligned with the front and one with the rear. We then measure the distance between the marks to determine the length of the car. Or we may judge how fast the car's dashboard clock is running by comparing its readings with those of synchronized clocks we have laid out along the roadway. A straightforward analysis would tell us that the rapidly moving car has shrunk and its clock slowed. The car driver would not agree with these measurements since they depend upon our judgment of the simultaneity of the placing of the marks and synchrony of the clocks. Indeed the car driver, carrying out an analogous measurement on us, would judge that our rods have shrunk and our clocks have slowed – and by the same factors, just as the principle of relativity demands.[6]

That we each judge the other's rods shrunk and clocks slowed is typical of relativistic effects. At first they seem paradoxical until we analyze them in terms of the relativity of simultaneity. Most complaints that relativity theory is paradoxical derive from a failure to accept the relativity of simultaneity.

The full complement of these kinematical effects is summarized in the equations of the *Lorentz transformation*. They describe what transpires when we view a system from two different inertial frames of reference; or, equivalently, what happens to one system when it is set into inertial motion. The body shrinks in length in the direction of motion; all its temporal processes slow; and the internal synchrony of its parts is dislocated according to the relativity of simultaneity. All these processes approach pathological limits as speeds approach c, which functions as an impassable barrier. The Lorentz transformation was not limited to spaces and times. Just as spaces and times transform in unexpected ways, Einstein's analysis of electrodynamical problems depended on an unexpected transformation for electric and magnetic fields. As we change inertial frames, a pure electric field or pure magnetic field may transform into a mixture of both.

The classical analog of the Lorentz transformation was later called the *Galilean transformation*. According to it, moving bodies behave just as you would formerly have expected: motion does not alter lengths, temporal processes, or internal synchrony and there is no upper limit to speeds.

A mathematically perspicuous representation of Einstein's kinematics was given by Hermann Minkowski in 1907 in terms of the geometry of a four-dimensional space-time. It lies outside the scope of this chapter.[7]

3. LORENTZ'S THEOREM OF CORRESPONDING STATES

3.1. Failing to See the Ether Wind

While Newton's physics had conformed to the principle of relativity, the revival of the wave theory of light in the early nineteenth century promised a change.[8] Light was now pictured as a wave propagating in a medium, the luminiferous ("light bearing") ether, which functioned as a carrier for light waves, much as the air does for sound waves. It seemed entirely reasonable to expect that this ether would provide the state of rest prohibited by the principle of relativity. As the Earth moves through space, a current of ether must surely blow past. A series of optical experiments were devised to detect the effects of this ether wind. The curious result in experiment after experiment was that no such result could be found. All "first-order" experiments, that is, ones that required the least sensitivity of the apparatus, yielded a null result.[9] This failure could be explained by a simple result, the Fresnel ether drag. The speed of light in an optically dense medium (like glass) with refractive index n is c/n. What would the speed of the light be if that medium moves with some speed v in the same direction? Will that speed be fully added to that of light? Fresnel proposed that only a portion would be added, precisely $v(1-1/n^2)$, imagining that the ether is partially dragged by the medium. It has to be just that factor. It turns out that if the ether is dragged by just that amount, then no first-order experiment can reveal the ether wind.

By the middle of the nineteenth century, the problem was enlarged by Maxwell's discovery that light was actually a wave propagating in the electromagnetic field. Maxwell's theory was also based on an ether that carried the electric and magnetic fields of his theory, and it too supplied a state of rest prohibited by the principle of relativity. The problem of explaining why no ether wind was detectable became part of a larger problem in electrodynamics. It became more acute when the Michelson-Morley experiment of 1887, the first second-order experiment, detected no ether wind. By 1903, Trouton and Noble had carried out a fully electrodynamic second-order experiment, again with a null result (Janssen 1995, ch. 1; 2009, 41–47).)

3.2. A Challenging Problem in Electrodynamics

The task of accommodating electrodynamics to these null results was undertaken by the great Dutch physicist, Hendrik A. Lorentz. In a series of papers in the 1890s and early 1900s, he was able to show that Maxwell's electrodynamics should not be expected to yield any positive result in these experiments. The computational task he faced was formidable. To arrive at his result, he needed a systematic comprehension of moving systems in electrodynamics. Motion immensely complicates electrodynamics. Take, for example, the basic entity of his electrodynamics, the electron, which he modeled as a sphere of electric charge surrounded by an electric field $\mathbf{E}$. As long as it is at rest in the ether, it could be analyzed merely by looking at the electrostatic forces between each of the parts of the electron. But once the electron is set in motion through the ether, each part becomes a moving charge; and a moving charge is an electric current; and an electric current generates a magnetic field $\mathbf{H}$; and that magnetic field acts on moving charges. See Figure 2.5. A thorough analysis is messy and eventually shows that the electron must be contracted slightly in its direction of motion.

The problem of computing the behavior of moving systems had been immeasurably easier in Newtonian physics since it conformed to the principle of relativity. The principle could be used to convert hard problems in moving systems into easy problems in systems at rest. Suppose,

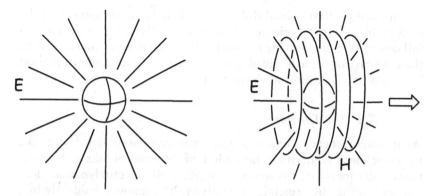

Figure 2.5. *Lorentz's electron at rest and in motion.*

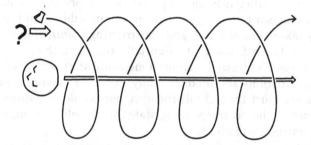

Figure 2.6. *A hard problem in Newtonian physics.*

Figure 2.7. *An easy problem in Newtonian physics.*

for example, that we want to know if a rapidly moving asteroid can gravitationally capture a satellite, as shown in Figure 2.6. What initial speed should we give the satellite so that capture is possible?

The problem is solved by first solving a much easier problem, shown in Figure 2.7: If the asteroid were at rest, is such a capture possible? Obviously, yes. What initial speed is needed? Computing it is the easiest problem in celestial mechanics.

But once we have solved the easy problem, we have also solved the hard problem, for the principle of relativity tells us that we recover a full description of a moving asteroid with its satellite by merely taking the easy case of the asteroid at rest and setting it into uniform motion by means of a Galilean transformation.

3.3. The Theorem

What Lorentz needed urgently was some computational device like the principle of relativity so he could find solutions of Maxwell's equations easily for moving systems.[10] But Maxwell's electrodynamics does not conform to the principle of relativity. Its equations hold only in a frame of reference at rest in the ether. Lorentz's ingenious discovery was a theorem in Maxwell's electrodynamics that mimicked the principle of relativity sufficiently for his purposes. The principle of relativity says that one can generate new systems compatible with the laws of nature by taking one solution and constructing *identical* uniformly moving copies. Lorentz saw that essentially the same thing could be done with Maxwell's electrodynamics. One could start with a solution of Maxwell's equations and produce oddly *distorted* moving copies of them. If one used just the right distortions, one would be assured that the new systems, the "corresponding state" of the old system, would also solve Maxwell's equations.

The rules Lorentz specified should not be a surprise. They are just the Lorentz transformations described in Section 2. But Lorentz did not give them Einstein's interpretation. They were merely artifices whose quite odd form was fixed by Maxwell's equations and justified solely by the fact that they enabled construction of new solutions from old. The largest (first-order) effect was a dislocation of the internal synchrony of the parts of the system that we now know as the relativity of simultaneity. For Lorentz, the rule was simply the assembly of a new system from the parts of the old, sampled at different times. The sampling rule was governed by his notion of "local time" – a sampling time that varied with the spatial location (hence "local"). Other first-order effects included odd transformations of fields: a pure electric field, such as the one surrounding an electron at rest, would become a mixture of electric and magnetic fields, just as shown in Figure 2.5. This first-order transformation was developed in Lorentz's (1895) *Versuch*. Higher-order effects soon followed and were codified in Lorentz (1904). They included the slowing of all temporal processes and the contraction of lengths in the direction of motion. (Einstein did not know of this later paper when he wrote his own on special relativity.)

With these rules and his theorem, Lorentz was able to compare systems moving and at rest in the ether and show that no existing experiment could decide which was at rest and which was moving. His device of local time was adequate for all first-order experiments, including the recovery of the Fresnel drag coefficient (though not the interpretation of a dragged ether). The higher-order contraction was sufficient for the Michelson-Morley experiment.

We can see just how Lorentz used these rules to describe electrons in motion. He solved the easy problem of electrons at rest and used the transformation to form its corresponding state, a contracted moving electron surrounded by a magnetic field. This example reveals an important complication. The electron at rest in Figure 2.5 cannot be governed solely by electromagnetic forces. Since like charges repel, another otherwise unknown, nonelectromagnetic force must be present in order to hold all the parts of the electron together and prevent it from blowing itself apart. How might this force transform? Lorentz made the natural supposition that it would transform just like electric and magnetic forces do under his Lorentz transformation. Only then could the contracted, moving electron of Figure 2.5 be recovered. This was a weak point of Lorentz's account for he was required to make presumptions about forces whose nature was quite unknown to him. The resulting contraction also happens to be the same length contraction used to explain the Michelson-Morley experiment, where it is sometimes called the *Lorentz-Fitzgerald contraction*. The awkwardness surrounding its introduction has led to suggestions that Lorentz's account is ad hoc. A better assessment is given by Janssen (2002a, 2002b), who urges that the superiority of Einstein's treatment lies in its giving a single explanation for what is otherwise an odd coincidence. Einstein shows us that forces of all types must transform alike because they inhabit the same space and time.

4. EINSTEIN'S PATH TO SPECIAL RELATIVITY[11]

4.1. The Magnet and Conductor Thought Experiment

The decisive moment in Einstein's path to special relativity came when he reflected on the interaction of a magnet and conductor in Maxwell's electrodynamics. The outcome was of such enduring importance that, years later when he wrote his 1905 paper on special relativity, this was the elementary consideration to which he gave pride of place in the paper's first paragraph.[12] As far as Maxwell's theory is concerned, the case of a magnet at rest in the ether is very different from that of one that moves. As shown in Figure 2.8, the magnet at rest is surrounded just by a static magnetic field **H**.

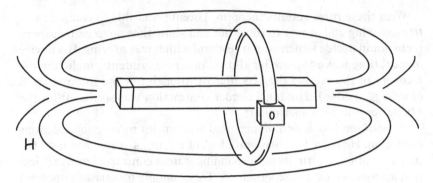

Figure 2.8. *Magnet and conductor at rest in the ether.*

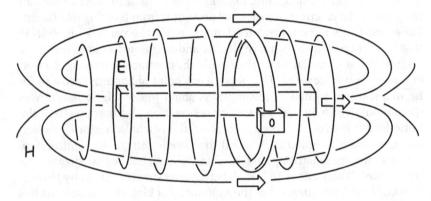

Figure 2.9. *Magnet and conductor moving in the ether.*

The moving magnet, however, is surrounded by both a magnetic field **H** and an electric field **E**, as shown in Figure 2.9. The latter arises from the complicated interactions between electric and magnetic fields in Maxwell's electrodynamics. At a point in space as the magnet moves past, the magnetic field will wax and wane. A time-varying magnetic field induces an electric field, a new entity not present in the first case.

Since the theory holds the two cases to be so distinct, one would expect that a simple measurement would distinguish them. The most straightforward would be to encircle the magnet with a conductor; that is, a wire with free charges in it that would be set in motion by the electric field to generate a measurable electric current. The conductor surrounding the magnet at rest would show no current; the conductor moving with the moving magnet would show a current and reveal its absolute motion. Or so one would expect. However, another electrodynamical interaction intervenes. Since the charges of the moving conductor are themselves moved through the magnetic field, that field also exerts a force on them and produces a current. The two currents – one due to the induced electric field, the other due to the motion of the charges in the magnetic field – are in opposite directions and turn out to cancel exactly. In both cases, there is no measurable current. Once again we have an experiment aimed at detecting motion in the ether, this time using a simple detector made from a magnet and a wire. And again we find a null result.

In an unpublished article in 1920 Einstein recalled how disturbed he was by the tension between the theoretical account and experimental outcome:

The idea, however, that these were two, in principle different cases was unbearable for me. The difference between the two, I was convinced, could only be a difference in choice of viewpoint and not a real difference. Judged from the [moving] magnet, there was certainly *no* electric field present. Judged from the [ether], there certainly was one present. Thus the existence of the electric field was a relative one, according to the state of motion of the coordinate system used, and only the electric and magnetic field *together* could be ascribed a kind of objective reality, apart from the state of motion of the observer or the coordinate system. The phenomenon of magneto-electric induction compelled me to postulate the (special) principle of relativity. (CPAE 7, Doc. 31, [p. 20]; Einstein's italics)

The principle of relativity, which prevailed among the observables, had to be extended to the full theory. This thought experiment gave Einstein the means to do it. The existence of the induced electric field was no longer the immutable mark of a magnet truly in motion; it was now merely an artifact of motion relative to the observer. Whatever may be the magnet's inertial motion, an observer moving with it will find a pure magnetic field; an observer in another state of inertial motion will find a mixture of magnetic and electric fields. That is just what moving magnetic fields look like, Einstein supposed – just as, in the later special theory of relativity, observers judge moving clocks to slow and rods to shrink, while comoving observers do not.

4.2. Field Transformations and the Relativity of Simultaneity

With this notion of field transformations, Einstein had created a potent device and it remained of central importance. For it was how Einstein would finally show in 1905 that Maxwell's electrodynamics conformed to the principle of relativity after all. The difficulty Einstein faced, however, was that no fully relativistic formulation of Maxwell's

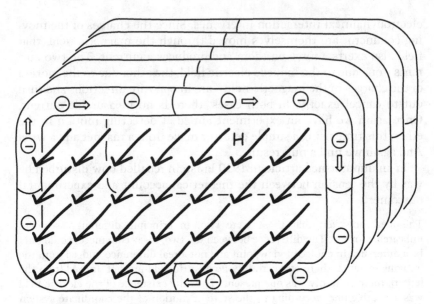

Figure 2.10. *Magnetic field inside a coil as seen by a comoving observer.*

electrodynamics was possible just using this new device of field transformations. It had to be coupled with the novel account of space and
time in special relativity. A simple thought experiment – *not* due to
Einstein – shows that the device of field transformations requires
Einstein's later notion of the relativity of simultaneity if it is to be
implemented in a relativized Maxwell's theory.[13]

Consider a very long coil of wire with a rectangular cross-section.
When a current is passed through the coil, a uniform magnetic field **H**
appears inside, with the magnetic field running along the axis of the
coil. The wire consists of a lattice of immobile positive charges, with
the current due to the motion of negatively charged electrons. The density of positive and negative charges will balance exactly so the wire
carries no net charge. A section through the coil is shown in Figure 2.10,
as it is seen by the "comoving observer," an observer who moves with
the coil.

We now set the coil into uniform motion. Figure 2.11 shows how
it will appear to a "resting observer," that is, one who remains at rest
while the coil moves past. Following Einstein's prescription, a resting
observer will find an induced electric field **E** associated with the magnetic field. From the case of the magnet and conductor (Figure 2.9) we
can see that the induced electric field will be perpendicular to both the

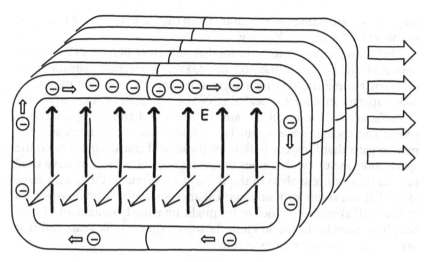

Figure 2.11. *Induced electric field inside a moving coil.*

magnetic field and the direction of motion. Since the magnetic field is uniform, the induced electric field will be uniform as well and it will run from the bottom of the coil to the top. Since there is no magnetic field outside an infinitely long coil, the electric field lines of force will terminate in the wire. Maxwell's theory is clear on what that means: electrical lines of force can only terminate in charges. The result is that the top of the coil carries a net negative charge and the bottom a net positive charge, both of which are not seen by the comoving observer.

How can this happen? The comoving observer judges the current-carrying electrons to take the same time to move from left to right as from right to left. The resting observer does not. They take more time to traverse the coil in the left to right direction (with the motion) and less in the other direction (against the motion). As a result, there is an accumulation of negative charges on the top and dilution of negative charges on the bottom of the coil, yielding net negative and positive charges respectively.[14]

This difference of traversal times cannot happen in classical (Galilean) kinematics. If one observer judges the two traversal times to be equal, then so must all observers. The difference can arise in special relativistic kinematics; it is a direct expression of the relativity of simultaneity. We have already seen this in Section 2.2 regarding light signals. The platform observer of Figure 2.1 judges the traversal times of light signals over the platform to be the same in both directions. The moving observer judges the traversal times to conform to Figure 2.2; they are not the same. This disagreement immediately leads to their differing judgments concerning the simultaneity of the events at A and B; that is, to the relativity of simultaneity.

This thought experiment shows that sufficient pursuit of Einstein's device of field transformations in Maxwell's electrodynamics must eventually force the relativity of simultaneity. The device cannot be used satisfactorily for the realizing of the principle of relativity until Einstein adopts the novel account of space and time of special relativity. The thought experiment is not Einstein's. We do not know the precise path that Einstein took from these field transformations to the space and time transformations of special relativity. He may have used physical reasoning such as in the thought experiment. Or he may have arrived at the result by mathematical analysis of the formal properties of Maxwell's equation, much as we might imagine Lorentz doing. Or he may have used both. We do know, however, that it took years and that several other considerations entered.

4.3. Einstein Considers an Emission Theory of Light

Einstein could not see how to formulate a fully relativistic electrodynamics merely using his new device of field transformations. So he considered the possibility of modifying Maxwell's electrodynamics in order to bring it into accord with an emission theory of light, such as Newton had originally conceived. There was some inevitability in these attempts, as long as he held to classical (Galilean) kinematics. Imagine that some emitter sends out a light beam at c. According to this kinematics, an observer who moves past at v in the opposite direction will judge the emitter moving at v and the light emitted at $c + v$. This last fact is the defining characteristic of an emission theory of light: the velocity of the emitter is added vectorially to the velocity of light emitted.

Einstein ran into numerous difficulties in his explorations of an emission theory. The principle difficulty, however, was this: if the emission theory was to be formulated as a field theory in which light is fully described as a propagating wave, then a light wave must somehow encode within it the velocity of its emitter, so that the theory could assign the correct velocity of propagation to each wave. No such encoding seemed possible, however, since experience showed that light waves were fully characterized simply by their intensity, color, and polarization.

That no field theory can do this is not immediately obvious. My conjecture (Norton 2004, Secs. 5–6) is that Einstein's objections to an emission theory of light can be made transparent through a celebrated thought experiment that he first hit upon at the age of sixteen and whose continuing cogency for Einstein would otherwise be unclear. As reported in his *Autobiographical Notes* (Einstein 1949a, 49–50) and elsewhere, he imagined chasing a beam of light at c. The result would be the observing of an electromagnetic waveform, frozen in space. "There seems to be no such thing, however," Einstein retorted, "neither on the basis of experience nor according to Maxwell's equations" (ibid.). Yet the retort is untroubling to an ether theorist. Maxwell's equations *do* entail quite directly that the observer would find a frozen waveform; and the ether theorist does not expect frozen waveforms in our experience since we do not move at the velocity of light in the ether. Why, then, was the thought experiment singled out for special attention in Einstein's recollections if its cogency is so doubtful?

The cogency becomes apparent if we place the thought experiment in Einstein's investigations of an emission theory of light. According to an emission theory, we should find frozen or slowed light waveforms if there are any sources of light moving sufficiently rapidly with respect to us. But we don't – just as Einstein remarked in his thought experiment. Even if we don't find them, the possibility of these static waveforms must be admitted by an emission theory if it is also a field theory. Now the sorts of static electric and magnetic fields possible were then well understood. Their investigation involved none of the relativistic complications of motion, rapid or otherwise. So an emission theory would have to agree with then current theories of electrostatics and magnetostatics, as Maxwell's theory did. In agreement with these theories, Maxwell's theory prohibits frozen waveforms – just as Einstein remarked in his thought experiment – and so also should a viable emission theory.

Finally, a field theory, patterned even loosely after Maxwell's theory, will use the present state of a wave to determine its velocity of propagation. This is what allows field theories to be deterministic, so that according to them the present can determine the future. Yet just such determination is denied by an emission theory if it is also a field theory. If the present state of a light wave is determined fully by its intensity, color, and polarization, it can have any velocity of propagation. As Einstein's thought experiment shows, it is even possible to have the extreme case of a completely frozen wave with no velocity of propagation; we merely need to move an observer at c with respect to the light's source. If an emission theory can be formulated as a field theory, it would seem to be unable to determine the future course of processes from their state in the present. As long as Einstein expected a viable theory of light, electricity, and magnetism to be a field theory, these sorts of objections would render an emission theory of light inadmissible.

4.4. Return to Maxwell's Theory

The early fruitlessness of Einstein's device of field transformations and his failed attempts to modify Maxwell's theory are just two episodes extracted from nearly a decade of thought on the problem of relative motion in electrodynamics. That thought must also have been entangled with his other investigations of what would become the light quantum hypothesis and the associated ebbing of his confidence in the exact validity of Maxwell's theory. Einstein recalled his reaction to these doubts and failures in his *Autobiographical Notes*: "Gradually I despaired of the possibility of discovering the true laws by means of constructive efforts based on known facts. The longer and more desperately I tried, the more I came to the conviction that only the discovery of a universal formal principle could lead us to assured results" (Einstein 1949a, 49). So he sought a theory that merely restricted the possibilities by means of principles whose grounding was secure. That decision brought special relativity to us as a theory founded on two postulates. In the light postulate, Einstein recorded one thing of which he had become sure. An emission theory fails. As he wrote in his 1905 paper "light is always propagated in empty space with a definite velocity *c which is independent of the state of motion of the emitting body*" (Einstein 1905r, 892; my emphasis).

As the walls closed in, Einstein was brought to his final crisis. In a story that often has been told (e.g., Stachel 2002a, 185), Einstein visited his friend Michele Besso some five or six weeks prior to the completion of the 1905 paper, bringing his struggle with him. The next day he reported with glee to his friend that he had found the solution, the relativity of simultaneity. In his *Autobiographical Notes*, Einstein (1949a, 51) recalled how his analysis had been decisively furthered by reading the philosophical writings of David Hume and Ernst Mach. While Einstein did not elaborate on how they assisted him, it is not hard to guess. Both Hume and Mach stress that concepts are only warranted insofar as they are anchored in experience. Einstein now saw that the classical notion of time incorporated a concept of absolute simultaneity that had no basis in experience. Emboldened by Hume and Mach's critiques, Einstein discarded the classical notion and the path to the completed theory was opened.[15]

4.5. Stellar Aberration

The analysis of stellar aberration provides a simple illustration of the different theories of light and their associated kinematics. It also supplies one of the most direct expressions of the relativity of simultaneity

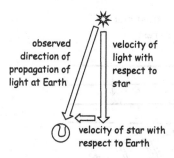

Figure 2.12. *Stellar aberration.*

in observables. Indeed the expression is so direct that I shall also suggest that it may have been important in the closing stages of Einstein's reflections.

In 1727, James Bradley observed that the motion of the Earth around the Sun affected the direction of starlight arriving at the Earth. The simple prescription for computing the change of direction is shown in Figure 2.12.[16]

The velocity of the light with respect to its emitter, the star, is added vectorially to the velocity of the star with respect to the Earth. The *direction* of the resulting compounded motion is the direction of the starlight observed on Earth. If this vectorial addition gave the correct direction, how could we avoid concluding that it also gave the correct velocity? To conclude that would be to accept an emission theory of light.

The passage to the emission theory is so natural that one might wonder how an ether-based, wave theory of light could possibly accommodate Bradley's result. Yet it turns out to be quite easy, as is shown in Figure 2.13. While the wave fronts of light propagating from a star are spherical, the small portion of the wave fronts reaching the Earth from a very distant star are virtually flat, so they become plane waves as depicted in the figure.

The wave fronts propagate toward a telescope on the Earth that moves from left to right. The telescope must be tilted as shown if a wave front that enters the front of the telescope is to pass along the barrel of the telescope to the observer's eyepiece. Otherwise the trailing telescope wall will intercept the wave before it reaches the eyepiece. The tilting of the telescope alters the apparent direction of the starlight in just the amount of Bradley's result.

This successful accommodation of aberration to the wave theory appears to fail completely, however, if we also demand that the wave theory respect the principle of relativity. For now we should expect the

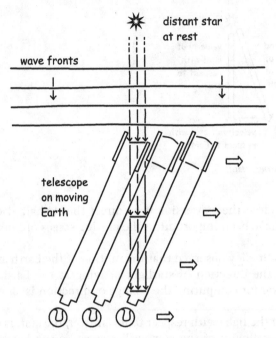

Figure 2.13. *Stellar aberration in an ether-based wave theory of light.*

same observable result if we conceive the star as at rest and the Earth moving (as in Figure 2.13), or if we conceive the star moving and the Earth at rest. According to the principle of relativity, the effect should only depend on the relative velocity of Earth and star and not on which is conceived as moving. Using classical notions of space and time, we arrive at the second case of a resting Earth by a Galilean transformation of the arrangement in Figure 2.13. The result is shown in Figure 2.14.

The transformation brings the Earth to rest and, at the same time, sets the star in motion in the opposite direction. The wave fronts remain perpendicular to the line joining the star and the Earth. With this arrangement, it is immediately apparent that the effect of stellar aberration is obliterated. The motion of the star no longer has an effect on how we must aim the telescope on Earth. If the telescope is pointed directly at the star, its light will pass to the eyepiece.[17] We seem to have a violation of the principle of relativity; whether the Earth or the star moves absolutely can be determined by checking for the presence or absence of stellar aberration.

One of the great achievements of Lorentz's (1895) *Versuch* was to show that, in Maxwell's electrodynamical theory of light, stellar aberration does depend solely on the relative velocity after all, or at least to

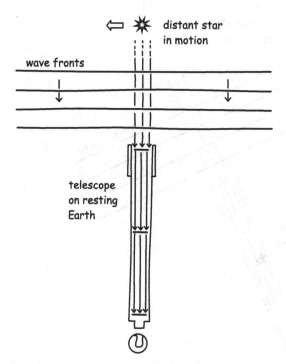

Figure 2.14. *Galilean transformation of Figure 2.13 to a resting Earth.*

the first-order quantities accessible to measurement. His demonstration depended upon the theorem of corresponding states. It requires us to use a Lorentz transformation, not a Galilean transformation, if we want to infer from the arrangement of Figure 2.13 how light would propagate were the Earth at rest. The effect of applying a Lorentz transformation is shown in Figure 2.15.

As before, the Earth is brought to rest and the star is set in motion in the opposite direction. In addition, the Lorentz transformation rotates the wave fronts so that they are no longer perpendicular to the line connecting the star and the Earth. That means that the telescope on Earth must be directed away from that line if the light from the star is to reach the eyepiece. The relative motion of the Earth and the star once again affects the apparent direction of propagation of the starlight just by the amount of Bradley's result. The observed result of stellar aberration is recovered and the outcome is in conformity with the principle of relativity. In the older analysis, the direction of propagation of a plane wave would be perpendicular to the wave fronts only for an observer at rest in the ether; for all others it would fail. Earthbound astronomers,

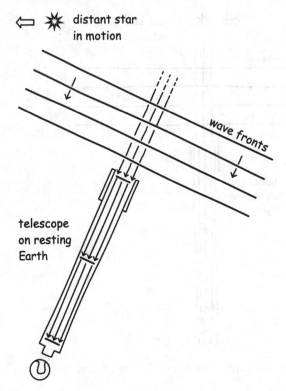

distant star
in motion

wave fronts

telescope
on resting
Earth

Figure 2.15. *Lorentz transformation of Figure 2.13 to a resting Earth.*

observing starlight, could use this failure to establish their absolute motion. Lorentz's analysis deprives them of this possibility, since, in his analysis, the direction of propagation is perpendicular to the wave fronts for all inertial observers.

What is intriguing about the core effect, the rotation of the wave fronts of Figure 2.15, is that it is due *entirely* to Lorentz's local time. Formally it is exactly the same effect as the rotation of the platform due to the relativity of simultaneity shown in Figure 2.4. That the vertically moving platform or wave fronts in the figures are oriented horizontally depends on a judgment of simultaneity. If observers change their state of motion and thus their judgments of simultaneity, the immediate outcome is the rotations shown in Figures 2.4 and 2.15.

Lorentz gives a very different interpretation of this effect than does Einstein's special theory of relativity. For Lorentz, it is simply a matter of consulting Maxwell's electrodynamics to find the effect of a star's motion on the light it emits. Such consultation is made through the theorem of corresponding states and, in this case, we find that the effect is a rotation of the wave fronts. For Einstein's special theory, the effect has nothing in particular to do with electrodynamics, but comes directly from space and time. It arises whenever a long object, be it a long platform as in Figures 2.3 and 2.4 or a wave front as in Figure 2.15, moves perpendicular to its length and we change the state of motion of the observer.

I'd like to suggest that this consideration might have been important to Einstein in the closing stages of the reflections leading to his 1905 paper. It might have provided a way to see how the relativity of simultaneity was grounded in experience, just as Hume and Mach had demanded of our scientific concepts. In the early 1950s, Robert Shankland visited Einstein with the express purpose of learning of the degree to which the Michelson-Morley experiment had influenced Einstein. To his dismay, he found Einstein barely able to recall whether he even knew of the experiment prior to his 1905 paper and suggesting that, if he did, he took the result for granted. That reaction is not hard to understand. The null outcome of the experiment is a direct result of the principle of relativity, and the experiment cannot decide between an emission theory of light and one based on the light postulate. Instead, Einstein volunteered, he had been more influenced by Fizeau's measurement of the speed of light in moving water and by stellar aberration. Shankland, unfortunately, did not ask Einstein to explain. We cannot be sure of what Einstein intended with these remarks. He may merely have meant that these experiments allowed a decision between a resting ether and a fully dragged ether. I lean toward another far more interesting explanation.

Einstein, I propose, interested in realizing a principle of relativity and in anchoring his theorizing in facts of experience, would have attended closely to what experience delivered. The Earth moves to new frames of reference as it slowly changes its direction of motion in its orbit around the Sun. In the course of a year, earthbound astronomers sample the light from a given star in many different frames of reference. They find the direction of deflection of stellar aberration and the associated rotation of the wave fronts to be different in each frame. So Einstein could take these views and read from them the transformation between the associated frames of reference that must figure in his principle of relativity. It must be one that can rotate vertically moving objects such as in Figures 2.3 and 2.4. The result would be a transformation that employs local time, that is, that embodies the sort of dislocation of temporal parts of the relativity of simultaneity. It could not be a Galilean transformation since that transformation does not rotate wave fronts.

What is important is that this dislocation of simultaneity would be read *from the observations* of stellar aberration and would be independent of Maxwell's electrodynamics and Lorentz's theorem of corresponding states. The only assumptions would be that starlight is a propagating waveform conforming to the principle of relativity.[18] If our concept of time was to be grounded in experience, here was experience calling for a concept of time that incorporated the relativity of simultaneity.

5. $E = MC^2$

5.1. The Result

Shortly after completing his paper on special relativity, Einstein found another consequence of the theory that he described in a short note "Does the Inertia of a Body Depend upon its Energy Content?" (Einstein 1905s). The basic notion was as simple as it was profound. Any quantity of energy, an amount "E" for example, also carries a mass "m" in direct proportion to the energy. The mass is computed by dividing the energy E by the number c^2. That number is so large that the associated mass is usually very tiny. Conversely, any mass m is also a quantity of energy E, where the conversion is effected by multiplying m by c^2. Because c^2 is so large, even a very small mass is associated with an enormous amount of energy.

This result of the inertia of energy can be applied whenever mass or energy transforms. Sometimes the effect is an imperceptible curiosity. When we talk on a battery-powered cell phone, the battery loses energy as it powers the phone. The accompanying, miniscule loss of mass of the battery is imperceptible to us. On other occasions, the effect is world changing. When uranium-235 undergoes fission, it breaks into other elements whose total mass turns out to be slightly less than that of the original uranium. That slight mass deficit manifests as an enormous quantity of energy in heat and radiation. As was discovered decades later, that process can power atom bombs or nuclear power plants.

To the casual reader, virtually all of Einstein's demonstrations of $E = mc^2$ seem curiously complicated, drawing on arcane results in electrodynamics, now generally regarded as more obscure than the result to be shown. Even a mid-century derivation (Einstein 1946b), offered as especially simple, takes the pressure of radiation as a primitive notion. The reasons for this obliqueness lie in the physics and in its history. Special relativity, as a theory of space and time, cannot make pronouncements by itself on energy, mass, and matter. It can only constrain the ways that they can manifest in space and time: they must be governed by laws that admit no absolute velocities. So some extra physical assumption must be supplied to determine which of the possibilities is realized. In Einstein's case, that extra assumption is conveyed by electrodynamics. The choice of electrodynamics for this purpose is entirely natural. The inertia of energy is a result already to be found in Maxwell's electrodynamics, just as the kinematics of special relativity were first discovered in Maxwell's theory. The real import of Einstein's demonstrations is to show that the inertia of energy cannot be localized to electrodynamics alone. Once it is secured there, relativity theory demands that it must hold for all forms of energy.[19]

5.2. A Demonstration

The following is a version of Einstein's (1905s) demonstration, simplified along the lines of Einstein (1946b).[20] It is designed to show that if the inertia of energy is realized in Maxwell's electrodynamics, it must be realized for all forms of mass and energy. The inertia of energy is expressed in Maxwell's theory for unidirectional radiation as follows: a quantity of radiant energy E carries momentum E/c in the direction of its motion. (To make the result familiar, assume that momentum has magnitude mc where m is the mass of radiation and we have $E/c = mc$ so that $E = mc^2$.)

A body with mass m' at rest emits two quantities of radiant energy $E'/2$ in opposite directions, as shown in Figure 2.16.

Because of the symmetry of the emission, the body remains at rest. We now view the process from a frame in which the body moves

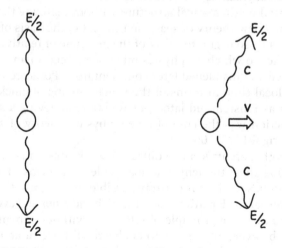

Figure 2.16. *A mass emits two quantities of radiation.*

perpendicularly to the direction of emission at v, and in which it has mass m. The quantities of radiant energy are now $E/2$ in the new frame. Thus they carry momentum $E/2c$ in the direction of propagation and a portion of that momentum in the ratio v/c lies in the direction of the body's motion. That portion is $(E/2c)(v/c) = (1/2)(E/c^2)v$. The law of conservation of momentum tells us that momentum gained by the radiation must equal that lost by the body. In the direction of the body's motion, the radiation has gained momentum $2 \times (1/2)(E/c^2)v = (E/c^2)v$. So the momentum of the body must be reduced by the same amount. The momentum of the body is mv and it must reduce by $(E/c^2)v$ as a result of the emission. Since the emission did not accelerate the body in its rest frame, the same will be true in this frame. Therefore the body's velocity remains v. So the decrease of momentum must come from a reduction in the mass m of the body. In sum, the body loses energy E and, as a result, loses momentum $(E/c^2)v$, which corresponds to a loss of mass of (E/c^2). This is the inertia of energy, now demonstrated for any body whatever that can emit radiation in the way shown.

6. CONCLUSION

The special theory of relativity owes its origins to Maxwell's equations of the electromagnetic field. (Einstein 1949a, 59)

In our brief review of the origins of Einstein's theory, we have seen much to affirm Einstein's judgment. The theory was already implicit in Maxwell's electrodynamics – so much so that Lorentz was able to discover its essential mathematical structure without realizing that he had chanced upon a new theory of space and time. On the basis of this theory, Poincaré had also begun to speak of the principle of relativity as one of the principles to which all physics must be subject. In an analysis that has excited and maddened later commentators, Poincaré interpreted Lorentz's local time in terms of the synchronizing of clocks by light signals, just as Einstein did later, but without conveying a sense that this construction was the core of a new physical theory of space and time (see Darrigol 1995, 2004).

In assessing both Lorentz's and Poincaré's work, one must guard against interpreting their thought and goals solely in terms of their proximity to Einstein's work. It is entirely possible to recognize that no experiment will reveal the Earth's motion through the ether and even to codify this expectation as a principle of relativity, without demanding that our theories be overturned so as to eradicate all trace of an ether and its state of rest. Another principle of physics, also discussed by Poincaré, illustrates this. The second law of thermodynamics assures us of an inexorable unidirectionality in time for all thermal processes and we routinely use it, while fully recognizing that its unidirectionality need not be reflected in the fundamental physics that governs thermal processes. The fundamental physics hides its time reversibility systematically from experimental detection, just as we might imagine that the physics of electricity and magnetism systematically hides the ether state of rest from us.[21] Poincaré seems not to have regarded his analysis of Lorentz's local time as a physical discovery. Rather it illustrated a persistent theme of his thought, the conventional character of parts of our science. Among many systems for synchronizing clocks we choose the one that we find most convenient, in that it makes the expression of laws of physics most simple. The choice is not imposed by nature but by our preferences and the point is illustrated by an odd synchrony scheme that gives simple expression for results in electrodynamics.[22]

Einstein's approach was quite different, characterized by an enduring conviction that the principle of relativity had to be realized throughout electrodynamics, even when years of investigation seemed to show the goal unrealistic and unachievable. While electromagnetism could not reveal the ether state of rest in the context of the magnet and conductor thought experiment, as Föppl had already pointed out, there were other equally simple experiments in electrodynamics that would, or so it seemed. The device Einstein learned from the thought experiment, field transformations, would have proved infertile as a means for realizing the principle of relativity. A thorough examination of its use in Maxwell's theory would have shown that different parts of the theory require different field transformations, as long as the kinematics remained Galilean (see Norton 2004, Sec. 2). Undaunted, Einstein was willing to sacrifice the greatest success of nineteenth-century physics, Maxwell's theory, seeking to replace it by one conforming to an emission theory of light, as the classical, Galilean kinematics demanded. With the failures mounting and his options exhausted, Einstein would entertain an extraordinary and desperate thought. Could he realize the principle of relativity in electrodynamics if he reshaped the very notion of time? This final gambit succeeded. When such stubbornness prevails, we wonder at its prescience; if it fails, we lament its folly.

Einstein is inseparably linked with analyses dense in light signals and the clocks they synchronize. What is notable in the preceding account is how little they figured in Einstein's path to special relativity. They were decisive in the final moments, some five to six weeks prior to the completion of the theory, when Einstein probably used them in his last, desperate gambit. But there is no evidence in the long years of investigation preceding that Einstein gave any serious thought to light signals and clocks. He did ponder light as it is judged by observers in different states of motion. But that was light as a propagating waveform in Maxwell's electrodynamics, not light as a signal, a point moving at c. The light of his original thought experiment at age sixteen was light as a propagating waveform, for he immediately recoiled at the resulting temporally frozen waveform. The optical experiments that Einstein singled out as important in his thought prior to the 1905 paper were stellar aberration and Fizeau's experiment. Both admit a very simple analysis in which the relativistic rule of velocity addition is applied to light signals. However, he seemed not to conceive them that way in 1905. His 1905 analysis of stellar aberration is given in terms of transforming light as a propagating waveform, much as in Section 4.5; and, in 1907, Einstein reported that he had only then learned from Laue of this perspicacious analysis of Fizeau's experiment.

The dominance of light signals and clock synchrony seems to be very much an artifact of Einstein's own final steps, their undeniable pedagogic value, and our own preoccupation with them.[23] They are not necessary to special relativity or to the relativity of simultaneity. They need not appear at all in a space-time formulation of special relativity, where the relativity of simultaneity arises naturally as our freedom to slice up the space-time into spaces in many different but equivalent ways.

ACKNOWLEDGMENTS

I am grateful to Tony Duncan, Allen Janis, Michel Janssen, and Robert Rynasiewicz for helpful comments.

NOTES

1 See Holton (1969), and for discussion informed by more recent discoveries in Einstein's correspondence, Stachel (1987).

2 Einstein wrote in his *Autobiographical Notes*: "Newton, forgive me: you found just about the only way possible in your age for a man of highest reasoning and creative power" (Einstein 1949a, 31).

3 Einstein (1959, 41) wrote that special relativity had been "developed from electrodynamics as an astoundingly simple combination and generalization of the hypotheses, formerly independent of each other, on which electrodynamics was built."

4 See Chapter 6.

5 A frame of reference is a system for assigning positions and times to events in association with a particular state of motion. It is conveniently realized by imagining space filled with a lattice of sticks and that every point of the lattice is equipped with a synchronized clock. Coordinates of time and space are assigned to events by a numbering system for the points of the lattice and by the readings of the clocks. An inertial motion is a uniform, straight-line motion naturally adopted by masses moving free of net forces. An inertial frame of reference is one that moves inertially.

6 See the Appendix, Section 1, for more detailed analysis.

7 See the Appendix, Section 2, for a basic introduction to Minkowki's geometry of space-time.

8 For a history of ether theories and electrodynamics in the nineteenth century, see Whittaker (1951) and Darrigol (2000b); and, for a treatment more narrowly focused on Einstein's 1905 paper, Miller (1981).

9 These experiments were expected to produce a measurable effect that is a function of v/c, where v is the speed of the Earth. Writing the effect as a power series, it is Effect = $A(v/c) + B(v/c)^2 + C(v/c)^3 + \ldots$. A first-order experiment seeks to measure $A(v/c)$. If that is zero, a second-order experiment would try to measure $B(v/c)^2$. Since (v/c) is very small, terms in $B(v/c)^2$ are still smaller and extraordinarily difficult to measure.

10 For a thorough discussion, see Janssen (1995).

11 This section draws on work of recent decades that had profited from extensive scrutiny of material in the Einstein archive. See CPAE 2, Stachel (2002a, Part IV), Rynasiewicz (2000), and Earman et al. (1983). My own attempts to extend these accounts are in Norton (2004), which expands on many of the points made in the text.

12 The version I will develop is a slight variant form given by August Föppl (1894, 309–10) in an electrodynamics text which Einstein probably read. For the version given by Einstein (1905r, 891), see Figure 6.2 in Chapter 6.

13 A simpler argument is that Maxwell's theory entails the constancy of the speed of light and that constancy, along with the principle of relativity, entails the relativity of simultaneity. This simpler argument, however, does not reveal how the kinematics of special relativity must permeate through even the simplest electrodynamical processes if the principle of relativity is to be respected; showing that is the function of this magnet and coil thought experiment.

14 An analogy: on a circular track, racing cars will accumulate on the slow side and dilute on the fast side.

15 For further discussion of what Einstein found in Hume and Mach's writings, see Norton (2010).

16 The figures that follow greatly exaggerate the change of direction. It is about twenty seconds of arc, which would be imperceptible in a properly scaled figure. The figures also show the special case of a star located in a direction perpendicular to the Earth's motion.

17 One might try to avoid the problem, as Born (1962, 141) suggests, by supposing that the direction of propagation is not perpendicular to the wave fronts. However this might be achieved, it is not a solution available to someone seeking to realize a principle of relativity. If the direction of propagation is perpendicular to the wave fronts in one frame, such as the one we designate as the ether frame, then, by the principle of relativity, that must also be true in any other inertial frame.

18 Fizeau's experiment measured the Fresnel drag for light propagating in moving water. In a similar analysis, it can be seen to give experimental support to Lorentz's local time, independently of electrodynamical theory. See Norton (2004, Sec. 7). According to a contemporary record of a lecture Einstein gave in Kyoto in 1922 (Akibo 2000, 13–14), Einstein read Lorentz's (1895) *Versuch*, noting how it solved the problem of the undetectability of

motion through the ether to first order. He then recalled how he proceeded to seek a relativistic account of Fizeau's experiment.

19 For another approach to this result, see Janssen (2003b).

20 Einstein's (1905s) derivation seems to have been complicated by an unfortunate definition of mass. His 1905 special relativity paper (Einstein 1905r) had defined *mass* as force/acceleration with the awkward outcome that mass has a different dependency on velocity according to whether the acceleration was parallel or transverse to the direction of its motion. Thus Einstein (1905s) demonstrated the inertia of energy for the rest mass only. The superior definition soon adopted by Einstein set mass equal to momentum/velocity and is used in the text.

21 The analogy is not perfect, of course, for at that time it was recognized by many, Einstein and Poincaré included, that certain microscopic processes might reveal the hidden processes that make the second law of thermodynamics true only with high probability.

22 See Poincaré (1898, 234; 1904, 306–8). Page references to reprints of these two papers are in Poincaré (1913).

23 For evidence of their enduring fascination, see Galison (2003).

3 Einstein on Statistical Physics

Fluctuations and Atomism

1. INTRODUCTION

If we want to summarize the ideas behind Einstein's early papers on molecular forces and in the field of what we now call statistical physics, we can say that this work was guided (1) by a strong belief and (2) by an important insight. The belief is atomism, the conviction that atoms and molecules really exist; the insight, which comes to play in the statistical physics papers, is that the study of fluctuations of physical quantities can lead to valuable new knowledge. Einstein's new approach in the field of fluctuations was to take them seriously instead of trying to show that they do not matter – as is usually done – and in some cases even studying situations where fluctuations are of the same order of magnitude as the physical quantity.

First, a few words about atomism. The idea that matter consists of small particles, invisible to the naked eye, indivisible and eternal, moving in vacuum, the properties of which determine the properties of matter as we know it, goes back to the Ancient Greeks. After having gone out of fashion for a long time – its mechanistic and deterministic implications were hard to reconcile with the existence of a higher being that was actively involved in the occurrences on Earth – it was revived in the seventeenth century by, among others, Gassendi, Descartes, Boyle, and Huygens. In Robert Boyle's words, "matter and motion" were the determining factors in nature. Through modifications of the older views, this atomism was made acceptable to the church authorities: one did not have to postulate the eternity of the particles, for instance, and neither did one have to assume that nature could exist without interference from outside. From that time on the existence of small elementary particles became an important explanatory concept in physics.

In the nineteenth century, the atomic view of matter led to important new developments, such as the kinetic theory of gases, the theory that describes gases as collections of large numbers of atoms that move with high speeds and collide with each other and the walls of the vessels in which they are enclosed.[1] Kinetic gas theory could give a fair, though not exact explanation of the relation between pressure, temperature, and volume of a dilute gas, for instance. But at the same time there were doubts: Did the successes of kinetic gas theory mean that the postulated atoms and molecules "really" existed? Even one of the great proponents of kinetic gas theory, the Austrian physicist Ludwig Boltzmann, was not always convinced of the reality of atoms: in a paper (Boltzmann 1897) he suggests that it might also be possible that atoms and molecules are just pictures that we form in our minds to become more familiar with the mathematical description of kinetic gas theory and to help us in the further development of this theory. This obviously sidesteps the question of the reality of the atoms. Such doubts were not uncommon among nineteenth-century physicists. One of the great champions of kinetic gas theory, the Dutch physicist Johannes Diderik van der Waals, admitted in his Nobel Prize acceptance lecture of 1912 that he, too, had been plagued by private doubts: "It will be perfectly clear that in all my studies I was quite convinced of the real existence of molecules ... yet still there often arose within me the question whether in the final analysis a molecule is a figment of the imagination and the entire molecular theory too" (Nobel Foundation 1967, 264–5).

In the presence of others, he stood firm: when his German colleague Woldemar Voigt visited him in 1899 and suggested that something was to be said for Boltzmann's view as sketched, Van der Waals became so annoyed that Voigt thought it wise to cut his visit short.[2]

Another Dutch proponent of the atomistic view was Hendrik Antoon Lorentz, who as early as 1878 unambiguously stated: "There will be hardly anybody nowadays, who does not know, that in the mind of physicists a material body is a system of very small particles, so called molecules, each of which may be composed of a number of still smaller particles, atoms" (Lorentz 1935–9, Vol. 9, 28). Throughout Lorentz's career atomism remained an important guiding principle, about which he showed no public doubts whatsoever. He made important contributions to kinetic gas theory and, more importantly, founded his influential theory of electrodynamics on the existence inside and outside of atoms of very small charged particles, later known as electrons.

In spite of the successes of kinetic gas theory and Lorentz's electron theory, at the end of the nineteenth century a strong antiatomistic sentiment can be found in the German-speaking world. One of the most important and influential critics was the Austrian physicist and philosopher Ernst Mach. In Mach's (positivistic) view, the atomic hypothesis was just that: a hypothesis that served to account for experimental facts. According to Mach, it would be foolish to conclude from the usefulness of the atomic hypothesis to the real existence of atoms and molecules. All that matters for him are statements about observations and nothing else.

Another important proponent of the antiatomistic attitude was the German physicist Wilhelm Ostwald, who also came with an alternative. For him the central concept in physics was energy in all its forms. This "energeticism" was not very long-lived, though: it was first formulated in the 1890s and already abandoned by the beginning of the twentieth century.

Although Einstein was strongly influenced by the work of Mach, he did not follow him on atomism. On the contrary, as will become clear in the following text, the real existence of molecules was so obvious for him that it did not need to be pointed out explicitly. At the same time, he searched for possibilities to derive experimentally testable consequences from the atomic hypothesis, as a means to convince its opponents.

Now to fluctuations. One of the key aspects of kinetic gas theory is that we are not interested in the motions of individual molecules, but only in average values. It would, in fact, be impossible to keep track of the rapid changes in the positions and speeds of the individual particles: there are just too many of them in an average volume (in the order of 10^{23}). But fortunately we do not have to: the measurable (macroscopic) properties of gases are connected with average values (to be precise, averages over time). For instance, pressure is the result of collisions of molecules with the walls during a certain period of time. And the temperature is proportional to the average kinetic energy (or the average quadratic speed) of the molecules. Interestingly, this does not mean that we cannot observe at least some trace of the motions of individual molecules. This is where the concept of fluctuations enters. Consider a particle, much larger than an atom (but visible with the help of a microscope) suspended in a fluid. This particle will constantly be hit on all sides by the atoms of the fluid, but because of the erratic motion of the atoms, the net force exerted by the atoms (and thus the speed of the suspended particle) will not be uniform but will vary rapidly in magnitude as well as direction, so that the particle will at one moment move in one direction and at the next moment in another direction. In this way the motion of the atoms becomes visible in an indirect way through the fluctuating motion of the suspended particles.[3]

2. THE FIRST TWO PAPERS: 1901–1902

Einstein's first two papers (Einstein 1901, 1902a) were later dismissed by him as "my two worthless beginners' works."[4] They may not be as groundbreaking as the famous papers published in 1905, but they are interesting nevertheless. Not only because they illustrate Einstein's commitment to atomism, but also because in both papers a link is made with experimental results. The latter aspect is characteristic for Einstein: throughout his life he remained interested in experimental matters and on several occasions he proposed or even performed experiments designed to provide information on fundamental matters. An example of the latter category is the work he did in collaboration with the Dutch physicist Wander de Haas on the microscopic origin of paramagnetism;[5] a typical example of the former category is an experiment he proposed in 1921 to decide between the particulate or wavelike character of radiation.[6]

In both of Einstein's first papers, thermodynamic arguments are combined with kinetic theory, in particular with certain assumptions about molecular forces, to obtain experimentally verifiable results. The first paper deals with capillary action in fluids, in particular the phenomenon of surface tension. Einstein first derives an expression for the surface potential energy of a fluid. He then makes the assumption that molecular forces behave in a similar way to the gravitational force: just as the force between two masses m_1 and m_2 can be written as $m_1 m_2 f(r)$ with $f(r)$ a universal function of the distance, so, according to Einstein, does the molecular force between a molecule of type a and one of type b have the form $c_a c_b \varphi(r)$, with c_a and c_b characteristic constants and $\varphi(r)$ a universal function of the distance. Using this assumption, Einstein succeeds in deriving two different relations that connect the quantities c with experimentally determined quantities (such as heat of evaporation). The values of c found in these two different ways are in reasonable agreement. In the paper he refrains from making explicit comments on the consequences of his results – although he speculates on a possible relation between intermolecular forces and gravitation – but in a letter to a friend he expressed his excitement about the possibility to explain seemingly unrelated phenomena with the help of one general hypothesis, stressing in particular the importance of his assumed analogy between molecular and gravitational forces.[7] It is interesting that Einstein never comments in the paper on the perhaps hypothetical nature of atoms and molecules and does not explicitly hail his results as evidence for their existence. Indeed, during his student years at the *Eidgenössische Technische Hochschule* in Zurich, Einstein had become familiar with the atomistic approach, for instance through the well-known physics textbook by Violle.[8]

As in the first paper, in the second one the real existence of atoms and molecules is presupposed without further comment. Here Einstein studies a complicated system, consisting of a substance in solution in a fluid, in which membranes are present that are permeable for the solvent molecules, but not for the solute particles. For the intermolecular forces the same assumption is made as in the first paper. There is no need to repeat the complicated arguments that Einstein uses to calculate quantities that can be experimentally measured, so as to test the postulated forces. Two more fundamental points are important, however. The first is the use of Van 't Hoff's law for the osmotic pressure, the pressure exerted by the solute particles. In analogy with the ideal gas law $pV = RT$ where p is the pressure, V the volume, R the gas constant, and T the temperature, Van 't Hoff had postulated that the pressure due to the solute molecules could be represented by a similar expression, provided that the solution is sufficiently diluted. In other words, the solute molecules behave like an ideal gas, and are not hindered by the presence of solvent molecules. A second point is that Einstein models the action of the semipermeable membranes by the introduction of fictitious external forces that act on the solute molecules only. In this way the calculations become much more tractable; as it turns out, the precise details of the nature of these fictitious forces have no influence on the final results.

3. THE YEAR 1905

We now leave the historical development for a moment and skip forward to the year 1905. Later on we will come back to the intervening years. As is well-known, 1905 is Einstein's *annus mirabilis* – in fact, a large part of this volume is devoted to this year – and one of the papers we will discuss here is part of the legend. It is the paper on Brownian motion (Einstein 1905k). But in order to understand this paper in its historical context, we will have to look at the work that preceded it, namely Einstein's doctoral dissertation on the dimensions of molecules (Einstein 1905j). The reason is that to a large extent the two follow the same method. It is interesting that the dissertation, although it appeared in 1905, is seldom included in the list of revolutionary papers of that year.[9]

In his dissertation, Einstein again turns to the theory of liquids to find experimental evidence for the existence of atoms and molecules. But now he does not look at molecular forces, which would give only indirect proof, but he sets himself the task to determine the actual dimensions of the particles. Here is what he does. He considers a two-component system, consisting of large molecules, suspended in a fluid. The suspended particles, the solute particles, are supposed to be much larger than the particles of the solvent; moreover, their concentration is supposed to be very low. These assumptions allow Einstein to make a few simplifications that facilitate the calculations he wants to make. In the first place, the low concentration of the solute particles allows him to use Van 't Hoff's law for the osmotic pressure. Here Einstein makes the implicit assumption that Van 't Hoff's law does not only hold for particles of atomic sizes, but also for particles that are much larger, though still invisible to the naked eye (but not necessarily to the microscope!). The other simplification is that the molecular structure of the solvent may be neglected, so that the solvent is treated as a continuous fluid, for which the laws of hydrodynamics hold. In the next step Einstein reaches back to a method used in his second paper: he introduces a fictitious external force that works only on the solute particles. That force will push the particles in one direction, causing differences in concentration. But difference in concentration means difference in osmotic pressure (through Van 't Hoff's law). For a system in thermal equilibrium the external force and the force of osmotic pressure have to balance each other, and that provides us with a relation between the external force and the density.

We can also look at the situation from a different point of view, namely the dynamical one. The external force will cause the solute particles to move in the solvent, whereby they experience a resistance that is caused by the fluid's viscosity. The speed with which the particles move is given by a relation derived by the English nineteenth-century physicist George Gabriel Stokes, in which the so-called viscosity coefficient of the fluid appears, as well as the diameter of the solute particles. But the concentration difference caused by the motion of the suspended particles – they all move in one direction – will be counteracted by the phenomenon of diffusion, the tendency of particles to move from places of higher concentration to those of lower concentration. Again, equilibrium will result, but this time a dynamical equilibrium between the particle flow caused by the external force and the diffusion flow. This gives us another equation in which the external force and the density gradient appear. Elimination of the external force leads to the following expression for the diffusion coefficient D:

$$D = \frac{RT}{6\pi\eta aN}$$

In this equation the gas constant R and the temperature T are known; for the viscosity coefficient η and the diffusion coefficient D experimental values are available for specific fluids; but Avogadro's number N (the number of molecules in a mole) and the diameter a of the suspended molecules are unknown. A second relation is thus needed to determine a and N.

In an earlier part of the dissertation Einstein had calculated the effect of the presence of suspended particles on the viscosity of the solvent.

For the ratio of the viscosity coefficients of fluid with suspension and pure fluid he found

$$\frac{\eta^{\cdot}}{\eta} = 1 + \frac{4\pi\rho Na^3}{3m}$$

in which the unknowns are, again, the diameter a and Avogadro's number N (ρ is the mass density of the fluid, and m its molar weight, both of which are known; $\eta^{\cdot}$ is the viscosity of the suspension). Because the ratio of the two viscosity coefficients is also experimentally known, we now have two equations for the two unknowns. Using data on sugar solutions in water, Einstein calculates Avogadro's number and finds $N = 2.1 \times 10^{23}$, which is a bit more than one-third of the current value (6.02×10^{23}).[10] Was Einstein's method flawed? No, but his algebra was. He had made a mistake in his calculation of the ratio of the two viscosity coefficients. Experiments done by the French physicist Jacques Bancelin in 1910 first suggested that something was wrong and not much later one of Einstein's students found the mistake.[11] The newly calculated value for N was 6.6×10^{23}, much closer to the value accepted today.

Why is a determination of Avogadro's number so important? It is because the existence of this constant is directly linked to atomism: it is a straightforward conclusion from the atomistic hypothesis that moles of all substances contain the same number of atoms or molecules, a conclusion that is hard to base on any other hypothesis. It was the French physicist Jean Perrin, who in 1909 named this number after the Italian physicist Amedeo Avogadro, who in 1811 had formulated the hypothesis that equal volumes of different gases at the same pressure and temperature contain equal numbers of molecules. Determinations of Avogadro's number with the help of different experimental methods provide strong evidence for the atomic hypothesis – but only when the same number is found in all cases.[12]

The next paper (Einstein 1905k), submitted eleven days after the date of the dissertation, is now known as Einstein's paper on Brownian motion. What is Brownian motion? In 1828 the Scottish biologist Robert Brown reported on a strange erratic motion of microscopically small particles (such as pollen or soot particles) suspended in water. At first these motions were interpreted as a sign of some life force present in the suspended particles, but later on several scientists speculated that the phenomenon was caused by the heat motion of the water molecules.

When he wrote his paper, Einstein was not sure that the phenomenon he studied was in fact Brownian motion. As he puts it in the first paragraph: "It is possible that the motions to be discussed here are identical with the so-called 'Brownian molecular motion'; however, the data available to me on the latter are so imprecise that I could not form a definite opinion on this matter."[13]

His uncertainty is also reflected in the title of the paper: "On the Movement of Small Particles Suspended in Stationary Liquids Required by the Molecular-Kinetic Theory of Heat." In any case, for Einstein the important point is his application of the kinetic theory of heat to the motion of suspended particles and the test of the correctness of this theory that his results allow.

In the paper on Brownian motion Einstein carries the method of his dissertation a step further. He starts with a suspension with the same properties as in the dissertation. But he now studies the motion of the suspended particles in some more detail. Why are the particles moving in the first place? Because they are constantly hit by the molecules of the solvent. Those hits are not regular, but reflect the disordered or random motion of the molecules: because of their collisions they will also move in an irregular way. If the particles were always hit at all sides by the same number of particles with the same speeds, they would obviously not move. But the momentum imparted to the particles will fluctuate, because of the irregular motion of the colliding molecules. Or the random motion of the solvent particles is reflected in the suspended particles. As mentioned before, the only thing one can say about this motion is that the mean kinetic energy – or the mean quadratic speed of the particles – is proportional to the temperature. Is it possible to make a similar statement about the irregular motion of the suspended particles? As Einstein shows, it is. With a simple probability argument he shows that the average density of the suspended particles is governed by the same diffusion equation he found in his dissertation and that the diffusion coefficient is proportional to the mean quadratic displacement of the molecules. If we now use our former expression for the diffusion coefficient in terms of the radius of the particles and Avogadro's number, an equation results that connects the mean square displacement with Avogadro's number: this yields yet another method to determine this crucial number. The mean square displacement can be measured by looking at a particle through a microscope and recording its position at regular intervals.[14]

Not long after the publication of the Brownian motion paper, experimenters started to test Einstein's theory. The most important of these was Perrin, who, in a series of careful experiments, confirmed Einstein's theory and obtained a very precise value of Avogadro's number.[15] The success of Einstein's theory played a very important role in convincing the skeptics of the reality of atoms and molecules. One of them was Ostwald. But one other unbeliever remained unconvinced: Ernst Mach. In 1909 Einstein sent him reprints of some of his papers, including one on Brownian motion. In the accompanying letter he drew Mach's particular attention to the connection between Brownian motion and the heat motion of molecules.[16] But, at least in print, Mach never admitted to the real existence of atoms.

In the years following 1905 Einstein published three more papers on Brownian motion, generalizing his theory (Einstein 1906b), clearing up a conceptual point (Einstein 1907c), and presenting an elementary theory of the phenomenon (Einstein 1908c). In the first paper, Einstein used the method of fluctuations; this approach made it possible to treat rotational Brownian motion as well.[17] He also made new applications of the phenomenon of fluctuations, for instance on voltage fluctuations in condensers (Einstein 1907b).[18]

4. FOUNDATIONS OF STATISTICAL PHYSICS, 1902–1904

Let us now go back to the years 1902–4 and to the three papers on the foundations of statistical physics Einstein published in those years (Einstein 1902b, 1903, 1904).[19] In the first paper, Einstein outlines an ambitious program. He wants to derive thermodynamics, in particular its second law, from kinetic theory. He claims that Boltzmann as well as the Scot James Clerk Maxwell have come close to this goal but have not reached it. In the introduction to his paper he mentions a "gap" ("eine Lücke") in those derivations that he wants to fill. From his correspondence with his fiancée Mileva Marić we know that he had in particular Boltzmann's work in mind, though it is not clear what exactly the "Lücke" was.

The ambition to find some kind of mechanical basis for the laws of thermodynamics, in particular its second law, had occupied many nineteenth-century physicists. Just as kinetic theory provided a molecular basis for the behavior of gases in equilibrium, one wanted to find a microscopic basis for the two laws of thermodynamics, in particular its second law. This law describes the behavior of systems in the course of time. The law states that this behavior is such that as time goes on, a certain magnitude, called *entropy*, will remain constant or increase, but never decrease. If one wants to "derive" this law from the laws of mechanics, one encounters a fundamental problem: whereas the second law indicates a "direction" in time through the behavior of the entropy, the mechanical laws that govern the motions of the individual molecules show no preferred direction in time. If at a given moment time is reversed, or the motions of all particles are reversed, a perfectly legitimate physical system results that does not violate any physical laws. So how can it be that if we go from the atomistic to the macroscopic level suddenly a directed time appears?

The solution Boltzmann developed was based on the idea that thermo-dynamics must be seen as a statistical theory. Just as the temperature is proportional to the average kinetic energy of the atoms and molecules, so the second law of thermodynamics should, in Boltzmann's view, be seen as a probability statement: the probability that a system goes from a state with a certain entropy to a state with equal or higher entropy is overwhelmingly much greater than the probability that it goes to a state with lower entropy. Now, of course, this reasoning hinges on the definition of probability, as well as the connection between this concept and the entropy S. Boltzmann conjectured the following connection: $S = k \log W$, where W is the probability of the state of the system and k a constant, later named Boltzmann's constant.[20] That leaves us with the problem of defining probability. For this Boltzmann used the following reasoning. If we look at a system of many particles, it is easy to see that the macroscopic properties we are interested in, such as pressure and temperature, are only very indirectly dependent on the precise microscopic properties of the individual atoms. For instance, exchanging two atoms, whereby atom 1 gets the speed and position of atom 2 and vice versa, will create a new "microstate" but will make no difference in macroscopic properties. Thus, the following definition of probability seems reasonable: the probability of some given macro-state is proportional to the number of microstates that correspond to it. This definition allows one to actually calculate W, and thus S for a given system.

In his series of three papers, Einstein took a different approach, one that was also based on ideas first developed by Boltzmann. To get a better grip on the number of microstates of a given system, Einstein now considers the system as a whole, and imagines it as a point in a multi-dimensional space, the phase space. To make this clearer, one should first consider a system consisting of one single particle. This system is fully described at a given moment when we know the position and the velocity of the particle. That is, we need six quantities to describe it: three spatial coordinates and three velocity components. These six quantities can be seen as the coordinates of a point in a six-dimensional space. In the same way, a system of N particles can be described by a point in a $6N$ dimensional space. Again, because many microstates correspond to the same macrostate, this macrostate will be represented by many points in phase space. Of course, because every system evolves in time, even when it is in equilibrium (because of collisions each micro-state changes into another one very rapidly) the points in phase space will travel through this space. One can visualize this evolution as the motion of a cloud of points through phase space, or, more technically, as the motion of an incompressible fluid through phase space.

In line with the earlier definition of probability in terms of number of microstates, we can look at the number of phase-space points (or the volume of phase space) corresponding to our macrostate as a measure for its probability. Or, again a bit more technical, we can take the density of our cloud of points in phase space to be proportional to the probability of our macrostate. Now comes a crucial assumption, but one that seems very natural: instead of taking time averages, Einstein takes phase-space averages to find the macroscopic properties we are interested in – he takes an average over phase space using the phase-space density as weight function. The admissibility of this assumption hinges on a deeper-lying assumption that is not made explicit by Einstein, namely the so-called ergodic hypothesis, the hypothesis that in its evolution a point in phase space will pass through every point of this space before it comes back to its point of departure. This hypothesis has been the subject of heated debate and is still under discussion.

Einstein now looks specifically first at systems of constant energy and then at systems of constant temperature. The cloud of points in phase space for a system of the latter kind is nowadays called a canonical ensemble, in the terminology introduced by the American physicist J. W. Gibbs, who had developed ideas similar to those of Einstein a few years earlier (see Gibbs 1902), but whose work was still very little known in Europe.[21] Introducing a probable form for the function that describes the density of the ensemble in phase space, Einstein calculates various average values. The fact that the temperature is constant leads to the identification of a certain constant in the density function with Boltzmann's constant.

One of the successes of Einstein's approach in these three papers is a derivation of an expression for the entropy of a system in terms of the ensemble density and a proof that this quantity obeys the second law of thermodynamics. (Einstein would use this expression in many later papers.) He also arrived at Boltzmann's connection between entropy and probability – but whereas Boltzmann had postulated it, Einstein found it from his derivation of the entropy and his considerations on probability of macroscopic states.

I have lumped the three papers together because they overlap to a large extent, at least as far as the general method and its results are concerned. The differences lie in certain refinements (such as the question whether it is necessary to split the total energy of a system in a kinetic and potential part) that do not concern us here. What does concern us are the final paragraphs in the third paper where Einstein almost as an afterthought presents some new results. He wants to give a physical interpretation of Boltzmann's constant k. A simple calculation leads him to the conclusion that k is a measure for the fluctuations in the energy of a system. More precisely, if the total energy of a system is given by $\langle E \rangle + e$ with $\langle E \rangle$ the mean energy and e its fluctuations, $\langle e^2 \rangle$ turns out to be proportional to k, independently of the specifics of the system under consideration. In characteristic Einsteinian fashion, the paper ends with an unexpected and adventurous application of this result: consider a space filled with radiation, such that the dimensions of the space are of the same magnitude as the wavelength of maximum energy of the radiation. In this case, Einstein argues, $\langle e^2 \rangle$ is of the same order of magnitude as $\langle E \rangle^2$. He then puts $\langle e^2 \rangle$ equal to $\langle E \rangle^2$ in the earlier found connection between $\langle e^2 \rangle$ and k and combines it with the Stefan-Boltzmann law, which states that the energy density of radiation is proportional to the fourth power of its temperature. In this way he finds that the cube root of the radiation volume (i.e., the linear dimension of the radiation space and thus the wavelength of maximum energy) is proportional to the temperature. Remarkably, such a relation also follows from radiation theory: it is Wien's displacement law. Even more remarkably, the proportionality constant Einstein found is of the same order of magnitude as the constant in the displacement law. That molecular statistics is able to reproduce such a fundamental property of radiation as the displacement law, Einstein says, cannot be purely coincidence.[22]

5. CRITICAL OPALESCENCE

Finally, a few words about a paper from 1910 (Einstein 1910d), in which yet another application is made of the phenomenon of fluctuations. The paper deals with the phenomenon of critical opalescence. What is critical opalescence? If light is scattered off a gas it is seen that the scattering increases enormously close to the critical point (the temperature and density at which a gas splits into two phases: liquid and gaseous). This phenomenon is called *critical opalescence*. Einstein showed in his paper that this increased scattering is due to large density fluctuations that occur close to the critical point.[23]

What makes the results of Einstein's paper especially interesting for a wider audience is a matter that is only mentioned in passing in the paper: the origin of the blue color of the sky. From ancient times on, people have asked themselves the question "why the sky is blue?" As more insight was gained in the structure of the world the question became more and more pressing. Why isn't the sky on Earth black (as it is outside of the atmosphere) except where the Sun is? The light of the Sun was shown to contain all colors, so why should the sky be blue? Somehow, it became clear, the blue light from the Sun was scattered more strongly by the atmosphere than the other colors, so that the blue component seemed to come from all directions, not just straight from the Sun. When the English physicist Lord Rayleigh calculated for the first time how light – that is, electromagnetic radiation – was scattered by atoms he found that short wavelengths were more strongly scattered than long ones: blue light is scattered more than red light. If one looks away from the Sun, the eye catches scattered sunlight, thus light in which the blue component is strongest. Thus the sky seems blue. For a while that seemed to solve the puzzle, and even today one can find this explanation in many popular books and on websites. But Einstein, and with him the Polish physicist Marian von Smoluchowski, realized that this scattering from individual particles would not have the desired effect if the atmospheric particles were arranged in an orderly pattern – it is crucial that they are distributed in an irregular way: density fluctuations in the atmosphere are crucial for the necessary scattering effect to appear. Interestingly, one can view this result as yet another proof of the atomistic constitution of matter. But, as Einstein points out, the mere existence of density fluctuations suffices for the explanation, irrespective of the underlying cause of the fluctuations.

Einstein's paper on critical opalescence was his last major contribution to classical statistical mechanics. His later work on quantum statistical physics is discussed in Chapter 4.

NOTES

1 See Brush (1976) for a historical overview of kinetic gas theory.
2 See Woldemar Voigt to Hendrik Antoon Lorentz, May 11, 1899 (Lorentz 2008, Letter 59).
3 See also Einstein's still very readable semipopular exposition of kinetic theory in Einstein (1915a), in which he calls it "amazing" that the molecular motions can be observed in this indirect way.
4 Einstein to Johannes Stark, December 7, 1907 (CPAE 5, Doc. 66). See also the editorial note, "Einstein on the Nature of Molecular Forces," CPAE 2, 3–8, for a historical discussion of the two papers.
5 See, e.g., Einstein and de Haas (1915a).
6 See Einstein (1922a). As was later pointed out by Paul Ehrenfest, the proposal was flawed (see the annotation to the paper in CPAE 7, Doc. 68).
7 Einstein to Marcel Grossmann, April, 141901 (CPAE 1, Doc. 100).
8 Violle (1892, 1893).
9 A possible explanation is that the journal version of the dissertation (Einstein 1906a) appeared in 1906, which obscured the fact that the paper on Brownian motion was, in fact, elaborating on the methods used in the dissertation. Interestingly, the dissertation is at the same time the most frequently cited paper by Einstein, because of the many practical applications of its results (see Pais 1982, 89–90).
10 In this context it should be noted that Einstein's terminology is often confusing: he frequently uses the term Grösse der Moleküle not for the actual size of the molecules, but for the number of molecules in a mole. This has

led the editors of the second volume of Einstein's *Collected Papers* astray in several instances (see, e.g., CPAE 2, 172, where a quotation by Einstein is mistranslated).

11 In the preceding equation, a factor of 2.5 is missing in the second term on the right-hand side. The student was Ludwig Hopf, who checked the calculations at Einstein's request, after he had unsuccessfully done so himself. The mistake was corrected in Einstein (1911e).

12 Apart from the method outlined in Einstein's dissertation, experiments on Brownian motion and critical opalescence provide values of Avogadro's number (see the following sections). Comparing Planck's radiation law with experimental data leads to yet another determination of the number, this time in a totally different field of physics.

13 Here, and in the following translations, are quoted from the translation of CPAE 2 by Anna Beck and Peter Havas (CPAE 2).

14 It was the invention of the ultramicroscope, just a few years earlier, that made the observation of suspended particles possible.

15 See Nye (1972) for a detailed historical discussion.

16 See Einstein to Ernst Mach, August 9, 1909 (CPAE 5, Doc. 174).

17 To be precise, Einstein derived general probabilistic expressions for the deviation of the values of physical quantities from their mean values. In rotational Brownian motion one looks at the variation of the angle of rotation of suspended particles that can rotate around a fixed axis.

18 Other important applications of the method of fluctuations are Einstein's calculations of energy and pressure fluctuations of radiation, which provided fundamental insights in the physical meaning of Planck's radiation formula (see Einstein 1909b).

19 Einstein's work in the field of statistical physics has been extensively discussed in the historical literature. A recent excellent technical analysis of Einstein's statistical papers from the years 1902–5 is given in Uffink (2006); see also Klein (1967, 1982), Pais (1982), Gearhart (1990), and Renn (2005c), and the editorial notes, "Einstein on the Foundations of Statistical Physics," CPAE 2, 41–55; "Einstein's Dissertation on the Determination of Molecular Dimensions," CPAE 2, 170–82; and "Einstein on Brownian Motion," CPAE 2, 206–22.

20 In fact, the special form of the relation between entropy and probability given here was introduced by Max Planck, who also coined the term *Boltzmann's constant.*

21 In fact, Einstein later wrote that he would never have written his papers had he known of Gibbs's work (see Einstein 1911c).

22 As is pointed out by Pais, Einstein's argument is flawed. I still mention it here because of the daring step to apply the concept of fluctuations to radiation, anticipating Einstein's introduction of the light quantum a year later (see Pais 1982, 70).

23 Results, similar to those of Einstein, were obtained by the Polish physicist Marian von Smoluchowski, who had earlier made important contributions to the theory of Brownian motion. See, e.g. Pais (1982, 100–3).

4 The Quantum Enigma

1. INTRODUCTION

Historians and physicists all agree that Albert Einstein significantly contributed to early quantum theory. They credit him with the light quantum concept, the concept of stimulated emission on which masers and lasers are based, and the Bose-Einstein condensation, which has meanwhile been observed on magnetically trapped, laser-cooled rubidium. There also is a consensus on the main heuristic device that produced this spectacular series of insights: statistical fluctuations of thermodynamic quantities regarded as indicators of the structure of the underlying microworld.[1]

Yet the nature of Einstein's contributions is clouded with three persistent myths, which the present article attempts to dissipate. According to the first of these myths, Einstein applied to radiation an idea of quantum discontinuity that Max Planck had already introduced for matter. We will see that the full awareness of a direct physical meaning of Planck's energy elements in fact belongs to Einstein, as Planck recognized. According to the second myth, Einstein's early arguments in favor of light quanta were so compelling that contemporary rejection must have resulted from conservatism or blindness. In reality, some of the arguments of Einstein's adversaries were truly disturbing. Einstein came to doubt the very existence of quanta for a couple of years around 1912. He only returned to them by some sort of pragmatism, with the discomfort of unsolved paradoxes. According to the third myth, the discovery of the Compton effect rapidly convinced all quantum physicists that Einstein's light quanta were real. In fact, Niels Bohr and his collaborators still rejected light quanta for some time. When they at last admitted a light quantum concept, they did so in a form that was unacceptable to Einstein.

The following account of Einstein's struggles with the quantum enigma is chronological. It first deals with various conceptions of statistical mechanics and their application to the problem of thermal radiation, which yielded the first quantum concepts at the turn of the nineteenth and twentieth centuries. After introducing the light quantum in 1905, Einstein soon left this narrow context and applied the quantum hypothesis to the thermal properties of matter. This successful part of his early quantum work largely contributed to the rise of a new quantum physics. In the same period, Einstein tried to construct the quanta in terms of singularities in the electromagnetic radiation field. By the end of 1911, he gave up this approach and instead tried to see how far quantum results could be obtained without discrete quanta. In 1914 he returned to the quanta, never to leave them despite their persistently paradoxical character. His new efforts led to a highly important description of radiation processes in terms of transition probabilities and to a new theoretical proof of the necessity of light quanta in 1916. They also led, in the 1920s, to fruitless speculation on ghostlike guiding fields for the light quanta and even to an erroneous proposal for a crucial experiment. Einstein's last important contribution to quantum theory, besides his later criticism of quantum mechanics, was his quantum theory of gas degeneracy of 1924, which strengthened the analogy between matter and light and confirmed Louis de Broglie's intuition of matter waves.

In the years that preceded Einstein's entry into the world of quantum paradoxes, physics was a most attractive, lively field that produced new phenomena, new entities, and new theories at a dizzying pace. In 1887–8, Heinrich Hertz demonstrated the superiority of James Clerk Maxwell's electromagnetic theory over its continental rivals and showed how to produce electromagnetic waves with a high-frequency electric oscillator. In the 1890s, the discoveries of X rays, the electron, and radioactivity signaled the emergence of an experimental microphysics in which the behavior of atomistic entities became more directly accessible. These discoveries lent credence to the kinetic molecular theory of matter that Rudolf Clausius, James Clerk Maxwell, and Ludwig Boltzmann had developed in the second half of the century. They also favored the atomistic reformulation of Maxwell's electromagnetic theory that the Dutch theorist Hendrik Antoon Lorentz was elaborating in order to explain electrolysis, optical dispersion, and the optics of moving bodies.

The foundations of these new theories were often unclear, their formulations multifarious and unstable. They were criticized, or even rejected by physicists who believed in the inexhaustible fertility of older, phenomenological methods. The young Einstein had no such reservations. His intense ambition prompted him to focus on the very questions' that the contemporary leaders of physics considered most challenging. One of them was the reconciliation of Lorentz's new electrodynamics with the principle of relativity. Another was the validity of Boltzmann's statistical methods for connecting the microworld with observable phenomena.[2]

Around 1900, the physics student Einstein focused on two problems that implied a combination of Lorentz's and Boltzmann's innovative methods: the explanation of the electrical, thermal, and optical properties of metals in terms of an electron gas, and the problem of thermal radiation. Planck's latest solution to the latter problem involved a novel combination of electrodynamical and thermostatistical arguments that attracted Einstein's attention. It also yielded the most precise determination of Avogadro's number of the time, a notable achievement for any believer in the world of atoms. Einstein reached his first quantum concepts in a criticism of Planck's approach, which we now proceed to examine.

2. MAX PLANCK ON BLACK-BODY RADIATION

In technical language from the late nineteenth century, a black body is a cavity whose walls can absorb and emit light of every possible frequency. The spectral density u_v of the radiation within the cavity is then defined so that $u_v \, dv$ represents the energy per unit volume of the part of the radiation whose frequency lies between v and $v + dv$. According to a thermodynamic theorem published by Gustav Kirchhoff in 1860, when a black body is maintained at a well-defined temperature, the spectral density of the radiation within the cavity depends on the temperature only. It does not depend on the nature of the walls or on the shape of the cavity. This remarkable result attracted the attention both of theorists and of experimenters. The development of electric lighting toward the end of the century, as well as the optical measurement of high temperatures, required knowledge of thermal radiation. With this industrial and metrological motivation, Berlin experimenters strove to accurately determine the black-body spectrum.[3]

At the same time, a Berlin expert on theoretical thermodynamics, Planck, sought to establish the law of this spectrum by theoretical means. His idea was that electromagnetic radiation interacting with a system of electric resonators at every possible frequency (miniature versions of Hertz's resonator) should irreversibly evolve toward black-body radiation. As resonators are relatively simple entities, Planck hoped that theory could determine this interaction and the final spectrum. As he had a special interest in the derivation of thermodynamic irreversibility that radiation theory seemed to permit, he focused on the entropy of the global system, which must indefinitely increase in time according to the second law of thermodynamics.[4]

Specifically, Planck determined the expression $S(U)$ of the entropy of a resonator of average energy U so as to make the total entropy of the system of resonators and the radiation an ever-increasing function of time. According to ordinary thermodynamics, the absolute temperature T of a resonator is related to its entropy through

$$\mathrm{d}S / \mathrm{d}U = 1 / T. \tag{1}$$

Once $S(U)$ is known, this relation yields U as a function of T. Planck then obtained the spectral density u_ν through the equilibrium relation

$$u_\nu = (8\pi\nu^2/c^3)\, U, \tag{2}$$

which he had earlier obtained by considering the interaction between "natural" (random phase) radiation and resonator.

After four years of labor, in 1899 Planck arrived at a derivation of the exponential law

$$u_\nu = \alpha e^{-\beta\nu/T}, \tag{3}$$

which Wilhelm Wien had already suggested and which earlier blackbody measurements seemed to corroborate. Unfortunately, toward the end of the same year Berlin spectroscopists found that the infrared part of the spectrum disagreed with Wien's law. A few months later, Planck recognized that his procedure for determining the resonator entropy was ambiguous; that there was an infinity of choices of the function $S(U)$ for which the global entropy increased in time.

In December 1900, Planck turned to another way of determining entropies that Boltzmann had invented in 1877. According to the Viennese theorist, the entropy of a given macrostate of a system is proportional to the logarithm of the number of "complexions" or microstates compatible with this macrostate

$$S = k \ln W \tag{4}$$

in Planck's notation. In order to determine W in the case of a resonator of average energy U, Planck counted the number of distributions of a number $P = NU/\varepsilon$ of energy elements ε over a large number N of replicas of this resonator. Through the relations (4), (1), and (2), this leads to the formula

$$u_\nu = \frac{8\pi\nu^2}{c^3} \frac{\varepsilon}{e^{\varepsilon/kT} - 1}. \tag{5}$$

Choosing $\varepsilon = h\nu$, wherein h is a new universal constant with the dimension of action, Planck obtained the law

$$u_v = \frac{8\pi h v^3}{c^3} \frac{1}{e^{hv/kT} - 1},$$ (6)

which agreed with Wien's law for large frequencies and with the recently observed spectrum for low frequencies (Planck 1900b, 1901a).[5]

Comparison of the new law with the latest Berlin measurements permitted a numerical determination of the two constants h and k. Planck emphasized that the resulting value of Avogadro's number $N = R/k$, where R is the constant of perfect gases, was the most accurate to date. The agreement with earlier values obtained through measurements on gases showed that Planck had successfully bridged the physics of two widely different sorts of systems: thermal radiation and gases.

The meaning of Planck's h constant was more obscure. As Planck noted, the choice $\varepsilon = hv$ of the energy elements was "the essential step" of his derivation. A closer analogy with Boltzmann's similar calculation in the case of a perfect gas would have required a quasi-infinitesimal value of the elements and led to the empirically inadequate

$$u_v = 8\pi v^2 kT/c^3,$$ (7)

the so-called Rayleigh-Jeans law. Planck did not offer any justification for the finite value of ε. Nor did he speculate on the implications for the dynamics of the resonators. In particular, he did not require the energy of a resonator to vary only by quantum jumps. He only knew that for some unknown reason, in the combinatorial computation of the resonator entropy, energy states differing only by a fraction of ε counted only as one state. As he later remembered in his autobiographical writings, in 1900 he did not know whether the energy elements were a mere computational fiction or should instead be ascribed direct physical significance. In his own opinion, the man who decided in favor of the second alternative was Einstein.[6]

Planck's lack of commitment regarding the physical meaning of the energy elements is apt to surprise the modern reader. It is now well-known that proper application of statistical mechanics to a thermalized resonator necessarily leads to the "equipartition of energy" among the various degrees of freedom of a mechanical system, which implies the average energy kT for a resonator and hence, through equation (2), the inadequate Rayleigh-Jeans law. In this light, the only way to justify Planck's calculation seems to be the introduction of discrete energy levels for the resonators. The inference is anachronistic, however. Planck shared the contemporary skepticism with regard to the validity of energy equipartition. Most important, he disagreed with Boltzmann's and Maxwell's belief that microdynamics entirely determined the thermodynamic behavior of physical systems. In his opinion, the micro-dynamics needed to be supplemented with special assumptions such as the molecular chaos and the "natural radiation" that forbade nonther-modynamic, entropy-decreasing processes. The finite energy elements could just be one of these assumptions.

Planck's peculiar understanding of the relation between micro- and macroworld has escaped the attention of most commentators. Yet the sources of this peculiarity are easily traceable. Like his mentor Hermann Helmholtz, Planck regarded the two laws of thermodynamics as orga-nizing principles with which any future theory had to comply. In partic-ular, an acceptable theory of microscale processes could not contradict the law of entropy increase. As Maxwell's and Boltzmann's kinetic the-ory of gases implied violations of this law in rare but principally observ-able fluctuations, Planck rejected it until he conceived, toward the close of the century, that some structural intricacy of the walls of the gas container could impede the entropy-violating fluctuations. Fifteen more years elapsed before he accepted the statistical understanding of the entropy law.[7]

3. EINSTEIN'S STATISTICAL MECHANICS

Planck was not the only theorist to believe, at the turn of the cen-tury, that the founders of statistical mechanics had left some leeway in its foundations. So did Maxwell's American admirer Josiah Willard Gibbs, Boltzmann's Austrian disciple Paul Ehrenfest, and a student at the Zürich Polytechnicum: Einstein. Although Boltzmann's writings contained pertinent answers to the foundational questions, these often eluded his readers for they were hidden in lengthy, complex memoirs and most of them did not appear in the more synthetic *Lectures on Gas Theory*. Einstein, who only knew Boltzmann's theory through that book, qualified his praise of "the magnificent Boltzmann" with the reproach of insufficiently justified foundations. In 1903–4, he published his own foundations of statistical mechanics, thus hoping to make it the most powerful tool for investigating the microworld (Einstein 1903, 1904).[8]

As is explained in Chapter 3, Einstein started with a physical defini-tion of the probability of a macrostate of a system as the fraction of time that the system spends in this state. For a sufficiently complex system that is left to itself, he assumed that this fraction reached a well-defined limit over the long run and that this limit did not depend on the initial microstate of the system, as long as this microstate had the same energy. Under this assumption, he could unambiguously derive the thermody-namic properties of the system from its microdynamics. In the case of

Hamiltonian systems, he thus retrieved the basic formulas of Gibbs's *Statistical Mechanics* (1902), of which he was still unaware.

An essential corollary of Einstein's foundations was the physical interpretation of thermostatistical probabilities in terms of the frequency with which the various configurations of the system occur in given macroscopic circumstances. Such was the case, for instance, of the combinatorial probability W occurring in the Boltzmann-Planck formula $S = k \ln W$. According to Einstein's interpretation of this formula, the state of highest entropy of an isolated system is the most likely to occur and it is the one observed at thermodynamic equilibrium. The scale of fluctuations around this state, or the thermal stability of the system, is given by the constant k.

Whereas Boltzmann and Gibbs focused on retrieving the results of macroscopic thermodynamics and therefore hurried to assert that perceptible fluctuations where too improbable to be seen in nature, Einstein regarded fluctuations and the concomitant violation of the second law of thermodynamics as the characteristic feature of statistical thermodynamics. Fluctuations depend on the microdynamics of the system, for instance on the molecular structure in the case of a gas. Einstein tried to imagine situations in which they would be observable. In 1904, the only case he could think of was thermal radiation in a cavity of dimensions comparable to the average wavelength. In this case, the energy fluctuation must be of the same order as the energy because of the beats of the waves. Through this consideration Einstein obtained a correct estimate of the average wavelength of black-body radiation (Einstein 1904) at a given temperature. The following year, he famously studied the case of Brownian motion (Einstein 1905k).[9]

To sum up, there was no consensus on the foundations of statistical mechanics at the beginning of the twentieth century. In particular, physicists disagreed on the physical significance of fluctuations. According to Planck, fluctuations were principally unobservable. Boltzmann and Gibbs regarded them as practically unobservable and therefore of little interest. Einstein imagined situations in which they were observable, and had them reveal the microstructure of physical systems. As we will see, much of his insights into the quantum world depended on this physical understanding of fluctuations.

4. THE REVOLUTION OF 1905

In a letter to his friend Conrad Habicht written in the spring of 1905,[10] Einstein announced a series of papers, one of which he judged to be revolutionary. This paper came out in the June issue of *Annalen der Physik* under the title "On a Heuristic Viewpoint about the Production and the Transformation of Light" (Einstein 1905i). In the first section, Einstein established that statistical mechanics, when combined with current electromagnetic theory, necessarily led to an absurd black-body law with infinite total energy. For this purpose, he imagined a set of resonators that interacted both with cavity radiation and with the molecules of a gas. On the one hand, the equipartition of energy between the oscillatory degree of freedom of a resonator and the translational degrees of freedom of the molecules implies the average energy value kT for every resonator. On the other hand, the kinetic equilibrium of the radiation with the resonators implies Planck's relation (2). The combination of these two results yields the Rayleigh-Jeans law (7).[11]

Einstein's reasoning departed from Planck's by introducing a gas as the thermalizing entity for the resonators. In Planck's system, the temperature of the resonators emerged only as a by-product of a special prescription for calculating their entropies. In contrast, Einstein judged that the radiation-resonators system did not in itself have a well-defined temperature, since the interaction between radiation and resonators did not alter the radiation spectrum and therefore did not permit mutual thermalization.[12]

Very likely, Einstein originally formulated this section of his paper as a criticism of Planck's procedure. He may have written something like: "Planck's radiation law is incompatible with the theoretical foundation on which he relied," as he actually dared to write in 1909 (Einstein 1909b, 187). A friend of his, however, advised him not to do so: Planck was then the acknowledged leader of German physics, as well as the chief editor of *Annalen der Physik*. This friendly intervention may be inferred from a letter that Michele Besso wrote to Einstein in January 1928: "For my part, in the years 1904 and 1905 I was your public. If in connection with the drafting of your papers on the quantum problems, I deprived you of a part of your fame, in return I secured a friend for you in Planck" (Einstein and Besso 1972).[13]

Besso probably meant that Planck came to be regarded as the initiator of the quantum revolution, whereas it was Einstein who first declared a radical failure of classical physics in the black-body problem. It was also Einstein who first introduced the idea of a sharp quantum discontinuity with direct physical implications. He did so in the middle section of the 1905 paper, through a clever inversion of Boltzmann's relation $S = k \ln W$. From Wien's law (3) for the spectral density in the high-frequency range, Einstein calculated the entropy variation of black-body radiation when its volume changes from V to v. The result is

$$S(v) - S(V) = k(E / h\nu)\ln(v / V). \tag{8}$$

Therefore, the probability of a fluctuation in which the whole radiation is confined within the fraction v/V of the total volume is

$$W = (v / V)^{E/hv}. \tag{9}$$

This is to be compared with the probability $(v/V)^N$ that the N molecules of a gas be found within a fraction v/V of the available volume. From this analogy and from his conviction that the probability W should be interpreted in terms of a really occurring fluctuation, Einstein concluded that the Wien part of black-body radiation behaved as spatially localized energy quanta hv: "Monochromatic radiation of small density (within the range of validity of Wien's formula) behaves, with respect to the theory of heat, as if it were made of independent energy quanta of the magnitude $[hv]$" (Einstein 1905i, 143).

Lastly, Einstein showed that this analogy, once extended to the absorption and emission of light, accounted for a number of heretofore unexplained facts. For example, the conservation of energy during the absorption and reemission of light by an originally unexcited molecule implies that the absorbed light quantum should be of greater energy than the emitted one and therefore that the frequency of fluorescence light should be smaller than the frequency of the incoming light, a law that George Gabriel Stokes had stated in the mid-nineteenth century. Similarly, the energy balance during photoelectric emission implies that the incoming light quantum hv should be greater than the work P necessary to extract an electron from the target, in conformity with Philipp Lenard's observation of a frequency threshold for the photoelectric effect (1902). The potential difference V that is needed to stop the photoelectron is then given by

$$hv = P + eV. \tag{10}$$

Einstein called for an experimental verification of the new law, as he often did in his early papers.

Einstein was perfectly aware of the revolutionary character of the light quantum hypothesis. He knew that Newton's emission theory of light had long ago been discarded and that the wave theory of light was the best confirmed theory of all physics. Accordingly, he announced the light quantum theory only as a "heuristic hypothesis." He also prepared the reader through an introduction in which he argued that the phenomena of wave optics could represent the average behavior of light when many light quanta were involved, just as the mechanics of continuous media only applied to large numbers of molecules.

One may still wonder whether Einstein had other, unexpressed reasons to tolerate the light quanta. It is well-known that sometime before 1905 he considered an emission theory of light in order to overcome the apparent incompatibility of electromagnetic theory with the principle of relativity. It is not clear, however, whether he literally meant corpuscular emission or merely an influence of the velocity of the source on the velocity of light. At any rate, he may have thought that the elimination of the ether made light more akin to a substance and therefore implied a partial return to Newton's concept. So he said in 1909 (Einstein 1909b). Also, in the relativity paper of 1905 he noted that the ratio E/v of the energy E of a "light complex" to its frequency was invariant under the Lorentz transformation. Presumably, he had in mind the consistency of the light quantum hypothesis with the theory of relativity (Einstein 1905r, 914).

5. QUANTIZED OSCILLATORS

In 1905, Einstein could not make sense of Planck's derivation of Planck's law. In fact, he seems to have deliberately avoided any reference to Planck's law in his reasoning. The light quantum argument required only Wien's law. Einstein even avoided Planck's constant h. For the energy of a light quantum, he did not write hv (as was done previously for the sake of transparency) but $(R/N)\beta v$, where R is the constant of perfect gases, N Avogadro's number, and β the logarithmic slope of Wien's law. He showed that Planck's most remarkable result, the new determination of Avogadro's number, did not depend on Planck's theory since it only required the comparison of the low-frequency part of the measured black-body spectrum with the Rayleigh-Jeans law (7).

The following year, Einstein ceased to avoid Planck's law as he discovered a way to justify Planck's formal steps toward this law (Einstein 1906d). If a resonator of frequency v can only absorb or emit full light quanta, Einstein reasoned, then its energy can only be an integral multiple of hv and Planck's characterization of the complexions for a set of resonators receives a dynamical justification. The only remaining difficulty is that Planck's derivation of the relation (2) between the average energy of a resonator and the spectral density of radiation becomes void. Einstein expressed the need of a new derivation based on some quantized dynamics for the interaction between matter and radiation. Ten years elapsed, however, before he filled the gap.[14]

For the moment, Einstein implemented an old idea of his instead: that the oscillators responsible for the specific heats of solids behaved like Planck's resonators, whether or not they carried electric charge.[15] In the simplest possible model, suppose that each molecule of the solid can vibrate at a unique frequency v independently of its neighbors. The average vibrational energy of a molecule is three times that given by Planck for a resonator, since the direction of the vibration is arbitrary. Calling N the number of molecules, the resulting specific heat of the solid is

$$C = 3N\frac{\mathrm{d}}{\mathrm{d}T}\left(\frac{hv}{e^{hv/kT}-1}\right). \tag{11}$$

For moderate temperatures, the ratio hv/kT is very small so that the specific heat approximately equals the constant $3Nk$ in conformity with the equipartition of energy and with the experimental law established by Laplace's disciples Pierre Louis Dulong and Alexis Petit in the early nineteenth century. For sufficiently small temperatures, the specific heat diminishes, and it goes exponentially to zero when the absolute zero is reached. As Einstein knew, contemporary experiments indicated such a violation of the Dulong-Petit law (Einstein 1907a). He hoped that further low-temperature experiments would confirm his formula, or a similar one based on a less simplistic representation of the solid's vibrations.[16]

6. CONVERSIONS TO THE QUANTUM

No major theorist took Einstein's light quantum seriously until the early 1920s. In contrast, Einstein's idea of quantized material energy quickly gained ground, if only as a working hypothesis. There were two reasons for that. The first was Walther Nernst's enthusiastic espousal of the quantum theory of specific heats, despite his original skepticism toward quantum assumptions. According to the third law of thermodynamics formulated by Nernst in 1906, both the entropy and the specific heat of any substance must go to zero when the temperature goes to zero, just as is the case in Einstein's quantum theory. Since Nernst had developed an experimental program for corroborating his theorem, his laboratory possessed the low-temperature techniques that were necessary to test Einstein's formula for the specific heat of solids, or improved formulas by himself and by Peter Debye, Max Born, and Theodor von Kármán. The results convinced Nernst that the quantum theory was worth pursuing. When in 1910 the Belgian industrialist and amateur-theorist Ernest Solvay offered to finance the first international congress of physics, Nernst convinced him to devote the discussions to this new theory.[17]

In the same period 1905–10, there appeared several proofs that the Rayleigh-Jeans law necessarily followed from ordinary electrodynamics applied to the interaction between matter and radiation. Most authoritative was the proof presented by Lorentz at the Rome congress of 1908 and based on an application of Gibbs's statistical mechanics to the interaction between a metal's electrons and radiation. The British theorist James Jeans gave still another, kinetic-theoretical proof in the same year. For anyone who could follow these learned considerations, there were only two escapes: either the observed black-body radiation did not represent a true equilibrium, or ordinary electrodynamics failed when applied to the interaction between matter and radiation. Although Jeans and Lorentz both flirted with the first possibility, most experts, including Planck, soon agreed that a drastic revision of electrodynamics was necessary. Einstein's suggestion of restricted energy levels was most simple and lent itself to easy thermostatistical derivations. But it was also most shocking for believers in the wavelike nature of radiation, because the absorption of a continuous wave could not possibly induce a quantum jump in a molecule. In his "second theory" of 1911, Planck imagined that the energy of a resonator grew continuously during the absorption of light, whereas emission occurred discontinuously after a full quantum had become available.[18]

At the Solvay Congress of 1911, Lorentz and Jeans demonstrated the impossibility of explaining black-body radiation classically, while Einstein, Nernst, Planck, and Arnold Sommerfeld showed how the quantum hypothesis explained the black-body spectrum, specific heats, and some features of the interaction between radiation and matter. The gathered elite agreed about the need to introduce in physics some basic discontinuity, although they disagreed on the precise form that discontinuity should take and on whether one should regard the discontinuity as primitive or as deducible from a more fundamental and less paradoxical theory.[19]

To sum up, in 1900 Planck introduced a formal element of discontinuity in physics, though with no intention to upset the fundamental continuity of physical interactions. In his peculiar understanding of statistical mechanics, finite energy elements could well occur in an entropy calculation without affecting the continuity of the underlying dynamics. In 1905, Einstein introduced a *physical* discontinuity in the most radical form of light quanta. In 1906, he reinterpreted Planck's energy elements in terms of quantum jumps of atomic entities during their interactions. From 1907 to 1911, various physicists consolidated Einstein's proof that ordinary electrodynamics necessarily failed in the black-body problem and corroborated his explanation of specific heats in terms of quantized oscillators.

The complexity of this historical process has brought about divergent views on the paternity of quantum discontinuity. Most physicists and a few respectable historians regard Planck as the originator of this concept. Thomas Kuhn (1978) famously made Einstein the true inventor of quantum discontinuity. The wisest attitude would be to admit that Planck, Einstein, and a number of other physicists including Lorentz, Ehrenfest, and Nernst all contributed to the emergence of quantum physics in different ways. Surely, Einstein was most radical; but he was not alone in treading the *terra incognita*.

7. WAVES AND CORPUSCLES

Whereas in the early 1910s a growing number of physicists agreed to quantize atomic entities or their interaction with radiation, almost no one accepted Einstein's light quantum. The only notable exception was the experimentalist Johannes Stark, who proposed a light quantum representation of X rays in 1907–8. Soon after their discovery, these rays had been assimilated to a kind of electromagnetic radiation. Some of their behavior nonetheless suggested a corpuscular interpretation, as Stark regularly observed in his laboratory. Most strangely, the electrons ejected during the ionization of molecules through these rays had energies comparable to those of the electrons that produced them. X rays thus seemed able to carry all their energy from the source to a distant point, as only a corpuscle could do. Yet theorists did not regard the argument as fatal to the electromagnetic theory of these rays, because one could still imagine the energy of the ejected electrons to be provided by the target. Since 1906 Lenard advocated this view in the case of the photoelectric effect. He countered Einstein's light quantum explanation with the idea that the incoming radiation acted as the trigger of a pistol to communicate to the photoelectron energy stored in the metal.[20]

Einstein does not seem to have given much weight to contemporary experimental evidence in favor of the light quanta. He rather tried to improve his theoretical argument based on the fluctuation of black-body radiation. One weakness of the original argument of 1905 was that it only applied to the Wien, high-frequency part of the spectrum. In 1909, Einstein removed this restriction by calculating, for radiation obeying Planck's law, the mean quadratic fluctuation $\overline{\Delta^2}$ of the energy E in a volume V and within a spectral interval dv around the frequency v. The result can be written in the form

$$\frac{\overline{\Delta^2}}{\overline{E}^2} = \frac{1}{N_v} + \frac{1}{Z_v}, \tag{12}$$

where N_v is the number of quanta hv contained in the average energy $\overline{E}$, and Z_v is the number $(8\pi v^2/c^3)Vdv$ of wave modes permitted in the volume V and in the frequency interval dv around v. The first term reflects the light quantum structure. The second corresponds to the interference of waves randomly crossing the volume V (Einstein 1909b).[21]

In the event that the argument would seem too formal, Einstein also offered a concrete reasoning inspired by his theory of Brownian motion. He considered a mirror that could freely move in a direction perpendicular to its faces and that perfectly reflected radiation on both faces, though only for frequencies within dv. When immersed in black-body radiation, the mirror acquires a kinetic energy owing to the fluctuating impulses of the radiation. Equating this energy with the value $kT/2$ given by the equipartition theorem, Einstein obtained an expression of the quadratic momentum-fluctuation of the mirror that again contained both a corpuscular term and a wave term. As a corollary to this reasoning, light quanta carry a definite momentum, just as any moving particle does (Einstein 1909b, 1909c).

Clever as they were, these arguments failed to convince the two highest authorities on radiation, Planck and Lorentz. In his comments on Einstein's (1909c) lecture in Salzburg in September 1909, Planck noted that the Brownian-mirror argument only proved the quantum character of the momentum exchange between radiation and matter, not the quantum nature of the radiation. As for the energy-fluctuation argument, it only bore on the structure of radiation when applied to a fictitious volume of radiation without material boundaries. Planck judged that the formula $S = k \ln W$ lacked foundation in this case, since pure radiation, unlike matter, was unable to reach thermal equilibrium by itself. In general, Planck believed that thermodynamic probability could not be defined without reference to matter. In addition, he believed that the probability of a given macrostate depended on some quantum-related blurring of the microstates, whereas Einstein's contended that this probability should always be defined in terms of the frequency of its occurrence in the course of time.[22]

Einstein corresponded with Planck on this matter for a while, with little success. He found a more congenial interlocutor in Lorentz. As he wrote to his friend Jakob Laub: "I am very busy with the question of the constitution of radiation, on which I am having an extensive correspondence with H. A. Lorentz and Planck. The first is an astonishingly deep and lovable man. Planck is also very pleasant in his correspondence. But he has difficulty entering alien ways of thought" (Einstein to Laub, May 17, 1909 [CPAE 5, Doc. 160]). In a letter to Ehrenfest,[23] Einstein denounced Planck's "metaphysical concept of probability," an allusion to his peculiar understanding of statistical mechanics. In contrast, Einstein applauded Lorentz's masterly exposition of the foundations of statistical mechanics and appreciated the provoking clarity of his criticism.

In a long letter,[24] Lorentz responded to Einstein's new fluctuation arguments by rejecting the idea of independent light quanta, despite their success in explaining the photoelectric effect and the properties of X rays. In order to account for interference with large path differences and for the resolving power of a large telescope, he argued, one would have to give to the light quanta a longitudinal and a transversal extension of the order of one meter, in which case their full absorption by a small target would become inconceivable.

Einstein replied that he had never believed in independent, localized light quanta, because, among other reasons, this concept was incompatible with the division of rays during refraction.[25] What he had in mind was a set of singularities driving a vector field, somewhat as electrons carry along an electrostatic field. He hoped to find a modification of Maxwell's equations that would allow for nondiffusing energy points traveling at the velocity of light (Einstein 1909b, 192–3). For the moment he was trying linear modifications, although he knew that a nonlinear equation would make unnecessary a separate equation of motion for the singularities. Despite the difficulty of this approach, he trusted the basic idea of a hybrid theory that would reflect the dual nature of fluctuations: "I think that the next phase of the development of theoretical physics will bring us a theory of light that may be conceived as a kind of fusion of wave and emission theories" (Einstein 1909c, 817).

8. THREE YEARS OF SKEPTICISM

A few months later, Einstein's confidence in this concept of light vacillated. In July 1910 he wrote to Sommerfeld: "Error is human; I would not be surprised if you found a weak point in my considerations. But that could well also be the clue that saves us from the dilemma.... To me the basic question is: 'Is there a way to unify the energy quanta and Huygens's principle?' Appearances are against it, but the good Lord has found the trick."[26] He soon retreated from this quantum view and examined another revolutionary possibility: "At present," he wrote to Laub in November, "I am very hopeful to solve the radiation problem, without light quanta. I am exceedingly curious to see how the thing evolves. Even the energy principle in its present form would have to be given up."[27] Perhaps he had in mind a virtual, wavelike radiation field correlating quantum jumps in distant molecules, as in the later theory of Bohr, Kramers, and Slater.[28]

A week later, Einstein renounced this new attempt: "Again, the solution of the radiation problem has come to naught. The devil has indulged in a rotten trick with me."[29] Six months later, Einstein confided to Besso his doubts on the existence of quanta in general: "I no longer ask myself if these quanta really exist. And I do not try any more to construct them because I now know that my brain is unable to do it. But I still explore the consequences as carefully as I can to learn the range of validity of this idea."[30] In February 1912 he wrote to Hopf: "Quanta certainly do what they ought to, but they do not exist, like the immovable ether. At the moment, the latter is turning diligently in its grave intending to come to life again – poor fellow."[31]

Einstein's skepticism should not be mistaken for an endorsement of theories such as Planck's second theory or Lenard's triggering hypothesis. As he wrote to Wien in May 1912: "I still prefer the 'honest' quantum theory to the compromises that have been *so far* propounded to replace it."[32] But he no longer sought to "construct" quanta and instead tried to see how far quantum results could be obtained without quantum discontinuity: "After many fruitless attempts," he wrote in the same letter, "I too have come to the conclusion that one will not be able to put the theory of radiation on its feet merely by constructing models. This is why I tried to formulate new questions through pure thermodynamics, without making use of a picture."

The first yield of this new strategy was a derivation of an obvious consequence of the light quantum hypothesis, the law of photochemical equivalence, without light quanta (Einstein 1912b). According to this law, first published by Stark in 1908, the amount of radiant, monochromatic energy needed to dissociate one mole of a photosensible compound is $Nh\nu$, where N is Avogadro's number and ν is the frequency of the impinging radiation. In his new derivation of this law, Einstein first wrote the condition for the kinetic equilibrium of radiation at density ρ_ν with the mass m_1 of the compound and the masses m_2 and m_3 of the dissociation products. Assuming that the dissociation probability of a molecule is proportional to $\rho_\nu m_1$, and that the recombination probability of the dissociation products is proportional to $m_2 m_3$, the equilibrium constant $m_2 m_3 / m_1$ must be proportional to the density ρ_ν. In order to comply with the laws of chemical thermodynamics, this constant must also be proportional to $e^{-\epsilon/kT}$, where ϵ is the dissociation energy per molecule and T the temperature of the radiation. In the domain of validity of Wien's law, this is only possible if $\epsilon = h\nu$, in conformity with the equivalence law.[33]

The following year, in a more pronounced retreat from quantum discontinuity, Einstein and Otto Stern derived Planck's law without quantization at all. They reasoned on a resonator-molecule globally free to move in one direction and immersed in thermal radiation, and they computed the velocity fluctuation of this resonator in two different manners: indirectly through the condition that this fluctuation should maintain the average translational energy $kT/2$ of the resonator (as for the Brownian mirror), and directly through a special assumption on the fluctuation mechanism. According to this assumption, fluctuation results partly from the pressure of random waves and partly from an intrinsic agitation corresponding to the zero-point energy $hv/2$ that Planck ascribed to his resonators in his second theory of radiation. Einstein and Stern thus obtained a differential equation for the spectral density, with Planck's law for solution. In this reasoning, the zero-point energy of resonators acts as a substitute for the quantization of resonators or radiation. Einstein was willing to admit this new concept in the name of possible connections with relativistic rest mass, radioactive decay, and specific-heat anomalies. Yet he did not believe in Planck's justification of it, and he had no illusion about the generality of his and Stern's reasoning (Einstein and Stern 1913).

9. BACK TO THE QUANTA

Einstein forcefully reasserted the reality of quanta in a publication of 1914 (Einstein 1914n). Perhaps Niels Bohr's new theory of spectra encouraged him to do so, although there is no trace in his writings of any reflections on this theory before 1916. Perhaps he realized that the zero-point energy failed to solve quantum difficulties that involved other entities than resonators. In any case, he now lent so much reality to the quantum states of a microentity as to compare them with different chemical species. This view induced the profound comment: "The concepts of physical and chemical change seem to lose their fundamental difference" (ibid., 822–3). For instance, a quantum jump in a resonator and the dissociation of a molecule are comparable processes, for they are both caused by the absorption of an energy quantum.[34]

Einstein immediately applied this analogy to a new, nonstatistical derivation of Planck's formula for the average energy of a harmonic oscillator at temperature T. If the various energy levels nhv of a resonator are identified with different chemical species, a thermalized set of resonators is comparable to a chemical mixture in equilibrium. According to the laws of chemical equilibrium, the free energy of the mixture must be a minimum, which implies that the concentration of the species nhv should be proportional to $e^{-nhv/kT}$. Consequently, the average energy of the resonators must be $U = 1/(e^{hv/kT} - 1)$, in conformity with Planck's result of 1900.

Through this reasoning, Einstein confirmed the quantum-theoretical version of Gibbs's canonical law, according to which the probability of the discrete energy value E_n is proportional to $e^{-E_n/kT}$ for a (nondegenerate) system in contact with a thermostat at temperature T. He still did not know how to proceed from the quantized resonators to the black-body law. As he had noted in 1906, Planck's derivation of the relation $u_v = (8\pi v^2/c^3)U$ between average resonator energy and radiation spectrum only applied to classical resonators. A new, quantum-theoretical picture of the interaction between matter and radiation was needed. Einstein found it in the summer of 1916, after the completion of his new theory of gravitation left him more time for quantum meditation (Einstein 1916j, 1917c).

The new picture presumably emerged from a combination of three elements: Einstein's derivation of the law of photochemical equivalence, his analogy between quantum states and chemical species, and Niels Bohr's theory of atomic spectra. According to Bohr, atoms and molecules can only exist in a series of quantum states $S_0, S_1, \ldots S_n, \ldots$ with well-defined energies $E_0, E_1, \ldots E_n, \ldots$. Their interaction with radiation occurs through quantum jumps with characteristic values of the frequency of the emitted or absorbed radiation. Regarding the quantum states as chemical species and remembering his photochemical reasoning, Einstein knew that he could derive Wien's law by balancing the absorption process $S_n + hv \rightarrow S_{n+1}$ with the emission process $S_{n+1} \rightarrow S_n + hv$ and by making the probability of the first reaction proportional to the density of radiation at frequency v. Something in this reasoning needed to be altered in order to get Planck's law instead of Wien's.[35]

At this point Einstein appealed to an analogy between classical and quantum theory. According to classical theory, an oscillating dipole spontaneously emits radiation, whether or not radiation is initially present in its surroundings. When external radiation encounters this dipole, it may either be absorbed if the phase of the incoming wave agrees with that of the oscillator, or it may be amplified in the contrary case. In the quantum theory of radiation, Einstein similarly admitted the existence of three kinds of processes: spontaneous emission (*Ausstrahlung*), absorption (*negative Einstrahlung*), and stimulated emission (*positive Einstrahlung*). The modern terminology is Bohr's. For the probability per time unit of the respective sorts of quantum jump, Einstein assumed the forms

$$A_m^n, \ \rho_v B_n^m, \ \rho_v B_m^n, \tag{13}$$

where n is the upper quantum state, m the lower one, and ρ_v is the density of radiation at the frequency v (Einstein 1916j).

Einstein did not say much on the nature of the probabilities he thus introduced. He only commented that his theory had the weakness to leave to chance the instant and direction of the spontaneous emission of light. He also noted the similarity between spontaneous emission and radioactive decay. Undoubtedly, he would have preferred a theory in which the emission and absorption probabilities were deduced from an underlying deterministic theory. He nonetheless expressed his "full trust in the present way of reasoning" (Einstein 1917c, 128). The probabilistic description of the interaction was a natural counterpart of the discrete character of quantum states: if a quantum system evolves mostly through quantum jumps, then the probability of a quantum jump obviously is the main quantity of physical interest. Instead of speculating on the precise timing and fine structure of the jumps, Einstein proceeded to show what could be done by means of the new probability coefficients.

At thermal equilibrium, Einstein reasoned, statistical mechanics requires the number of atoms in a quantum state n to be proportional to $e^{-E_n/kT}$. The kinetic equilibrium between the atoms and surrounding radiation further requires that the number of quantum jumps from m to n should be equal to the number of reverse jumps:

$$B_n^m \rho_\nu e^{-E_m/kT} = (B_m^n \rho_\nu + A_m^n) e^{-E_n/kT}. \qquad (14)$$

In the high temperature limit for which $\rho_\nu \to \infty$, this condition gives

$$B_m^n = B_n^m. \qquad (15)$$

Therefore, the equilibrium value u_ν of the density ρ_ν has the form

$$u_\nu = \frac{A_m^n / B_m^n}{e^{-(E_n - E_m)/kT} - 1}. \qquad (16)$$

According to a thermodynamic theorem by Wien, u_ν/ν^3 must be a function of ν/T only. Hence $E_n - E_m$ must be proportional to ν. Einstein thus derived Bohr's strange frequency rule $\Delta E = h\nu$ with complete generality and without recourse to any of the empirical laws of spectra. He then required the expression of u_ν to agree with the Rayleigh-Jeans law (7) in the low-frequency limit. The outcome was Planck's law (see Eq. 6), as well as the relation

$$A_m^n / B_m^n = 8\pi h\nu^3 / c^3 \qquad (17)$$

between Einstein's two probability coefficients (Einstein 1916j). To Besso, Einstein announced "an amazingly simple derivation of Planck's formula, I should like to say *the* derivation."[36]

This is not all. In a sequel to this argument, Einstein examined the momentum fluctuation of Bohr atoms surrounded by thermal radiation. He first inferred the mean quadratic fluctuation by indirect reasoning similar to that of the Brownian mirror. Then he computed the same quantity directly, in terms of the transition probabilities A and B. The two methods agree only if the elementary emission process is directed in space and involves a momentum exchange $h\nu/c$, in conformity with the light quantum idea (Einstein 1917c).

Einstein's new theory of radiation is now remembered for the introduction of stimulated emission, which famously permitted the conception of masers and lasers. For Einstein and for his contemporaries, the importance of these memoirs lay elsewhere. First, Einstein filled an important gap in the derivation of Planck's law by means of a simple, statistical description of radiation processes. Second, he corroborated two basic assumptions of Bohr's atomic theory: the existence of stationary states and the frequency rule. In this regard, it should be emphasized that before Einstein's and Sommerfeld's contributions of 1916, Bohr believed that his frequency rule only applied to strictly periodic systems. For instance, he regarded the Zeeman effect as a violation of this rule. Einstein's new considerations established its complete generality.[37]

Third, and most important for Einstein, the new theory of radiation required the directed character of emitted radiation and thus vindicated the light quantum. This argument was the most compelling Einstein had offered so far. In the earlier Brownian-mirror argument, the form of the momentum fluctuation could still be imputed to some intricacy of the interaction between mirror and radiation. As for the energy fluctuation of thermal radiation far from any source, its derivation required a questionable application of Boltzmann's relation to parts of the radiation field. A partial coherence of this radiation possibly threatened the statistical independence of these parts, as Lorentz noted in 1912 (Lorentz 1916, 76). In contrast, Einstein's new argument did not depend on Boltzmann's relation; it only required minimal assumptions on the interaction between matter and radiation, such as the existence of quantum states and the conservation of energy.

10. CRUCIAL EXPERIMENTS

Anyone who could follow Einstein's not-so-simple argument was compelled to admit one of three uncomfortable possibilities: the existence of light quanta, the nonexistence of discrete quantum states, or the violation of energy conservation for elementary radiation processes. Among the contemporary leaders of quantum theory, Einstein opted for the first possibility, Planck for the second, and Bohr for the third. Sommerfeld was agnostic. Einstein's own preference for the light quantum option did not go without worries and anxieties.

In the early 1920s, Einstein renewed his efforts to construct a hybrid theory of light, involving both light quanta and a wave field. He now regarded the wave field as a ghost entity, carrying no energy but some-how able to guide the quanta in their motion. This vague attempt never reached the stage of publication. Einstein did, however, propose a cru-cial experiment for deciding between the classical and the quantum-theoretical derivation of the frequency of the field emitted by a moving atom (Einstein 1922a). Einstein expected the frequency to depend on the direction of emission in the classical case (Doppler effect), and to have the constant value given by Bohr's frequency rule in the quan-tum case. Walther Bothe and Hans Geiger performed the experiment and concluded that the frequency of the emitted light did not depend on the direction of emission. In a letter to Born of December 30, 1921, Einstein commented: "This is the certain proof that the wave field does not have a real existence and that the Bohr emission is an instanta-neous process in the proper sense. This is the greatest scientific event in my life that has happened for years" (Einstein, Born, and Born 1969). Ehrenfest and Laue soon chilled Einstein's enthusiasm by pointing to an error in his classical analysis of the optical device that was supposed to reveal a frequency spread. With this correction, Bothe and Geiger's result became equally compatible with the classical and quantum con-cepts of emission.[38]

As is recounted in Chapter 5, X-ray experiments performed in the early 1920s by Maurice de Broglie and by Arthur Holly Compton even-tually decided in favor of the light quantum. In these experiments, the object interacting with radiation was simpler and better understood than in earlier photoelectric experiments. Instead of the macroscopic metal surface of Millikan's photoelectric experiments, de Broglie dealt with the well-understood inner shells of a Bohr atom, Compton with quasifree electrons. This simplicity permitted the complete determina-tion of the energy or the momentum of the object before and after its interaction with light, and left no room for triggering mechanisms à la Lenard.[39]

These experiments were widely regarded as definitive proofs of the reality of Einstein's light quanta. Accordingly, historians often pre-sent the episode as closing a long period of universal misapprehension of Einstein's insight. Yet it was not so. Bohr still preferred to give up energy conservation rather than accepting the light quantum. With the collaboration of Hendrik Kramers and John Slater, he proposed a theory in which virtual, energy-less wave fields provided statistical correla-tions between quantum jumps in distant atoms. When Bothe and Geiger refuted a consequence of this theory, Bohr still refused to accept the light quantum as a real entity with well-defined space-time behavior.[40]

The standard form of modern quantum mechanics follows Bohr in this respect: the trajectory of quantum objects such as light quanta and electrons is ill-defined, because their position and momentum cannot be simultaneously determined. In this perspective, light quanta are only good to express conservation laws in experimental situations in which the corpuscular aspect of light is dominant. On his side, Einstein never claimed to know what the light quanta actually were. On December 12, 1951, he wrote to Besso (Einstein and Besso 1972, 453): "Those fifty years of conscious brooding have not brought me any closer to an answer to the question: What are light quanta? Today the first rascal thinks he knows what they are, but he deludes himself."[41]

11. QUANTUM DEGENERACY AND MATTER WAVES

Einstein did not contribute to the later evolution of the concept of light quantum. Generally speaking, his contributions to quantum theory after the advent of quantum mechanics were of a critical nature, as is recounted in Chapter 10. Yet his theory of radiation of 1916 was not his last important contribution to quantum theory. In 1924, Einstein received a strange manuscript from an unknown Indian theorist named Satyendra Nath Bose. Bose somehow managed a purely corpuscular derivation of Planck's law, even though the Boltzmann statistics of independent light quanta was known to lead to Wien's law since Einstein's paper of 1905. This achievement depended on a peculiar way of computing the number of complexions that define the thermodynamic probability. Through blind use of an old combinatorial formula by Boltzmann, Bose implicitly assumed that complexions that only differ with respect to the identity of the distributed light quanta count only as one complexion. In other words, he treated the light quanta of a given frequency as indistinguishable entities.[42]

Instead of worrying about Bose's implicit assumptions, Einstein saluted his achievement and proceeded to apply the same method to a gas of material molecules. He thus obtained the so-called Bose-Einstein theory of the quantum gas, with the characteristic phenomenon of low-temperature "degeneracy" as well as the Bose-Einstein condensation. Degeneracy, or the gradual vanishing of entropy and specific heat when the absolute temperature reaches zero, was believed to be required by Nernst's third law of thermodynamics when applied to substances that remain gaseous at very low temperature. Planck and Erwin Schrödinger had proposed various theories of this hypothetical phenomenon, based on the quantization of the translational motion of molecules and ad hoc statistical techniques. Although Einstein's own theory was not much better founded, it led to simpler formulas and it had the great advantage, from Einstein's point of view, to derive from a very close analogy between light and matter (Einstein 1925a, 1925b).[43]

The most extreme supporter of this analogy was the French theorist Louis de Broglie. In 1923 de Broglie had discovered a relativistic invariant way to associate a monochromatic plane wave to a moving particle, thus extending the wave-corpuscle duality to matter. He had transposed the approximation of geometrical optics to the new matter waves in order to retrieve the classical equations of motion of a corpuscle. He had also used the picture of corpuscles sliding on an associated wave as a basis for statistical calculations, and thus obtained Planck's law in the case of radiation and approximately the Maxwell-Boltzmann law in the case of a material gas. He had retrieved the Bohr-Sommerfeld condition for quantized orbits by requiring the synchronicity of the wave and corpuscle motions around the atomic nucleus. Last but not least, he had suggested that the wave nature of matter might lead to observable diffraction phenomena.[44]

Einstein received a copy of de Broglie's dissertation from Paul Langevin, while he was completing his new theory of gas degeneracy. In his thank-you letter, he exclaimed: "He [Louis de Broglie] has lifted a corner of the great veil."[45] Following his usual procedure, Einstein computed the density fluctuation of the Bose-Einstein gas and found two terms, as he had in 1909 in the case of thermal radiation. One term clearly corresponded to the corpuscular structure of the gas. The other, Einstein showed, derived from the random interference of de Broglie's waves. Just as the fluctuation argument of 1905 had revealed the corpuscular properties of light, twenty years later a similar argument revealed the wavelike properties of matter. The rest of the story is well-known. Schrödinger, seduced by the statistical aspects of de Broglie's dissertation and by the connection with Einstein's new theory of the quantum gas, sought a differential equation for de Broglie's waves in a Coulomb potential. After a few failed trials, he found one whose bounded solutions correctly represented the stationary states of the hydrogen atom in the nonrelativistic approximation. Thus was born one of the early forms of quantum mechanics.

12. CONCLUSION: EINSTEIN AND QUANTUM MECHANICS

Although Einstein did not directly contribute to the construction of quantum mechanics, he provided several of the conceptual resources that permitted this construction. First of all, he introduced quantum discontinuity in the sharp, physical form that Bohr needed to develop his atomic theory. Second, his radiation theory of 1916 solidly established Bohr's two postulates and permitted a quantitative – though probabilistic – description of the quantum jumping that occurred during the emission and absorption of radiation. Third, Einstein developed a theory of the quantum gas that implicitly contained essential features of quantum mechanics, including quantum indistinguishability and the wave behavior of matter. Last (firstly in chronological order), Einstein introduced a light quantum concept that anticipated the modern photon.

This last, revolutionary step played a more ambiguous role in the genesis of quantum mechanics. The form of quantum mechanics that emerged from Bohr's and Sommerfeld's programs in 1925 owed almost nothing to Einstein's light quantum (although it definitely benefited from Einstein's statistical description of radiation processes through the A and B coefficients). On the contrary, Bohr and his collaborators could progress only by leaving the radiation mechanism largely undetermined, focusing on atomic structure, and deploying formal, partial analogies with classical electrodynamics. When the Compton effect seemingly confirmed the light quantum, Bohr and Heisenberg only used it to justify their abandonment of the concept of atomic orbits. In contrast, the other form of quantum mechanics, the wave mechanics of de Broglie and Schrödinger, emerged as a bold extension of the analogy between light and matter that Einstein's light quantum initiated in 1905. The light quantum also played a role in the final synthesis of the two early forms of quantum mechanics by inspiring the role that Bohr gave to wave-particle duality in his complementarity interpretation.

Ironically, Einstein never accepted the interpretation of quantum mechanics that his own unveiling of the paradoxes of light suggested. This rejection of Bohr's complementarity and of any interpretation that admits irreducible probabilities was based on the very principle that permitted Einstein's success in forging early quantum concepts: *Probabilitatem esse deducendam*.[46] For the young Einstein, the principle meant that thermodynamic probabilities should be reduced to temporal frequencies in a perfectly determined evolution. In Einstein's later years, the principle further implied that quantum-mechanical probabilities should be derived from a more fundamental deterministic theory. As Einstein could never do that to his satisfaction, the quantum paradoxes of his revolutionary youth haunted him till his last days.

NOTES

1 Cf. Klein (1963b, 1964, 1965, 1967, 1977, 1979, 1982).
2 Cf. CPAE 2, Introduction, and Renn (1997).
3 Cf. Jammer (1966), Kangro (1970), Kuhn (1978), Darrigol (1992), Hoffmann (2001).
4 In fact, the resonators can only alter the spatial distribution of radiation, not its spectrum, as Planck partially came to realize. Hence Planck could not determine the equilibrium spectrum as the final spectrum in his model.

Instead he had recourse to the entropy consideration that is summarized below. Cf. Klein (1962), Kuhn (1978), and Darrigol (1992).

5 Planck had earlier obtained the law (6) by interpolation at the level of entropies. The proportionality between ε and ν was required by Wien's "displacement law" (a consequence of thermodynamics and of Maxwell's radiation pressure), according to which u_ν must have the form $\nu^3 f(\nu/T)$.

6 Planck (1920, 127). Cf. Kuhn (1978), Needell (1980), Darrigol (1988, 1992, 2000a, 2001), and Gearhart (2002).

7 Cf. Needell (1980).

8 Cf. Klein (1967), Renn (1997), Darrigol (1988), and Uffink (2006).

9 Cf. Klein (1967).

10 Einstein to Habicht, May 18 or 25, 1905 (CPAE 5, Doc. 27).

11 Cf. Klein (1963b).

12 Cf. Büttner, Renn, and Schemmel (2003) and Darrigol (1988).

13 Cf. Darrigol (1988).

14 Cf. Kuhn (1978).

15 This suggestion can be found in Einstein to Marić, March 23, 1901 (CPAE 1, Doc. 93).

16 Cf. Klein (1965) and Kuhn (1978).

17 Cf. Klein (1965).

18 Cf. Hiebert (1978), Kuhn (1978), and Needell (1980). There was also the proof of Einstein and Hopf (1910), based on the Brownian motion of a resonator immersed in thermal radiation and free to globally move in one direction: the fluctuating momentum exchange of the resonator with natural (random-phase) radiation obeying Maxwell's equations only agrees with the value $kT/2$ of its (global) translational kinetic energy (required by equipartition) if the radiation is distributed according to the Rayleigh-Jeans law.

19 Cf. Barkan (1993) and Jammer (1966).

20 Cf. Wheaton (1983).

21 Cf. Klein (1964, 1982). On the later history of Einstein's fluctuation formula, cf. Duncan and Janssen (2008).

22 Planck, discussion following Einstein (1909c, 825–6). See also Planck (1912, 101–2).

23 Einstein to Ehrenfest, September 4, 1918 (CPAE 8, Doc. 608).

24 Lorentz to Einstein, May 6, 1909 (CPAE 5, Doc. 153).

25 Einstein to Lorentz, May 23, 1909 (CPAE 5, Doc. 163).

26 Einstein to Sommerfeld, July 1910 (CPAE 5, Doc. 211).

27 Einstein to Laub, November 4, 1910 (CPAE 5, Doc. 231).

28 On the Bohr-Kramers-Slater (BKS) theory and Einstein's reaction to it, cf. Klein (1970a), Darrigol (1992, 246–7), and Duncan and Janssen (2007, 597–617).

29 Einstein to Laub, November 11, 1910 (CPAE 5, Doc. 233).

30 Einstein to Besso, May 13, 1911 (CPAE 5, Doc. 267).

31 Einstein to Hopf, after February 20, 1912 (CPAE 5, Doc. 364). Cf. Darrigol (1988, 76–7).

32 Einstein to Wien, September 17, 1912 (CPAE 5, Doc. 395).

33 Cf. the editorial note, "Einstein on the Law of Photochemical Equivalence" (CPAE 4, 109–13). Einstein's reasoning bizarrely required a different temperature for the gas mixture and for the radiation.

34 According to Hevesy's letter to Bohr of September 23, 1913 (Bohr 1981, 532),
Einstein first reaction to Bohr's theory was one of admiration. However, in a
letter to Sommerfeld, he pointed to the importance of Sommerfeld's recent
extension of Bohr's theory in making Bohr's ideas "entirely convincing"
(Einstein to Sommerfeld, August 3, 1916 [CPAE 8, Doc. 246]).
35 Cf. Bergia and Navarro (1988) and Darrigol (1988).
36 Einstein to Besso, August 11, 1916 (CPAE 8, Doc. 250).
37 Cf. Darrigol (1992, 118–21).
38 Cf. editorial comments in Einstein (1989, 150–5). On Einstein's ghost field,
cf. Lorentz (1927, Secs. 50–3) and Klein (1970a).
39 Cf. Stuewer (1975) and Wheaton (1983).
40 On BKS see note 28.
41 Cf. Stachel (1986b).
42 Cf. Jammer (1966, 248–9) and Darrigol (1991, 256–7).
43 Cf. Darrigol (1991).
44 Cf. Darrigol (1993).
45 Einstein to Langevin, December 16, 1924 (Einstein Archive, Jerusalem
15–376).
46 "Probabilities should be deduced," from Einstein to Pauli, January 22, 1932,
in Pauli (1985, 109).

5 The Experimental Challenge of Light Quanta

1. INTRODUCTION

When Robert A. Millikan published his autobiography in 1950 at the age of eighty-two, he included a chapter on "The Experimental Proof of the Existence of the Photon." In it he recalled that at the Washington meeting of the American Physical Society in April 1915 he "presented my complete verification of the validity of Einstein's equation." He then added:

This seemed to me, as it did to many others, a matter of very great importance, for it ... proved simply and irrefutably I thought, *that the emitted electron that escapes with the energy hv gets that energy by the direct transfer of hv units of energy from the light to the electron* and hence scarcely permits of any other interpretation than that which Einstein had originally suggested, namely that of the semi-corpuscular or photon theory of light itself. (Millikan 1950, 101–2; Millikan's italics)

Millikan's photoelectric-effect experiments thus settled the issue in 1915, although a few pages later he did allow that Arthur H. Compton's "great discovery" of the Compton effect in 1923 "helped in the removal of all doubts, if any still existed, as to the necessity of the foregoing conclusion ..." (ibid., 105).

Millikan's account is a striking example of linear history.[1] It also reflects Millikan's propensity to rewrite history as it suited him. The earliest instance I found dates to 1906 when he and his younger colleague at the University of Chicago, Henry G. Gale, published their textbook, *A First Course in Physics.* In it, Millikan reproduced a picture taken in 1899 of J. J. Thomson reading in his study while sitting in a chair once owned by James Clerk Maxwell.[2] The original picture and Millikan's reproduction of it are identical – except that Millikan carefully etched out the cigarette in Thomson's left hand,[3] presumably because he did not wish to corrupt young and impressionable students. This, and his previously mentioned account, beautifully illustrate what I like to call Millikan's philosophy of history: if the facts don't fit your theory, change the facts.[4]

143

We shall see that the experimental testing of Albert Einstein's equation of the photoelectric effect – for the "discovery" of which Einstein would win the Nobel Prize for Physics in 1921 – was far more problematic than Millikan would have us believe.

2. THE PHOTOELECTRIC EFFECT AND BEYOND

Heinrich Hertz, *ordentlicher* Professor of Physics at the Technische Hochschule in Karlsruhe, first observed the photoelectric effect at the end of 1886, a few months after he had embarked on his famous experiments on electromagnetic waves (Stuewer 1975; Buchwald 1994, 243–4). He enclosed the secondary spark gap in a dark case to increase the visibility of the sparks, observed that their maximum length then decreased, and proved by the middle of 1887 that this occurred only when ultraviolet light emitted at the primary spark gap was not allowed to strike the secondary spark gap. Full enlightenment came a decade later: ultraviolet light expels electrons from the metal knobs forming the secondary spark gap, allowing its length to increase. By 1902, when Philipp Lenard carried out his pioneering experiments (Lenard 1902), more than 160 papers had been published on the photoelectric effect.[5]

Lenard, Hertz's former assistant at the University of Bonn and since 1898 *ordentlicher* Professor of Physics at the University of Kiel, projected ultraviolet light onto a small aluminum plate in an evacuated chamber, established a retarding potential difference between it and an electrode a few centimeters away, and used a quadrant electrometer to measure the total charge Q of the electrons expelled from it.[6] He found, first, that the total charge Q (or the total number) of expelled electrons was proportional to the intensity I of the ultraviolet light; second, that the initial velocity v_i of the expelled electrons did not depend on the intensity I; and third, that three different sources of ultraviolet light of different frequencies expelled electrons with three different initial velocities v_i (Lenard 1902; in Lenard 1944, 255–6, 264–6). He took his second finding to mean that prior to their expulsion the electrons had been in rapid motion inside their parent atoms, so that, as Arthur L. Hughes put it, "the light acts as a kind of trigger, precipitating a state of instability ... resulting in the ejection of an electron" (Hughes 1914, 48; see also Lenard 1944, 267).

Three years later, Einstein (1905i) introduced his "very revolutionary" light quantum hypothesis into physics,[7] citing three phenomena that offered experimental support for it. The most significant of these was the photoelectric effect whose explanation on the "usual conception" of light, he said, "meets with especially great difficulties" as "presented in a pioneering work by Mr. Lenard" (CPAE 2E, 99). These difficulties vanished, however, if one assumed that "a light quantum transfers its entire energy to a single electron" whose maximum energy then is given by

$$\Pi e = (R/N)\beta v - P,$$

where Π is the retarding potential difference, R is the ideal gas constant, N is Avogadro's number, β is the constant that appears in Wilhelm Wien's black-body radiation law, v is the frequency of the incident light, and P is the energy lost by the electron (charge e) in reaching the surface of the irradiated metal, the so-called work function. Thus "Π, presented as a function of the frequency of the exciting light in Cartesian coordinates, must be a straight line whose slope is independent of the nature of the substance investigated" (CPAE 2E, 101), which, Einstein added, "does not conflict with the properties of the photoelectric effect observed by Mr. Lenard" (ibid.). Einstein did not neglect to point out that if the incident light quantum did not transfer all of its energy to the electron, then an additional term would be present on the left-hand side of the equation – an insight that would be vindicated almost two decades later.

Four years after his *annus mirabilis* of 1905, Einstein again was occupied "incessantly with the question of the constitution of radiation," as he remarked in a letter to his friend Jakob J. Laub on May 17, 1909.[8] Four months earlier, he had gained entirely new insight into the question by analyzing the energy and momentum fluctuations in black-body radiation (Einstein 1909b),[9] now assuming Max Planck's law to be valid, and found that the expressions for them split naturally into a sum of two terms, one of which arose from the interference of waves, the other "if the radiation consisted of point quanta of energy hv moving independently of each other" (CPAE 2E, 366), where he now replaced his earlier combination of constants with Planck's constant h.

This constituted Einstein's introduction of the wave-particle duality into physics, and in September 1909, soon after he had become *ausserordentlicher* Professor at the University of Zurich, he reiterated his analysis in a lecture at a meeting of the Gesellschaft Deutscher Naturforscher und Ärtze in Salzburg, Austria (Einstein 1909c). By then he was convinced "that the next stage in the development of theoretical physics will bring us a theory of light that can be understood as a kind of fusion of the wave and emission theories of light" (CPAE 2E, 379). Among the "extensive group of facts" that could be understood "far more readily from the standpoint of Newton's emission theory" than from that of the wave theory was the striking observation that when electrons struck a metal plate and produced X rays, which in turn struck a second metal plate and expelled electrons, their velocity was close to that of the primary electrons and did not depend on the separation of the metal plates. The intervening X rays thus did not behave like spreading spherical waves but like localized light quanta that were emitted from the first plate in a "directed" elementary process (ibid., 388).

That was the first scientific meeting that Einstein attended, and his audience included many prominent or soon-to-be prominent physicists. In the discussion, Planck, the doyen of German physicists, left no doubt where he stood. To Planck, accepting light quanta meant rejecting electromagnetic waves, and that, he said, "seems to me to be a step which in my opinion is not yet necessary" (ibid., 395).

The only physicist who sided with Einstein was Johannes Stark, who just five months earlier had been appointed *ordentlicher* Professor of Experimental Physics at the Technische Hochschule in Aachen. Picking up on Einstein's example, Stark asked how, other than by adopting Einstein's light quantum hypothesis, it was possible to understand how X rays could travel through "great distances, up to 10 meters," and "still achieve concentrated action on a single electron" (ibid., 397). Planck responded that, "X rays are a special case; I would not assert too much about them." The great difficulty was to understand how light quanta could undergo interference "at the enormous phase differences of hundreds of thousands of wavelengths" (ibid.). Einstein had the final word, saying that, "It probably wouldn't be as difficult to incorporate the interference phenomena as one thinks," because "it must not be assumed that radiations consist of noninteracting quanta" (ibid., 398). He suggested that a light quantum possibly could be pictured as a "singularity surrounded by a large vector field," and that a superposition of such fields might account for interference.

Stark, inspired by Einstein's lecture, returned to Aachen and published a series of remarkable papers in 1909–10. In his first one (Stark 1909), he stated explicitly, for the first time in the literature, that the momentum of a light quantum of frequency v and velocity c is hv/c, and he then analyzed the production of an X-ray quantum when one electron struck another one in a metal plate, drawing the correct vector diagram describing the collision process. He also discussed the inverse process, the collision of an X-ray quantum with an electron in a metal plate. He was immediately challenged by Arnold Sommerfeld (1909), however, who argued that an electron striking a metal plate would undergo a rapid deceleration, so that the X rays produced would not consist of quanta but of electromagnetic pulses – and he then went on to develop his well-known theory of *Bremsstrahlung*. Stark responded to Sommerfeld both in correspondence and in print (Stark 1910a, 1910b), but Sommerfeld's views soon prevailed. At the end of 1910, the Japanese physicist Hantaro Nagaoka visited Stark in Aachen, and on February 22,

1911, he wrote a letter to Ernest Rutherford in Manchester, informing Rutherford that Stark was "propounding his 'Lichtquantentheorie'; there is some doubt whether he will succeed in explaining the interference phenomena, or not. The Germans say that he is full of phantasies, which may be partly true" (quoted by Badash 1967, 59).

3. CLASSICAL THEORIES OF THE PHOTOELECTRIC EFFECT

That was the common attitude of physicists toward Einstein's light quantum hypothesis. The only exceptions were Stark and Einstein, who in early 1909 had initiated a correspondence with H. A. Lorentz in Leiden, seeking Lorentz's views on his developing ideas on the constitution of radiation.[10] That drew the great Dutch theoretical physicist into the debate, who then became the first prominent theoretical physicist after Planck to discuss the shortcomings of Einstein's light quantum hypothesis in print – and to propose a classical, non-Einsteinian theory of the photoelectric effect.

Lorentz's most general criticism was the same as Planck's, that independently moving, localized light quanta could not account for interference and diffraction, but in 1910 he drove it home incisively (Lorentz 1910a, esp. 354–5; Lorentz 1910b, esp. 1249). He argued that Otto Lummer and Ernst Gehrcke's interference experiments in Berlin, which involved path differences of up to eighty centimeters, proved that that had to be the minimum longitudinal extension of light quanta, while George Ellery Hale's observations with his new reflecting telescope on Mount Wilson in southern California, with its mirror of 150 centimeters in diameter, proved that that had to be the minimum lateral extension of light quanta. How then could such monstrously large light quanta pass through the small pupil of the human eye, or strike a single electron, without being subdivided into light quanta of lower frequency? Or how could a light quantum be split up into reflected and refracted light quanta at the interface between two different media without experiencing a change in its frequency?

Still, Lorentz "would not like to quarrel with the heuristic value" of Einstein's light quantum hypothesis, but he "would like to defend the old theory [of Maxwell] as long as possible" (Lorentz 1910b, 1250). He admitted that an Einsteinian light quantum striking an electron in a metal surface would cause it to be ejected immediately, as observed, while a spreading electromagnetic wave would require a large amount of time to deliver sufficient energy to it for ejection. Nevertheless, the disadvantages of Einstein's hypothesis outweighed its advantages. He therefore proposed an alternate explanation of the photoelectric effect based upon Arthur E. Haas's "somewhat risky" atomic model, which envisioned an electron possessing energy hv at the surface of a Thomson positively charged plum-pudding atom. An incident electromagnetic wave then would set an electron inside the atom into oscillation, and if it acquired energy hv it would be ejected. "In Haas's hypothesis," Lorentz thus concluded, "the riddle of the energy elements is combined with the question of the nature and action of positive electricity, and it may be that these different questions will, for the first time, find their complete solution" (ibid., 1253). It was far more likely that the photoelectric effect would be understood on the basis of atomic-structure considerations than on the basis of Einstein's light quantum hypothesis. That approach also was advocated between 1910 and 1913 by J. J. Thomson, Arnold Sommerfeld, and Owen W. Richardson.

Thomson, Cavendish Professor of Experimental Physics, initially accepted Lenard's "trigger" explanation of the photoelectric effect, but by 1908 he regarded it as "exceedingly improbable" after learning that Erich Ladenburg (1907) in Berlin had found a definite relationship (actually a *quadratic* one) between the energy E of the photoelectrons and the frequency v of the incident ultraviolet light (Thomson 1908, 422). Two years later, Thomson (1910a) published the first of two new theories of the photoelectric effect, both of which assumed a *linear* relationship between E and v even though, as we shall see, that remained in doubt experimentally for several more years.

Thomson assumed that positive-negative doublets are dispersed throughout the atoms in a photosensitive surface, and that some of them have electrons (corpuscles) circulating below their positive ends with various kinetic energies T proportional to their angular frequencies $d\varphi/dt$. Thus an incident electromagnetic wave of frequency v would resonate with one or more of them, doing "work upon the doublet, twisting its axis" and causing the "liberation" of the electron with a kinetic energy T proportional to the frequency v, where Thomson took the proportionality constant to be Planck's constant h. Thus, Thomson wrote, his "theory enables us to explain the electrical effects produced by light, without assuming that light is made up of unalterable units, ... a view which it is exceedingly difficult to reconcile with well-known optical phenomena" (ibid., 245–6).

James H. Jeans (1910) in Cambridge criticized Thomson's theory, but Thomson (1910b) immediately rebutted Jeans's criticism. In any case, Thomson rejected his theory in 1913 and proposed an entirely new one (Thomson 1913) based upon an entirely new atomic model – a move that reflected his general attitude toward atomic models. As his biographer Lord Rayleigh remarked, "J. J. was not inclined to be dogmatic about his atomic theories, and indeed he was quite prepared to change them, sometimes without making it altogether clear that he had wiped the slate clean ..." (Rayleigh 1943, 141). Thomson focused on what a particular atomic model *would* explain and not on what it would *not*.

Thomson's new atomic model embodied two forces, a radial inverse-cube repulsive force "diffused throughout the whole of the atom" and a radial inverse-square attractive force "*confined to a limited number of radial tubes in the atom*" (Thomson 1913, 793; Thomson's italics). The two forces would cancel at some distance $r = a$ from the center of the atom, so that a small displacement of the electron would cause it to oscillate about its equilibrium position with simple harmonic motion of frequency f. An incident electromagnetic wave of frequency $v = f$ thus would resonate with the electron, so that, Thomson asserted, if some "casual magnetic force" moved the electron laterally out of the attractive radial tube, it would experience the "uncontrolled action" of the repulsive force and be expelled from the atom with a kinetic energy T proportional to f, where Thomson again took the proportionality constant to be Planck's constant h. Thus, Thomson wrote, we again have "the well-known law of Photo-Electricity" (ibid., 795).

Thomson's theories seem contrived and ad hoc in retrospect, but we must remember that little of a definite nature was known at this time about the structure of the atom, which was a major reason that his theories were viewed sympathetically by some of his prominent contemporaries (Hughes 1914, 50–1; Hallwachs 1916, 460–2, 530–2). They demonstrated, above all, that it was possible to envision atomic models on which a quantitative theory of the photoelectric effect could be based, so that Einstein's light quantum hypothesis need not be accepted. That same point was made in a more sophisticated way by Sommerfeld (1911a).

Sommerfeld, Professor of Theoretical Physics at the University of Munich, shared Lorentz's and Thomson's skepticism toward Einstein's light quantum hypothesis but was convinced that a quantum theory of *matter* was unavoidable and that physicists should "pursue the h-hypothesis in its various consequences and trace other phenomena back to it" (ibid., 1092). One of those was the photoelectric effect, and together with his assistant and former student, Peter Debye, he presented a new theory of it in September 1911. It would stand, Sommerfeld explained, "approximately midway between ... Lenard's idea, according to which the ripping off of the electron is a resonance phenomenon in the atom ... and ... Einstein's light-quantum hypothesis ..." (ibid., 1088). He postulated that in "every purely molecular process a definite and universal amount of action" is taken up or given up in time τ by an electron of kinetic energy T and potential energy V according to the condition

$$\int_0^\tau (T - V)dt = \frac{h}{2\pi}.$$

Setting up the electron's equation of motion, he then showed that at complete resonance between the natural frequency v_0 of the electron and the frequency v of the incident electromagnetic radiation, $T = hv_0 = hv$, which, he said, is "Einstein's law" (ibid., 1089).

So it was, except for the absence of the work function, which, however, could be ignored because, Sommerfeld later wrote, that is "obviously foreign to the pure molecular process and would appear only if we were to follow the photoelectron along on its subsequent path through the surface of the metal" (Debye and Sommerfeld 1913, 884; Sommerfeld 1968, 89). The true difference between his and Einstein's theories, of course, was a fundamental one: Sommerfeld assumed that the photoelectron accumulated its energy over a finite period of time τ in a resonance process which, he showed, meant that a plot of v versus T was not linear, as Einstein would have it, but "should possess for each natural frequency [v_0] of the atom, a maximum." Whose relationship fit the existing experimental data better? Sommerfeld had the answer: J. R. Wright, working in Millikan's laboratory at the University of Chicago, had just reported experiments that showed "with certainty," Sommerfeld said, first, that the "maximum photoelectron energy does not vary approximately linearly with the frequency," and second, in agreement with recent experiments of Robert W. Pohl and Peter Pringsheim in Berlin, that the magnitude of the photoelectron current depends on the plane of polarization of the incident radiation. "With respect to both points," Sommerfeld concluded, "our theory is in better accord with Wright's measurements than Einstein's light-quantum theory" (Sommerfeld 1911a, 1091).

At the end of October 1911, one month after Sommerfeld had reached that conclusion, he left for the Solvay Congress in Brussels where he presented a paper in which he extended his theory to the case of near resonance, that is, when $v \approx v_0$ (Sommerfeld 1911b). He derived a more complicated expression for the kinetic energy T of the photoelectrons but showed that it again reduced to $T = hv_0 = hv$ at resonance. Einstein, attending from Prague, was the first to respond in the discussion, and he later intervened in it repeatedly. Among his objections to Sommerfeld's theory was its prediction that "the number of electrons liberated per second would not be proportional to the intensity of the [incident] light," since they would have to accumulate their energy over time before emission (ibid., 390). Thus, he remained convinced, as he said in the discussion following his own paper, that "it is necessary ... for us to introduce a hypothesis such as that of quanta along side the indispensable equations of Maxwell" (Einstein 1911h, 443; Einstein 1914a, 358; Eucken 1914, 359, reprinted in CPAE 3, Doc. 27, 556).

Einstein bolstered his arguments for light quanta soon after he returned to Prague. He wrote two papers in January and May 1912 in which he proved, first, *"that a gas molecule that decomposes under the absorption of radiation of frequency v_0 absorbs (on average) the radiation energy hv_0 in the course of its decomposition"* (Einstein 1912b; CPAE 5E, 94; Einstein's italics), and second, that "the energy absorbed per molecular decomposition does not depend on the proper frequency of the absorbing molecule but on the frequency of the radiation that brings about the decomposition" (Einstein 1912f; CPAE 5E, 123). These conclusions conflicted directly with Sommerfeld's theory, but Sommerfeld did not abandon it. Assisted by Debye, he continued to pursue his theory of the photoelectric effect, for instance, analyzing in 1913 the effect of a damping term in the electron's equation of motion and estimating the accumulation time τ to be on the order of 10^{-5} second (Debye and Sommerfeld 1913). That no doubt was the most troubling feature of his theory, since Wilhelm Hallwachs (1916, 469) soon cited experiments in which the intensity of the incident radiation was so low that it should have taken between sixty seconds and six hundred years for the photoelectrons to accumulate enough energy to be ejected. Yet, they were ejected immediately, as Einstein's theory required.

In 1912, however, Owen W. Richardson, Professor of Physics at Princeton University, had proposed yet another, very different theory of the photoelectric effect that, he wrote, would "avoid … the vexed question of the nature of the interaction between the material parts of the system and the aethereal radiation" and would assume "the existence of a statistically steady condition of the aethereal and electronic radiations" (Richardson 1912a, 617). He would adopt a thermodynamic, macroscopic approach.

Richardson carried out his analysis by envisioning a cylinder with a piston in it that was completely opaque to electrons, which he took to behave like an ideal gas, and completely transparent to black-body radiation, which he took to obey Wien's law as a sufficiently close approximation to Planck's law. By moving the piston slowly up and down, electrons then would be emitted and absorbed by the bottom of the cylinder in a reversible manner. Equating their rates of emission and absorption, he found in a second paper a general equilibrium equation that he solved, showing that no electrons would be emitted if the energy hv of the radiation were less than some value w_0, the "latent heat of evaporation" per electron at absolute zero (which he took to play the same role as Einstein's work function), while a steady emission would occur if hv were equal to or greater than w_0. These results, Richardson concluded, "have been derived without making use of the hypothesis that radiant energy exists in the form of 'Licht-quanten'.... It appears therefore that the confirmation of [my theory] ... by experiment would not necessarily involve the acceptance of the unitary theory of light" (ibid., 574).

That was a highly significant conclusion, because Richardson and his graduate student, Karl T. Compton, were concurrently carrying out experiments on the photoelectric effect. In Richardson's judgment, all earlier experiments were "very contradictory," because while most physicists believed that the maximum initial kinetic energy of the photoelectrons varied linearly with the frequency of the incident radiation, some had found that it varied quadratically (Richardson and Compton 1912, 575).

Richardson and Compton set out to resolve this uncertainty by making measurements on eight different photosensitive metals when irradiated by three to five different ultraviolet wavelengths, usually between 2,000 and 2,700 ångstroms. Theirs were the most refined and accurate experiments to date. They determined both the maximum and the mean kinetic energies of the photoelectrons and found a linear relationship between both and the frequency of the incident radiation. But whose *theory*, Richardson's or Einstein's, did their experiments support? Richardson concluded forthrightly that they could be taken to "confirm the theory of photoelectric action which was recently developed by one of the writers" (Compton and Richardson 1913, 530).

Richardson's conclusion, nonetheless, was open to criticism because, as Arthur L. Hughes pointed out, his theory was "developed under the assumption that the [photosensitive] substance is in thermal equilibrium with complete [black-body] radiation travelling with equal intensities in all directions," while his and Compton's experiments were carried out with "directed monochromatic light" (Hughes 1914, 47). Moreover, in 1913 Pohl and Pringsheim (1913, 1018) in Berlin questioned whether Richardson and Compton's experiments, in fact, had decided the issue of a linear or quadratic relationship, since both could fit their experimental points owing to the narrow range of frequencies they had used. This criticism stimulated Richardson to develop his theory further before he left Princeton in early 1914 to become Wheatstone Professor of Physics at King's College, London (Richardson 1914), and to carry out further experiments there with F. J. Rogers (Richardson and Rogers 1915), but these, too, were inconclusive on this question.

4. MILLIKAN'S EXPERIMENTS AND THEORY

Robert Andrews Millikan joined the faculty of the University of Chicago in 1896, and for around a decade he devoted his energies largely to teaching and writing textbooks before embarking on his measurements of the charge of the electron that culminated in his pioneering oil-drop experiments in 1913 (Millikan 1913a). Concurrently, since 1905, he had worked intermittently on the photoelectric effect,[11] and in 1907 he and his student George Winchester published their first results, reporting that the energy of the photoelectrons is independent of the temperature of the photosensitive metal. In yet another four years, as we have seen, his student J. R. Wright reported that he had not found "anything approaching a linear relationship" between the photoelectron energy and the frequency of the incident light. Millikan's doubts about the validity of Einstein's equation and light quantum hypothesis were soon reinforced in the spring and summer of 1912 when he visited Berlin and attended the lectures of Planck who, Millikan recalled, "very definitely rejected the notion that light travels through space in the form of bunches of localized energy" (Millikan 1950, 67). He thus came to realize "how difficult it would be to get a convincing answer to the problem of this Einstein photoelectric equation" and, he recalled, "I scarcely expected ... that the answer, when and if it came, would be positive; but the question was very vital and an answer of some sort had to be found" (ibid., 100). He began working in earnest in October 1912 and, he said, "it occupied practically all of my individual research time for the next three years" (ibid.)

Millikan revealed his ignorance of the origin of Einstein's light quantum hypothesis in December 1912 in a lecture he delivered on "Atomic Theories of Radiation" at a meeting of the American Association for the Advancement of Science in Cleveland, Ohio (Millikan 1913b). He asserted that J. J. Thomson, in his Silliman Lectures at Yale University in 1903, had proposed the "first form of the modern atomistic theories of radiation," and that Einstein's hypothesis was "simply the J. J. Thomson theory of the discontinuous distribution of radiant energy in space, assumed still to be electromagnetic ..., with the addition of Planck's original assumption that a given source emits and absorbs energy in units which are multiples of $h\nu$" (ibid., 129–30). Millikan then asked: "Why not adopt it?" His answer: "*Simply because no one has thus far seen any way of reconciling such a theory with the facts of diffraction and interference so completely in harmony in every particular with the old theory of ether waves.... That we shall ever return to a corpuscular theory of radiation I hold to be quite unthinkable*" (ibid., 132; Millikan's italics).

Still, the experimental challenge remained and eventually, by "an element of great good fortune," he found "the key to the whole problem." He discovered that radiation over a wide range of frequencies ejected photoelectrons from the highly electropositive alkali metals, lithium, sodium, and potassium. But numerous experimental problems had to be solved: he had to eliminate reflected and scattered light, measure small photoelectron currents accurately, and correct for contact potential differences, all under high vacuum. His final apparatus, however, worked beautifully. He could irradiate one end of a cylinder of sodium, for example, then rotate it through 120° with an external electromagnet and measure its contact potential difference, then rotate it another 120° and scrape its surface clean with a rotating knife, and finally rotate it another 120° back to its original position and irradiate it once again. It was, Millikan (1916, 361) said, "a machine shop in vacuo."

Millikan later summarized his and his students' experiments by declaring that, "After ten years of testing and changing and learning and sometimes blundering ... this work resulted, contrary to my own expectation, in the first direct experimental proof in 1914 of the exact validity ... of the Einstein equation, and the first direct photoelectric determination of Planck's h" (Millikan 1923, 61–2).[12] He reported his results in April 1915 at a meeting of the American Physical Society in Washington, D.C.; they were published in March 1916 in *The Physical Review*.

Millikan wrote Einstein's equation as

$$\frac{1}{2}mv^2 = Ve = h\nu - p$$

and listed its predictions: (1) that for each "exciting frequency" ν above a "certain critical value" ν_o, there is a "definitely determinable" maximum velocity v_{max} of the photoelectrons; (2) that the stopping potential V varies linearly with the frequency ν; (3) that the slope of the line is "numerically equal" to h/e; (4) that when the velocity $v = V = 0$, then $p = h\nu_o$, that is, the intercept of the line on the frequency axis is the critical frequency ν_o; and (5) that the contact electromotive potential difference E between any two photosensitive metals is given by $E = (h/e)(\nu_o - \nu_o') = (V_o - V_o')$.

Of these predictions, Millikan declared that only the first "had been tested even roughly" by Lenard in 1902, but that Carl Ramsauer (1914a, 1914b) in Heidelberg recently had "vigorously denied" its correctness. The second prediction, despite Richardson and Compton's "exceptionally fine work," was "not deducible from existing data," while the third and fourth had "never been subjected to careful experimental test," and the fifth had "never been tested at all" (Millikan 1916, 356–7). Now, however, Millikan had confirmed all five predictions. In fact, the accuracy with which his experimental points fell on a linear plot of V versus v with slope h/e when sodium was irradiated with light of wavelengths between 2,535 and 5,461 ångstroms, and lithium between 2,535 and 4,339 ångstroms, was remarkable. No longer could anyone doubt the validity of Einstein's equation which, Millikan soon claimed, "is destined to play in the future a scarcely less important rôle than Maxwell's equations have played in the past, for it must govern the transformation of all short-wave-length electromagnetic energy into heat energy" (Millikan 1917, 227–9).

That left its theoretical interpretation. "We are confronted," Millikan wrote, "by the astonishing situation that these facts were correctly and exactly predicted nine years ago by a form of quantum theory which has now been pretty generally abandoned" (Millikan 1916, 355). Einstein had brought "forward the bold, not to say the reckless" light quantum hypothesis – "reckless" because it "seems [to be] a violation of the very conception of an electromagnetic disturbance" and thus "flies in the face of the thoroughly established facts of interference." Einstein had proposed it in 1905 "apparently ... solely because it furnished a ready explanation" of Lenard's observation that the photoelectron's energy "is independent of the intensity of the [incident] light while it depends on its frequency." Einstein's assumption, however, "was but a very particular form of the ether-string theory advanced by J. J. Thomson two years earlier," which "seems at present to be wholly untenable" (ibid., 383). Indeed, his oil-drop experiments, in which he had varied the electric field continuously, proved that the ether does not have a discontinuous or "fibrous structure," so that "we must abandon the Thomson-Einstein hypothesis of localized energy."

The question therefore was, "how else can the equation be obtained?" Clearly, Millikan wrote, we must search for "a substitute for Einstein's theory" (ibid., 385). The energy of the photoelectron, he soon pointed out, "must have come either from the energy stored up inside of the atom or else from the light. There is no third possibility" (Millikan 1917, 231). Since the photoelectron acquires the same energy whether the light source is "an inch ... or a mile away," it might seem, as Lenard had suggested, "that the light simply pulls a trigger in the atom." That, however, could not be true since its energy then would not depend on the "kind of a wave-length [that] pulls the trigger, while it ought to make a difference what kind of a gun, that is, what kind of an atom, is set off." "*The energy of the escaping corpuscle must come then*," Millikan concluded, "*in some way or other, from the incident light*" (ibid.; Millikan's italics).

On the "spreading-wave theory," the great problem was that the tiny photoelectron would receive only "a very minute fraction" of its observed energy. Thus Millikan calculated that an atom of cross-sectional area 10^{-15} cm^2 three meters away from a light source of wavelength 5,000 ångstroms, or of energy $hv = 4 \times 10^{-12}$ erg, would have to be illuminated for four hours for one of its electrons to absorb the energy with which it was ejected, "yet the corpuscle is observed to shoot out the instant the light is turned on." Millikan therefore concluded that "*there is no alternative but to assume that the corpuscles which are ejected are already possessed of an energy almost equal to hv*" (Millikan 1916, 385; Millikan's italics), which meant that there are "oscillators of all frequencies within the absorbing body" that "are at all times in all stages of energy loading up to the value hv." Most would have a "frequency or frequencies characteristic of the substance," but a few of "every conceivable frequency" would be mixed in with them. Those therefore that "are in tune with the impressed waves" will absorb energy until it reaches a "critical value," at which time "an explosion occurs and a corpuscle is shot out [of the atom] with an energy hv," part of which it will "fritter away" in striking other atoms before it emerges from the photosensitive metal. In sum, Millikan wrote:

The Thomson-Einstein theory throws the whole burden of accounting for the new facts upon the unknown nature of the ether and makes radical assumptions about its structure. The loading theory leaves the ether as it was and puts the burden of an explanation upon the unknown conditions and laws which exist inside the atom, and have to do with the nature of the electron. (Millikan 1917, 237)

Millikan thus fell completely in line with Lorentz, Thomson, Sommerfeld, and Richardson and proposed a classical, non-Einsteinian theory of the photoelectric effect. No one expressed his views on Einstein's light quantum hypothesis more clearly than he did in 1917 when he wrote that:

Despite ... the apparently complete success of the Einstein equation, the physical theory of which it was designed to be the symbolic expression is found so untenable that Einstein himself, I believe, no longer holds to it, and we are in the position of having built a very perfect structure and then knocked out entirely the underpinning without causing the building to fall. It [Einstein's equation] stands complete and apparently well tested, but without any visible means of support. These supports must obviously exist, and the most fascinating problem of modern physics is to find them. Experiment has outrun theory, or, better, *guided by erroneous theory*, it has discovered relationships which seem to be of the greatest interest and importance, but the reasons for them are as yet not at all understood. (ibid., 230; my italics)

We recall that this is the same man who in 1950 wrote that his experiments "proved simply and irrefutably I thought," that they scarcely permit "any other interpretation than that which Einstein had originally suggested, namely that of the semi-corpuscular or photon theory of light ..." (Millikan 1950, 101–2).

5. THE COMPTON EFFECT

Millikan, of course, also missed the mark when he alleged that Einstein had abandoned his light quantum hypothesis. Quite the contrary. Although Einstein did doubt the existence of light quanta between 1910 and 1913, he submitted two new papers for publication in 1916 that reaffirmed their existence and offered penetrating new insights into the emission and absorption of radiation (Einstein 1916j, 1916n). These papers are best known for his introduction of his famous A and B coefficients, by which he gave an "amazingly simple and general" derivation of Planck's black-body radiation law, but he regarded as their "most important" result his proof that the emission or absorption of radiation of energy hv by a molecule involves the transfer of momentum hv/c. These, therefore, are directed processes: "There is no emission of radiation in the form of spherical waves." Thus, Einstein wrote, the "establishment of a quantumlike theory of radiation" appears to be "almost unavoidable" (CPAE 6E, Doc. 38, 232).

Einstein's contemporaries disagreed, and they had another weighty reason for doing so. In the spring of 1912, Max Laue, *Privatdozent* in Sommerfeld's Institute for Theoretical Physics in Munich, concluded that X rays could be diffracted by a crystal, and soon thereafter Sommerfeld's assistant Walter Friedrich and his doctoral student Paul Knipping confirmed Laue's prediction using a copper-sulfate crystal (Friedrich, Knipping, and Laue 1912; Laue 1912). As Ernest Rutherford (1918, 81) wrote, this dramatic discovery provided a "complete proof" that X rays are "a type of light wave" of short wavelength.

That was virtually the universal conviction of physicists when Arthur Holly Compton began his researches. He received his PhD degree in physics at Princeton University in 1916 and subsequently spent one year at the University of Minnesota, two years at the Westinghouse Lamp Company in Pittsburgh, and one year as a National Research Council Fellow at the Cavendish Laboratory before being appointed in 1920 as Wayman Crow Professor and Head of the Department of Physics at Washington University in St. Louis. Two years later he discovered the Compton effect.[13]

The most striking feature of Compton's long and arduous route to his discovery was its autonomy: his thoughts evolved largely owing to his own experiments and theoretical interpretations of them. At Minnesota he carried out X-ray diffraction experiments that proved that the electron, not the atom, is the "ultimate magnetic particle." At Westinghouse he came across a paper by Charles G. Barkla of the University of Edinburgh who reported experiments indicating that the mass-absorption coefficient of 0.145 ångstrom X rays in aluminum is smaller than the Thomson mass-scattering coefficient (Barkla and White 1917), an apparently impossible result, which Compton explained by assuming that these X rays were being diffracted by exceedingly large electrons in the aluminum atom. At the Cavendish he then carried out experiments that proved that γ rays become "softer" or of longer wavelength when scattered by thin metal plates, and that the softening increased with increasing scattering angle – results he explained by assuming that the γ rays were striking electrons in the scatterer, which then were propelled forward at high velocities and emitted Doppler-shifted γ rays of longer wavelength. At Washington University he carried out similar experiments with X rays, eventually using a Bragg spectrometer to measure the increase in wavelength of the scattered X rays, which he again interpreted as Doppler-shifted radiation.

Thus, during no less than six years of intense experimental and theoretical work, Compton never doubted that γ rays and X rays are electromagnetic waves. And only once, in a long National Research Council report of October 1922, did he cite Albert Einstein, but then only to note that Einstein's equation of the photoelectric effect had been "shown by Richardson to be a direct consequence of Planck's radiation formula" (Compton 1922, 25; Shankland 1973, 346). His epiphany occurred one month later. He saw that he could reinterpret his X-ray scattering experiments quantitatively by assuming that an X-ray quantum of energy hv and momentum hv/c was striking an electron in the scatterer in a billiard-ball collision process in which both energy and momentum are conserved. He reported his discovery on December 1 or 2, 1922, at a meeting of the American Physical Society in Chicago; his paper was published in The Physical Review in May 1923 (Compton 1923a). Nowhere in it did he cite Einstein's light quantum paper of 1905 or even mention Einstein's name.

Compton, in fact, never seems to have read Einstein's paper. He may have learned of Einstein's light quantum hypothesis at the Cavendish Laboratory in 1919–20, but he probably first recognized its significance for his experiments during conversations he had with his colleague at Washington University, G. E. M. Jauncey, whom he acknowledged explicitly as contributing to "many of the ideas involved" in his paper (Compton 1923a, 502; Shankland 1973, 401).[14] His earlier ambivalence toward light quanta was actually reinforced in early 1922 when he discovered that X rays can be totally internally reflected by glass and silver mirrors, which he took to be clear proof of their wave nature (Compton 1923b). He sent off a full report of these experiments to the *Philosophical Magazine* on December 6, 1922, precisely one week before he sent off his paper on his quantum theory of scattering to *The Physical Review* on December 13, 1922. Thus, within a single week, Compton reported conclusive experimental evidence for *both* the wave *and* the particle natures of X rays. Nothing can better symbolize the dilemma that Compton and his contemporaries faced in trying to understand the perplexing dual nature of radiation.

Prior to Compton's discovery, Einstein's light quantum hypothesis had been taken seriously by a few European physicists, among them Maurice de Broglie in Paris and Charles D. Ellis in Cambridge. In 1921-2 de Broglie had sent X rays and Ellis γ rays into various substances and determined the energy spectra of the expelled electrons by magnetic-spectroscopic techniques. De Broglie interpreted his experiments by assuming that the incident X rays possess energy hv and expel electrons from different atomic energy levels. He reported his results in April 1921 at the third Solvay Congress in Brussels, where Rutherford noted that they agreed with Ellis's γ-ray experiments.[15] Soon thereafter Maurice's younger brother Louis reviewed Einstein's papers and in November 1922 gave a derivation of Planck's law on the assumption that black-body radiation consists of "light molecules" whose energies are multiples of hv - the first step en route to his discovery of "matter waves" (Broglie 1922).

Einstein had reentered the picture in December 1921 by proposing what he thought would be a crucial experiment to decide between the wave and quantum theories of light (Einstein 1922a).[16] He suggested that a beam of positive ions (canal rays) should be sent perpendicular to the optical axis of a simple optical system, so that the light emitted by them would emerge as collimated monochromatic light and then enter a dispersive medium, where it would be deviated upward by a few degrees if it consisted of Doppler-shifted plane waves but would not be deviated if it consisted of light quanta. His colleagues in Berlin, Hans Geiger and Walther Bothe, immediately tried the experiment and observed no deviation - a result that startled his friend Paul Ehrenfest in Leiden and prompted him to carefully reexamine Einstein's analysis. Ehrenfest concluded that while the normal to the individual plane waves should be deviated, the normal to the energy-carrying wave group should not. Thus Geiger and Bothe's observation did not constitute the decisive proof for light quanta that Einstein had envisioned - a conclusion with which Einstein fully agreed by the end of January 1922.

Several months later, however, in June 1922, Erwin Schrödinger, *ordentlicher* Professor of Theoretical Physics at the University of Zurich, provided other theoretical support for Einstein's light quantum hypothesis. He assumed that a moving molecule emits a light quantum and proved that its frequency, as given by both the nonrelativistic and relativistic Doppler effects, could be recovered provided that "the emitted quantum hv always – and in every co-ordinate system – must possess linear momentum hv/c ..." (Schrödinger 1922, 301; 1984, 11).

Still, no one took Einstein's light quantum hypothesis more seriously than Peter Debye, now Professor of Experimental Physics and Director of the Physical Institute at the Eidgenössische Technische Hochschule in Zurich. Like Compton, Debye recognized the puzzling nature of Barkla's experimental result of 1917, but he did not pursue it until after he saw Compton's X-ray spectra of October 1922. He then analyzed the scattering problem in detail, explicitly adopting Einstein's concept of "needle radiation" (Debye 1923). His results agreed with Compton's, but he placed less confidence in them. Compton had concluded that the "remarkable agreement between our formulas and the experiments can leave but little doubt that the scattering of X rays is a quantum phenomenon" (Compton 1923a, 501; Shankland 1973, 400). Debye, by contrast, concluded that his analysis should be

considered as nothing but an attempt to derive as detailed conclusions as possible from the two assumptions: "energy quanta" and "radiation quanta," taking recourse only to general laws: "law of conservation of energy" and "principle of conservation of momentum." By means of experiments *which reveal characteristic deviations from this scheme* we may hope to secure deeper insight into the laws of quantum theory.... (Debye 1954, 88; my italics)

Debye's and Compton's work were independent and virtually simultaneous. Much later, however, Debye declared that since Compton had carried out the experiments *and* had provided the theory, it should be called the Compton effect.[17]

Compton, in fact, never stopped working. He found further experimental support for his quantum theory of scattering both at Washington University and at the University of Chicago, where he moved in the summer of 1923, replacing Millikan who had become head of the new California Institute of Technology. Others also confirmed Compton's experiments (Stuewer 1975, 237–46), but at the end of October 1923 George L. Clark, a National Research Council Fellow working in William Duane's laboratory at Harvard University, announced that he could not (Clark and Duane 1923). Duane had rejected Einstein's light quantum hypothesis as early as 1915 as an interpretation of the so-called Duane-Hunt law, which he and his graduate student, Franklin L. Hunt, had established to high precision, namely, that the maximum frequency ν_{max} of X rays produced by electrons of energy E in a Coolidge X-ray tube was given by $E = h\nu_{max}$ (Duane and Hunt 1915). Duane now supported Clark, precipitating what Compton later called "the most lively scientific controversy that I have ever known."[18] It lasted more than a year and survived two face-to-face debates at two scientific meetings between Compton and Duane as well as visits to each others' laboratories. Only after Samuel K. Allison, another National Research Council Fellow, carried out further experiments in Duane's laboratory was agreement reached. Duane withdrew his objections "at a memorable meeting of the American Physical Society" in Washington after Christmas in 1924. The Compton effect was an experimental fact of nature.

6. THE BOHR-KRAMERS-SLATER THEORY

That did not resolve its theoretical interpretation, however. During 1923–4 some physicists tried to explain the Compton effect classically by invoking the Doppler shift, but these theories foundered because they could not account either for the sharpness of the shifted line or for the ranges of the recoil electrons (Stuewer 1975, 290–1). Others accepted the Compton effect and drew significant theoretical conclusions from it (ibid., 288–90). Thus Wolfgang Pauli in Hamburg proved on its basis that a state of equilibrium could exist between black-body radiation and free electrons, and Louis de Broglie saw it as proof of "the actual reality of light quanta," which supported his developing ideas on "matter waves." At root, however, the fundamental problem was, as Einstein put it in April 1924, that there now were two theories of light, "both indispensable and – as one has to admit today in spite of twenty years of immense efforts by theoretical physicists – without any logical connection" (Einstein 1924c).

Niels Bohr, Hendrik A. Kramers, and John C. Slater (1924) proposed to fill that lacuna. Slater received his PhD degree from Harvard University in June 1923 and left that fall for Europe on a traveling fellowship, first visiting the Cavendish Laboratory in Cambridge (Slater 1975, 6). There, greatly impressed with Compton's discovery, he became convinced that the "two apparently contradictory aspects of the mechanism" governing the interaction of radiation and matter "must be really consistent, but [that] the discontinuous side is apparently the more fundamental ..." (Slater 1924, 307). He therefore set out to develop a "picture of optical phenomena ... by associating the essentially continuous radiation field with the continuity of existence in stationary states [in an atom], and the discontinuous changes of energy and momentum with the discontinuous transitions from one state to another." Thus, any "atom may ... be supposed to communicate with other atoms all the time it is in a stationary state, by means of a virtual field of radiation, originating from oscillators having the frequency of possible quantum transitions ..." (ibid., 307–8). The virtual radiation field would carry no energy, but its associated Poynting vector would guide light quanta between two atoms in accordance with the conservation laws, with the emission and absorption processes being described probabilistically by Einstein's A and B coefficients. Slater's ideas were received enthusiastically by Ralph H. Fowler, the leading Cambridge mathematical physicist (Slater 1975, 11).[19]

Leaving Cambridge, Slater arrived in Niels Bohr's institute in Copenhagen around Christmas, where his ideas were accorded an entirely different reception. Bohr had never accepted the physical reality of light quanta. In 1913 he had insisted that when an electron makes a transition from a higher to a lower energy level in a hydrogen atom, electromagnetic waves, not light quanta, are emitted, and during the years that followed he never changed his mind on this question. Thus, most recently, in his Nobel Lecture on December 11, 1922, he admitted that Einstein's equation of the photoelectric effect had been confirmed precisely, but he then added: "In spite of its heuristic value, ... the hypothesis of light-quanta, which is quite irreconcilable with so-called interference phenomena, is not able to throw light on the nature of radiation" (Bohr 1922, 14). Bohr was convinced that his correspondence principle provided the essential bridge between the classical and quantum realms, so that light quanta had only a "formal" significance. And he permitted no opposition on that score from his Dutch assistant, Kramers. In fact, according to his biographer, as early as 1921 "Kramers actually sketched a derivation of ... the Compton effect," but "Bohr immediately set out to convince Kramers of the errors of his ways," extolling "the 'wondrous' wave theory of light" (Dresden 1987, 289–90). That was the brick wall that Slater faced in Copenhagen.

Slater recalled that Bohr and Kramers were "enthusiastic about the idea of the electromagnetic waves emitted by oscillators during the stationary states" and "at once coined the name 'virtual oscillators' for them," but "to my consternation ... they completely refused to admit the real existence of the [light quanta]" (Slater 1975, 11). At least as early as 1919 Bohr had entertained the possibility that energy was not conserved in individual atomic interactions (Klein 1970a, 19–21; Stuewer 1975, 293), and now he and Kramers opposed the strict conservation of energy and momentum "so vigorously that I saw that the only way to keep peace and get the main part of the suggestion published was to go along with them with the statistical idea."[20] They, in fact, excluded Slater entirely in the writing of their joint paper.[21]

As applied to the Compton effect, Bohr and Kramers insisted that the radiation of increased wavelength had to be viewed as Doppler-shifted radiation emitted by an imaginary recoiling electron having "a certain probability of taking up in unit time a finite amount of momentum in any given direction," so that "a statistical conservation of momentum is secured in a way quite analogous to the statistical conservation of energy in the phenomena of absorption of light" à la Einstein (Bohr, Kramers, and Slater 1924, 799–800; Bohr 1984, 115–16). Thus there could be no correlation between the emission of the scattered radiation and the recoil of the electron.

When the Bohr-Kramers-Slater theory became known in January 1924 (four months prior to its publication), negative reactions abounded (Stuewer 1975, 294–7). Pauli told Bohr in a letter of February 21, 1924, that while he knew what both of the words *communicate* and *virtual* meant, he could not guess the meaning of Bohr's theory.[22] Two months later, in a letter of April 29, 1924, Einstein told Max Born's wife Hedwig that if "an electron ejected by a light ray can choose *of its own free will* the moment and direction in which it will fly off," then he "would rather be a shoemaker or even an employee in a gambling casino than a physicist."[23] One month later, in a letter of May 31, 1924, to his friend Ehrenfest, Einstein reported that he had just reviewed the Bohr-Kramers-Slater paper at the Berlin colloquium, and that "the idea is an old acquaintance of mine," but that he did not "consider it to be the real thing."[24] He gave five main reasons, among them that a "final abandonment of strict causality is very hard for me to tolerate" (Klein 1970a, 33). Four months later, in a letter of October 2, 1924, Pauli told Bohr that "as a physicist" he was "totally opposed" to the theory, a view that was shared by "very many other physicists, perhaps even most,"[25] foremost among whom was Einstein, whom he had recently met in Innsbruck. Pauli listed Einstein's reasons, which partially overlapped with those Einstein had given Ehrenfest four months earlier, and then Pauli added a couple more of his own, including his conviction that it was not meaningful to distinguish between Compton transitions and ordinary atomic transitions because their total probability of occurrence had to be given by Einstein's B coefficient.

A few physicists, however, reacted positively to the Bohr-Kramers-Slater theory and attempted to develop it further (Stuewer 1975, 297–9). Most prominent among them was Erwin Schrödinger, who wrote a long letter to Bohr on May 24, 1924,[26] explaining that ever since he had been a student of Franz Exner in Vienna,[27] he had been friendly to the idea that the conservation laws were valid only statistically. Then, four months later, Schrödinger (1924) published a thorough explication of the Bohr-Kramers-Slater theory, discussing its applicability to the Compton effect and other phenomena, and asserting that its "most exciting" aspect, which "claims equally great interest from the physical and philosophical points of view," was its "fundamental violation of the laws of conservation of energy and momentum in each radiation process."

That was the central point at issue, and as Charles D. Ellis (1926) observed, "it must be held greatly to the credit of this theory that it was sufficiently precise in its statements to be disproved definitely by experiment." In fact, soon after it was published in May 1924, Bothe and Geiger (1924) announced that they had an experiment underway to test it. Their idea was to position two detectors, each covering an angular range of 180°, opposite to each other in an attempt to record coincidences between them arising respectively from the scattered light and recoil electrons – which would occur only if energy and momentum were strictly conserved. By January 15, 1925, Max Born could write to Bohr that "everyone" in Berlin was taking their results to mean that "Einstein triumphs," even though they "do not yet regard their results as final."[28] Three months later they did. They sent off two papers for publication on April 18 and 25, 1925 (Bothe and Geiger 1925a, 1925b), reporting that they had recorded a coincidence rate that was "not in accord with Bohr's interpretation of the Compton effect," and therefore that one had to "probably assume that the light-quantum concept possesses a high degree of reality ..." (Bothe and Geiger 1925b, 662–3). Geiger also informed Bohr of their conclusion in a letter of April 17, 1925,[29] to which Bohr responded four days later.[30] That same day he added a postscript to a letter to Ralph H. Fowler in Cambridge, saying that, "It seems therefore that there is nothing else to do than to give our revolutionary efforts as honourable a funeral as possible."[31]

Still, when Bohr had responded to Born's letter of January 15, 1925, he had pointed out that it would be difficult to come to a definite conclusion until an experiment was carried out in which two detectors were positioned exactly where the scattered light quantum and recoil electron were expected to appear.[32] That experiment, in fact, was currently underway by Compton and his student Alfred W. Simon at the University of Chicago. They reported their results on June 23, 1925 (Compton and Simon 1925), two months after Bothe and Geiger had reported theirs. Compton and Simon had sent a collimated beam of X rays into a cloud chamber and had positioned one counter at an angle φ to detect the recoil electron and a second counter at an angle ϑ where they had calculated that the scattered X-ray quantum should appear. Of the many stereoscopic photographs they had taken, thirty-eight had captured both recoil electrons and scattered X rays, and of those, eighteen had shown that the scattered X rays had appeared within 20° of the calculated angle. That number was much too high to be ascribed to chance but was, they concluded, "in direct support" of the conservation laws and did not "appear to be reconcilable with the view of the statistical production of recoil and photo-electrons proposed by Bohr, Kramers and Slater" (ibid., 299; Shankland 1973, 518).

The Bothe-Geiger and Compton-Simon experiments thus conclusively established, as Werner Heisenberg put it, "the reality of light quanta" (Heisenberg 1926, 993; 1984, 56). Of course, as Einstein wrote to Ehrenfest, "We both had no doubts about it."[33] Slater, too, was vindicated, but for the rest of his life he remained bitter over his treatment by Bohr and Kramers (Slater 1975, 11). Pauli was jubilant, writing to Kramers on July 27, 1925, that, "In general, I think it was a magnificent stroke of luck that the theory of Bohr, Kramers and Slater was so rapidly refuted by the beautiful experiments of Geiger and Bothe, as well as the recently published ones of Compton [and Simon]." Then, with characteristic insightfulness, he added that, "It can now be taken for granted by every unprejudiced physicist that light quanta are as much (and as little) physically real as electrons. Classical concepts in general, however, should not be applied to either."[34]

NOTES

1 For other examples, see Stuewer (1999).
2 The original picture is reproduced in G. P. Thomson (1964, Figure 7, facing p. 53).
3 Millikan's reproduction of the picture appears in Millikan and Gale (1906, facing p. 482).
4 I also have discussed this case and reproduced the photographs in question in Stuewer (2000, 977–8).
5 A bibliography of articles that appeared in 1887–1902 is given by Hallwachs (1916, 536–45).
6 For a recent discussion, see Hon (2003).
7 For a full discussion, see Klein (1963b). For Einstein's work on quantum theory, see also Chapter 4.
8 Einstein to Laub, May 17, 1909 (CPAE 5, Doc. 160).
9 For a full discussion, see Klein (1964).
10 Einstein to Lorentz, March 30, 1909 (CPAE 5, Doc. 146).
11 Millikan (1916, 359–61) summarized all of this work, with references to the literature.
12 He misquoted this slightly in his autobiography (Millikan 1950, 102–3).
13 For a full account, with references to the literature, see Stuewer (1975).
14 For an insightful account of Jauncey's contributions, see Jenkin (2002).
15 For an account of de Broglie's and Ellis's work, see Wheaton (1983, 264–74).
16 For a full account, see Klein (1970a, esp. 8–13).

17 Debye stated this explicitly in an interview with T. S. Kuhn and G. E. Uhlenbeck on May 3, 1962, that is deposited in the Archive for History of Quantum Physics.
18 For a full account of the Duane-Compton controversy, see Stuewer (1975, 249–73).
19 Also see Slater to Bohr, December 8, 1923 (Bohr 1984, 492–3).
20 Slater to B. L. van der Waerden, November 4, 1964, quoted by van der Waerden (1967, 11).
21 Slater to J. H. Van Vleck, July 27, 1924, cited in Klein (1970a, 24).
22 Pauli to Bohr, February 21, 1924 (Pauli 1979, 147–8; Bohr 1984, 410–12, 412–14 [English translation]).
23 Einstein to Hedwig Born, April 29, 1924; quoted by Klein (1970a, 32).
24 Einstein to Ehrenfest, May 31, 1924; quoted by Klein (1970a, 32).
25 Pauli to Bohr, October 2, 1924 (Pauli 1979, 163–4; Bohr 1984, 415, 418 [English translation]).
26 Schrödinger to Bohr, May 24, 1924 (Bohr 1984, 490–2).
27 To appreciate how thoroughly Exner was committed to statistical views, see Stöltzner (2002).
28 Reproduced in Bohr (1984, 302–4, esp. 303–4).
29 Geiger to Bohr, April 17, 1925 (Bohr 1984, 352–3).
30 Bohr to Geiger, April 21, 1925 (Bohr 1984, 353–4).
31 Bohr to Fowler, April 21, 1925 (Bohr 1984, 81–4; quote on p. 82).
32 Bohr to Born, January 18, 1925 (Bohr 1984, 304–6; see 306).
33 Einstein to Ehrenfest, August 18, 1925, quoted by Klein (1970a, 35).
34 Pauli to Kramers, July 27, 1925 (Pauli 1979, 232–4; quote on p. 233).

6 "No Success Like Failure ..."

Einstein's Quest for General Relativity, 1907–1920

1. INTRODUCTION

In 1905, Einstein published what came to be known as the special theory of relativity, extending the Galilean-Newtonian principle of relativity for uniform motion from mechanics to all branches of physics.[1] Two years later he was ready to extend the principle to arbitrary motion. He felt strongly that there can only be relative motion, as is evidenced, for instance, by his opening remarks in a series of lectures in Princeton in 1921, published in heavily revised form the following year (Einstein 1922c). A typescript based on a stenographer's notes survives for the first two, nontechnical lectures. On the first page of this presumably verbatim transcript we find Einstein belaboring the issue of the relativity of motion in a way he never would in writing:[2]

Whenever we talk about the motion of a body, we always mean by the very concept of motion relative motion ... we might as well say "the street moves with respect to the car" as "the car moves with respect to the street."... These conditions are really quite trivial ... we can only conceive of motion as relative motion; as far as the purely geometrical acceleration is concerned, it does not matter from the point of view of which body we talk about it. All this goes without saying and does not need any further discussion. (CPAE 7, Appendix C [p. 1])

Although Einstein insists that these points are trivial, we shall see that they are not even true. What makes his comments all the more remarkable is that by 1921 Einstein had already conceded, however grudgingly, that his general theory of relativity, worked out between 1907 and 1918, does *not* make all motion relative.

It has widely been recognized that general relativity is a misnomer for the new theory of gravity that was the crowning achievement of Einstein's career. In a paper entitled "Is 'General Relativity' Necessary for Einstein's Theory of Gravitation?" in one of the many volumes marking the centenary of Einstein's birth, for instance, the prominent relativist Sir Hermann Bondi (1979) wrote: "It is rather late to change the name of Einstein's theory of gravitation, but general relativity is a physically meaningless phrase that can only be viewed as a historical memento of a curious philosophical observation" (181).[3]

Einstein obviously realized from the beginning that there is a difference between uniform and nonuniform motion. Think of a passenger sitting in a train in a railway station looking at the train next to hers. Suppose that – with respect to the station – one train is moving while the other is at rest. If the motion is uniform and if the only thing our passenger sees as she looks out the window is the other train, there is no way for her to tell which one is which. This changes the moment the motion is nonuniform. Our passenger can now use, say, the cup of coffee in her hand to tell which train is moving: if nothing happens to the coffee, the other train is; if the coffee spills, her train is.

This is where gravity comes in. The key observation on the basis of which Einstein sought to extend the relativity principle to nonuniform motion is that, at least locally, the effects of acceleration are indistinguishable from those of gravity. Invoking this general observation, a passenger can maintain that her train is at rest, even if her coffee spills. She can, if she is so inclined, blame the spill on a gravitational field that suddenly came into being to produce a gravitational acceleration equal and opposite to what she would otherwise have to accept is the acceleration of her own train.[4] This was the idea that launched Einstein on his path to general relativity (Einstein 1907j, Part V). A few years later, he introduced a special name for it: the *equivalence principle* (Einstein 1912c, 360, 366).

This principle by itself does not make nonuniform motion relative. As Einstein came to realize in the course of the work that led him toward the new theory, two further conditions must be satisfied.

The first of these is that it should be possible to ascribe the gravitational field substituted for an object's acceleration on the basis of the equivalence principle to a material source – anything from the object's immediate surroundings to the distant stars. Otherwise, acceleration with respect to absolute space would simply be replaced by the equally objectionable notion of fictitious gravitational fields. If this further condition is met, however, the gravitational field can be seen as an epiphenomenon of matter and all talk about motion of matter in that field can be interpreted as shorthand for motion with respect to its material sources (Maudlin 1990, 561). This condition was inspired by Einstein's reading of the work of the nineteenth-century Austrian philosopher-physicist Ernst Mach (Hoefer 1994, Barbour and Pfister 1995, Renn 2007c).

The second condition is that all physical laws have the same form for all observers, regardless of their state of motion. In particular, this should be true for the gravitational *field equations*, the equations that govern what field configuration is produced by a given distribution of sources. This form-invariance is called *general covariance*.[5] Einstein had great difficulty finding field equations that are both generally covariant and satisfactory on all other counts (Renn 2007a, Vols. 1-2). He originally settled for field equations of severely limited covariance. He published these equations in a paper coauthored with the mathematician Marcel Grossmann (Einstein and Grossmann 1913). They are known among historians of physics as the *Entwurf* (German for *outline*) field equations after the title of this paper. The precursor to general relativity with these field equations is likewise known as the *Entwurf* theory. In the course of 1913, Einstein convinced himself that the restricted covariance of the *Entwurf* field equations was still broad enough to make all motion relative. In a vintage Einstein maneuver, he even cooked up an ingenious argument, known as the *hole argument* (see Section 3 and note 70), purporting to show that generally covariant gravitational field equations are inadmissible (Einstein and Grossmann 1914a). By the end of 1914 he felt so sure about the *Entwurf* theory that he published a lengthy self-contained exposition of it (Einstein 1914o).[6] In the fall of 1915, however, he got dissatisfied with the *Entwurf* field equations. In November 1915, with the Göttingen mathematician David Hilbert breathing down his neck,[7] Einstein dashed off a flurry of short papers to the Berlin Academy, in which he proposed, in rapid succession, three new field equations of broad and eventually general covariance (Einstein 1915f, 1915g, 1915h, 1915i).[8] The final generally covariant equations are known today as the *Einstein field equations*. Einstein subsequently replaced the premature review article of 1914 by a new one (Einstein 1916e). This article, submitted in March and published in May 1916, is the first systematic exposition of general relativity.[9] When Einstein wrote it, he was laboring under the illusion that, simply by virtue of its general covariance, the new theory made all motion relative.

The other condition, however, was not satisfied: general covariance does not guarantee that all gravitational fields can be attributed to material sources. In the fall of 1916, in the course of an exchange with the Dutch astronomer Willem de Sitter, Einstein was forced to admit this. He thereupon modified his field equations (without compromising their general covariance) by adding a term with the so-called *cosmological constant* (Einstein 1917b). Einstein's hope was that these new field equations would not allow any gravitational fields without material sources.

In a short paper submitted in March 1918, in which Einstein silently corrected some of his pronouncements on the foundations of general relativity in the 1916 review article, he introduced a special name for this requirement: *Mach's principle* (Einstein 1918e, 241). This was one of three principles Einstein identified in this paper as the cornerstones of his theory, the other two being the equivalence principle and the relativity principle. The characterization of Einstein's project in terms of the equivalence principle, Mach's principle, and general covariance follows this 1918 paper. This was the first time that Einstein explicitly separated the three notions involved. In a footnote he conceded that he had not clearly distinguished the relativity principle – identified essentially with general covariance (cf. note 73) – from Mach's principle before (ibid.).[10] A few months later, it became clear that even the field equations with the cosmological term do not satisfy Mach's principle. Within another year or so, Einstein came to accept that general relativity does not banish absolute motion from physics after all.

This, in a nutshell, is the story of Einstein's quest for general relativity from 1907 to about 1920. His frustrations were many. He had to readjust his approach and his objectives at almost every step along the way. He got himself seriously confused at times, especially over the status of general covariance (see Section 3).[11] He fooled himself with fallacious arguments and sloppy calculations (Janssen 2007). And he later allegedly called the introduction of the cosmological constant the biggest blunder of his career (Gamow 1970, 149–50).[12] There is an uplifting moral to this somber tale. Although he never reached his original destination, the bounty of Einstein's thirteen-year odyssey was rich by any measure.[13]

First of all, what is left of absolute motion in general relativity is much more palatable than the absolute motion of special relativity or Newtonian theory.[14] Einstein had implemented the equivalence principle by making a single field represent both gravity and the structure of space-time. He had rendered the effects of gravity and acceleration (i.e., the deviation from inertial motion) indistinguishable, at least locally, by making them manifestations of one and the same entity, now often called the inertio-gravitational field. If Mach's principle were satisfied, this field could be fully reduced to its material sources and all motion would be relative. But Mach's principle is not satisfied and the inertio-gravitational field exists in addition to its sources. When two objects are in relative nonuniform motion, this additional structure allows us to determine whether the first, the second, or both are actually moving nonuniformly. In this sense, motion in general relativity is as absolute as it was in special relativity. In his Princeton lectures, however, Einstein (1956a, 55–6) argued that there is an important difference between the two theories: in general relativity, the additional structure is a bona fide physical entity that not only acts but is also acted upon. As Misner, Thorne, and Wheeler (1973, 5) put it in their textbook on general relativity: *"Space acts on matter, telling it how to move. In turn, matter reacts back on space, telling it how to curve."*

By 1920, Einstein had probably recognized that Mach's principle was predicated on an antiquated nineteenth-century billiard-ball ontology (Hoefer 1994, Renn 2007b). In the field ontology of the early twentieth century, in which matter was ultimately thought of as a manifestation of the electromagnetic and perhaps other fields, it amounts to the requirement that the gravitational field be reduced to these other fields. A recognition of this state of affairs[15] may have helped Einstein make his peace with the persistence of absolute motion in general relativity. Instead of trying to reduce one field to another, he now tried to unify the two. This can be gleaned from "Ether and Relativity," the inaugural lecture Einstein gave upon accepting a visiting professorship in Leyden in 1920. Einstein was not pandering to his revered senior Dutch colleague Hendrik Antoon Lorentz when he presented the inertio-gravitational field in this lecture as a new relativistic incarnation of the ether eliminated by special relativity (Einstein 1920j).

Special relativity combines the electric and the magnetic field into one electromagnetic field, and space and time into space-time. General relativity combines the gravitational field and the space-time structure into one inertio-gravitational field. It thus made sense to try to combine the electromagnetic field and the inertio-gravitational field into one unified field. Einstein spent the better part of the second half of his career searching in vain for a theory along the lines of general relativity that would accomplish this.[16]

Even though general relativity does not eliminate absolute motion, the case can be made that it does eliminate absolute space(-time). In the classic debate between Newton (through his spokesperson Clarke) and Leibniz (Alexander 1956), these two notions seemed to stand or fall together. Modern philosophy of space and time has made it clear that they do not. The appearance that they do is due to a conflation of two related but separate issues (Earman 1989, 12–15).

The first issue is the one we have been considering so far: is all motion relative or is some motion absolute? This question, as we just saw, ultimately boils down to the question whether or not the space-time structure is something over and above the contents of space-time. To the extent that it is still meaningful to distinguish space-time from its contents once the former has been identified with a physical field (Rynasiewicz 1996), one would have to answer this question affirmatively. This, in turn, implies that absolute motion persists in general relativity.

The second issue concerns the ontological status of space-time. Is the space-time structure supported by a substance, some sort of container, or is it a set of relational properties, like the marriage of me and my wife?[17] The two views thus loosely characterized go by the names of *substantivalism* and *relationism* (or *relationalism*), respectively. Fairly or unfairly, Newton's name has been associated with substantivalism as well as with absolutism about motion, Leibniz's name with relationism as well as with relativism about motion. It is possible, however, to be an absolutist about motion while being a relationist about the ontology of space-time. Although the jury is still out on the latter count, the ontology of space-time, represented by the inertio-gravitational field in general relativity, is probably best understood in relational rather than substantival terms. In that case, however, the causal efficacy implied by the slogan that space-time both acts and is acted upon cannot be that of a substance.

If the verdicts on these two issues stand as final, the centuries-old debate between Newtonians and Leibnizians will have ended in a draw: Newtonians were right that there is absolute motion, Leibnizians were right that there is no absolute space. Accordingly, the best arguments in support of their respective positions would both be correct. Newton's rotating-bucket experiment (see Section 4) shows that rotation is absolute; Leibniz's mirror or shift argument (see Section 3) shows that space is relational. One can argue, however, that the terms of the debate have changed so drastically since the seventeenth century that it does not make much sense to belatedly declare winners and losers (Rynasiewicz 1996).[18]

The central argument for the claim that general relativity vindicates relationism can be seen as a modern version of Leibniz's shift argument and is based on Einstein's resolution of the hole argument through the so-called *point-coincidence argument* (see Section 3). Originally, the hole argument was nothing but a fig leaf to cover up the embarrassing lack of covariance of the *Entwurf* field equations (Janssen 2007). The point-coincidence argument likewise started out as an expedient to silence two correspondents, who took Einstein to task for publishing generally covariant field equations without explaining what was wrong with the hole argument (see note 71). Despite their inauspicious origins, both arguments have enjoyed a rich afterlife in the literature on the philosophy of space and time. This illustrates my general point that Einstein's quest for general relativity was anything but fruitless.

This becomes even clearer when we shift our attention from foundational issues to physics proper. Even though the equivalence principle could not be used for its original purpose of making all motion relative, Einstein did make it one of the cornerstones of a spectacular new theory of gravity that is still with us today. The insight that space-time and gravity should be represented by one and the same structure may well turn out to be one of the most enduring elements of Einstein's legacy (Janssen 2002a, 511–12). In addition to laying the foundation for his theory, Einstein, among other things, explained the anomalous advance of the perihelion of Mercury (Einstein 1915h);[19] successfully predicted both the bending of light in gravitational fields and its gravitational redshift (Einstein 1907j, 1911h, 1915h);[20] launched relativistic cosmology (Einstein 1917b);[21] suggested the possibility of gravitational waves (Einstein 1916g, 1918a),[22] gravitational lensing,[23] and frame dragging (Einstein 1913c, 1261–2);[24] came up with the first sensible definition of a space-time singularity (Einstein 1918c);[25] and caught on to the intimate connection between covariance and energy-momentum conservation (Einstein 1914o, 1916o) well before Emmy Noether (1918) formulated her celebrated theorems connecting symmetries and conservation laws inspired by this particular application in general relativity (Rowe 1999, Janssen and Renn 2007).[26] Even Einstein's "biggest blunder" – the cosmological constant – has made a spectacular comeback in recent years. It can be used to give a simple account of the accelerated expansion of the universe. These successes more than compensate for Einstein's failure in his quest for general relativity.

It is this quest, however, that will be the main focus of this chapter.[27] In Sections 2–5, I examine four different ways in which Einstein, between 1907 and 1918, tried to relativize all motion and I explain how and why each of these attempts failed. In the course of this quest for general relativity, Einstein formulated and reformulated the three principles that he identified in 1918 as providing the foundation for his theory – the equivalence principle, the relativity principle, and Mach's principle.

The failure of Einstein's quest raises an obvious question: how do we make sense of the success of Einstein's theory of gravity given that some of the main considerations that led him to it turned out to be misguided (Renn 2007b, 21–3)? In the concluding Section 6, I discuss three factors that may help answer this question. First, throughout the pursuit of his lofty philosophical goals, Einstein never lost sight of the more mundane physics problem at hand, namely how to reconcile the basic insight of the equivalence principle, the intimate connection between inertia and gravity, with the results of special relativity.[28] Second, in developing his new theory, Einstein relied not only on his philosophical ideas but also on an elaborate analogy between the electromagnetic field, covered by the well-established theory of electrodynamics, and the gravitational field, for which he sought a similar theory (Janssen and Renn 2007, Renn and Sauer 2007). Finally, as we shall see in the course of Sections 2–5, Einstein, who could be exceptionally stubborn, displayed a remarkable flexibility at several key junctures where his philosophical predilections led to results that clashed with sound physical principles, such as the conservation laws for energy and momentum. None of this is to say that Einstein's philosophical objectives only served as a hindrance in the end. Without them Einstein would probably have taken a more conservative approach, making gravity just another field in the Minkowski space-time of special relativity rather than part of the fabric of space-time. As we shall also see in Section 6, however, Einstein showed, through his contributions to a theory first proposed by the Finnish theorist Gunnar Nordström, that even this more conservative approach eventually leads to a connection between gravity and space-time curvature as in general relativity (Einstein and Fokker 1914).[29]

2. FIRST ATTEMPT: THE EQUIVALENCE PRINCIPLE

One day in 1907, at the Patent Office in Berne, while working on a review article on his original theory of relativity (Einstein 1907j), it suddenly hit Einstein: someone falling from the roof of a house does not feel his own weight. As he wrote in a long unpublished article intended for *Nature* on the conceptual development of both his relativity theories, this triggered "the best idea of [his] life."[30] It is illustrated in Figure 6.1. The upper half shows Einstein looking out the window and meeting the eyes of a man who moments earlier fell off his scaffold as he was cleaning windows a few floors up. Einstein is at rest in the gravitational field of the Earth, the man is in free fall in this field, accelerating toward the pavement. For the duration of the fall, he is experiencing something close to weightlessness.

Although to this day few have experienced this condition firsthand, we have all at least experienced it vicariously through footage of astronauts in free fall toward Earth as they orbit the planet in a space shuttle. Einstein only had his imagination to go on. If it were not for air resistance, the unfortunate window cleaner, like the astronaut in orbit around the Earth, would, during his fall, feel as if he were hovering in outer space, far removed from any gravitating matter. Moreover, on Galileo's principle that all bodies fall alike, he would fall with the same acceleration as his bucket and his squeegee. These objects would thus appear to be hovering with him. In short, at least in the observer's immediate vicinity, moving with the acceleration of free fall in a gravitational field seems to be physically equivalent to being at rest without a gravitational field. Likewise, being at rest at one's desk, resisting the downward pull of gravity, seems to be physically equivalent to sitting at the same desk in the absence of a gravitational field but moving upward with an acceleration equal and opposite to that of free fall on Earth. An astronaut firing up the engines of her rocket ship in outer space will be

Figure 6.1. *The equivalence principle.* Drawn by Laurent Taudin.

pinned to her seat as if by a gravitational field (an experience similar to the one we have during takeoff on a plane). Einstein used observations like this for an extension of sorts of the relativity principle for uniform motion to nonuniform motion.

Figure 6.1 depicts the four physical states described. Both in situation (I), near the surface of the Earth, and in situation (II), somewhere in outer space, the man on the left (a) and the man on the right (b) can both claim to be at rest *as long as they agree to disagree on whether or not there is a gravitational field present.* For Einstein sitting at his desk in situation (I), there is a gravitational field, he is at rest, and the other man is accelerating downward. For the falling man, there is no gravitational field, he is at rest, and Einstein is accelerating upward. Situation (II) fits the exact same description.

This extended relativity principle, however, is very different from the relativity principle for uniform motion. The situations of two observers in uniform motion with respect to one another are physically fully equivalent. This is not true for nonuniform motion. Resisting and giving in to the pull of gravity (Ia and Ib, respectively) feel differently; so do accelerating and hovering in outer space (IIa and IIb, respectively). In fact, the equivalence captured in Figure 6.1 is not between different observers in the same situation – that is, between observers (a) and (b) in situation (I) or (II) – but between different situations for the same observers – that is, between situations (I) and (II) for observer (a) or (b).

We call the uniform motion of one observer with respect to another relative because the situation is completely symmetric. It is therefore arbitrary in the final analysis (even if hardly ever in practice) which one we label "at rest" and which one we label "in motion." There is no such symmetry in the case of nonuniform motion. Nonuniform motion is thus not relative in the sense that uniform motion is. What is relative in this sense in the situations illustrated in Figure 6.1 is *the presence or absence of a gravitational field.* Situations (I) and (II) can both be accounted for with or without a gravitational field. From the perspective of observer (a), both situations involve a gravitational field; from the perspective of observer (b), there is none in either.

If we try to extend the descriptions of situations (I) and (II) to include all of space, the equivalence of the description with and the description without a gravitational field breaks down. Contrary to what Einstein thought in 1907, we cannot fully reduce inertial effects, the effects of acceleration, to gravitational effects. As mentioned in the introduction, however, general relativity in its final form does trace local inertial and gravitational effects to the same structure, the inertio-gravitational field.

In Newtonian physics particles get their marching orders, figuratively speaking, from the space-time structure and from forces acting on them. According to Newton's first law (the law of inertia), a particle moves in a straight line at constant speed as long as there are no forces acting on it. This is true regardless of its size, shape, or other properties.

Forces cause a particle to deviate from its inertial path. By how much depends on its susceptibility to the particular force (e.g., an electric force will only affect charged particles) and on its resistance to acceleration. The marching orders issued by forces are thus specific to the particles receiving them. There is one force in Newtonian physics, however, giving marching orders that are as indiscriminate and universal as those issued by the space-time structure: gravity. Newton accounted for this universality by setting *inertial mass*, a measure for a particle's resistance to acceleration, equal to *gravitational mass*, a measure for its susceptibility to gravity.[31] Newton did some pendulum experiments to test this equality, now often called the *weak equivalence principle*. It was tested with much greater accuracy in a celebrated experiment of the Hungarian physicist Baron Loránd von Eötvös (1890). Einstein was still unaware of this experiment in July 1912. He first cited it in his 1913 paper with Grossmann (CPAE 4, 340, note 3).

The equality of inertial and gravitational mass, without which Galileo's principle that all bodies fall alike would not hold, is an unexplained coincidence in Newtonian physics. To Einstein it suggested that there is an intimate connection between inertia and gravity (see Section 6 for further discussion). The universality of gravity's marching orders makes it possible to move gravity from the column of assorted forces to the column of the space-time structure. General relativity combines the space-time structure (more accurately: the inertial structure of space-time) and the gravitational field into one inertio-gravitational field. This field specifies the trajectories of particles on which no additional forces are acting. Einstein thus removed the mystery of the equality of inertial and gravitational mass in Newton's theory by making inertia and gravity two sides of the same coin.

In the passage from the unpublished *Nature* article of 1920 referred to at the beginning of this section, Einstein drew an analogy with electromagnetism to explain the situation: "Like the electric field generated by electromagnetic induction, the gravitational field only has a relative existence. *Because, for an observer freely falling from the roof of a house, no gravitational field exists while he is falling, at least not in his immediate surroundings*" (CPAE 7, Doc. 31 [p. 21], Einstein's italics).[32] He had explained the example from electromagnetism in the preceding paragraph.[33] It is the thought experiment, illustrated in Figure 6.2, with which Einstein (1905r) opened his first paper on special relativity.

Consider a bar magnet and a conductor – say, a wire loop with an ammeter – in uniform motion with respect to one another. In prerelativistic electrodynamics, it made a difference whether the conductor or the magnet is at rest with respect to the ether, the medium thought to carry electric and magnetic fields. In case (a) – with the conductor at rest

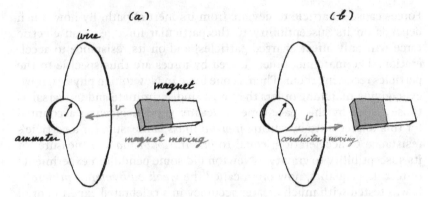

Figure 6.2. *The magnet-conductor thought experiment.* Drawn by Laurent Taudin.

in the ether – the magnetic field at the location of the wire loop is grow-ing stronger as the magnet approaches. Faraday's induction law tells us that this induces an electric field, producing a current in the wire, which is registered by the ammeter. In case (b) – with the magnet at rest in the ether – the magnetic field is not changing and there is no induced electric field. The ammeter, however, still registers a current. This is because the electrons in the wire are moving in the magnetic field and experience a Lorentz force that drives them around in the wire. It turns out that the currents in cases (a) and (b) are exactly the same, even though their explanations are very different in prerelativistic theory.

Einstein found this unacceptable. He insisted that situations (a) and (b) are one and the same situation looked at from two different perspec-tives. It follows that the electric field and the magnetic field cannot be two separate fields. After all, there is both a magnetic and an elec-tric field in situation (a), while there is no electric field in situation (b). Einstein concluded that there is only an electromagnetic field that breaks down differently into electric and magnetic components depend-ing on whether the person making the call is at rest with respect to the magnet or with respect to the conductor. The equivalence principle in its mature form can be formulated in the exact same way. There is only an inertio-gravitational field that breaks down differently into inertial and gravitational components depending on the state of motion of the person making the call.[34] This is what Einstein meant when he wrote in 1920 that "the gravitational field only has a relative existence." This statement must sound decidedly odd to the ears of many modern relativ-ists. Their criterion for the presence or absence of a gravitational field – does the so-called *curvature tensor* have nonvanishing components or

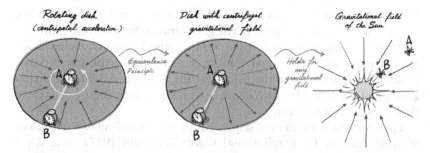

Figure 6.3. *The rotating disk*. Drawn by Laurent Taudin.

not? – leaves no room for disagreement between different observers (see Section 3).

It took Einstein more than a decade to articulate the mature version of the equivalence principle (Einstein 1918e).[35] In the meantime, the general insight that acceleration and gravity are intimately linked had guided Einstein on his path to the new theory. The equivalence principle, seen now as a heuristic principle, allowed him to infer effects of gravity from effects of acceleration in Minkowski space-time (Norton 1985).[36] The mature equivalence principle retroactively sanctioned such inferences, at least qualitatively. From the point of view of general relativity, the space-time structure of special relativity is nothing but a specific inertio-gravitational field.

Einstein first applied this type of reasoning to linear acceleration in Minkowski space-time and used it to derive the gravitational redshift and the bending of light in a gravitational field (Einstein 1907j, Einstein 1911h). I will illustrate it using the case of rotation in Minkowski space-time, which played an important role in the development of Einstein's theory (see Section 4). The inertial effects due to rotation (such as those one experiences when trying not to be thrown off a merry-go-round) can, in the spirit of the equivalence principle, be reinterpreted as due to a gravitational field.

The situation is illustrated in Figure 6.3. The first drawing shows a circular disk rotating in Minkowski space-time.[37] The inward pointing arrows represent the centripetal acceleration of points of the rotating disk. They give the direction in which the velocity of these points is *changing*. Now consider an observer on and at rest with respect to the rotating disk. Appealing to the equivalence principle, this observer can consider herself at rest in the centrifugal gravitational field represented by the outward pointing arrows shown in the second drawing in Figure 6.3.[38] Special relativity tells us what happens in the situation in the first drawing. The equivalence principle tells us the same things will happen in the peculiar gravitational field in the second. By determining in this manner what happens in the gravitational field corresponding to rotation in Minkowski space-time, we can expect to gain insights about what happens in gravitational fields in general, such as the gravitational field of the Sun shown in the third drawing in Figure 6.3. Such insights gave Einstein valuable clues about features of a new theory of gravity that goes beyond Newton's.

First, we examine the consequences of the special-relativistic effect of time dilation for gravitational theory (Einstein 1917a, Sec. 23).[39] Attach two clocks to the rotating disk, one, A, at the center, the other, B, at the circumference. B is moving, while A is practically at rest (it is spinning on its own axis with a velocity much smaller than that of B). According to special relativity, moving clocks tick at a lower rate than clocks at rest. One revolution of the disk thus takes less time on B than it takes on A. This is just a variant of the famous twin paradox in special relativity.[40] The equivalence principle tells us that the gravitational field pointing from A to B in the second drawing likewise causes clock B to tick at a lower rate than clock A. The same will be true for the gravitational field of the Sun pictured in the third drawing. The ticking of a clock will slow down as it is lowered in the Sun's gravitational field. The frequency of light emitted by atoms will be subject to this same effect. Hence, the frequency of light emitted by atom B close to the Sun will be lower than the frequency of light emitted in the same process by the identical atom A farther away from the Sun. If an atom is lowered in a gravitational field, the frequency of the light it emits will shift to the red end of the spectrum. This phenomenon is known as *gravitational redshift*. The conclusion of this simple argument based on the equivalence principle is confirmed by general relativity in its final form.[41]

The rotating disk can also be used to establish that gravity will bend the path of light, although Einstein always used the more straightforward example of a linearly accelerated elevator for this purpose (see, e.g., Einstein and Infeld 1938, 233–4). The first two drawings of Figure 6.3 show a white line painted on the disk connecting A and B. Consider the rotating disk in the first drawing and suppose a light signal is sent from A in the direction of this painted line. The light will travel in a straight line, but, since the disk is rotating under it, it will *not* follow the line AB. The light will cross the circumference slightly behind B. The equivalence principle tells us that the light will follow this exact same path across the disk at rest with a gravitational field as depicted in the second drawing. It will start out in the direction AB but veer off to the right (i.e., in the direction opposite to that of the disk's rotation in the first drawing).[42] The light will thus travel along a path that is bent. We should expect this to be the case in gravitational fields in general. This conclusion is confirmed, at least qualitatively, by general relativity in its final form. The phenomenon is known as *light bending*. When British astronomers announced in 1919 that the effect had been detected during a solar eclipse,[43] it made headlines on both sides of the Atlantic. Einstein became an overnight sensation, the world's first scientific superstar.

Einstein also considered the consequences of the special-relativistic effect of length contraction for gravitational theory. Suppose we put measuring rods on the radius and on the circumference of the rotating disk. According to special relativity, moving objects contract in the direction of motion. This does not affect the length of the rods on the radius since the radius is perpendicular to the motion of the disk. The length of the rods on the circumference, however, will be affected. On the circumference of a rotating disk one can thus fit more rods than on the circumference of a nonrotating disk of the same diameter. An observer on the rotating disk, Einstein argued, will therefore find that the ratio of the circumference and the diameter of the disk is greater than π, the value of this ratio in Euclidean geometry. The equivalence principle tells us that someone in the gravitational field in the second drawing in Figure 6.3 will likewise find a value greater than π. It follows that the spatial geometry in this and presumably other gravitational fields would be non-Euclidean. This argument can be found in four of Einstein's best-known expositions of general relativity (Einstein 1916e, 774–5; 1917a, Sec. 23; 1922c, 38–9; Einstein and Infeld 1938, 239–44).[44] Stachel (1989a)[45] argues that this is what first suggested to Einstein that gravity should be represented by curved space-time. The argument, however, is trickier than it looks at first glance. For an observer *next to the disk* the rods placed on the circumference of the rotating disk are clearly contracted. But is this also true for an observer *on the disk*? It is if she uses the same time coordinate as the observer standing next to the disk. It is not if she uses the time coordinates of the instantaneous rest frames of those rods. The latter option may seem more natural, but is not unproblematic. For one thing, when we use the instantaneous rest frames of the rods on the rotating disk, the curves in space-time representing the circumference of the disk for the observer on the rotating disk will not be closed (they will be spiraling up through space-time). Whatever one makes of Einstein's argument,[46] however, the notion that the spatial geometry on a rotating disk would not be Euclidean played an important heuristic role for him. It suggested a new way of making all motion relative. Before turning to this new attempt, I briefly discuss how Einstein came to abandon his original idea of reducing all nonuniform motion to gravity.

In 1912, partly in response to a special-relativistic theory of gravity published by Max Abraham (1912),[47] Einstein proposed his first formal new theory of gravity based on the equivalence principle. Up to that point, he had only explored isolated applications of the principle. The centerpiece of Einstein's theory was its gravitational field equation. One requirement the equation had to fulfill was that the static homogeneous gravitational field corresponding to constant linear acceleration (so-called Born acceleration) in Minkowski space-time be a vacuum solution (i.e., a solution for the case without any gravitating matter). The equation that Einstein (1912c) initially published met this requirement. As Einstein quickly discovered, however, the equation violated energy conservation. The equivalence of energy and mass, expressed in special relativity's most famous equation, $E = mc^2$, demands that all energy, including the energy of the gravitational field, acts as a source of gravity. In the original field equations of Einstein's 1912 theory only the mass-energy of matter entered as a source. Einstein (1912d) thus had to add a term representing the mass-energy of the gravitational field. Unfortunately, the gravitational field corresponding to Born acceleration is only locally a vacuum solution of these amended equations. This made Einstein reluctant to add the extra term (ibid., 455–6). It meant that the equivalence principle, even for constant acceleration and static homogeneous gravitational fields, only held in infinitesimally small regions of space (Norton 1984, 106).[48] Einstein faced a choice between the philosophical promise of the equivalence principle to make all motion relative and the physical requirement of energy conservation. He opted for the latter: physics trumped philosophy.

3. SECOND ATTEMPT: GENERAL COVARIANCE

To implement the insight that gravity is intimately connected with the geometry of space(-time) in a formal theory, Einstein turned to the mathematics of curved surfaces developed by Gauss.[49] As a student at the *Eidgenössische Technische Hochschule* (ETH) in Zurich, he had studied this subject relying on notes of his classmate Grossmann. As luck would have it, when Einstein realized that this was the kind of mathematics he needed, the two of them were about to be reunited at their alma mater. In early 1912, Einstein was appointed professor of theoretical physics at the ETH, where Grossmann was professor of mathematics. Grossmann familiarized Einstein with the extension of Gauss's theory to higher dimensions by Riemann, Christoffel, and others.[50] Einstein reportedly told his friend: "You must help me or else I'll go crazy" (Pais 1982, 212; Stachel 2002b, 107).

The central quantity in the geometry of Gauss and Riemann is the *metric tensor* or *metric* for short. In general relativity it does double duty. It gives the geometry of space-time – or, to be more precise, its

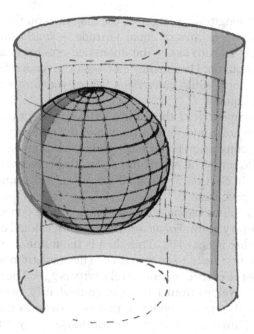

Figure 6.4. *Mapping the Earth.* Drawn by Laurent Taudin.

chrono-geometry – and the potential for the gravitational field. The description of a 3+1D locally Minkowskian curved space-time (three spatial and one temporal dimension) with the help of a metric is completely analogous to that of a two-dimensional locally Euclidean curved surface, such as the surface of the Earth.

Figure 6.4 shows a simple way of making a map of a miniature copy of this surface. A sheet of paper is rolled around the equator of a globe, forming a snug-fitting cylinder (as indicated by the dashed lines in the figure). The surface of the globe is projected horizontally on this cylinder mantle. The sheet is rolled out and a grid of regularly spaced horizontal and vertical lines is drawn on the part containing the image of the globe. With the help of this grid a unique pair of coordinates can be assigned to every point of the globe except for the two poles. To turn this grid into a useful map, instructions must be provided for converting distances in terms of (fractions of) steps on the grid to actual distances on the globe. In standard terminology, *coordinate distances* must be converted to *proper distances*. The conversion factors are given by the metric. They vary with direction and they vary from point to point. Right at the equator, where the map touches the globe, the conversion factors are equal to 1 in all directions. Everywhere else, the distance between lines of equal longitude is *larger* on the map than on the globe, while the distance between lines of equal latitude is *smaller* on the map than on the globe. In both cases, the difference gets larger as one moves away from the equator. Hence, the "east-west conversion factor" gets *smaller* and the "north-south conversion factor" gets *larger* as one moves away from the equator.

The "east-west" component of the metric will vanish at the poles. Since all points on the horizontal line at the top of the grid correspond to the North Pole, the conversion factor multiplying the finite distances between them must be zero. The metric has a so-called *coordinate singularity* at the poles. In Section 5, we shall encounter an example of such a singularity in space-time.

For an arbitrary two-dimensional curved surface, three conversion factors are needed at every point. For an arbitrary n-dimensional curved space(-time) this number is $n(n+1)/2$. This then is the number of components of the metric that need to be specified. The standard notation for the components of the metric in general relativity is $g_{\mu\nu}$.[51] The Greek indices take on integer values from 1 to n (or, equivalently, from 0 to $n-1$). So $g_{\mu\nu}$ has a total of n^2 components, that is, sixteen in the case of 3+1D space-time. However, since the metric tensor is *symmetric* (i.e., for all values of μ and ν, $g_{\mu\nu} = g_{\nu\mu}$), only $n(n+1)/2$ of those components are independent, that is, ten for 3+1D space-time. This means that the gravitational potential in Einstein's theory likewise has ten components.[52]

The metric field $g_{\mu\nu}(x^\rho)$ assigns values to the components $g_{\mu\nu}$ of the metric to points labeled with coordinates $x^\rho \equiv (x^1, \ldots, x^n)$. In three-dimensional Euclidean space these could be the familiar Cartesian coordinates, $(x^1, x^2, x^3) = (x, y, z)$. In the case of the two-dimensional surface in Figure 6.4, the coordinates (x^1, x^2) refer to the grid drawn on the sheet. There are infinitely many other grids that can be used to assign a unique pair of coordinates to points of this or any other surface. It is not necessary (and often impossible) to cover the entire surface with one map. An atlas of partly overlapping maps will do. Any one-to-one mapping from a region of the surface to a region of the plane $\mathbb{R}^2 = \mathbb{R} \times \mathbb{R}$ (where $\mathbb{R}$ is the set of real numbers) will do as a map. With any map a metric field $g_{\mu\nu}(x^1, x^2)$ needs to be specified that gives the corresponding conversion factors from coordinate distances to proper distances.

Gauss made the remarkable discovery that at every point of an arbitrary two-dimensional surface one can define curvature without reference to the surface's three-dimensional Euclidean embedding space.[53] Moreover, he found that this intrinsic so-called Gaussian curvature is the same function of the components of the metric field and their first- and second-order derivatives with respect to the coordinates in all coordinate systems. The same is true for the components of the Riemann curvature tensor describing the intrinsic curvature of higher-dimensional spaces.

The transformation rules for translating the metric field and other quantities encoding the geometry of the surface from one coordinate system to another are also the same for all coordinate systems. The geometry of any curved surface can thus be described in the exact same way regardless of the choice of coordinates. The Gaussian theory of curved surfaces is *generally covariant*. All this holds for the Riemannian extension of the theory to higher-dimensional spaces, such as 3+1D space-time, as well.[54]

Once a metric has been introduced, the length of lines in space(-time) can be computed. The lines of extremal length – the shortest ones in ordinary space, the longest ones in space-time – are called (metric) *geodesics*. In Riemannian geometry these are also the *straightest* lines, called *affine geodesics*. Which lines are the geodesics in a given Riemannian space is determined by the *geodesic equation*. This equation involves the *Christoffel symbols*, a sum of three terms, each of which is a gradient of the metric. In electricity theory, the field is the gradient of the potential. Since the components of the metric double as the gravitational potentials in Einstein's theory, the Christoffel symbols are natural candidates for representing the components of the gravitational field. It was only in 1915 that Einstein adopted this definition of the gravitational field. Originally, he had simply set the gravitational field equal to the gradient of the metric field, that is, to one of the three terms in the Christoffel symbols (see Section 6 for further discussion).

In the following years, the mathematicians Gerhard Hessenberg, Tullio Levi-Civita, and Hermann Weyl worked out the general concept of an *(affine) connection* (Stachel 2007, 1044–6).[55] This quantity allows one to pick out the straightest lines directly, without the detour through the metric and lines of extremal length. In Riemannian geometry, the connection is given by the Christoffel symbols, but it can be defined more generally and independently of the metric. Since what matters for the equivalence principle are the straightest rather than the longest lines in space-time, one can argue that general relativity is most naturally formulated in terms of the connection (Stachel 2007, 1041). Since Einstein formulated his theory in terms of the metric (and to this day textbooks tend to follow his lead), it looks as if the mathematical tools he needed were right at hand. In hindsight, it may be more accurate to say that he made do with the tools he had (Stachel 2002b, 86).

With the help of the notion of a geodesic, metric or affine, the situations illustrating the equivalence principle in Figure 6.1 can readily be characterized in geometrical language. The *world lines*, the trajectories through space-time, of an observer hovering freely in outer space far away from gravitating matter (IIb) or in free fall on Earth (Ib) are (time-like)[56] geodesics, the world lines of an observer accelerating in outer space (IIa) or resisting the pull of gravity on Earth (Ia) are nongeodesics. As the examples illustrate, moving on a geodesic is physically different from moving on a nongeodesic.

In both situations, flat Minkowski space-time (II) and curved space-time (I), both observers, nongeodesic (a) and geodesic (b), can use their own world line as the time axis of a coordinate system providing a map of the space-time region in their immediate vicinity. The metric field will be given by different functions of the coordinates for the two observers, but, because of the general covariance of Riemannian geometry, they will use the same equations involving the same functions of the metric field to describe the situation. This suggested to Einstein that the property of general covariance itself could be used to extend the principle of relativity from uniform to accelerated motion. In special relativity in its standard form, two inertial observers in uniform motion with respect to one another can use the same equations if they use special coordinate systems related to one another through special coordinate transformations called Lorentz transformations. By allowing arbitrary coordinates and arbitrary coordinate transformations, Einstein thought, one automatically extends the principle of relativity from uniform to arbitrary motion. Unlike Lorentz transformations in Minkowski space-time, however, the transformations between the coordinate systems of observers like (a) and (b) in situations (I) and (II) in Figure 6.1 are not between *physically equivalent* states of motion. We already saw this in Section 2. The point can be made succinctly in terms of the geometrical language introduced in this section: no coordinate transformation turns a geodesic into a nongeodesic or vice versa.

Erich Kretschmann (1917), a former student of Max Planck who had become a high school teacher, took Einstein to task for his conflation of general covariance and general relativity.[57] Given enough mathematical ingenuity, Kretschmann pointed out, just about any space-time theory, with or without absolute motion, can be written in generally covariant form. Einstein (1918e) granted this criticism but predicted that the generally covariant version of, say, Newtonian theory would look highly artificial compared to a theory such as general relativity that is naturally expressed in generally covariant form. This expectation was proven wrong when generally covariant formulations of Newtonian theory were produced in the 1920s (Norton 1993b, Sec. 5.3).

Kretschmann also put his finger on the crucial difference between the invariance under Lorentz transformations of the standard description of Minkowski space-time in special relativity and the invariance under arbitrary coordinate transformations of the standard description of curved space-times in general relativity. Only the former transformations capture a symmetry of the space-time. They map the set of all inertial states – in geometrical terms: the set of all geodesics representing all possible inertial paths – back onto itself. The state of rest in one coordinate system will be mapped onto a state of uniform motion in another, but, since all such states are physically equivalent, that does not make any difference. This, then, is how Lorentz invariance expresses the relativity of uniform motion. General relativity allows many different space-times depending on the matter distribution. The set of all geodesics of all these space-times has no nontrivial symmetries. The theory's general covariance, therefore, is not associated with any such relativity-of-motion principle whatsoever.[58]

General covariance, however, is important for the relativity of the gravitational field expressed in the mature version of Einstein's equivalence principle.[59] Once again consider Figure 6.1. Both in situation (I) and in situation (II), observer (a) – Einstein, sitting at his desk, moving on a nongeodesic – will say that there is a gravitational field while observer (b) – the falling-window-cleaner/hovering-astronaut moving on a geodesic – will say that there is none. If we want to insist that there are no grounds for preferring the judgment of one over the other, it had better be the case that the laws of physics are the same for both of them. General covariance guarantees that this is true for all observers.[60]

Both in the Minkowski space-time of situation (II) and in the curved space-time of situation (I), observer (b) can, at least in his immediate vicinity, use special relativity in standard coordinates, using his own world line as the time axis. This is because locally curved space-time is indistinguishable from flat Minkowski space-time, just as the surface of the Earth or any other curved surface is locally indistinguishable from a flat Euclidean plane. In Minkowski space-time in standard coordinates the components of the metric are constants, so all gradients and hence the Christoffel symbols are zero. Representing the gravitational field by the Christoffel symbols, observer (b) concludes, in situation (I) as well as in situation (II), that the gravitational field is zero and that the inertio-gravitational effects experienced by observer (a) are due to inertial forces. For observer (a), the Christoffel symbols do not vanish, neither in situation (I) nor in situation (II), and he will ascribe the inertio-gravitational effects he experiences to gravitational forces. General covariance and the identification of the Christoffel symbols as the gravitational field thus implement the relativity of the gravitational field of the mature equivalence principle.

Most modern relativists see things differently. They would say that there is only a gravitational field in situation (I) and not in the flat Minkowski space-time of situation (II). They would also object to having the presence or absence of a gravitational field depend on which observer is making the call. In the spirit of general covariance, they would prohibit such coordinate-dependent notions and insist that only quantities transforming as tensors be used to represent physically meaningful quantities. One consequence of the transformation rules for tensors is that, if all components of a tensor vanish in one coordinate system, they vanish in all of them. The Christoffel symbols then are clearly not tensors. For many modern relativists, this disqualifies them as candidates for the mathematical representation of the gravitational field. Instead, as mentioned in Section 2, the nonvanishing of the curvature tensor is used as a coordinate-independent criterion for the presence of a gravitational field.[61] To the end of his life, however, Einstein preferred to use the Christoffel symbols instead.[62]

By late 1912, for reasons good and bad, general covariance, or at least a covariance broad enough to cover arbitrary states of motion, had become central to Einstein's quest for general relativity. That winter he set out to find field equations for his new theory. He hoped to extract field equations of broad covariance from generally covariant equations. The fruits of his labor, in which he was assisted by Grossmann, have been preserved in what is known as the Zurich Notebook (CPAE 4, Doc. 10).[63] Despite considerable effort, he could not find physically sensible field equations of broad covariance and ruefully settled for equations of severely limited covariance. They were first published in May 1913 (Einstein and Grossmann 1913). It was only in November 1915 that Einstein replaced these *Entwurf* field equations by the generally covariant field equations named after him. The Zurich Notebook shows that almost three years earlier he had come within a hair's breadth of these generally covariant equations based on the Riemann curvature tensor. As he told some of his colleagues in 1915,[64] he had rejected them at the time because they did not seem to be compatible with energy-momentum conservation or reduce to the equations of Newtonian gravitational theory for weak static fields. In 1913, Einstein thus saw another attempt to generalize the principle of relativity foiled because he could not get the physics to work out.

The restricted covariance of the *Entwurf* field equations continued to bother Einstein until, in late August 1913, he convinced himself using the "hole argument" that such restrictions are unavoidable.[65] Generally, covariant field equations, the argument was supposed to show, cannot

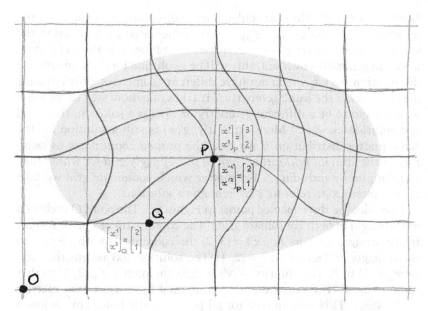

Figure 6.5. *The hole argument.* Drawn by Laurent Taudin.

do the basic job of uniquely determining the space-time geometry once the matter distribution has been specified. After his return to general covariance in November 1915, Einstein produced an escape from the hole argument, known as the "point-coincidence argument."[66]

Figure 6.5 illustrates how Einstein's hole argument works. It shows a 1+1D space-time (one spatial and one temporal dimension) with two coordinate systems, one with unprimed coordinates, (x^1, x^2), referring to the (lighter) grid with straight lines and one with primed coordinates, (x'^1, x'^2), referring to the (darker) grid with squiggly lines. The two grids coincide except in the shaded oval-shaped region. This region, devoid of matter, is the hole from which the hole argument derives its name. All candidate field equations are *local* in the sense that they set functions of the metric field and its derivatives, all evaluated at the same point, equal to functions describing the field's material sources evaluated at that same point. If such equations are generally covariant, the hole argument purports to show, the matter distribution does not uniquely determine the geometry inside the hole.

The functions describing the matter distribution in this case are the same in both coordinate systems. This is because, outside the hole, the two coordinate systems coincide, and, inside the hole, these functions are identically zero. Let $g_{\mu\nu}(x^1, x^2)$, abbreviated $g(x)$, be a solution of the field equations for this particular matter distribution in terms of the unprimed coordinates. Let $g'_{\mu\nu}(x'^1, x'^2)$, abbreviated $g'(x')$, describe the same geometry in terms of the primed coordinates. If the field equations are generally covariant, this will be a solution for the same matter distribution. So far, we do not have different geometries, only different *descriptions* of the same geometry.[67] It takes one more step to get what would seem to be a different geometry: if $g'(x')$ is a solution, then $g'(x)$ is a solution as well.[68] More explicitly, $g'(x')$ is still a solution for the same matter distribution *if we read the primed coordinates as referring to the straight grid rather than to the squiggly grid for which they were originally introduced.* No matter which coordinate grid we take its arguments x' to refer to, $g'(x')$ remains a solution.

Consider the three labeled points in Figure 6.5. The point O is chosen as the origin of both coordinate grids. The coordinates of P with respect to the straight grid are $(x^1, x^2) = (3, 2)$. Its coordinates with respect to the squiggly grid are $(x'^1, x'^2) = (2, 1)$. The solution $g(x)$ assigns the metric $g_{\mu\nu}(3, 2)$ to P. The solution $g'(x')$ assigns the metric $g'_{\mu\nu}(2, 1)$ to that same point P. The curvature at P computed from those two metrics is the same. This will be true for all points in the hole. This is just a different way of saying that $g(x)$ and $g'(x')$ describe the same geometry. The solution $g'(x)$, it seems, does not. This solution assigns the metric $g'_{\mu\nu}$ to the point Q with coordinates $(x^1, x^2) = (2, 1)$ *with respect to the straight grid.* So the curvature assigned to one point (P) by both $g(x)$ and $g'(x')$ is assigned to another point (Q) by $g'(x)$. The solutions $g(x)$ and $g'(x)$ thus do seem to describe different geometries inside the hole. To block this violation of determinism, Einstein argued, the covariance of the field equations needs to be restricted. Field equations that preserve their form under coordinate transformations affecting only matter-free regions must be ruled out.[69]

Einstein used this argument in print on several occasions to defend the restricted covariance of the *Entwurf* field equations.[70] In November 1915, however, he published generally covariant field equations without losing a word about the hole argument. Einstein (1915f, 1915g, 1915i) focused instead on demonstrating that his new field equations respect energy-momentum conservation and are compatible with Newton's theory in the appropriate limit. Problems on these two counts had made him forego general covariance in the first place. When his friends Michele Besso and Paul Ehrenfest pressed him on the hole argument, Einstein rolled out a new argument, the point-coincidence argument.[71] The hole argument was never mentioned in print again, but a version of this new argument was included in the first systematic presentation of the new theory a few months later (Einstein 1916e, 776–7).

The printed version of the point-coincidence argument is disappointing. Its premise is that all we ever observe are spatiotemporal coincidences, such as the intersections of world lines.[72] Since there is no reason to privilege one coordinatization of a set of point coincidences over any other, the argument continues, all physical laws, including the field equations, should be generally covariant.[73] This does provide an escape from the hole argument. The different geometries found for the same matter distribution agree on all point coincidences. If that exhausts all we can ever observe, we have no empirical means of telling these geometries apart. We still have indeterminism but of a benign kind. If we deny reality to anything but point coincidences, there is no indeterminism at all. This way of avoiding indeterminism, however, comes at the price of "a crude verificationism and an impoverished conception of physical reality" (Earman 1989, 186).[74]

The letters to Besso and Ehrenfest suggest a more charitable interpretation of Einstein's resolution of the hole argument. In these letters, it seems, Einstein used point coincidences to put his finger on an unwarranted implicit assumption in the hole argument without which indeterminism cannot be inferred in the first place. Consider, once again, Figure 6.5. Suppose that, in the solution $g(x)$, two world lines cross at P. In the solution $g'(x)$, the corresponding world lines cross at Q. This is a different state of affairs only if there is some way of identifying Q other than by referring to it as the point where these two world lines meet. It is at this juncture that the hole argument starts to unravel.

The identity of a point, one can argue, though the issue remains controversial, lies in the sum total of the properties assigned to that point by the metric field and all matter fields. It cannot be identified or individuated independently of those properties. It only has *suchness* and no *primitive thisness* or *haecceity*. Since candidate field equations are local in the sense previously specified, *all* properties assigned to P by $g(x)$ are assigned to Q by $g'(x)$. But then P and Q are only different labels for one and the same space-time point, and $g(x)$ and $g'(x)$ are only different descriptions of one and the same geometry. Generally covariant field equations can be perfectly deterministic after all.

In modern terms, all fields are defined on a so-called *differentiable manifold*, which, for our purposes, one can think of as an amorphous set of points. The manifold still needs to be "dressed up" by a metric field if it is to represent space-time. Metric fields such as $g(x)$ and $g'(x)$ generated in the hole argument dress up different points of the bare manifold to become a particular space-time point. If points of the bare manifold could be individuated independently of the fields defined on it, these differently dressed-up manifolds would represent distinct though empirically indistinguishable space-times and we would have (a benign form of) indeterminism. We can avoid this consequence by denying that bare manifold points can be individuated in this way. That, in turn, means that we cannot think of the bare manifold as some kind of container. The combination of the hole argument and (the charitable interpretation of) the point-coincidence argument thus amounts to an argument against a substantival and in favor of a relational account of the ontology of space-time.[75]

This argument for relationism can be seen as a modern version of a classic argument against absolute space given by Leibniz in the course of his correspondence with Clarke (Alexander 1956, 26).[76] One way to make the argument is the following: Newtonian space is the same everywhere, so the location of the world's center of mass makes no observable difference. This seems to violate Leibniz's *principle of sufficient reason*. For no reason whatsoever, God had to put the center of mass of the universe at one point rather than another. To avoid such consequences, Leibniz insisted on his *principle of the identity of indiscernibles*. Since it is impossible to tell two worlds apart that differ only in the position of their center of mass, they must be one and the same world. But then Newtonian space cannot be some kind of container. In the hole argument, a violation of determinism replaces the deity's violation of the principle of sufficient reason that so vexed Leibniz. In the point-coincidence argument, determinism is restored through an account of the identity and individuation of space-time points in the spirit of the principle of the identity of indiscernibles with which Leibniz restored the principle of sufficient reason. So, even though general covariance does not eliminate absolute motion, Einstein's struggles with general covariance did produce what would appear to be a strong argument against absolute space(-time).

4. THIRD ATTEMPT: A MACHIAN ACCOUNT OF NEWTON'S BUCKET

When it looked as if general covariance was not to be had, Einstein explored another strategy for eliminating absolute motion. This one was directly inspired by his reading of Mach's attempt to get around a classic argument for the absolute character of acceleration, an argument based on Newton's thought experiment of the rotating bucket in the *Scholium* on space and time in the *Principia* (Cohen and Whitman 1999, 412–13). Looking back on this period, Einstein wrote: "Psychologically, this conception [that a body's inertia is due to its interaction with all other matter in the universe][77] played an important role for me, since it gave me the courage to continue to work on the problem when I absolutely could not find covariant field equations" (Einstein to De Sitter,

Figure 6.6. *The rotating-bucket experiment.* Drawn by Laurent Taudin.

November 4, 1916 [CPAE 8, Doc. 273]).[78] Consider a bucket of water set spinning. As the water catches up with the rotation of the bucket, it will climb up the side of the bucket. Since the effect increases as the relative rotation between water and bucket decreases and is maximal when both are rotating with the same angular velocity, Newton argued, the effect cannot be due to this relative rotation.[79]

Figure 6.6 illustrates a different way of making the same point. The bucket experiment is broken down into four stages, the fourth being a flourish added by later authors (Laymon 1978, 405). In stage (I) the bucket and the water are at rest. In stage (II) the bucket has started to rotate but the water has yet to catch up with it. In stage (III) it has. In stage (IV) the bucket is abruptly stopped while the water continues to rotate. Comparison of these four stages shows that the shape of the water surface cannot be due to the relative rotation of the water with respect to the bucket. In stages (I) and (III) there is no relative rotation, yet the surface is flat in one case and concave in the other. In stages (II) and (IV) there is relative rotation, yet, once again, the surface is flat in one case and concave in the other.

The concave shape of the spinning water, Newton argued, is due to its rotation with respect to absolute space. Three centuries later, Mach resurrected another option briefly considered but rejected by Newton: rotation with respect to other matter in the universe. "Try to fix Newton's bucket and rotate the heaven of fixed stars," Mach (1960, 279) asked his readers to imagine, "and then prove the absence of centrifugal forces."

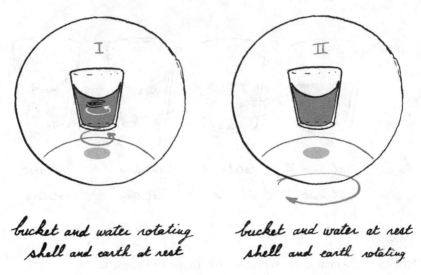

bucket and water rotating
shell and earth at rest

bucket and water at rest
shell and earth rotating

Figure 6.7. *Machian account of Newton's rotating-bucket experiment.* Drawn by Laurent Taudin.

The implication is that it should make no difference whether the bucket or the "heaven of fixed stars" is rotating: in both cases the water surface should become concave. This idea is illustrated in Figure 6.7, depicting the Earth, the bucket, and the water at the center of a spherical shell, much larger than shown in the figure, representing all other matter in the universe. On the left (situation I), the bucket and the water are rotating and the Earth and the shell are at rest. On the right (situation II), it is the other way around.

The problem with Mach's proposal is that, according to Newtonian theory, the rotation of the shell will have no effect whatsoever on the water in the bucket, *so the water surface on the right in Figure 6.7 (situation II) should have been drawn flat.* For most of the reign of the *Entwurf* theory and beyond, Einstein was convinced that this was a problem not for Mach's analysis but for Newton's theory and that his own theory vindicated a Machian account of the bucket experiment.

Einstein thought, mistakenly, that his theory reduced the two situations pictured in Figure 6.7 to one and the same situation viewed from the point of view of two different observers, one at rest with respect to the shell, the other at rest with respect to the bucket. He thought this followed from two more specific claims. First, the metric field of Minkowski space-time in rotating coordinates is a vacuum solution of the field equations, that is, a solution in which there is no gravitating matter at all. Second, this is the metric field that a spherical shell rotating in the opposite direction with the same angular speed would produce near its center. We need to take a closer look at both claims as well as at the conclusion Einstein drew from them.

We can take the space-time in which we perform the bucket experiment to be Minkowskian even though the telltale shape of the water surface obviously depends on the gravitational field of the Earth (cf. note 37). The metric field of Minkowski space-time in the standard coordinates for an observer at rest with respect to the shell is a vacuum solution of the field equations. This is true both for the *Entwurf* field equations of 1913 and the Einstein field equations of 1915. For the two situations in Figure 6.7 to be equivalent, it is necessary – though not sufficient – that this metric field also be a vacuum solution, at least near the center of the shell, in the coordinates used by the observer at rest with respect to the bucket. The Einstein field equations automatically satisfy this requirement. Their general covariance guarantees that an arbitrary solution in some coordinate system remains a solution under arbitrary transformations to other coordinate systems. This is not true for the *Entwurf* field equations. Einstein had to check whether this specific solution, the Minkowski metric in standard coordinates, remains a solution under the specific transformation to a rotating coordinate system. In this context, Einstein and Grossmann (1914b, 221) talked about "justified transformations" between "adapted coordinate systems" (i.e., adapted to the metric field). Earlier, Einstein had distinguished such transformations for specific solutions from the usual transformations for arbitrary solutions by labeling them "nonautonomous" and "autonomous," respectively.[80] This terminology reflects that the former depend on the metric field that is being transformed while the latter do not. Already in the Zurich Notebook, Einstein had retreated to field equations invariant under nonautonomous transformations whenever he could not find equations invariant under ordinary autonomous transformations (Renn 2007a, Vol. 2, 495–6, 533–5).

Einstein went back and forth for more than two years on whether or not the transformation to rotating coordinates in the special case of Minkowski space-time is a justified transformation in the *Entwurf* theory. To settle the issue, he had to determine whether or not the *rotation metric*, the metric field of Minkowski space-time in rotating coordinates, is a vacuum solution of the *Entwurf* field equations (Janssen 2007). A sloppy calculation preserved in the Einstein-Besso manuscript (cf. note 19) and probably dating from early 1913 reassured him that it is (CPAE 4, Doc. 14 [pp. 41–2]). In a letter to Lorentz of August 1913, Ehrenfest reported that Einstein had meanwhile done this calculation "five or six times," finding "a different result almost every time" (Janssen 2007, 833). Einstein appears to have accepted for a few months late in 1913 that the rotation metric is not a solution, but eventually he convinced himself on general grounds that it had to be.[81] In the authoritative exposition of the *Entwurf* theory of late 1914, this result, erroneous as it turns out, is hailed as a vindication of a Machian account of the bucket experiment (Einstein 1914o, 1031). In September 1915, Einstein redid the calculation of 1913 once more, this time without making any errors, and discovered to his dismay that the rotation metric is not a solution (Janssen 1999). He thereupon carefully reexamined the *Entwurf* theory, discovered a flaw in a uniqueness argument for the *Entwurf* field equations that he had published the year before, and used the leeway this gave him to introduce new field equations of broad covariance preserving their form under ordinary autonomous transformations to rotating coordinates (Janssen and Renn 2007).

The rotation metric was now a vacuum solution of the field equations. Is it also the metric field that a rotating shell produces near its center? It is not, neither according to the *Entwurf* field equations nor according to the Einstein field equations. To calculate the metric field for a given matter distribution one typically needs boundary conditions, the values of the metric field at spatial infinity. When Einstein calculated the metric field of a rotating shell in 1913, he uncritically took those values to be Minkowskian (CPAE 4, Doc. 14 [pp. 36–7]; Einstein 1913c, 1260–1). He thus started with Minkowski space-time and calculated only how the rotating shell would curve this Minkowski space-time in its interior. This curvature, it turns out, is much too small to make the water surface concave. More importantly, treating the effect of the rotating shell as a small perturbation of the metric field of Minkowski space-time defeats the purpose of producing a Machian account of the bucket experiment. Only a small part of the metric field is due to the rotating shell this way; most of it is due to absolute space-time, albeit of the Minkowskian rather than the Newtonian variety. To put it differently, only a small part of the inertia of particles near the center of the shell is due to their interaction with the rest of the matter in the universe, represented by the shell; most of it is determined by the absolute Minkowskian space-time. The theory thus fails to satisfy what Einstein (1913c, 1261) called the "hypothesis of the relativity of inertia" (see also Einstein 1912e and Einstein 1917b, 147).[82] This problem will arise for any physically plausible boundary conditions. At this point, Einstein clearly had a blind spot for the role of boundary conditions in his theory, something that would come back to haunt him (see Section 5).

As long as the rotation metric is a solution of the field equations, however, the relativity of the gravitational field expressed by the mature equivalence principle does hold for a bucket rotating in Minkowski space-time. Consider situation (I) on the left in Figure 6.7 from the perspective of two observers, one at rest with respect to the shell and one at rest with respect to the bucket. As we just saw, the latter perspective on situation (I) is *not* the same as situation (II) on the right in Figure 6.7. For one thing, the water surface should have been drawn almost flat in situation (II). Furthermore, the metric field, which is not represented in Figure 6.7, is very different in the two situations.

Focus on situation (I). For an observer at rest with respect to the shell, the components of the metric field are constants, there is no gravitational field, the concave shape of the water surface is due to inertial forces, and the particles forming the shell are hovering freely in outer space. For an observer at rest with respect to the bucket, the components of the metric vary from point to point, there is a gravitational field, the concave shape of the water surface is due to forces exerted by this gravitational field, and the particles forming the rotating shell are falling freely in this gravitational field.[83]

Note, once again, how different the latter perspective on situation (I) is from situation (II). First, in situation (I), the values of the metric field become infinite as we go to infinity, while they are assumed to remain perfectly finite in situation (II).[84] Second, in situation (I), there is no need for cohesive forces keeping the particles of the rotating shell together, whereas such forces are required in situation (II). Finally, in situation (I), the gravitational field does *not* have the shell as its source, whereas the shell is supposed to be the source of the field in situation (II).

Einstein conflated the situation on the left in Figure 6.7, redescribed in a coordinate system in which the bucket is at rest, with the very different situation on the right. He believed accordingly that the metric field of a rotating shell would automatically be the rotation metric as long as the field equations used to compute this field preserve their form under the transformation to a frame rotating with the bucket. As he told Besso in July 1916, it is "obvious given the general covariance of the [field] equations," that the metric field near the center of a rotating ring, a case analogous to that of a rotating shell, is just the rotation metric. It is therefore, he added, "of no further interest whatsoever to actually do the calculation. This is of interest only if one does not know whether rotation-transformations are among the 'allowed' ones, i.e., if one is not clear about the transformation properties of the equations, a stage which, thank God, has definitively been overcome" (Einstein to Besso, July 31, 1916 [CPAE 8, Doc. 245]). Correspondence between Einstein and the Austrian physicist Hans Thirring in 1917 reveals that this misconception persisted for at least another year and a half. When Thirring first calculated the metric field inside a rotating shell, he was puzzled, as he told Einstein,[85] that he did not recover the rotation metric, as he expected on the basis of remarks in the introduction of Einstein's

(1914o) exposition of the *Entwurf* theory. In his reply Einstein failed to straighten out Thirring and in a follow-up letter he explicitly confirmed Thirring's expectation.[86] By the time he published his final results, Thirring (1918, 33, 38) had realized that the metric field inside a rotating shell and the rotation metric correspond to completely different boundary conditions. He cited Einstein (1917b) and De Sitter (1916b) in this context. As we shall see shortly, the role of boundary conditions was at the heart of the debate between Einstein and De Sitter. Yet, Einstein did not breathe a word about them in his letters to Thirring.[87]

Thirring's work serves as a reminder that, as with Einstein's first two attempts, something good came of Einstein's third failed attempt to eliminate absolute motion. Following up on his study of the effect of a rotating hollow shell on the metric field *inside* of it, Thirring studied the effect of a rotating solid sphere on the metric field *outside* of it (Lense and Thirring 1918). Einstein (1913c, 1261) had also pioneered calculations of this effect, now known as "frame dragging."[88] In April 2004, NASA launched a satellite carrying the special gyroscopes of a complicated experiment called *Gravity Probe B* aimed at detecting it. The data analysis took many years but the final results fully confirmed the predictions of general relativity (Everitt et al. 2011).

5. FOURTH ATTEMPT: MACH'S PRINCIPLE AND COSMOLOGICAL CONSTANT

The period from late 1915 to the fall of 1916 can be seen as an idyllic interlude in Einstein's quest for general relativity. The first systematic exposition of the theory dates from this period (Einstein 1916e). This widely read article is probably one of the reasons that the impression has lingered that with general relativity Einstein succeeded in banishing absolute motion from physics.[89] With the new field equations of November 1915, the entire theory was generally covariant at last. Einstein believed that this automatically extended the relativity principle for uniform motion, associated with Lorentz invariance, to arbitrary motion (see Section 3). He also believed that it sufficed for a Machian account of Newton's bucket experiment (see Section 4). Kretschmann disabused him of the first illusion in 1917; De Sitter of the second in the fall of 1916.[90]

General relativity retains vestiges of absolute motion through the boundary conditions at infinity needed to determine the metric field for a given matter distribution. During a visit to Leyden in the fall of 1916, Einstein was confronted with this problem by De Sitter. The solution he initially proposed was so far-fetched that he never put it in print. We only know of it through the ensuing correspondence[91] and through two papers of De Sitter (1916b, 1916c).[92]

To ensure that the metric field has the same boundary conditions for every observer, Einstein argued, the value of all its components at spatial infinity must be either 0 or ∞. He imagined there to be masses outside the visible part of the universe that would contribute to the metric field in such a way that these degenerate values turn into Minkowskian values at the edge of the observable universe. De Sitter derided this proposal. This was a cure worse than the disease. It just replaced Newton's absolute space by invisible masses. What if better telescopes made more of the universe visible? Would these special masses then have to be pushed even farther out?

Einstein came to accept these criticisms. As he told De Sitter in February 1917: "I have completely abandoned my views, rightfully contested by you, on the degeneration of the $g_{\mu\nu}$. I am curious to hear what you will have to say about the somewhat crazy idea I am considering now."[93] This "crazy idea" was actually quite ingenious: if boundary conditions at spatial infinity are the problem, why not eliminate spatial infinity? Einstein thus explored the possibility that the universe is spatially closed.[94] He considered the simplest example that he could think of. In the *Einstein universe*, as this first relativistic cosmological model came to be known, the spatial geometry is that of the three-dimensional hypersurface of a hypersphere in 4D Euclidean space. This hypersurface is analogous to the ordinary two-dimensional surface of an ordinary sphere in 3D Euclidean space. It is also analogous to a circle, the 1D boundary of a round disk in 2D Euclidean space.

Suppressing two spatial dimensions, we can likewise visualize the spatially closed 1+1D Einstein universe as a circle of some large radius R persisting through all eternity, forming an unbounded cylinder mantle, as illustrated in Figure 6.8. The Einstein universe is therefore also known as the *cylinder universe*. It is a *static* world. The diameter R of the cylinder does not change over time. De Sitter emphasized a few months later that our universe is almost certainly *not* static.[95] Einstein ignored these warnings.

Before Einstein could use the cylinder universe as a new solution to the problem of boundary conditions, he had to check whether it is allowed by his theory, and, if so, for what matter distribution. He computed the components of the metric field of the cylinder universe in a convenient coordinate system and inserted them into the field equations. In this coordinate system the matter distribution is at rest and fully characterized by its mass density ρ.

The result of Einstein's calculation was that the metric field of the cylinder universe is *not* a solution of the field equations as they stood. It *is* a solution, however, of slightly altered equations. A term

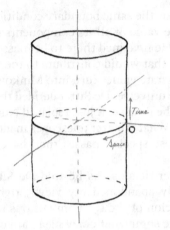

Figure 6.8. *Einstein's cylinder universe.* Drawn by Laurent Taudin.

proportional to $g_{\mu\nu}$, the so-called *cosmological term*, needs to be added. The proportionality constant lambda – nowadays used both in lower (λ) and in upper (Λ) case – is the infamous *cosmological constant*. It has to be exceedingly small so as not to disturb general relativity's agreement with Newton's theory of gravity in the limit of slow motion and weak fields. The cosmological constant determines both the radius R and the mass density ρ of the cylinder universe: $\lambda = 1/R^2 = \kappa\rho/2$ (where kappa is Einstein's gravitational constant). The radius of the cylinder universe is thus fixed for all times and large; its mass density constant, the same everywhere and small.

When Einstein first considered tinkering with his field equations, he must have anticipated renewed criticism from De Sitter. While abandoning his nebulous distant masses, he was now helping himself to an arbitrary new constant of nature. Mathematically, it turns out, the cosmological term is a natural addition to the Einstein field equations, but that was not immediately clear. In the paper in which he introduced the cosmological constant, however, Einstein (1917b) masterfully preempted the predictable charge of arbitrariness.

The title of the paper, "Cosmological Considerations on the General Theory of Relativity," suggests that Einstein's aim was simply to apply his new theory to cosmology. Today the paper is remembered and celebrated for launching modern relativistic cosmology. It did have a hidden agenda, however, which Einstein revealed in a letter to De Sitter about a month after its publication:

From the standpoint of astronomy, I have, of course, built nothing but a spacious castle in the sky. It was a burning question for me, however, whether the relativity thought can be carried all the way through or whether it leads to contradictions. I am satisfied now that I can pursue the thought to its conclusion, without running into contradictions. Now the problem does not bother me anymore, whereas before it did so incessantly. Whether the model I worked out corresponds to reality is a different question. (Einstein to De Sitter, March 12, 1917 [CPAE 8, Doc. 311])

This hidden agenda explains why the order of presentation in the paper is the opposite of the order in which the results presented had presumably been found. In the *context of discovery*, to borrow Hans Reichenbach's (1938, 6–7) terminology, Einstein, in all likelihood, conceived of the cylinder universe first, added the cosmological term to make sure the model is allowed by the field equations, and only then started to worry about making the extra term plausible. In the *context of justification*, preempting the kind of criticism he could expect from De Sitter, Einstein argued for the extra term first and then showed that the field equations with the cosmological term do allow the cylinder universe.

Einstein's justification for adding the cosmological term turned on an analogy with Newtonian cosmology.[96] To prevent a static universe from collapsing, he argued, a gravitational repulsion needs to be added, both in Newtonian theory and in general relativity. The cosmological term provides this repulsion. Arthur S. Eddington (1930) was the first to point out in print that the equilibrium thus produced in Einstein's cylinder universe is unstable. Much to the surprise of modern commentators (Weinberg 2005, 31), Einstein failed to recognize this.

What did De Sitter make of Einstein's new proposal? In response to the letter from which I quoted, he wrote: "As long as you do not want to force your conception on reality, we are in agreement. As a consistent train of thought, I have nothing against it and I admire it. I cannot give you my final approval before I have had a chance to calculate with it" (De Sitter to Einstein, March 15, 1917 [CPAE 8, Doc. 312]). Five days later, De Sitter had done his calculations. They had led him to an alternative solution of Einstein's amended field equations. He communicated this result in a letter to Einstein,[97] which served as the blueprint for a paper submitted to the Amsterdam academy shortly thereafter (De Sitter 1917a).

Following a suggestion by Ehrenfest, De Sitter considered a natural analogue of the cylinder universe in which time is treated in a similar way as the three spatial dimensions (ibid., 1219, note 1). This *De Sitter universe* has the space-time geometry of the 3+1D hypersurface of a hyper-hyperboloid in 4+1D Minkowski space-time. It is therefore also known as the *hyperboloid universe*. Figure 6.9 shows a lower-dimensional version of this space-time, the 1+1D surface of a

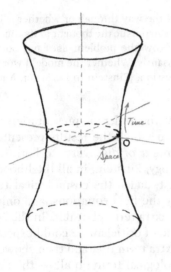

Figure 6.9. *De Sitter's hyperboloid universe.* Drawn by Laurent Taudin.

hyperboloid embedded in 2+1D Minkowski space-time. All points on the hyperboloid have the same spatiotemporal distance to its center in the embedding space (the origin of the coordinate axes shown in the figure). A hyperboloid in 2+1D Minkowski space-time is thus the analogue of a sphere in three-dimensional Euclidean space.[98]

As Einstein had done for the cylinder universe, De Sitter checked whether the hyperboloid universe is allowed by the field equations with the cosmological term. He found that it is, provided that the radius R of the "waist" of the hyperboloid satisfies the relation $\lambda = 3/R^2$ and the mass density ρ equals zero everywhere. De Sitter had thus found a vacuum solution of the new field equations.

This defeated the purpose of Einstein's introduction of the cosmological term. The inertia of test particles in De Sitter's hyperboloid universe is due to space-time rather than to their interaction with distant matter. It was crucial for Einstein's new attempt to implement the relativity of arbitrary motion that this be impossible. As he wrote to De Sitter: "It would be unsatisfactory, in my opinion, if a world without matter were possible. Rather, it should be the case that the $g_{\mu\nu}$-field is *fully determined by matter and cannot exist without the latter.* This is the core of what I mean by the requirement of the relativity of inertia" (Einstein to De Sitter, March 24, 1917 [CPAE 8, Doc. 317]). De Sitter got Einstein's permission to quote this passage in a postscript to his

paper (De Sitter 1917a). The second sentence is the first explicit statement of what Einstein (1918e) dubbed "Mach's principle" the following
year. If this principle were satisfied, absolute motion would finally be
eradicated. A body's motion is defined with respect to the metric field.
If Mach's principle is true, this field is nothing but an epiphenomenon
of matter and all talk about motion with respect to it is nothing but a
façon de parler about motion with respect to the matter generating it
(Maudlin 1990, 561). Vacuum solutions were therefore anathema and
Einstein immediately set out to find grounds to dismiss the one De
Sitter had purportedly found.

Einstein eventually fastened onto the so-called *static form* of the
solution, an alternative way of mapping the hyperboloid universe in
which it can more readily be compared to Einstein's cylinder universe
(De Sitter 1917b, 1917c).[99] The hyperboloid universe looks anything but
static in Figure 6.9. Consider horizontal cross-sections of the hyperboloid. These circles represent space at different times. Going from the
distant past to the distant future, we see that these circles get smaller
until we reach the waist of the hyperboloid and then get larger again. It
thus looks as if the hyperboloid universe contracts and then reexpands.
One has to keep in mind, however, that this conclusion is based on an
arbitrary choice of space and time coordinates.

Figure 6.10 shows an alternative coordinatization of the hyperboloid
universe. In these *static coordinates*, De Sitter's universe, shown on the
right, is represented by a cylinder, just as Einstein's, shown on the left.
In both worlds, space is represented by a circle of radius R at all times.
In these coordinates, the spatial part of the metric field of the De Sitter
universe is exactly the same as that of the metric field of the Einstein
universe in its standard coordinatization. Only the temporal parts are
different.

Compare the component g_{44} of the two metric fields, the square of
the conversion factor $\sqrt{g_{44}}$ from coordinate time to proper time, at $t = 0$.
The situation will be the same for any other value of t. Space at $t = 0$ is
represented by the circles through O and P, the positions at that time of
an observer and of the "horizon" or "equator" for that observer, respectively. In the Einstein universe, the time conversion factor is the same
everywhere: $\sqrt{g_{44}} = 1$. For all points on the circle, one unit of proper time,
represented by the vertical line segments in Figure 6.10, corresponds to
one unit of coordinate time. In the De Sitter universe, the time conversion factor varies from point to point: $\sqrt{g_{44}} = \cos(r / R)$ (where the
distance r from point O runs from 0 to πR). It is equal to 1 for $r = 0$,
then steadily decreases until it vanishes at the horizon P at distance
$r = (\pi/2)R$. As indicated on the right in Figure 6.10, when we go from
O to P, segments of coordinate time of increasing length correspond to

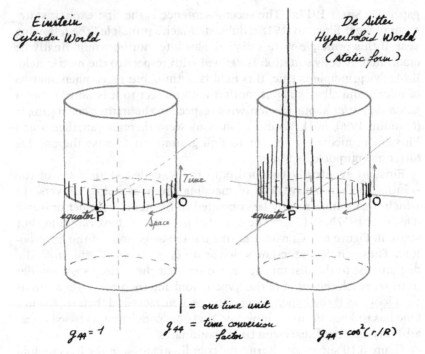

Figure 6.10. *Comparing Einstein's cylinder universe and De Sitter's hyperboloid universe.* Drawn by Laurent Taudin.

one unit of proper time. At the horizon *P* we need a segment of infinite length.

Einstein used this odd behavior of the temporal component of the metric to argue that the De Sitter world is not empty after all. That the vertical line segments in the drawing on the right in Figure 6.10 get longer and longer as we go from *O* to *P* means that it takes an increasing amount of coordinate time for a clock to advance one unit of proper time. It thus *looks as if* clocks are slowing to a crawl as they approach the horizon. This is reminiscent of the gravitational redshift experienced by clocks brought ever closer to some massive object (see Section 2). Einstein concluded that a large amount of matter must be tucked away at the horizon in the De Sitter universe. The main difference between the Einstein universe and the De Sitter universe, he thought, was that in the former matter was spread out evenly, while in the latter it was concentrated at the horizon.

On the postcard on which Einstein first spelled out this line of reasoning, De Sitter scribbled in the margin: "That would be distant masses yet again!"[100] And on the back he elaborated: "How large does the 'mass' of this matter have to be? I suspect ∞! I do *not* adopt such matter as ordinary matter. It is *materia ex machina* to save Mach's dogma" (ibid., my italics; the pun – ex *Mach*ina – was probably unintended). For all the exasperation one senses in these comments, De Sitter could not put his finger on the error in Einstein's argument.

The analysis of the static form of the De Sitter solution strengthened Einstein in his belief that the field equations with cosmological term do not allow vacuum solutions. In March 1918, he submitted two short papers in response to De Sitter's challenge to his latest attempt to eliminate absolute motion. In the first, Einstein (1918e) reworked the foundations of his theory[101] and officially introduced Mach's principle. In the second, he conjectured that the De Sitter solution, an apparent counterexample to Mach's principle, "may not correspond to the case of a matter-free world at all, but rather to that of a world, in which all matter is concentrated on the surface $r = (\pi/2)R$:[102] this could well be proven by considering the limit of a spatial matter distribution turning into a surface distribution" (Einstein 1918c, 272). When he wrote these lines, Einstein may have known that such a proof was already in the works. It appeared a few months later in Weyl's book, *Space-time-matter* (Weyl 1918b, Sec. 33), galleys of which Einstein was receiving in installments.[103]

At the end of May 1918, on the very same day that Einstein sent Weyl a letter in which he expressed his satisfaction over the latest version of this proof, another mathematician, Felix Klein, sent Einstein a letter in which he showed that the singular behavior of the metric field of the De Sitter world in static coordinates is just an artifact of those coordinates.[104] It may come as a surprise that this had not been clear to all parties involved right away. As we saw, De Sitter had found his solution by considering a completely regular hypersurface embedded in a 4+1D Minkowski space-time. It follows that any singularity in any coordinate representation of the solution has to be a coordinate singularity and cannot be an intrinsic singularity (cf. the poles in the example in Figure 6.4). Einstein and De Sitter had, in fact, recognized the degeneration of the metric field of the hyperboloid universe in *other* coordinates as pathologies of those coordinates.[105] And in his paper on the De Sitter solution, Einstein (1918c) had taken a significant first step toward formulating a sensible criterion to distinguish intrinsic singularities from coordinate singularities.[106] Yet, despite all of this, Einstein did not immediately appreciate Klein's point. In his response he wrote that Weyl had just furnished the proof for his conjecture that there must be a large amount of mass at the horizon of the hyperboloid universe.[107]

In his next letter, Klein reiterated the point of the previous one in simpler terms.[108] Klein's reasoning is illustrated in Figure 6.11.[109] The figure shows geometrically how to get from the original hyperboloid (Figure 6.9) to the static form of the De Sitter solution (Figure 6.10). This is done through a clever choice of time slices of the hyperboloid. Imagine that the plane cutting the hyperboloid horizontally at the waist, that is, the plane through the circle with O and P on the left in Figure 6.11, can pivot around the coordinate axis of the embedding space-time going through P. Rotate this plane from −45° to +45° around this axis and let its successive cross-sections with the hyperboloid represent time slices from past to future infinity. In the figure, these cross-sections look like ellipses that get ever more elongated as their angle with the horizontal plane increases until they degenerate into a pair of parallel lines for angles of ±45°. In terms of the metric of Minkowski space-time, however, for all angles between −45° and +45°, they have the exact same shape as the circle that forms the hyperboloid's waist (recall that all points on the hyperboloid have the same spatiotemporal distance to its center in the embedding space-time). Stacking up these circles, we arrive at the cylinder mantle on the right in Figure 6.11, which is just the static form of the De Sitter solution.

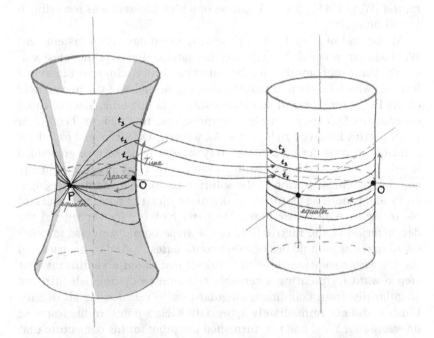

Figure 6.11. *Mapping a double-wedge-shaped region of De Sitter's hyperboloid universe onto a cylinder.* Drawn by Laurent Taudin.

As Figure 6.11 shows, these static coordinates only cover the shaded double-wedge-shaped region of the hyperboloid. More importantly for our purposes, we can now see why the time conversion factor in these coordinates vanishes at the two points on the edge of these wedges. The time slices all intersect at P (like the lines of equal longitude on Earth at the poles). This one point on the hyperboloid thus gets mapped onto a vertical line on the cylinder mantle (like the horizontal line representing the poles on the map in Figure 6.4). The distance between different points on this line needs to be multiplied by zero to reflect that they all represent the same point P on the hyperboloid. This is why $g_{44} = 0$ at P. There is nothing special about P. We could go through the exact same argument using a different set of axes in the embedding spacetime and g_{44} would be zero at some other pair of points. Contrary to what Einstein and Weyl believed at the time, there is no mass anywhere in the De Sitter universe.

To Einstein's credit, he immediately accepted this dire consequence of Klein's analysis once Klein had explained it to him in terms he understood. On the half-empty verso of Klein's letter, Einstein drafted his response. Testifying to his supreme surefootedness as a writer, the draft does not contain a single deletion and is virtually identical to the actual letter sent a few days later. The letter begins:

You are completely right. The De Sitter world in and of itself is free of singularities and all its points are equivalent. A singularity only arises from the substitution which gives the transition to the static form of the line element.... My critical comment on the De Sitter solution stands in need of a correction; there actually is a singularity-free solution of the gravitational equations without matter. (Einstein to Klein, June 20, 1918 [CPAE 8, Doc. 567])

Einstein then retreated to the position that the De Sitter solution could still be ruled out as a model of our universe precisely because it cannot be turned into a static model without the introduction of a singularity.

Einstein never did publish a correction to his critical note on the De Sitter solution. But he lost his enthusiasm for Mach's principle – and for the cosmological constant that had been the price he paid for it – once he had been forced to admit that the De Sitter solution is a counterexample. Looking back at this period the year before he died, Einstein wrote:

In my view one should no longer speak of Mach's principle at all. It dates back to the time in which one thought that the "ponderable bodies" are the only physically real entities and that all elements of the theory which are not completely determined by them should be avoided. (I am well aware of the fact that I myself was long influenced by this *idée fixe*). (Einstein to Felix Pirani, February 2, 1954)[110]

Although this statement dates from a much later period, the disenchantment with Mach's principle can already be discerned in "Ether and Relativity," in which Einstein (1920j) presented the metric field as a new kind of ether, thus abandoning the requirement that the metric field be reducible to matter. This development was greeted enthusiastically by De Sitter (Hoefer 1994, 329).

This marks the end of Einstein's crusade against absolute motion. After four failed attempts he finally threw in the towel. Around 1920, he embarked on a new project, the unification of the inertio-gravitational field and the electromagnetic field through the extension of general relativity in various different directions.[111] This project he pursued until his dying days (Pais 1982, 479).

6. POSTMORTEM: HOW EINSTEIN'S PHYSICS KEPT HIS PHILOSOPHY IN CHECK

It should be clear by now that general relativity does not generalize the relativity principle of special relativity from uniform to nonuniform motion. The combination of the equivalence principle and general covariance leads to what can be called the relativity of the gravitational field – the recognition that an effect due to gravity for one observer can be due to inertia for another – not to the relativity of arbitrary motion. Einstein's theory also does not vindicate Mach's suggestion that Newton's bucket experiment can be accounted for in terms of relative motion with respect to distant matter. Nor is the theory such that the metric field can be reduced to its material sources, as demanded by what Einstein called Mach's principle. General relativity thus failed to fulfill many of the high hopes Einstein had nourished during the long years he had spent in search of it. The consoling thought in all of this is that Einstein had found a tremendously successful new theory of gravity.

The analysis so far may have left the impression that it was purely serendipitous that Einstein arrived at this theory at the end of his journey. Many of the guideposts he had relied on along the way had, after all, listed a destination that was nowhere to be found. The aim of this concluding section is to dispel this impression. I want to highlight three factors that help explain the success of Einstein's search for a new theory of gravity despite the failure of many of his philosophical objectives. First, Einstein did not just want to eliminate absolute motion, he also wanted to reconcile some fundamental insights about gravity with the results of special relativity and integrate them in a new broader framework. Second, when these efforts led him to the introduction of the metric field, he carefully modeled its theory on the successful theory of the electromagnetic field of Maxwell and Lorentz. Third, whenever his philosophical agenda clashed with sound physical principles, Einstein jettisoned parts of the former instead of compromising the latter. In short, throughout his quest for general relativity, Einstein checked whether the philosophical goals he had set himself could be realized in a physically sensible theory.

Einstein's later recollections, especially those in the lecture in Glasgow mentioned at the beginning of this chapter (Einstein 1933c, cf. note 2), leave little doubt that his interest in gravity predated his hope that the subject might hold the key to the relativity of arbitrary motion. Special relativity made Newton's theory of gravity unacceptable. Like other physicists around 1905, Einstein sought to replace this theory, based on instantaneous action-at-a-distance, by a new theory in which, as in the electrodynamics of Maxwell and Lorentz, action is mediated by fields propagating with the speed of light. Working out the law for the force this field exerts on a test particle, Einstein arrived at "a result which raised my strong suspicions. According to classical mechanics, the vertical acceleration of a body in the vertical gravitational field is independent of the horizontal component of its velocity.... But in the theory I advanced, the acceleration of a falling body was not independent of its horizontal velocity" (Einstein 1933c, 286–7).[112] The acceleration of a falling body will likewise depend on the horizontal velocities of its constituent parts and thus on "the internal energy of a system" (ibid.). This is at odds with Galileo's principle that the acceleration of free fall is the same for all bodies. Recognizing this conflict, Einstein seems to have had an epiphany: "This law, which may also be formulated as the law of the equality of inertial and gravitational mass, was now brought home to me in all its significance. I was *in the highest degree amazed* at its existence and guessed that *in it must lie the key to a deeper understanding of inertia and gravitation*"[113] (Einstein 1933c, 287, my emphasis). Einstein's interest thus shifted from the conflict between special relativity and Newtonian action-at-a-distance, on which his contemporaries continued to focus, to the conflict between special relativity and Galileo's principle (Renn 2007b, 61). Einstein quickly gave up on the attempt to develop a theory of the gravitational field within the framework of special relativity. Such a theory, he felt "clearly failed to do justice to the most fundamental property of gravitation" (Einstein 1933c, 287). What he would come to call the equivalence principle would have to be the cornerstone of a truly satisfactory new theory of gravity. In his Glasgow lecture, as in the article intended for *Nature* of 1920 (CPAE 7, Doc. 31; cf. notes 30 and 32), Einstein still presented the equivalence principle as intimately connected with the relativity of arbitrary motion, but that persistent and unfortunate association does not diminish its value as a constraint on Einstein's theorizing about gravity.

In Einstein's 1912 theory for static gravitational fields, a variable speed of light plays the role of the gravitational potential. Einstein thus gave up one of the two postulates of special relativity, the light postulate, in his effort to extend the other, the relativity postulate (Einstein 1912h, 1062). From the point of view of the *Entwurf* theory, the precursor to general relativity proposed the following year, the variable speed of light of the 1912 theory is one of the components of the metric field. From this point of view, gravity had thus already become part of the fabric of space-time in the 1912 theory. Space-time is no longer the Minkowski space-time of special relativity.

That same year, Nordström (1912) published a paper in which he proposed a theory of gravity that stays within the confines of special relativity. In this theory, gravitational interaction, like electromagnetic interaction, is conceived of in terms of a field in Minkowski space-time. In a note added in proof, Nordström (1912, 1129) informs the reader that Einstein had told him (in a letter that is no longer extant) that this theory runs afoul of the general problem with horizontal velocities mentioned: the acceleration of free fall of a body rotating in a horizontal plane would be less than that of the same body without such rotation. Nordström initially shrugged off the objection, insisting the effect would be too small to measure.

Einstein took an active part in the further development of Nordström's theory.[114] A sizable fraction of the first two papers by Nordström (1912, 1913a) on his new theory went into deciding on the quantity that should represent the material source of the gravitational field. Nordström settled on the energy density. In his part of the *Entwurf* paper, Einstein approvingly passed on the suggestion of Max Laue, then at the University of Zurich, that it should be the so-called *trace* of the energy-momentum tensor instead (Einstein and Grossmann 1913, 21).[115] Acknowledging both Einstein and Laue, Nordström (1913b, 533) adopted this suggestion. This established a first parallel between the theories of Einstein and Nordström. In Einstein's theory, the ten independent components of the energy-momentum tensor act as the material source for the ten independent components of the metric field. In Nordström's amended theory, a scalar constructed out of the energy-momentum tensor acts as the material source for the one-component gravitational potential.

Even in the modified version of the Nordström theory, the acceleration of free fall of a body depends on its horizontal velocity, as it must in any special-relativistic theory of gravity (see note 112). The Einstein-Laue amendment, however, did remove the dependence of the acceleration on a body's rotation and on the kinetic energy of its constituent particles. The general treatment of stressed bodies by Laue (1911a, 1911b), which Nordström (1913a) had already made extensive use of, shows that such dependence disappears once the internal forces that keep a body from flying apart are taken into account.[116] This illustrates a more general point. A new theory of gravity had to incorporate the insights of special relativity and these insights went well beyond the prohibition against instantaneous action-at-a-distance or the inclusion of gravity among its own sources required by the equivalence of mass and energy. The work by Laue and others on the relativistic mechanics of continua, in which the (stress-)energy-momentum tensor takes center stage, was especially important in this regard. In an unpublished review article on special relativity written in 1912, Einstein aptly called this development "the most important new advance in the theory of relativity" (CPAE 4, Doc. 1 [p. 63]).[117]

Based on one gravitational potential and flat space-time, Nordström's theory was much simpler than the *Entwurf* theory with its ten gravitational potentials and curved space-time. In defense of his own more complicated theory, Einstein concocted a clever thought experiment showing that Nordström's theory violated energy conservation, albeit only under highly artificial circumstances (Einstein and Grossmann 1913, 21–2).[118] Einstein conceded, however, that his main reason for preferring the *Entwurf* theory was that it generalized the relativity principle to arbitrary motion (ibid.). This was a remarkable admission. To generalize the relativity principle, Einstein thought, a theory of broad covariance was needed (see Section 3). Yet so far he had been unable to establish whether the limited covariance of the *Entwurf* theory was broad enough for his purposes. That he nonetheless preferred his own theory over Nordström's shows that he had the courage of his convictions; that he carefully examined and even contributed to the strengthening of Nordström's theory shows that he was not dogmatic about them.

This same open-mindedness is on display in a lecture that Einstein (1913c) gave in Vienna in September 1913. Einstein compared and contrasted the Nordström theory and the *Entwurf* theory, giving roughly equal time to both. He had meanwhile found a way to restore energy conservation in his competitor's theory. This made it a perfectly viable alternative to the *Entwurf* theory. True to his belief that Galileo's principle held the key to a new theory of gravity, Einstein had no interest in theories in which this principle does not hold. For this reason, he made no mention of the gravitational theory proposed by Mie (1913), who predictably took exception to the omission in question time (Einstein et al. 1913, 1262–3).[119]

To decide between the Nordström theory and the *Entwurf* theory empirically, Einstein (1913c, 1262) told his audience in Vienna, one had to wait for a solar eclipse. In Nordström's special-relativistic theory, light propagates in straight lines at constant speed. It is not bent by gravity. Nordström's theory thus respects the equality of inertial and gravitational mass but does not implement the equivalence principle.[120] After all, it follows directly from the latter that gravity does bend light (see Section 2). The *Entwurf* theory predicts an effect half the size of that predicted by general relativity (Einstein 1915h, 834). Einstein expressed the hope that the solar eclipse of August 1914 would bring the decision between the two theories.[121] In the meantime, other arguments would have to do. Whereas in the *Entwurf* paper, Einstein had presented his theory's broader covariance as its main advantage, he now pointed to the relativity of inertia, which, he argued, was realized in the *Entwurf* theory but not in the Nordström theory (Einstein 1913c, 1260–1). This reflects the shift in Einstein's strategy for eliminating absolute motion between May and September 1913 (cf. the quotation at the beginning of Section 4). In the present context, the important point is that, both in the *Entwurf* paper in May and in the Vienna lecture in September, Einstein argued that philosophical considerations gave his theory the edge over Nordström's while acknowledging that in terms of more mundane physical considerations it was a toss-up.

The following year, Einstein produced a much stronger argument in favor of moving beyond special relativity. He showed that Nordström's theory could readily be reformulated as a theory in which, as in his own theory, gravity is incorporated into the space-time structure. This possibility was first brought out by the escape Einstein found from his own argument against the Nordström theory in the *Entwurf* paper. Einstein communicated this escape to Nordström, who presented it in his next paper on his theory, dutifully acknowledging his source (Nordström 1913b, 543–5).[122] Einstein argued that the only way to guarantee energy conservation in the Nordström theory was to assume a universal dependence of the dimensions of physical systems and the duration of physical processes on the gravitational potential.[123] Because of this universality, clocks and rods would no longer measure times and distances in the flat Minkowski space-time posited by the theory but times and distances in some curved space-time.[124]

Early in 1914, in a joint paper with Lorentz's former student Adriaan D. Fokker, Einstein reformulated Nordström's theory using Riemannian geometry (Einstein and Fokker 1914). In this reformulation of Nordström's theory, as in Einstein's own theory, the metric field describes both the gravitational potential and the space-time geometry. The metric field in the Nordström theory is determined by a generally covariant equation and an additional condition.[125] The generally covariant equation sets the so-called curvature scalar, a quantity constructed out of the Riemann curvature tensor involving first- and second-order derivatives of the metric field, equal to the trace of the energy-momentum tensor. The structure of this equation is thus similar both to the *Entwurf* field equations and to the Einstein field equations. Unlike these equations, however, the equation in the Nordström theory only has one component. To determine the ten independent components of the metric, one needs the additional condition that for any metric field $g_{\mu\nu}(x^\rho)$ allowed by the theory there is a coordinate system in which it can be written in the simple form $\varphi(x^\rho)\eta_{\mu\nu}$, where $\eta_{\mu\nu}$ are the constant components of the metric for Minkowski space-time in the standard form used by inertial observers in special relativity and $\varphi(x^\rho)$ is a so-called *conformal factor*. This extra condition guarantees that the velocity of light is a constant, as it should be in a special-relativistic theory. It is also what rules out the light bending required by the equivalence principle. The conformal factor is just the gravitational potential in the original formulation of the Nordström theory. The field equation of the original formulation is recovered if this special form of the metric field is inserted into the generally covariant equation.

The reformulation of the Nordström theory by Einstein and Fokker shows that even this most satisfactory of special-relativistic theories of gravity eventually leads beyond special relativity. As Norton (1992b, 1993a) as well as Giulini and Straumann (2006, Sec. 5) emphasize, the new formulation turns gravity from a field in flat Minkowski space-time to part of the fabric of curved space-time. As Einstein and Fokker (1914, 321) put it themselves, their reformulation shows that the Nordström theory is covariant under a group of transformations broader than the class of Lorentz transformations characterizing special relativity. In Einstein's thinking at the time, this was tantamount to a generalization of the relativity principle. The main difference between Nordström's theory and his own *Entwurf* theory then was that the latter not only extended the relativity principle but also implemented the equivalence principle, reducing the equality of inertial and gravitational mass to an essential unity – a *Wesensgleichheit* (see note 35) – of gravity and inertia.

Regardless of how the point is made, the recasting of Nordström's theory in terms of Riemannian geometry bolstered Einstein's confidence that he was on the right track with a theory like the *Entwurf* theory based on the metric tensor. Considerations of how to reconcile the physical insights represented by Galileo's principle and special relativity, which had led to Einstein's interest in Nordström's theory in the first place, ended up pointing in the same direction as the considerations about extending the relativity principle that had guided Einstein in his formulation of the *Entwurf* theory.

As noted in Section 3, Einstein gave up the search for field equations of broad covariance in 1913 because he could not find any that were compatible both with energy-momentum conservation and with the results of Newtonian theory in the case of weak static fields. When he finally did publish field equations of broad and eventually general covariance in 1915, Einstein accordingly made sure that they passed muster on both counts. What I did not mention so far is that Einstein used these requirements not just to check whether they were met by candidate field equations he was considering but also to generate candidates specifically designed to meet them. This is how Einstein arrived at the *Entwurf* field equations in the Zurich Notebook (Renn 2007a, Vol. 2, 706–11). Like the relativity principle and the equivalence principle, these physical principles thus guided Einstein in his theory building.

In a similar vein, Einstein relied strongly on the analogy with electrodynamics, both for the further elaboration of the *Entwurf* theory and for the transition to the new theory in November 1915.[126] Much of Einstein's work on the *Entwurf* theory in 1913–14 went into recasting it in a form in which it could readily be compared with electrodynamics. This is nicely illustrated by the Vienna lecture. Einstein (1913c, 1249–50) began by explaining that one should expect the transition from Newton's theory to a new theory of gravity to be similar to the transition from Coulomb's electrostatics to Maxwell's electrodynamics. In the body of the lecture, Einstein presented the *Entwurf* field equations in a form that closely matches the field equations for the electromagnetic field. He consistently used the equations in this form in subsequent publications (Janssen and Renn 2007, 847). Like Maxwell's equations, the *Entwurf* field equations in this new form set the divergence of the field equal to its source.[127] The equations governing the transfer of energy-momentum between the gravitational field and matter can likewise be written in a form that is similar to the corresponding equation in the case of the electromagnetic field. It was on the basis of these parallels that Einstein initially identified the gravitational field as the gradient of the metric field (see Section 3).

The following year, Einstein developed a more general formalism to analyze various properties of the *Entwurf* field equations (Einstein 1914o, Einstein and Grossmann 1914b).[128] He derived a set of conditions in this formalism that, on the one hand, determine under which (nonautonomous) transformations the field equations are invariant and, on the other, ensure that the field equations imply energy-momentum conservation.[129] The central quantity in this formalism is the so-called Lagrangian. Specification of the Lagrangian for the gravitational field is tantamount to the specification of the vacuum field equations. Einstein modeled the Lagrangian for the gravitational field in the *Entwurf* theory on the Lagrangian for the electromagnetic field in Maxwell's theory. It is essentially the same quadratic expression in the components of the field in both cases.

When, sometime in October 1915, Einstein finally came to accept that the rotation metric is not a vacuum solution of the *Entwurf* field equations (see Section 4), he held on to his general formalism, including the expression for the Lagrangian in terms of the gravitational field. He only changed the definition of the field from the gradient of the metric to the Christoffel symbols (see Section 3). The resulting new field equations were of broad covariance. Purely mathematical considerations had already led Einstein to consider these equations three years earlier. They can be found in the Zurich Notebook. At that time, physical considerations had steered Einstein away from these equations and toward the *Entwurf* field equations. Now the formalism that Einstein had developed for the *Entwurf* theory, relying heavily on the analogy with electrodynamics, not only led him back to the equations rejected earlier, but also provided him with all the guidance he needed to demonstrate that they are compatible with energy-momentum conservation after all. Moreover, the connection between energy-momentum conservation and the covariance of the field equations, one of the key insights enshrined in his general formalism, gave Einstein the decisive clue for solving the other problem that had defeated him before, namely to show that these field equations reproduce the results of Newtonian theory in the case of weak static fields. With both these problems taken care of, Einstein rushed his rediscovered field equations into print (Einstein 1915f). Within days he realized that they were still not quite right. Guided once again by his general formalism, Einstein fixed the remaining problems in two further communications to the Berlin academy in November 1915 (Einstein 1915g, 1915i).

This whole chain of events was triggered by Einstein's redefinition of the gravitational field. One can thus understand his assessment at the time that the old definition had been a "fateful prejudice" (Einstein 1915f, 782) and that the new one had been the "key to the solution."[130] Einstein later downplayed the importance of the physical considerations encoded in his general formalism for the transition from the *Entwurf* field equations to the Einstein field equations. The way he came to remember it was that he had chosen the new equations purely on grounds of mathematical elegance (Janssen and Renn 2007, Sec. 10).

Ultimately, it was probably the convergence of physical and mathematical lines of reasoning that reassured Einstein that the field equations of his fourth communication of November 1915 were the right ones. Confident that no further corrections would be needed, he could afford to poke fun at the way victory had at long last been achieved.

As he told Ehrenfest in late December: "It is convenient with that fellow Einstein, every year he retracts what he wrote the year before."[131] This self-deprecating comment nicely captures the flexibility we have seen Einstein exhibit at several junctures on his road to the new theory. Three days later, Einstein likewise told his Polish colleague Władysław Natanson: "I once again toppled my house of cards and built a new one."[132] In terms of the philosophical objectives that had guided Einstein in the search for his theory, the new structure turned out to be yet another house of cards.[133] As a physical theory, however, it has proved to be remarkably sturdy and durable.

ACKNOWLEDGMENTS

This essay builds on a couple of earlier attempts to provide a concise account of the crusade against absolute motion and absolute space that fueled the development of Einstein's general theory of relativity (Janssen 2004, 2005). I also drew heavily on my contributions to *The Genesis of General Relativity* (Renn 2007a, Vols. 1–2) and the Einstein edition (CPAE 4, 7, and 8). For discussion, comments, references, and encouragement, I want to thank Mark Borrello, John Earman, Michael Friedman, Hubert Goenner, Lee Gohlike, Jan Guichelaar, Geoffrey Hellman, David Hillman, Don Howard, Ted Jacobson, Christian Joas, Dan Kennefick, Anne Kox, Dennis Lehmkuhl, Christoph Lehner, Charles Midwinter, John Norton, Antigone Nounou, Oliver Pooley, David Rowe, Rob "Ryno" Rynasiewicz, Catharine Saint Croix, Tilman Sauer, Robert Schulmann, Chris Smeenk, Jim Smoak, John Stachel, Roberto Torretti, Bill Unruh, Jeroen van Dongen, Christian Wüthrich, and, especially, Jürgen Renn. Special thanks are also due to Laurent Taudin, who patiently drew all the diagrams. Generous support for work on this essay was provided by the *Max-Planck-Institut für Wissenschaftsgeschichte* in Berlin.

NOTES

1 See Chapter 2 as well as the Appendix.
2 Otherwise, Einstein's introduction of general relativity here is similar to that in the George A. Gibson foundation lecture that he gave at the University of Glasgow over a decade later on June 20, 1933. In the published text of this lecture we read: "only a relative meaning can be assigned to the concept of velocity" and "[f]rom the purely kinematical point of view there was no doubt about the relativity of all motions whatsoever" (Einstein 1933c, 286). Page references are to the reprint in Einstein (1954). Cf. note 14.
3 See also Fock (1959, xviii). Bondi's remarks and similar ones by Synge, another leading relativist of the same era (see note 36), are quoted and discussed by Schücking and Surowitz (2007, 19). In his article Bondi tried to preempt charges of sacrilegiousness: "One may surely admire and embrace

Einstein's theory of gravitation while rejecting his route to it, however heuristically useful he himself found it" (Bondi 1979, 180). A few years after the publication of general relativity, Ernst Reichenbächer (1920) had already published an article with a title similar to Bondi's: "To What Extent Can Modern Gravitational Theory Be Established without Relativity?" In his response, Einstein (1920k) essentially argued not for the relativity of arbitrary motion but for something more appropriately called the *relativity of the gravitational field* (see Section 2, note 59, and Janssen 2012).

4 Einstein (1918k) produced an account of the twin paradox along these lines (Janssen 2012, 162–9). Gustav Mie used the example of the passenger in the accelerating train to criticize this way of extending the relativity of uniform motion to accelerated motion in the discussion following a lecture by Einstein in Vienna in 1913 (Einstein et al. 1913, 1264; cf. note 59). See also Weyl (1924, 199; cf. note 34).

5 In Section 3, we shall see how Einstein came to equate general covariance with general relativity. For an insightful review of the decades-long debate over the status of general covariance, see Norton (1993b).

6 For expressions of his strong confidence in the theory at this point, see Einstein to Heinrich Zangger [after December 27, 1914], and January 11, 1915 (CPAE 10, Vol. 8, Doc. 41a and Vol. 8, Doc. 45a).

7 See Sauer (1999, 2005b), Renn and Stachel (2007), and Brading and Ryckman (2008) for comparisons of the relevant contributions of Einstein and Hilbert.

8 For a reconstruction of these developments, see Janssen and Renn (2007).

9 Along with the Princeton lectures (Einstein 1922c) and his popular book on relativity (Einstein 1917a), this is Einstein's best-known exposition of general relativity. It is included in *The Principle of Relativity*, an anthology still in print today (Einstein et al. 1952). For detailed commentary, see Janssen (2005) and Sauer (2005a).

10 See Lehner (2005) for a somewhat different take on the changes in the status of and the relation between these three principles in Einstein's thinking in this period.

11 This is nicely captured in the title of a paper by Earman and Glymour (1978), "Lost in the Tensors," even though the paper was quickly superseded by a talk by Stachel in 1980 (eventually published as Stachel 1989b) and a paper by Norton (1984).

12 For more on the checkered history of the cosmological constant, see Chapter 7.

13 The story of Einstein's quest for general relativity thus simultaneously confirms the first and refutes the second part of the observation in Bob Dylan's 1965 song *Love Minus Zero/No Limit* that "there's no success like failure and that failure is no success at all." I used the first part as the title of this chapter.

14 One can argue that absolute motion is already less objectionable in special relativity than it was in Newtonian theory (Dorling 1978). Once again, consider the two passengers whose trains are in nonuniform motion with respect to one another. According to Newtonian theory, these two observers, using ideal rods and clocks (i.e., ideal in the sense of measuring intervals in the space and time posited by Newtonian theory), will arrive at

equivalent *descriptions* of the motion of the other observer, the only difference being the direction of the motion. Yet, the *effects* of the motion (e.g., whether or not the coffee in their cups spills) is different for the two observers. This, Einstein (1916e, 771–3) pointed out, amounts to a violation of the principle of sufficient reason: two motions that look the same have different effects. This, Einstein suggested, is what makes absolute motion so objectionable. If this were all there is to the problem of absolute motion, Einstein had already solved it in 1905 (Dorling 1978). According to special relativity, given the behavior of ideal clocks and rods posited by the theory, the two observers will describe the motion of the other observer differently. Contrary to what Einstein claims in the passages from the lectures in Princeton and Glasgow quoted at the beginning of this chapter, in both special and general relativity it *does* matter, even "from the purely kinematic point of view" or in terms of the "purely geometrical acceleration" (i.e., acceleration as determined by ideal rods and clocks in the space-time posited by the relevant theory) "from the point of view of which body we talk about [nonuniform motion]." Since the two motions under consideration here *look* different according to special relativity, it is not surprising that their effects are different as well. There is no violation of the principle of sufficient reason. For further discussion, see Janssen (2005, 62–6; 2012, 165).

15 As Einstein wrote, for instance, to Vilhelm Bjerknes, November 12, 1920 (CPAE 10, Doc. 201): "the electromagnetic field has turned out to be more fundamental than ponderable masses."

16 For discussion of Einstein's work on unified field theory, see Chapter 9.

17 This analogy nicely illustrates that being a set of relational properties does not make a structure any less real. One need only think of adultery.

18 See, e.g., DiSalle (2006) for an attempt to parse the philosophical debate over these issues in a different way.

19 For accounts of Einstein's work on the perihelion problem, see the editorial note, "The Einstein-Besso Manuscript on the Motion of the Perihelion of Mercury" (CPAE 4, 344–59), Earman and Janssen (1993), and Janssen (2003a).

20 See Earman and Glymour (1980a, 1980b), Kennefick (2009, 2012) (cf. note 43), and CPAE 9 (Introduction, Secs. III–V) for discussions of these two classical tests of general relativity (the third being the prediction of an additional advance of the perihelion of Mercury of some forty-three seconds of arc per century [see note 19]). For a more comprehensive discussion of the response of astronomers to general relativity, see Crelinsten (2006).

21 See Chapter 7.

22 See Chapter 8.

23 See Renn et al. (1997) and Renn and Sauer (2003).

24 Pfister (2007) convincingly argues that Einstein actually deserves most of the credit for what is usually referred to as the Lense-Thirring effect (Lense and Thirring 1918).

25 See Earman and Eisenstaedt (1999) and Earman (1995).

26 See, e.g., Brading (2002) for analysis of Noether's theorems.

27 In Chapter 13, Friedman places the development of general relativity in the context of the history of the philosophy of geometry. For other accounts of

the development of general relativity written for a general audience, see Renn (2006, chs. 5–6) or Eisenstaedt (2006).

28 For a concise overview of the development of general relativity that largely focuses on this strand of the story rather than on the failed quest for general relativity, see Giulini and Straumann (2006).

29 See Norton (1992b, 1993a) and, drawing on this work, Giulini and Straumann (2006, Sec. 5, 145–51). The first of Norton's two papers is reprinted in Renn (2007a, Vol. 3, 413–542) along with translations of the original papers by Nordström (1912, 1913a, 1913b).

30 CPAE 7, Doc. 31 [p. 21]. A literal translation of the German original ("der glücklichste Gedanke meines Lebens") would be: "the happiest thought of my life," where "happy" is to be taken in the sense of "fortunate." Einstein also told this story in a lecture in Kyoto on December 14, 1922 (Abiko 2000, 15; CPAE 13, Doc. 399; and the editorial note, "Einstein's Lecture at the University of Kyoto").

31 To be more precise, this is a particle's *passive* gravitational mass, a measure of how strongly it is attracted by other particles. Its *active* gravitational mass measures how strongly it attracts other particles. These two quantities also have the same numerical value.

32 In the next sentence, Einstein admittedly still suggested that this consideration leads to an extension of the relativity principle to nonuniform motion. Old habits die hard. For further discussion, see Janssen (2002a, 507–8; 2012, 160).

33 The relevant passage is quoted and discussed in Section 4.1 of Chapter 2, where a variant of the thought experiment shown in Figure 6.2 is analyzed.

34 In a remarkable semipopular article, Hermann Weyl (1924, 198–9) put the notion of what he called a "guiding field" that cannot be split uniquely into inertial and gravitational components at the center of his discussion of the foundations of general relativity.

35 As Einstein (1918e, 241) put it, "Inertia and gravity are of the exact same nature." Six years earlier, Einstein (1912h, 1063) had already written about the equivalence of inertial and gravitational mass and the equivalence of a static gravitational field and the acceleration of a frame of reference using the same term, *wesensgleich*, which I translated as "of the exact same nature" (Norton 1992b, 447, note 42). Page references to Norton (1992b) are to the reprint in Renn (2007a, Vol. 3).

36 In the preface of his textbook on general relativity, J. L. Synge admitted that he had never been able to define the equivalence principle in a way that would not make it either trivial or false, but he still recognized its heuristic value: "The Principle of Equivalence performed the essential office of midwife at the birth of general relativity.... I suggest that the midwife be now buried with appropriate honours and the facts of absolute space-time faced" (Synge 1960, ix–x). He spoke for many when he observed that "the word 'relativity' now means primarily Einstein's theory and only secondarily the obscure philosophy which may have suggested it originally" (ibid., ix). Cf. the comment by Bondi quoted in the introduction.

37 For our purposes, we can still think of a merry-go-round even though that involves an additional gravitational field perpendicular to the plane of the disk. Since we only consider what happens in the plane of the disk, this does not matter.

38 For objects *moving* with respect to the rotating disk, the centrifugal force is not the only inertial force of rotation. There is also the so-called Coriolis force. In the case of the rotation of the Earth, this is the force responsible, for instance, for the whirling around of air in hurricanes and tornadoes. In addition to the field corresponding to the centrifugal force, the gravitational field corresponding to rotation will thus have components corresponding to the Coriolis force that are not shown in the second drawing of Figure 6.3. From notes taken by Hans Reichenbach, one of his students at the time, we know that Einstein discussed the role of the Coriolis force when he covered the example of the rotating disk in lectures in Berlin in 1919 (CPAE 7, Doc. 19, note 6). As Einstein noted and as can be seen from this example, if the equivalence principle holds, the gravitational field cannot be represented by the gradient of an ordinary one-component potential.

39 See also Einstein 1959, Appendix III, first added to the English translation of Einstein (1917a) in 1920 (CPAE 6, 538, note 83).

40 See Section 2.7 in the Appendix for an analysis of the twin paradox. The appendix also provides elementary explanations of time dilation and length contraction.

41 The experimental verification of the effect was much more contentious. See Hentschel (1993), Pound (2000, 2001), and CPAE 9, xxxvii–xl.

42 The effect is caused by the component of the gravitational field corresponding to the Coriolis force (cf. note 38).

43 Kennefick (2009, 2012) convincingly argues against earlier scholars – notably, Earman and Glymour (1980a) – that the data obtained did bear out Einstein's prediction.

44 See also, e.g., Einstein's lectures in Madrid in March 1923 (CPAE 13, Appendix H, 880–1).

45 Since Stachel's paper, additional documents pertaining to Einstein and the example of the rotating disk have come to light (Sauer 2008).

46 Although they do not mention the argument explicitly, Salzman and Taub (1954, 1662–3) side with Einstein noting that something called the *"intrinsic circumference* of the rotating disk" does not have the Euclidean value of 2π times the disk's radius.

47 For discussion of Abraham's theory and Einstein's criticism of it, see the editorial note, "Einstein on Gravitation and Relativity: The Static Field" (CPAE 4, 122–8) and Renn (2007d).

48 Page references to this paper are to the reprint in Howard and Stachel (1989). Three years earlier, in response to criticism by Max Planck of the definition of constant acceleration in his 1907 review article, Einstein (1908b) had already been forced to accept a restriction of the principle to bodies at rest in the accelerated frame (Schücking and Surowitz 2007, 7). This did not stop Einstein, however, from applying the equivalence principle to the motion of light.

49 In addition to the "rotating disk" argument mentioned in the preceding section, there was one other element that may have pointed Einstein to Gauss's theory of curved surfaces. In a note added in proof to the second paper on his theory for static gravitational fields, Einstein (1912d, 458) noted that the equation of motion for a particle moving in a gravitational field can be given in the form of a simple extremal principle. From the point of view

of the metric theory to be discussed in this section, the path length has a maximum for the path actually taken by the particle. This result may well have suggested to Einstein that motion in a gravitational field is analogous to motion along geodesics (extremal paths) of a curved surface (see CPAE 4, the headnote, "Research Notes on Gravity," 193–4, and, for further analysis, Blum et al. 2012). There is also a page in the so-called Zurich Notebook (more on this notebook below) devoted to geodesic motion along a surface (CPAE 4, 208–9; Renn 2007a, Vol. 2, 592–6).

50 The Göttingen mathematician Felix Klein later noted that this had been a rather one-sided introduction to the field (Renn 2007a, Vol. 2, 611, note 212). For an account of how his collaboration with Grossmann began, see Einstein's Kyoto lecture (Abiko 2000, 16, cf. note 30).

51 Richard Feynman once boasted that he had found his way to the 1957 conference on general relativity at Chapel Hill by asking a cab driver to take him to the same place the man had taken others going "gee-mu-nu, gee-mu-nu" (Feynman and Leighton 1985, 258–9).

52 We already saw in Section 2 that, if the equivalence principle is to hold, the potential must have more than one component (see note 38).

53 Gaussian curvature would be meaningful to critters constrained to the surface, such as ants crawling along on it. Without leaving the surface, they could ascertain that they live on a curved surface by measuring the angles of triangles drawn on it and noting that these angles do not add up to π.

54 Since space-time is locally Minkowskian or pseudo-Euclidean, it is, strictly speaking, only *pseudo-Riemannian*.

55 For an English translation of the key parts of Levi-Civita's (1917) paper on the subject, see Renn (2007a, Vol. 4, 1081–8). With the help of the connection, the old Gaussian interpretation of curvature in terms of *angular excess* of geodetic triangles (cf. note 53) was replaced by the modern interpretation in terms of *parallel displacement* (Janssen 1992). In curved spaces a vector transported parallel to itself around a closed loop will no longer point in the same direction as the original vector.

56 Cf. Appendix, Section 2.5.

57 Norton (1999a), building on some of his earlier work (Norton 1992a), argues that the conflation was the result of the collision in Einstein's work of two different traditions in geometry, one going back to Klein's famous Erlangen program, the other going back to Riemann (Janssen 2005, 61–2).

58 For further discussion of Kretschmann's paper, see Norton (1992a, 1993b) and Rynasiewicz (1999). See Anderson (1967, Secs. 4.2–4.4) for a classic discussion of covariance groups and symmetry groups.

59 Mie (1917) defended a similar view of the role of general covariance in Einstein's theory (see CPAE 8, Doc. 346, note 3). Mie completely agreed with Kretschmann (1917) and recommended the latter's paper to Einstein (Mie to Einstein, February 17, 1918 [CPAE 8, Doc. 465]). Einstein continued to connect general covariance with relativity of motion, both in his correspondence with Mie in 1918 and in his response to Reichenbächer (1920) who presented arguments similar to Mie's (cf. note 3). What Einstein (1920k) argued for in his reply to Reichenbächer, however, is more accurately described as the relativity of the gravitational field than as the relativity of arbitrary motion. See Janssen (2012) for further discussion.

60 Dieks (2006) defends Einstein against the charge of conflating general covariance and general relativity by arguing that his goal was to eliminate preferred frames of references in the sense of laws taking on a special form in them rather than in the sense of their special states of motion. See Janssen (2012) for a critique of Dieks's proposal.

61 The curvature of Minkowski space-time, for instance, is zero regardless of whether we use an inertial or a rotating frame of reference. As Einstein pointed out in response to criticism of his claim that the spatial geometry in a rotating frame in Minkowski space-time is curved, it does not follow that the curvature of any three-dimensional subspace of Minkowski space-time must also have zero curvature (Stachel 1989a, 54–5).

62 See, e.g., Einstein to Max von Laue, September 12, 1950 (EA 16–148), quoted by Stachel (2002a, 256) in a supplementary note to a reprint of his paper on the rotating disk (Stachel 1989a). Einstein likewise saw no problem representing the energy and momentum of the gravitational field by a quantity that is not a tensor. Since the mature equivalence principle makes the presence of a gravitational field coordinate-dependent, it is only natural that its energy and momentum are too. Einstein (1918f) defended his pseudotensor of gravitational energy-momentum against criticism of various colleagues, including Levi-Civita, Lorentz, and Klein. Part of this debate over the pseudotensor is covered by Cattani and De Maria (1993). Trautman (1962) provides a concise overview of subsequent work on energy and momentum conservation in general relativity.

63 This notebook is the centerpiece of Renn (2007a, Vols. 1–2) where it is presented in facsimile with a transcription and a detailed commentary. High-quality scans of the notebook are available at the Albert Einstein Archives.

64 See Janssen and Renn (2007, 913–14) for the relevant passages from letters to Michele Besso, Hilbert, and Arnold Sommerfeld.

65 An embryonic version of the hole argument can be found on a page in Besso's hand dated August 28, 1913 (Janssen 2007). It was first published as part of the comments Einstein (1914d) added to the reprint of the *Entwurf* paper in a mathematics journal (Einstein and Grossmann 1914a).

66 For historical discussion of these arguments, see, e.g., Stachel (1989b), Norton (1984, 1987), Howard (1999), Janssen (2007), and Engler and Renn (2013). For philosophical debate, see, e.g., Earman and Norton (1987), Stachel (1986a, 2002c), Earman (1989), Maudlin (1990), Rynasiewicz (1992, 1994), Brighouse (1994), Howard (1999), Saunders (2003), Rickles and French (2006), and Pooley (2006). For a good introduction to the debate and further references, see the entry on the hole argument in the online *Stanford Encyclopedia of Philosophy* (Norton 2008). For a more recent (but also more selective) review, see the article on the hole argument for the online *Living Reviews in Relativity* (Stachel 2013).

67 Before Stachel (1989b) and Norton (1984), commentators, including Pais (1982, 222) thought that *this* was the indeterminism lurking in the hole argument, and they therefore dismissed Einstein's argument as a beginner's blunder of someone who has just learned Riemannian geometry.

68 This extra step is already present in the embryonic version of the hole argument mentioned in note 65 (Janssen 2007, 821–3).

69 That it is far from trivial to spot the flaw in this argument was forcefully demonstrated by the discovery of page proofs of Hilbert's (1916) first paper on general relativity, which show that even the great mathematician had originally fallen for it (Corry et al. 1997).

70 Einstein (1914d, 260; cf. note 65), Einstein and Grossmann (1914b, 217–18), Einstein (1914e, 178; 1914o, 1067).

71 See Einstein to Ehrenfest, December 26, 1915 and January 5, 1916 (CPAE 8, Docs. 173 and 180) and Einstein to Besso, January 3, 1916 (CPAE 8, Doc. 178).

72 Einstein in all likelihood got the notion of point coincidences from a paper by Kretschmann (1915) that was published just days before Einstein wrote the letter in which the new argument makes it first appearance (Howard and Norton 1993, 54). As can be inferred from the manuscript mentioned in note 65, Einstein had rejected a similar escape from the hole argument two years earlier (Janssen 2007, Sec. 4).

73 Two years later, Einstein elevated this observation to the statement of the relativity principle: "The laws of nature are nothing but statements about spatio-temporal coincidences; they therefore find their only natural expression in generally-covariant equations" (Einstein 1918e, 241).

74 Howard (1999) draws attention to the harmful influence of this reading of the point-coincidence argument in philosophy of science. It was not just readers of Einstein's 1916 review article, however, who interpreted the argument this way. It was also how Lorentz interpreted the original argument in Einstein's letters to Ehrenfest, Lorentz's successor in Leyden. It was this version of the argument that convinced Lorentz of the need of general covariance (Kox 1988, Janssen 1992).

75 The argument, however, does leave the determined substantivalist plenty of wiggle room. First, the account of identity and individuation that it is based on remains controversial: can identity truly be a matter of *suchness* alone or does it always involve some *thisness* as well (Maudlin 1990)? Second, the argument specifically targets substantivalists committed to the reality of bare manifold points as the ultimate carriers of all physical properties. One can argue that this is not the right way to assign physical meaning to bare manifold points (Wilson 1993). Or one could opt for a more sophisticated form of substantivalism that avoids commitment to the reality of bare manifold points. Both moves, however, would seem to end up blurring the distinction between substantivalism and relationism (Rickles and French 2006, 3–4).

76 See Earman (1989, ch. 6) for discussion.

77 This notion had already been explored in Einstein (1912e) and in the Einstein-Besso manuscript (CPAE 4, Doc. 14 [pp. 36–8]; cf. note 19). See also Einstein (1913c, 1261).

78 For other discussions of Einstein's efforts to implement Machian ideas in his new theory of gravity, see, e.g., Barbour (1992, 2007), Hoefer (1994, 1995), Barbour and Pfister (1995), and Renn (2007c).

79 For Newton, the bucket experiment was first and foremost an argument against the Cartesian concept of motion rather than an argument for absolute acceleration (Laymon 1978; Huggett 2000, ch. 7). For discussion of

the responses of Huygens, Leibniz, Berkeley, Kant, Maxwell, Mach, and Poincaré to Newton's bucket experiment, see Earman (1989, ch. 4).

80   Einstein to Lorentz, August 14, 1913 (CPAE 5, Doc. 467).

81   Einstein to Lorentz, January 23, 1915 (CPAE 8, Doc. 47).

82   In the early 1960s, the small increase of inertia that Einstein did find as a result of interaction with distant matter was shown to be an artifact of the coordinates he used (Torretti 1978, 20).

83   The gravitational field will exert both centrifugal and Coriolis forces in this case (cf. note 38). The latter are twice the size of the former and point in the opposite direction, thus keeping the particles in orbit (Janssen 2005, notes 24 and 44; 2012, 170).

84   The degenerate nonphysical values in situation (I) need not bother us here, since we are only interested in the large but finite region occupied by the shell.

85   Thirring to Einstein, July 11–17, 1917 (CPAE 8, Doc. 361).

86   Einstein to Thirring, August 2, 1917 and December 7, 1917 (CPAE 8, Docs. 369 and 405). See also Einstein to Eduard Hartmann [April 27, 1917] (CPAE 8, Doc. 330).

87   What is also puzzling is that Einstein mentioned that he was working on the problem of boundary conditions in a letter to Besso of May 14, 1916 (CPAE 8, Doc. 219), i.e., several months before the exchange with De Sitter.

88   See also the Einstein-Besso manuscript (CPAE 4, Doc. 14 [pp. 18–24, 32–5, 41–2, 45–9]). For further discussion, see Pfister (2007).

89   Another factor, I suspect, is that early critics may have been reluctant to take Einstein to task on this score for fear of being lumped in with antirelativists, whose attacks on Einstein had more nefarious motives (see CPAE 7, 101–13; Rowe 2006; Rowe and Schulmann 2007, ch. 3; and Wazeck 2009, 2013).

90   See the correspondence between Einstein and De Sitter in 1916–18 published in CPAE 8 and the editorial note, "The Einstein-De Sitter-Weyl-Klein Debate" (ibid., 351–7). For other accounts of the developments discussed in this section, see, e.g., Röhle (2002) and Giulini and Straumann (2006, Sec. 6.6).

91   De Sitter to Einstein, November 1, 1916; Einstein to De Sitter, November 4, 1916 (CPAE 8, Docs. 272 and 273).

92   The second of these was the second installment of a trilogy in the Monthly Notices of the Royal Astronomical Society that first introduced British scientists to Einstein's new theory (De Sitter 1916a, 1916c, 1917c).

93   Einstein to De Sitter, February 2, 1917 (CPAE 8, Doc. 293).

94   He had mentioned this possibility the year before in the letter to Besso cited in note 87.

95   De Sitter to Einstein, April 1, 1917 (CPAE 8, Doc. 321).

96   For more detailed discussion of these considerations, see Norton (1999b) and Chapter 7. Drawing on his historical analysis of the difficulties with Newtonian cosmology, Norton (1995, 2003) shows that there is a class of cosmological models in which the arbitrariness of the split between inertial and gravitational effects expressed in the mature equivalence principle amounts to a true relativity of acceleration. The particles in relative acceleration toward one another in these models all move on geodesics.

97   De Sitter to Einstein, March 20, 1917 (CPAE 8, Doc. 313).

98   Cf. Appendix, Sections 2.5 and 2.6 and Figure A.27.

99 See also De Sitter to Einstein, June 20, 1917 (CPAE 8, Doc. 355).

100 Einstein to De Sitter, August 8, 1917 (CPAE 8, Doc. 370).

101 This may be the reason that this paper was published in *Annalen der Physik*, in which Einstein (1916e) had published his big review article, while most of his papers on general relativity during this period appeared in the Proceedings of the Berlin academy (Janssen 2005, 60).

102 In the 1+1D version of the model the horizon consists of two points rather than a two-dimensional surface.

103 Einstein first mentioned these galleys the day after submitting his paper on the De Sitter solution (Einstein to Weyl, March 8, 1918 [CPAE 8, Doc. 476]). Einstein and Weyl only discussed the relevant passage in correspondence in April and May (see, e.g., Einstein to Weyl, April 18, 1918 [CPAE 8, Doc. 511]). Although Weyl thus helped Einstein defend Mach's principle, he later explicitly distanced himself from it (see, in particular, the popular article mentioned in note 34). For further discussion of Weyl's contributions to cosmology, see Goenner (2001).

104 Einstein to Weyl, May 31, 1918, and Klein to Einstein, May 31, 1918 (CPAE 8, Docs. 551 and 552).

105 Einstein to De Sitter, March 24, 1917, and De Sitter to Einstein, April 1, 1917 (CPAE 8, Docs. 317 and 321).

106 For discussion, see, e.g., Earman (1995) or Earman and Eisenstaedt (1999, Sec. 3).

107 Einstein to Klein (before June 3, 1918), (CPAE 8, Doc. 556).

108 Klein to Einstein, June 16, 1918 (CPAE 8, Doc. 566).

109 The following analysis follows Schrödinger (1956).

110 This passage and a similar passage from the autobiographical notes (Einstein 1949a, 29) are quoted and discussed by Hoefer (1994, 330) and Renn (2007c, 61).

111 See Chapter 9.

112 See Norton (1993a, 6–11) for a reconstruction of how Einstein presumably derived the result that the acceleration of a falling body decreases when it is moving sideways. The paper also provides diagrams illustrating the effect (ibid., Figs. 1 and 2). Einstein first alluded to this problem with special-relativistic theories of gravity in print in the course of his polemic with Abraham (Einstein 1912h, 1062–3). Renn (2007b, 55) presents a more elementary argument that shows that the acceleration of free fall must depend on a body's horizontal velocity in any special-relativistic theory of gravity (see also Giulini 2006, 16). Consider two trains traveling in opposite directions on parallel tracks with constant speeds. The moment a passenger in one train comes face to face with a passenger in the other train, they each drop some object from the same height. The relativity of simultaneity implies that, if the objects were to hit the floor of the respective trains simultaneously according to the passenger in one train, they would not do so according to the passenger in the other train. Since the situation of the two passengers is completely symmetric, it follows that the objects must hit the floor one after the other for both passengers. One easily verifies that each observer will claim that the object he or she dropped hit the floor first (consider Fig. A.26 in the Appendix and let the events P and Q represent the two objects hitting the floor of their respective trains).

113 In an article Einstein is known to have read (see Einstein to Mileva Marić, September 28, 1899 [CPAE 1, Doc. 57]), Wilhelm Wien (1898) had noted that the equality of inertial and gravitational mass strongly suggests that there is a connection between acceleration and gravity.

114 My discussion of Einstein's engagement with the Nordström theory follows Norton (1992b, 1993a). For insightful further discussion, see Giulini (2006).

115 The trace of the energy-momentum tensor $T_{\mu\nu}$ is the sum of its diagonal components, $T_{11}, T_{22}, \ldots$ This quantity is invariant under Lorentz transformations. It transforms as a scalar under such transformations.

116 For detailed discussion, see Norton (1992b, Secs. 9–10, 437–50) and Giulini (2006, Sec. 6).

117 For further discussion of these developments, see, e.g., Janssen and Mecklenburg (2006, 107–11) and Walter (1999). Einstein's 1912 manuscript is available in a facsimile edition (Einstein 1996).

118 See Norton (1993a, Sec. 5) for discussion of Einstein's objection and a helpful diagram (ibid., Fig. 5).

119 The following year, Mie (1914) published a sharp critique of the Entwurf theory. By 1917, however, he had abandoned his own theory and accepted general relativity, though not Einstein's interpretation of general covariance in terms of relativity of motion (see note 59). See Smeenk and Martin (2007) for an introduction to some of Mie's papers on gravity presented in English translation in Renn (2007a, Vol. 4, 633–743).

120 In modern terms, Nordström's theory satisfies the weak but not the strong equivalence principle.

121 The following year, Erwin Freundlich, a Berlin astronomer and Einstein's protégé, set out for the Crimea to observe this eclipse, but then World War I broke out and Freundlich was interned by the Russians. Another expedition was rained out (Earman and Glymour 1980a, 60–2). In a sense, this was a fortunate turn of events for Einstein since the effect predicted by the Entwurf theory was too small (Earman and Janssen 1993, 129).

122 The paper was submitted from Zurich, so Nordström would have had the opportunity to discuss his theory in person with both Einstein and Laue (Norton 1992b, 455).

123 See Norton (1993a, Sec. 6) for discussion and another helpful diagram (ibid., Fig. 6). Giulini (2006, 26) disputes Einstein's claim that this is the only way to save energy conservation in the Nordström theory.

124 This is similar to the situation in Lorentz's ether theory of electrodynamics out of which special relativity grew. By 1899, Lorentz assumed that any moving system contracts by a factor depending only on the ratio of the system's velocity with respect to the ether and the velocity of light and that any process in a moving system takes longer than the same process in the system at rest by that same factor (Janssen 2002b, 425). As a result, clocks and rods in Lorentz's theory measure times and distances in the Minkowski space-time posited by special relativity rather than times and distances in the Newtonian space-time posited by Lorentz's own theory.

125 See Norton (2007, 775–6) for discussion. Einstein thought that the Entwurf field equations could likewise be recovered from the combination of generally covariant equations and additional conditions (Einstein and Fokker

1914, 328). In the case of the *Entwurf* theory, Einstein knew the additional conditions (these were the conditions for "adapted coordinates" mentioned in Section 4) but not the generally covariant equations that together with these conditions would give the *Entwurf* field equations. See Janssen and Renn (2007, 842–3, 866–7) for further discussion of Einstein's understanding during this period of how field equations could be extracted from generally covariant equations, which by themselves were inadmissible as field equations on account of the hole argument.

126 Einstein had already used the electrodynamical analogy when he was looking for a generalization of his 1912 theory for static gravitational fields to stationary fields. See, in particular, Einstein (1912e) where the electrodynamical analogy is mentioned in the title, Einstein to Besso, March 26, 1912, and Einstein to Ehrenfest (before June 20, 1912) (CPAE 5, Docs. 377 and 409). We already saw how Einstein used an analogy with electrodynamics in 1920 to explain the relativity of the gravitational field (cf. Fig. 6.2).

127 There is still an important difference between the two cases: the source of the electromagnetic field is the electric charge-current density, while the source of the gravitational field is the energy-momentum density, both of matter *and of the gravitational field*. In Chapter 8, Kennefick emphasizes the importance both of analogies and of disanalogies between gravity and electromagnetism in the later debate over gravitational waves.

128 The discussion of the developments of 1914–15 is based on Janssen and Renn (2007). Our account of how Einstein found the generally covariant field equations now named after him deviates at key points from the one given in a classic paper by Norton (1984). For a short version of our new still controversial account, see Janssen (2005, 75–82). Ryckman briefly discusses the controversy in Section 2 of Chapter 12. For more detailed discussion, see van Dongen (2010, ch. 1).

129 Two years later, Einstein (1916o) used this same formalism to show that the general covariance of the Einstein field equations is directly related to energy-momentum conservation (Janssen and Renn 2007, Sec. 9).

130 Einstein to Sommerfeld, November 28, 1915 (CPAE 8, Doc. 153).

131 Einstein to Ehrenfest, December 26, 1915 (CPAE 8, Doc. 173).

132 Einstein to Natanson, December 29, 1915 (CPAE 8, Doc. 175).

133 As the contributions by Howard, Ryckman, Friedman, and Lehner to this volume show, Einstein's theory was philosophically fruitful in many other ways.

7   Einstein's Role in the Creation of Relativistic Cosmology

1. INTRODUCTION

Einstein's paper, "Cosmological Considerations in the General Theory of Relativity" (Einstein 1917b), is rightly regarded as the first step in modern theoretical cosmology. Perhaps the most striking novelty introduced by Einstein was the very idea of a cosmological model, an exact solution to his new gravitational field equations that gives a global description of the universe in its entirety. Einstein's paper inspired a small group of theorists to study cosmological models using his new gravitational theory, and the ideas developed during these early days have been a crucial part of cosmology ever since. We will see that understanding the physical properties of these models and their possible connections to astronomical observations was the central problem facing relativistic cosmology in the 1920s. By the early 1930s, there was widespread consensus that a class of models describing the expanding universe was in at least rough agreement with astronomical observations. But this achievement was certainly not what Einstein had in mind in introducing the first cosmological model. Einstein's seminal paper was not simply a straightforward application of his new theory to an area where one would expect the greatest differences from Newtonian theory. Instead, Einstein's foray into cosmology was a final attempt to guarantee that a version of "Mach's principle" holds. The Machian idea that inertia is due only to matter shaped Einstein's work on a new theory of gravity, but he soon realized that this might not hold in his "final" theory of November 1915. The 1917 paper should thus be read as part of Einstein's ongoing struggle to clarify the conceptual foundations of his new theory and the role of Mach's principle, rather than treating it only as the first step in relativistic cosmology.

Einstein's work in cosmology illustrates the payoff of focusing on foundational questions such as the status of Mach's principle. But it also illustrates the risks. In the course of an exchange with the Dutch astronomer Willem De Sitter, Einstein came to insist that on the largest scales the universe should not evolve over time – in other words, that

228

it is static. Although he originally treated this as only a simplifying assumption, Einstein later brandished the requirement that any reasonable solution must be static to rule out an anti-Machian cosmological model discovered by De Sitter. Thus Einstein's concern with Mach's principle led him to introduce the first cosmological model, but he was also blind to the more dramatic result that his new gravitational theory naturally leads to *dynamical* models. Even when expanding universe models had been described by Alexander Friedmann and Georges Lemaître, Einstein rejected them as physically unreasonable. Einstein's work in cosmology was also not informed by a thorough understanding of contemporary empirical work. The shift in theoretical cosmology brought about by Einstein's work occurred at the same time as a shift in the observational astronomer's understanding of the nature of spiral nebulae and large-scale structure of the cosmos. Others with greater knowledge of contemporary astrophysics, including Arthur Eddington, De Sitter, Lemaître, and Richard Tolman, made many of the more productive contributions to relativistic cosmology.

This chapter describes the early days of relativistic cosmology, focusing on Einstein's contributions to the field. Section 2 gives an overview of the "Great Debate" in observational astronomy during the 1910s and 1920s. Section 3 describes Einstein's "rough and winding road" to the first cosmological model, with an emphasis on the difficulties with Newtonian cosmology and the importance of Mach's principle. Conversations and correspondence with De Sitter forced Einstein to consider the status of Mach's principle in his new theory, leading to the introduction of his cosmological model. Although Einstein hoped that this would insure the validity of Mach's principle, De Sitter quickly produced an apparent counterexample, a new model in which Mach's principle did not hold. Einstein's attempts to rule out this counterexample drew him into a tangle of subtle problems regarding the nature of singularities that also ensnared his correspondents Hermann Weyl and Felix Klein. Throughout the 1920s, theoretical cosmology focused for the most part on understanding the features of the models due to Einstein and De Sitter and attempting to find some grounds on which to prefer one solution to the other. Section 4 takes up these debates, which came to an end with the discovery that several other viable solutions had been overlooked. This oversight can be blamed on Einstein's influential assumption that the universe does not vary with time, which was elevated from a simplifying assumption to a constraint on any physically reasonable cosmological model. This assumption kept Einstein, and many of the other leading lights of theoretical cosmology, from discovering expanding universe models. Section 5 describes the expanding models and their relation to observational evidence in favor of expansion. In the concluding section I briefly discuss the fate of the consensus formed in the early days of relativistic cosmology in the ensuing decades, focusing on debates regarding the status of the so-called cosmological principle.

2. THE GREAT DEBATE

The first three decades of the twentieth century were a period of active debate within the astronomical community regarding the overall structure of the cosmos.[1] At the turn of the century, two leading observational cosmologists, the influential Dutch astronomer Jacobus Kapteyn and the German Hugo von Seeliger, had embarked on the ambitious project of determining the architecture of the Milky Way. Astronomers working in this line of research used sophisticated statistical techniques to overcome the fundamental problem facing the project, namely that of determining distances to the stars, in order to convert observational data into a three-dimensional map of the stellar system. The program culminated in the "Kapteyn Universe," according to which the galaxy is a roughly ellipsoidal distribution of stars, with the Sun near the center.[2] Starting in 1918, Kapteyn and others had to contend with the alternative "Big Galaxy" view introduced by the American astronomer Harlow Shapley. Based on studies of large, dense groups of stars called *globular clusters*, Shapley argued that the Milky Way was a factor of ten larger than in Kapteyn's model, with the Sun placed some distance from the center. Partially as a result of Shapley's challenge, the focus of observational cosmology shifted from determining stellar distributions in the Kapteyn tradition to using other objects (globular clusters and nebulae) as indicators of large-scale structure. This era has been called the "second astronomical revolution" to reflect the impact of the new ideas and techniques introduced during this period. I will set the stage for Einstein's work by focusing on two aspects of this revolution, the debates regarding the spiral nebulae and related debates in cosmogony.

By 1930 astronomers had abandoned much of the turn-of-the-century conventional wisdom regarding "spiral nebulae." Around 1900 many astronomers held that the Milky Way galaxy was a unique system encompassing all observed objects, including the enigmatic spiral nebulae. The English astronomer Agnes Clerke, for example, confidently asserted that "no competent thinker" could accept that any spiral nebula is comparable to the Milky Way (Clerke 1890, 368). But by 1930, few competent thinkers still held Clerke's opinion. A majority of astronomers instead accepted the "island universe" theory, according to which the spiral nebulae are similar in nature to the Milky Way.[3]

The nature of the spirals had been the subject of speculation and debate for more than 150 years. The speculative proposal of Kant, Lambert, and Wright that the nebulae are "island universes," composed of stars and similar to the Milky Way in structure, first garnered support from the observations of William Herschel and Lord Rosse in the early to mid-nineteenth century. Yet by the turn of the century, there were three major objections to this idea.[4] In 1885 a nova flared in the Andromeda Nebula, and at its brightest this star reached a luminosity of roughly one-tenth that of the entire nebula. This led to the first objection: faced with the implausible idea that a single star could outshine millions of others, many astronomers concluded instead that the Andromeda Nebula was not a group of stars like the Milky Way. Second, the technique of spectroscopy revealed bright emission lines in the spectra of some nebulae, characteristic of luminous gas rather than starlight. This suggested that the nebulae are regions of hot gas within the Milky Way, rather than groups of unresolved stars at a great distance. Although some nebulae have starlike spectra, this could be explained as the consequence of reflected starlight. Finally, the positions of the nebulae seemed to be related to the galaxy: the nebulae shunned the plane of the galaxy and clustered near the poles. Such a correlation would be unsurprising if the spirals were within the galaxy, but it appeared to be inexplicable based on the island universe theory.

The fortunes of the island universe theory improved dramatically by the mid-1910s. In his influential *Stellar Movements and the Structure of the Universe* (1914), the British astrophysicist Arthur S. Eddington argued that, despite the paucity of direct evidence in its favor, the island universe theory should be accepted as a useful "working hypothesis," whereas the idea that the spirals lie within the galaxy should be rejected because it led to an unproductive dead end. As Eddington emphasized, the classification of distinct types of nebulae undercut the second criticism: the spiral nebulae might lie well beyond the galaxy even though other nebulae are part of the Milky Way. However, the other criticisms remained unanswered. Island universe advocates had to admit that novae could somehow produce tremendous energy, yet at least the novae discovered in spirals appeared to be significantly dimmer than those within the Milky Way.[5] The clustering of spirals near the galactic poles remained a mystery throughout the 1920s, but it was eventually explained as an observational selection effect due to the absorption of light by interstellar dust (Trumpler 1930).[6]

Critics of the island universe theory marshalled two new and apparently quite damaging results in the late 1910s. Two of the most prominent critics, Shapley and Adriaan van Maanen, attempted to put the final nails in the coffin of the island universe theory. As Shapley put it in a letter to van Maanen, "Between us we have put a crimp in the island universe, it seems – you in bringing the spirals in and I by pushing the Galaxy out" (quoted in Smith 1982, 105). Starting in 1914, van Maanen measured what he took to be the rotational motion of objects within several spiral nebulae. These measurements only made sense if the nebulae were relatively small, nearby objects; if the nebulae were as large as the Milky Way and far away, as the island universe theory required, the motions would dramatically exceed reasonable physical limits. Shapley, for his part, argued that the Milky Way was much larger than had previously been assumed (by a factor of ten), leaving plenty of room for the spirals within the galaxy. Shapley reached the "Big Galaxy" view based on a novel astronomical yardstick: he used Cepheid variable stars as "standard candles." Henrietta Leavitt had established that the period of the variation in the brightness of these stars bears a fixed relationship to their intrinsic brightness (absolute magnitude). After calibrating the scale using nearby Cepheids, Shapley could then directly calculate the distances of remote Cepheids by determining their period.[7]

The decade closed with the so-called Great Debate between Shapley and a defender of the island universe theory, Heber Curtis. The event did not live up to its billing, since Shapley chose to avoid tackling the island universe theory directly.[8] In fact, Shapley's "Big Galaxy" view was not directly incompatible with the island universe theory, as Shapley's own shifting views on the matter in the years leading up to the debate indicate.[9] Shapley did not present a detailed attack on the island universe theory in the published exchange (Shapley 1921), although he briefly mentioned the clustering of the spirals around the galactic poles and the incompatibility of the island universe theory with van Maanen's measurements. The appeal to van Maanen's measurements of internal motion in the spiral nebulae had little chance of persuading Curtis, who responded that van Maanen's measurements would only be convincing once they had been successfully reproduced (Curtis 1921, 214). Controversy regarding van Maanen's measurements continued throughout the 1920s, until they were eventually rejected as artifacts of subtle systematic errors.

For his part in the debate, Curtis focused on the crucial issue of how to reliably measure distance to astronomical objects. Shapley had pioneered the use of Cepheid variables to establish distances to globular clusters, but Curtis questioned the internal consistency of this method as well as its relative reliability compared to other distance measures. Ironically, this controversial method would bring about the resolution of the debate within a few short years of Curtis and Shapley's exchange. Edwin Hubble observed a Cepheid variable in the Andromeda Nebula in 1923 with the powerful 100-inch telescope on Mount Wilson. Based on observations of variables in a number of galaxies, Hubble concluded that the spiral nebulae were much farther away than opponents of the island universe theory allowed.[10] This groundbreaking observational work convinced almost all astronomers of the validity of the island universe theory by the mid-1920s. Hubble's work provided what had been lacking in earlier stages of the debate: a way of measuring distance to the spirals that island universe advocates and opponents both considered reliable.[11]

The great debate concerning the nature of the spiral nebulae was entangled with controversies in cosmogony.[12] This field focused on describing the origins and evolution of various astronomical structures, ranging in scale from the Earth-Moon system to the solar system and beyond. After the turn of the century, the Laplacian nebular hypothesis was under attack. Laplace had suggested that structures such as the solar system resulted from the condensation of gaseous nebulae, an account augmented by Herschel's claim that the spiral nebulae are protosolar systems at an earlier stage of evolution. Theoretical development of the Laplacian idea drew on increasingly sophisticated mathematical treatments of a problem that is simple to state, if not to solve: what are the stable configurations of a rotating fluid according to Newtonian gravitational theory, and how do different configurations evolve over time?[13] The Americans Thomas Chamberlin and Francis Moulton coupled forceful criticisms of the nebular hypothesis with a new theory of the formation of the solar system proposed in 1905. Their account had two striking features: first, the planets were formed by the accretion of smaller hard fragments (planetesimals) rather than by condensation of a gaseous cloud, and second, an encounter with another star triggered formation of the planets. The Chamberlin-Moulton theory triggered refinements of the nebular hypothesis. James Jeans incorporated the second idea within a modified nebular hypothesis, according to which an encounter with a nearby star triggered the condensation of the hot gaseous cloud into the solar system.

The debate between the nebular hypothesis and Chamberlin-Moulton's alternative continued throughout the 1910s and 1920s. Both theories appealed to a wide variety of astronomical phenomena to justify a number of speculative assumptions. In a 1913 review of two books on cosmogony, Karl Schwarzschild complained that the field was "heterogeneous" and "impenetrable" due to "prolixity, pretension, confusion, and the general lack of mathematical control" (Schwarzschild 1913, 294). Many of the cosmogonists followed Herschel in appealing to observations of spiral nebulae, taken to be protosolar systems at an earlier stage of evolution, to vindicate aspects of their accounts.[14] Schwarzschild singled out this idea as particularly dubious, and the developments described vindicated his skepticism. But more significantly, the new understanding of the spiral nebulae dramatically increased the scale of the known universe. The concerns of cosmogony appear more parochial, although still important, once these differences of scale are recognized. Rather than appealing to results regarding the evolution of rotating fluids or the interactions of planetesimals, progress in the study of the structure and evolution of the universe at these largest scales came from an unexpected direction: a new theory of gravitation.

3. RELATIVISTIC COSMOLOGY

The debates just described form the background for the reception and further development of relativistic cosmology. However, Einstein's foray into cosmology was not motivated by the problems described, and he apparently had not kept abreast of new results regarding spiral nebulae and the ensuing controversies. Instead, his groundbreaking 1917 paper began by highlighting an apparently embarrassing dilemma for Newtonian cosmology. Einstein emphasized this alleged flaw to cast his own newly minted gravitational theory in a better light, but the motivations for his first foray into cosmology lay elsewhere. Einstein's "rough and winding road" to cosmology was a continuation of the path he followed to the discovery of general relativity. Einstein created the field of relativistic cosmology as a final effort to guarantee that his new gravitational theory satisfied Machian ideas.[15]

3.1. Paradoxes of Newtonian Cosmology

Einstein (1917b) opened with the following dilemma. If matter is uniformly distributed throughout an infinite universe, Newtonian gravitational theory does not consistently apply, for reasons that we will consider shortly. Newtonian cosmology seems to allow only an alternative picture: an "island" of stars clumped together in an otherwise empty universe. Einstein argued that even this possibility can be ruled out, since the island would be unstable. Einstein presented this as an inescapable dilemma for Newtonian cosmology, and his discussion of it served the dual purpose of exposing the inconsistencies of a preceding theory and preparing readers of his paper for a modification of his own gravitational theory.[16]

The problem Einstein noted regarding Newtonian cosmology can be stated quite simply. Einstein discovered the problem independently, but it has a long history stretching back to shortly after the publication of Newton's *Principia*, and Hugo von Seeliger had recently given a

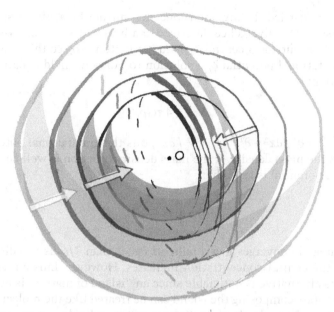

Figure 7.1. *The total gravitational force on the point mass* O *is given by the sum of forces due to each hemispherical shell. Each half-shell contributes a force of equal magnitude,* $F = G\rho\pi d$, *and the sum does not converge. Drawn by Laurent Taudin.*

systematic exposition of the problem (Seeliger 1895).[17] Einstein gave a characteristically straightforward formulation of the problem in a popular work (Einstein 1917a, 71–2): the number of lines of gravitational force passing through a sphere around a point mass O enclosing a uniform matter distribution is given by $\rho_0 V$, where ρ_0 is the mass density and V the volume of the sphere. The force per unit area is then proportional to $\rho_0 R$, where R is the radius of the sphere. As $R \to \infty$, the force also diverges – a result Einstein called "impossible." Although Einstein did not elaborate further, the failure of convergence can be elucidated more clearly by considering cases such as that illustrated in Figure 7.1 (following Norton 1999b). The gravitational force on O at the origin is given by the sum of forces due to hemispherical shells of thickness d surrounding O. Since the increase of mass ($\propto r^2$) cancels the decrease in gravitational force with distance ($\propto r^{-2}$), the force due to each shell is the same – it is given by $F = G\rho\pi d$, where G is Newton's gravitational constant and ρ is the mass density. The total gravitational force is then given by an infinite sum of equal terms with opposite signs, and this sum fails to converge.

Einstein carried the argument one step further by applying statistical physics. The divergence described can be avoided if the mass density falls off outside a central "island" of matter. To see this, consider the gravitational potential ϕ, a solution to Poisson's field equation for gravitation:

$$\nabla^2 \phi = 4\pi G\rho, \tag{1}$$

where $\nabla^2 \equiv \partial^2 / \partial x^2 + \partial^2 / \partial y^2 + \partial^2 / \partial z^2$, ϕ is the gravitational potential, and ρ is the mass density. If the mass density function is well behaved, then the potential is given by:

$$\phi(r) = \int G \frac{\rho(r)}{r} dV. \tag{2}$$

This integral converges if $\rho(r)$ falls off faster than $1/r^2$ as the distance r from the central concentration increases. However, Einstein argued that this alternative is not viable since an "island of matter" is not stable. The stars composing the island can be treated like the molecules of an ideal gas. The island would "evaporate" as individual stars acquired enough kinetic energy to escape the gravitational attraction of the other stars, just as water molecules evaporate into the air. If the potential ϕ converged to a small, finite value at infinity, at thermal equilibrium the ratio between mass at the center and mass at infinity must take a finite value. Einstein concluded that requiring the mass distribution to vanish at infinity implied that it also vanished at the center, ruling out an "island" of concentrated matter.

Seeliger and his contemporaries had considered several ways to avoid the dilemma, from denying the universality of gravitation to ruling out an infinite, uniform matter distribution by *fiat*. Einstein and Seeliger both advocated a modification of Newton's inverse square law, effected in Einstein's case by simply adding a term to Eq. (1):[18]

$$\nabla^2 \phi - \Lambda\phi = 4\pi G\rho. \tag{3}$$

The modified field equations admit a uniform field as a solution for a constant mass density ρ_0,

$$\phi = -\frac{4\pi}{\Lambda} G\rho_0. \tag{4}$$

However, Einstein and Seeliger did not have the same attitude toward a solution of the dilemma along these lines. Seeliger took the proposal

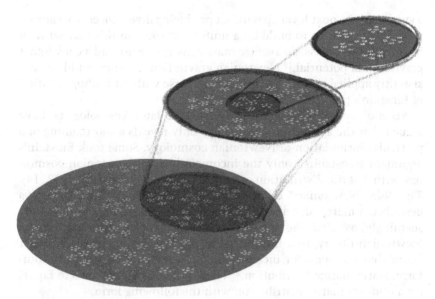

Figure 7.2. *Charlier's hierarchical universe has a fractal-like structure, and it avoids divergences without introducing a preferred central region.* Drawn by Laurent Taudin.

seriously enough to suggest that astronomical measurements could constrain his modification of Newton's law. Einstein, however, commented that his solution "does not in itself claim to be taken seriously; it merely serves as a foil for what is to follow" (Einstein 1917b, 543). Einstein was not primarily concerned with resolving the paradoxes of Newtonian cosmological theory, but the addition of a constant to Poisson's equation paves the way for the introduction of the infamous cosmological constant a few pages later.

Einstein's presentation of the dilemma facing Newtonian cosmology is at first compelling, but on closer examination the argument falls apart due to two significant oversights. First, a way between the horns of the dilemma had already been found. A decade before Einstein's paper, the Swedish astronomer Carl Charlier had explicitly constructed a "hierarchic cosmos," inspired by the speculations of Fournier d'Albe, that neatly avoided both problems discussed (Charlier 1908). The crucial trick was to produce a mass distribution which both avoided the divergences and lacked a preferred center. Charlier's model had a fractal structure: stars are grouped into spherical galaxies, galaxies into spherical meta-galaxies, and so on (see Figure 7.2). Charlier derived constraints on how densely the systems at one level could be "packed" into a system of the next level up without producing divergences. He showed that it was possible to build up a uniform cosmos in this way with an infinity of total mass, an average mass density of zero, and a convergent gravitational potential.[19] Newtonian gravitational theory could be consistently applied to such a hierarchic universe without leading to either of Einstein's problems.

Second, and more significantly, several later cosmologists have argued that the apparent inconsistency only reveals a shortcoming of a particular formulation of Newtonian cosmology. Some took Einstein's argument to establish only the incompatiblity of Newtonian cosmology with a *static* distribution of matter (see, e.g., Heckmann 1942, 14). The "Neo-Newtonian" cosmological models developed in the 1930s described a matter distribution that changed over time, and thus they seemingly avoided the inconsistency.[20] Geometric formulations of Newtonian theory, first introduced in the 1920s, support the stronger claim that the apparent inconsistency can be avoided even for a uniform, static matter distribution.[21] There is a class of solutions to Eq. (1) for a uniform matter distribution with the following form,

$$\phi(\mathbf{r}) = \frac{2}{3}\pi G\rho |\mathbf{r} - \mathbf{r}_0|^2, \tag{5}$$

with distinct solutions corresponding to different choices of an arbitrary point r_0.[22] The force $\mathbf{F}$ acting on a test particle with unit mass is then given by the gradient of the potential:

$$\mathbf{F} = -\nabla\phi = -\frac{4}{3}\pi G\rho(\mathbf{r} - \mathbf{r}_0), \tag{6}$$

which represents a spherically symmetric field of force directed at the "center point" r_0, where the force vanishes. The original problem resurfaces, in that a completely *uniform* matter distribution mysteriously leads to a *nonuniform* force field with an arbitrary preferred central point at r_0. However, these peculiar properties of the solutions have no directly observable consequences. The geometric formulation modifies Newtonian theory to incorporate what Norton (1995) aptly called the "relativity of acceleration": the choice of how to decompose a given free-fall trajectory into inertial motion and gravitational deflection is conventional. The indeterminacy noted by Einstein is just a consequence of this feature, since there appears to be a physical difference between different choices of the decomposition. In the geometrical formulation the choice of a particular decomposition has as little significance as the choice of a rest frame within special relativity.

Einstein's response to the unraveling of his dilemma for Newtonian cosmology illustrates its relatively minor role in his thinking. Franz Selety (1922) spelled out how Charlier's hierarchic model avoided the dilemma in painstaking detail. In his brief response, Einstein (1922e) conceded the point before reiterating the line of reasoning leading to his own cosmological model, giving Mach's principle the leading role.[23]

3.2. Mach's Principle

Einstein turned to cosmology after conversations with De Sitter in the fall of 1916 convinced him that an idea that had played a crucial role in his discovery of general relativity was at stake. "Mach's principle" (as Einstein later dubbed it) is difficult to formulate precisely, but Einstein hoped to capture Mach's idea that inertia derives from the distribution of matter rather than space itself. Einstein also called this the "relativity of inertia," which he put as follows: "In a consistent theory of relativity, there can be no inertia *relative* to 'space,' but only an inertia of masses *relative to one another*" (Einstein 1917b, 145, original emphasis).[24] There are problems with both formulations; in particular, it is not clear how to characterize the matter distribution without employing the metric. In what follows, we will focus almost exclusively on the connection between Mach's principle and the status of boundary conditions and vacuum solutions. For an entry into the extensive literature on Mach's principle, its historical role in Einstein's path to general relativity, and its relation to other foundational principles, such as the requirement of general covariance and the equivalence principle, see Chapter 6, Torretti (2000), and Hoefer (1994). The discussions and correspondence with De Sitter forced Einstein to consider the status of this principle more carefully, and he struggled to give it a sharper formulation and to insure that it held in his new gravitational theory.[25] Einstein's famous cosmological model and the infamous cosmological constant were both by-products of this effort.

One threat to Mach's principle came from the need to stipulate boundary conditions in order to solve the gravitational field equations.[26] For example, Schwarzschild (1916) derived his expression for the gravitational field of a single body such as the Sun by requiring (among other things) that the solution resembles Minkowski space-time, the flat space-time of special relativity, at infinity. This boundary condition captures the natural requirement that at great distances from the source the gravitational field should approach the "empty space" solution to the field equations. There is nothing unusual about imposing boundary conditions in order to find a solution to a set of field equations. But for Einstein, introducing boundary conditions meant allowing a vestige of

absolute motion to creep back into the theory. The inertia of a particle far from the Sun in Schwarzschild's solution would be fixed by the
space-time structure imposed at infinity as a boundary condition rather
than just its relation with other matter. For Einstein this was contrary
to the spirit, if not the letter, of his theory. The structure at infinity
could be used to distinguish states of motion absolutely, without reference to other matter. On Einstein's view, there should be no such
structures attributed to space itself in his new theory. Vacuum solutions clearly violated this requirement, because they possess structure
that cannot be due to matter. But boundary conditions also violated this
requirement by introducing spatial structure "at infinity" that is independent of the matter distribution.[27]

Einstein proposed two ways to banish this last remnant of "absolute
space" in letters to De Sitter from 1916 and 1917. His first proposal
was to stipulate that at infinity the metric field takes on "degenerate
values" (all components equal to 0 or ∞) rather than approaching a flat,
Minkowski metric. Einstein (1917b) argued that the degenerate boundary values follow from requiring that the inertia of a body drops to zero
as it approaches infinity. This was intended to capture the idea that there
would be no "inertia" relative to space itself.[28] A fundamental problem
with this proposal is the difficulty with specifying the nature of a solution "at infinity" in a manner that does not depend on the choice of
coordinates. A more obvious problem led Einstein to abandon this line
of thought fairly quickly. A flat Minkowski metric appeared to be compatible with stellar motions on the largest scale yet observed. Einstein's
proposal required postulating "distant masses" that would somehow
reconcile degenerate values of the metric field "at infinity" with these
observations. De Sitter did not share Einstein's Machian intuitions,
and the lack of a Machian explanation of inertia did not bother him as
much as this ad hoc introduction of such unobserved distant masses.[29]
Einstein soon dropped this first proposal in favor of a more radical way
of handling the problem of boundary conditions.

Einstein's ingenious suggestion was to do away with the problem of boundary conditions by getting rid of boundaries entirely. He
introduced a cosmological model built up from spatial sections that
describe the universe at a given cosmic time. These spatial sections
are three-dimensional analogs of a two-dimensional sphere such as the
Earth's surface. Like the Earth's surface, each spatial section is finite
yet unbounded, leaving no place to assign boundary conditions. This
cosmological model satisfied two strong idealizations. First, the model
describes a simple uniform mass distribution without the lumps and
bumps observed by astronomers.[30] Second, Einstein argued that the
model should be static, which means that the properties of the model

do not vary with time; each spatial section "looks like" any other.[31] This second requirement was justified with an appeal to astronomical observations. Since the observed relative velocities of the stars are much smaller than the speed of light, Einstein argued that there is a coordinate system such that all the stars are permanently at rest (at least approximately). This is a weak justification for such a crucial assumption. As De Sitter promptly objected in correspondence, "we cannot and must *not* conclude from the fact that we do not see any large changes on this photograph [our observational "snapshot" of stellar distributions] that everything will always remain as at that instant when the picture was taken."[32] With the benefit of hindsight it is easy to fault Einstein for so quickly ruling out solutions that change over time on such weak grounds. Yet many of his contemporaries followed his lead in limiting consideration to static models. Even De Sitter, despite recognizing the weakness of Einstein's argument, nonetheless failed to study dynamical models until 1930. In any case, Einstein was not primarily interested in a detailed comparison between his cosmological model and observations. He frankly explained his motives in introducing the model in a letter to De Sitter: he called it "nothing but a spacious castle in the air," built to see whether his Machian ideas could be consistently implemented.[33]

The castle in the air came at a price. Einstein was forced to modify the field equations of general relativity in order to admit this cosmological model as a solution. All that was required was the straightforward addition of a cosmological constant term, Λ, a modification Einstein had foreshadowed with the similar change to Poisson's equation discussed previously:[34]

$$R_{\mu\nu} - \Lambda g_{\mu\nu} = -\kappa\left(T_{\mu\nu} - \frac{1}{2}g_{\mu\nu}T\right). \tag{7}$$

This equation specifies the relationship between the space-time geometry (the left-hand side) and the distribution of matter and energy (the right-hand side).[35] Einstein showed that if these modified field equations are to hold in his model, the cosmological constant Λ is fixed by the mean density of matter ρ, namely $\Lambda = \kappa\rho/2$. The value of Λ is also related to the radius of the spatial sections R by the equation $\Lambda = 1/R^2$. Combining the two equations made it possible to estimate the "size of the universe" based on the mean density of matter.

The addition of the cosmological constant is a fairly natural generalization of the original field equations, and it satisfies the formal and physical constraints that Einstein had used in deriving the original equations.[36] Physically, a positive Λ term represents a "repulsive force" that counteracts the attraction of gravitation.[37] Observations available to Einstein could have been used to show that Λ must be *incredibly* close to zero, since even a slight deviation from zero affects the general relativistic predictions for motions within the solar system. De Sitter argued that Λ was certainly less than 10^{-45} cm^{-2} and probably less than 10^{-50} cm^{-2},[38] and later calculations such as those reported in Tolman (1934) gave even tighter constraints. However, even a very small non-zero value of Λ changes the space of solutions to the field equations. Einstein's static universe is a solution of the modified field equations, and a nonzero Λ term also rules out solutions of the original field equations such as Minkowski space-time, a welcome result for Einstein.[39]

Einstein did not have long to rest easy with the modified field equations and his ingenious way of saving Mach's principle. The introduction of Λ was not actually necessary for Einstein's way of eliminating boundary conditions to work; it was only necessary given the additional assumption that the model must be static. Einstein (1917b) concluded with a remark suggesting that it might be possible to construct time-varying models with closed spatial sections without the Λ term.[40] Such models would do away with the need to specify boundary conditions without Λ, undermining the rationale for its introduction. But Einstein must have been surprised to discover that the introduction of Λ was also not sufficient to guarantee that Mach's principle holds in general relativity. After learning of Einstein's solution, De Sitter promptly produced a second solution to the modified field equations. Since it was a vacuum solution (with nonzero Λ) it apparently violated Mach's principle. As we will see in the next section, until 1930 much of the literature in relativistic cosmology focused on understanding the properties of these two models.

4. COSMOLOGICAL MODELS: *A* VERSUS *B*

Einstein and De Sitter's published papers brought their debate to the wider scientific community. Eddington had invited De Sitter to provide a précis of Einstein's new gravitational theory in the *Monthly Notices*, and De Sitter responded with a series of three clear articles that introduced Einstein's theory – along with De Sitter's own cosmological model – to the English-speaking world. By the mid-1920s a handful of mathematical physicists had begun to study the properties of the Einstein and De Sitter solutions. Part of this literature focused on the question of whether the De Sitter solution could be dismissed on the grounds that it included a "singularity," with a debate regarding both the content of that charge and whether it applied. This topic was part of the broader debate regarding what properties a cosmological model should have to qualify as physically reasonable. The major conceptual innovation introduced by Einstein was the very possibility of a mathematical description of the universe as a whole, but it was not immediately clear what observational and physical content these abstract models possessed. With the benefit of hindsight, this early exploration of cosmological models was limited by illicit assumptions regarding what counts as physically reasonable. In particular, it was more than a decade before the community overcame Einstein's insistence on static models. Several theorists followed De Sitter's lead in seeking out a stronger connection between cosmological models and astronomical observations. With his strong empiricist bent, De Sitter actively explored the connection between his model and observations, suggesting in particular an explanation of the redshift in the spectral lines of the spiral nebulae. This point of contact became far more important on the heels of Hubble's observational results published in 1929.

In the course of these explorations of cosmological models, this first generation of relativists encountered a number of the most striking novelties introduced by Einstein's theory of gravitation. To determine a particular solution's physical properties, and to find out whether it harbored any singularities, relativists had to differentiate artificial features due to the coordinates being used from genuine features of the space-time. This was no simple matter, given that they lacked the mathematical tools developed later for isolating the intrinsic features of a space-time. In addition, as Eisenstaedt (1989) has emphasized, in cosmology theorists faced a variety of situations far removed from familiar problems studied in Newtonian theory. (In fact, Newtonian cosmology and Newtonian analogs of the expanding universe models were studied systematically several years later, based on insights from general relativity.)

4.1. The Many Faces of Model B

De Sitter derived his solution as a variation on Einstein's theme: rather than taking only the spatial sections to be closed spheres, de Sitter treated the entire space-time manifold as a closed space.[41] In his letter to Einstein introducing the solution[42] as well as his publications (De Sitter 1917a, 1917c), De Sitter laid out the two solutions side by side – labeling Einstein's model "A" and his own model "B" in the published papers – to emphasize their similarities. However, B differed from A in several crucial respects. Perhaps most importantly, De Sitter's solution proved to be much harder to grasp, leaving most theorists to rely too heavily on particular coordinate representations; Eddington (1923) characterized Einstein's solution as "commonplace" compared to the complex geometry of De Sitter's solution.[43] De Sitter's solution was a vacuum solution of the modified field equations, with $\Lambda = 3/R^2$ and $\rho = 0$. Unlike Einstein's solution, the cosmological constant is *not* related to the matter density. For De Sitter this counted as an advantage of his solution, since he took Einstein's model to include "world matter" (responsible for producing the Λ term) in addition to ordinary matter. However, Einstein insisted that the uniform matter distribution in his own model was merely an approximation to ordinary matter. A more realistic model would be inhomogeneous but still static, and such a model would be to the cylinder universe as "the surface of a potato to the surface of a sphere."[44] De Sitter's attitude toward his own solution was at first fairly skeptical, and he professed a preference for model *C*, his label for flat Minkowski space-time, over both *A* and *B*, since model *C* was a solution of the original field equations without Λ. This initial skepticism eventually faded somewhat, and De Sitter took his model seriously enough to study its observational consequences, as we will see.

The De Sitter solution appeared in a variety of guises over the following decade, as Einstein, De Sitter, and others used different coordinates to study its properties.[45] In retrospect, it is easy to fault these discussions for failing to disentangle properties of the De Sitter solution from properties that reflect a particular coordinate representation of it. The participants in the debates were aware of the need to focus on invariant quantities that do not depend upon the choice of coordinates, but in practice this fundamental principle was often not heeded. The issue is a more subtle analogue of a familiar problem with maps of the Earth's surface. Representing the Earth's surface on a two-dimensional map inevitably produces distortions. Knowledge of the underlying geometry of the Earth's surface makes it easy to avoid confusing these distortions due to the map with actual features of the surface. For example, a Mercator projection breaks down at two points (such as the North and South Poles), but it is easy to see that this is due to the projection and not to some special feature of the surface at those points. Einstein, De Sitter, and other cosmologists did not have a similar grasp of the geometry of the De Sitter solution to fall back on in assessing odd features of the solution revealed in different coordinate systems.

Einstein and De Sitter's discussion of model *B* eventually focused on a particular, and quite misleading, choice of coordinates. After originally presenting the solution geometrically, De Sitter wrote his solution in a static form to facilitate comparison with Einstein's model.[46] The transition to this set of coordinates appears to have been driven by Einstein's preference for a static model, and in this guise the De Sitter solution appears to represent, like Einstein's model, a static space-time with uniform spatial sections. But this impression is misleading. Suppose that we consider freely falling particles that are initially at rest with respect to the background matter distribution. In Einstein's model the distance between such particles measured on successive spatial sections remains constant. This is just to say that the trajectories followed by the freely falling particles (timelike geodesics) coincide with the curves defined by fixed values of the spatial coordinates. This is not the case in De Sitter space-time. In De Sitter space-time, the distance between the particles changes over time – they either scatter or move closer together rather than staying at a fixed distance. This is a consequence of the fact that the curves defined by fixed values of the spatial coordinate, using static coordinates, are not timelike geodesics.[47]

Unlike Einstein, De Sitter and later Lemaître (1925) took the non-static character of the De Sitter solution to be a positive feature, since this might explain the receding motion of the spiral nebulae (an idea we will return to). Lemaître's paper introduced a set of coordinates independently discovered by Robertson (1928). The De Sitter solution appears strikingly different in these two coordinate systems. The contrast between the static coordinates and the new coordinates is illustrated in Figure 7.3. The global geometry of the De Sitter solution can be represented as the surface of a hyper-hyperboloid embedded in a five-dimensional Minkowski space (see Figure 6.9).[48] The static coordinates used by De Sitter and Einstein only covered two wedge-shaped portions of the surface (see Figure 7.3a). The spatial sections are bounded and overlap at two points in the diagram (corresponding to a two-dimensional surface in the full solution), called the "mass horizon." The resulting degeneracy of the time coordinate reflects on the choice of coordinates rather than on the underlying geometry of the solution – the hyperboloid is entirely regular at these points. The solution takes on a very different guise in the Lemaître-Robertson coordinates. These coordinates cover only the upper half of the hyperboloid (see Figure 7.3b). The diagram illustrates a striking contrast between the properties of spatial sections for the two cases: in the static coordinates, the spatial sections have finite volume, whereas in Lemaître and Robertson's coordinates the spatial sections have *infinite* volume. Historically, the Lemaître-Robertson coordinate system served as much more than a reminder of the need to focus on invariant quantities in assessing the properties of a solution. For both Lemaître and Robertson, careful study of the De Sitter solution played a part in the discovery of more general nonstatic cosmological models.

4.2. Singularities

Starting with his first response to the new model, Einstein gave a series of arguments intended to rule out the De Sitter solution on physical grounds. Clearly Einstein hoped to do away with this counterexample

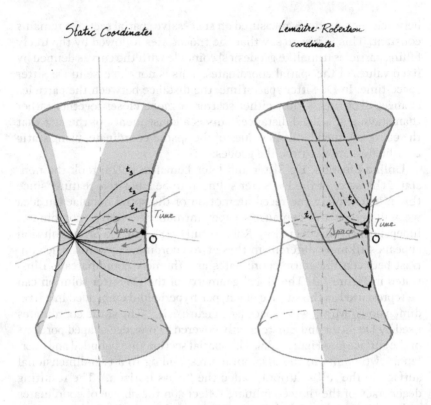

Static Coordinates

Lemaître-Robertson coordinates

Figure 7.3. *Coordinate representations of the De Sitter solution. (a) Static coordinates. These cover only the two wedge-shaped regions of the hyperboloid. The spatial slices have finite volume, and they overlap at the "mass horizon," the point p (and a point on the opposite side of the hyperbola). (b) Lemaître-Robertson coordinates. These cover only the upper half of the hyperboloid, and the spatial sections have infinite volume.* **Drawn by Laurent Taudin.**

to Mach's principle by revealing some fatal flaw unrelated to the fact that it was a vacuum solution. The most important and interesting charge was that model *B* harbored a "singularity," roughly a point or region where the metric field is ill-behaved.[49] If the metric field "blows up" to an infinite value or drops to zero, quantities appearing in the field equations would no longer be well defined. In this sense, a singularity would signal the breakdown of the theory, a result Einstein regarded as clearly unacceptable.

In order for the existence of singularities to be used as a criterion for separating physically reasonable cosmological models from the unclean, the concept of a "singularity" had to be defined more precisely. Giving such a definition has turned out to be a surprisingly subtle problem, and a widely accepted and useful definition of singularity was not in place until the mid-1960s.[50] A difficult first step toward a satisfactory definition involves distinguishing coordinate effects from true singularities. For example, the apparent singularity at the origin of polar coordinates is clearly an artifact of the coordinates. Einstein took this first step (Einstein 1918c): for a space-time to qualify as nonsingular, he required that for every point lying in regions within a "finite distance," there must be a coordinate system in which the metric is well behaved.[51] More intuitively, for a space-time satisfying the definition, all the regions within a finite distance from some starting point could be covered by a patchwork of coordinate systems in which the metric is continuous. The metric for the De Sitter solution in static coordinates fails to satisfy this definition because it breaks down at the "mass horizon," the points in Figure 7.3a (surfaces in the full solution) where the spatial sections intersect.[52] Einstein argued that the De Sitter solution should be rejected as unphysical due to this alleged singularity.

Einstein called this surface the "mass horizon" since he expected that the singular behavior of the metric indicated the presence of matter. Anticipating results about to appear in Weyl's book *Raum-Zeit-Materie*, which he was reading in proof, Einstein suggested that the De Sitter solution was after all *not* a vacuum solution (Einstein 1918c, 272). Weyl constructed a general solution by combining three different solutions, including one representing an incompressible fluid, and argued that the original De Sitter solution could be recovered in the limit as the fluid zone goes to zero volume.[53] Einstein welcomed this result, since it insured that the De Sitter solution could no longer be regarded as a counterexample to Mach's principle. The De Sitter solution was no vacuum solution – the mass had just been "swept away into unobserved corners," as Eddington (1923) put it.

Einstein's claims regarding the singularity in the De Sitter solution provoked a number of critical responses. Klein's analysis brought out the central defect of Einstein's assessment quite clearly (at least for those familiar with projective geometry): in Klein's geometrical representation of the solution, the hyperboloid is an entirely regular surface without any kinks or discontinuities that would correspond to a genuine singularity. Thus the alleged singularity had to be an artifact of a poor choice of coordinates. This point was not clear to Klein's contemporaries. Eddington, for example, gave a geometrical description of the de Sitter solution similar to Klein's, followed with an inconclusive discussion of the mass horizon idea and singularities (Eddington 1923, 164–6). Eddington suggested incorrectly that the De Sitter solution could not be fully covered by a single, entirely regular coordinate system. The young Hungarian physicist Cornelius Lanczos, drawing on Klein's work, showed that Eddington was mistaken by explicitly writing the De Sitter solution in a form that was globally singularity free (Lanczos 1922).[54] Klein did convince Einstein of the error of his ways, and although Einstein admitted the mistake in letters to Klein and to De Sitter's colleague Paul Ehrenfest,[55] he never publicly retracted his earlier criticism.

However, accepting that the De Sitter solution was free of singularities did not mean that Einstein accepted it as a physically reasonable model. Instead he rejected the De Sitter solution because it was nonstatic. The requirement that any physically reasonable model must be static seems to have emerged opportunistically as a way to rule out De Sitter's anti-Machian solution. At least, Einstein gives no indication of a deeper or more plausible motivation for this requirement. What started as a weakly motivated simplifying assumption in deriving his own model had thus turned into a substantive constraint, a constraint that Einstein did not critically examine for the next decade. If he had, he would have discovered that nonstatic models are much more natural in general relativity than the static model he preferred.

This early study of singularities shows the first generation of relativists grappling with difficult problems raised by Einstein's new theory, and the study of the De Sitter solution shaped other areas of research. Lemaître (1932) gave an insightful study of the alleged singularity in the Schwarzschild solution that drew on his earlier work regarding the De Sitter solution, as Eisenstaedt (1993) has emphasized, and the study of alternate coordinate representations of the De Sitter solutions was also connected with Lemaître's discovery of expanding models, as it was for H. P. Robertson. Furthermore, with the benefit of hindsight we can see in these early debates the first steps toward differentiating two distinct features of cosmological models. The odd behavior of the metric that Einstein took to signal the presence of a singularity instead shows that there is an "event horizon" in the De Sitter solution. De Sitter's initial response to Einstein, in which he emphasized that the alleged singularity could be ignored since it was physically inaccessible, contained a kernel of truth. The event horizon demarcates regions from which light signals can and cannot reach a given observer; regions beyond the event horizon are thus physically inaccessible to the observer. The event horizon is not an intrinsic feature of space-time, since it is defined relative to a given observer.[56] The study of horizons has been an important part of cosmology since its earliest days, and the presence of horizons in the standard cosmological models is one of the puzzles used to motivate a recent modification of the standard cosmological models known as *inflationary cosmology.*

4.3. Redshift

The previous sections focused on general criteria for what might count as a reasonable cosmological model. De Sitter, with his strong empiricist bent, added another way of comparing models *A* and *B* in one of his early discussions (De Sitter 1917b): he predicted a redshift in the spectral lines of light emitted from distant objects, with the amount of redshift a function of the distance. In addition to calculating the effect, De Sitter took an important step in suggesting that the movements of the spiral nebulae rather than the stars (which Einstein had focused on) should be used as gauges of cosmic structure on the largest scales. The nature of the redshift effect and the precise functional dependence of redshift on distance for the De Sitter model were both matters of substantial controversy for the following decade and a half. But by 1930, Hubble's observational work had both convinced many astronomers that a linear relationship between redshift and distance obtained and promoted the idea that De Sitter's model explained the effect.

De Sitter recognized that the redshift effect would be a useful way to compare observations with the new cosmological models. Spectral lines have precise wavelengths that can be measured in the laboratory, making it possible to detect a systematic displacement toward either the red or blue end of the spectrum in light received from distant objects. One cause of such displacement is the Doppler effect, due to the relative motion of the emitter and receiver of a light signal. Light received from an object moving toward the receiver is shifted toward the blue end of the spectrum (by an amount that depends on the relative velocity of the emitter), whereas light from an object moving away from the receiver is shifted toward the red end of the spectrum. (The changing frequency of the siren of an ambulance approaching and receding illustrates this effect for sound waves.) Cosmology textbooks typically include a stern warning not to confuse Doppler shifts with cosmological redshift. The Doppler shift is caused by the motion of the emitter, whereas the cosmological redshift results from the stretching of the wavelength of light with the expansion of space (see Figure 7.4). The nature of the cosmological redshift was, however, not so clear to De Sitter and his contemporaries.

The disputes regarding cosmological redshifts can be traced back to two basic problems.[57] The first, and most important, problem is that calculations of the redshift effect require choosing the world lines representing the observer and the observed object, or, more generally, a set

Doppler Effect   *Cosmological Redshift*

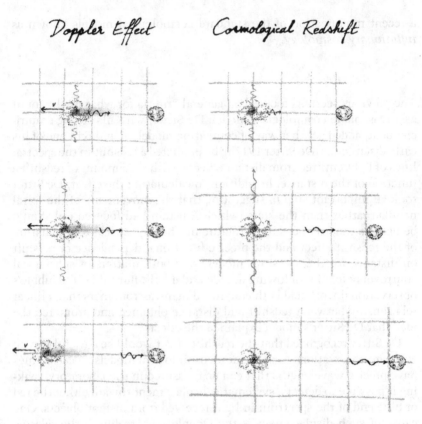

Figure 7.4. *Contrast between Doppler effect and cosmological redshift. The Doppler effect is due to the motion of the emitter, which effectively "stretches" the wavelength of the emitted light. Cosmological expansion also leads to redshift, because light emitted from distant objects is "stretched" by the expanding space.* Drawn by Laurent Taudin.

of world lines for several objects. Without this specification a cosmological model is fundamentally incomplete. From a modern point of view, choosing the appropriate world lines is often completely straightforward; one might choose, for example, world lines for freely falling test particles. These particles follow geodesics, lines of extremal length, through the space-time.[58] Weyl recognized the need to specify the class of curves traced out by objects in a cosmological model, and proposed using a set of geodesics that converge at a point in the past to represent the world lines of galaxies, based on a fairly obscure causal argument (see Goenner 2001). In the early debates these choices were not always clearly stated, and the confusion was exacerbated by the different coordinate representations of the models being considered. For example, in the static coordinates frequently used by De Sitter, the world lines for particles at rest with respect to the coordinates are not geodesics. Particles following world lines that appeared to be "at rest" with respect to static coordinates were in fact undergoing acceleration as they departed from geodesic motion. Second, within general relativity the cosmological redshift (or blueshift) can be calculated for a specific cosmological model given a choice of curves traversed by the observer and emitter. While the result of this calculation is independent of the choice of coordinates, as it should be, identifying the causes of the calculated redshift does depend upon this choice. There are six distinct causes of redshift in a cosmological model.[59] In different coordinate frames, the calculated redshift effect will be traced back to different causes. Theorists at the time often did not take all of these different causes into account, failed to draw the appropriate distinctions among them, and did not appreciate how their arguments and calculations depended on the choice of coordinates. Calculations of the redshift effect in the same model often led to conflicting results due to these two problems. De Sitter initially argued that the redshift should increase quadratically with distance in his model B, whereas Weyl and Robertson (1928) later derived a linear redshift-distance relation.

These disputes regarding how to calculate and understand the redshift occurred even among those who accepted general relativity and its cosmological models. There were also many astronomers who threw out cosmological redshifts with general relativity. The relativistic account of cosmological redshifts, such as it was, met with widespread skepticism among astronomers, leading to a variety of alternative explanations (such as Fritz Zwicky's "tired light" hypothesis). None of these alternative explanations has been successful.

On the observational side, the main difficulty was finding a way to reliably measure distances to the spiral nebulae. Vesto Slipher had begun measuring redshifts in the 1910s, and Eddington (1923) reported that of Slipher's forty nebulae, thirty-six exhibited redshifts, on average quite large. Slipher made his observations with a small twenty-four-inch refractor at the Lowell Observatory, and in 1928 Hubble and his collaborator M. L. Humason turned the 100-inch Mount Wilson telescope to the task of measuring redshifts. But in addition to the telescope power, Hubble was able to add another crucial component – reliable distances to the nebulae based on observations of Cepheid variables. By the time he published his results Hubble had redshift and distance measurements (of varying quality) for forty-six nebulae, and the data fit a linear redshift-distance relation (Hubble 1929).[60] Hubble had made one of the outstanding discoveries of twentieth-century astronomy.

5. THE EXPANDING UNIVERSE

After a decade of study and debate focused mainly on the Einstein and De Sitter models, several cosmologists belatedly reexamined their options. De Sitter's presentation to the Royal Astronomical Society in January 1930 aptly summed up the sorry state of affairs (Royal Astronomical Society 1930). Rather than making a decisive case for one model or another, De Sitter concluded that *neither* model was completely satisfactory.

On the one hand, Einstein's static model failed to account for the observed redshifts of the spiral nebulae. Eddington had recently uncovered another serious defect of the model. Einstein's model treated the distribution of matter (represented by the cosmological constant) as perfectly uniform, and Einstein had assumed that it would be possible to construct a more realistic model that treated "normal" matter as lumps or condensations within this uniform background.[61] However, Eddington showed that Einstein's solution was unstable: a slight departure from uniformity would trigger runaway expansion or contraction (Eddington 1930). Departures of the mass density from the value in Einstein's model ($\Lambda = \kappa\rho/2$) are enhanced through dynamical evolution, because the mass density changes while the cosmological constant remains fixed under expansion or contraction. A local concentration of matter triggers contraction; the effect of the contraction is to increase the matter density, which then differs even more from the (fixed) value of the cosmological constant, leading to further contraction. (Similarly for the case of a local deficit in density, triggering expansion.) As Eddington emphasized, the result follows directly from an equation Lemaître had derived from Eq. (7), and it is surprising that the instability escaped notice for so long.[62] Eddington did not immediately discard Einstein's solution due to the stability problem; instead, he followed Lemaître in suggesting that Einstein's model might describe the initial state of the universe, with some event triggering a transition to the De Sitter model. But once this instability was recognized, it was no longer possible to treat the model as Einstein had, as an approximation to the real, lumpy universe.

On the other hand, De Sitter's model predicted a redshift effect apparently compatible with observations. But it also failed to give a realistic approximation to the observed universe. De Sitter calculated the mean density of matter required in his model based on observed redshifts, and concluded that it was too high to be approximated as a vacuum.[63] As Eddington (1933, 46) put the question, "Shall we put a little motion into Einstein's world of inert matter, or shall we put a little matter into de Sitter's Primum Mobile?"

One answer to the dilemma was to drop one of the assumptions that led to such a short list of viable models. By the time of De Sitter's presentation, both Eddington and De Sitter had finally begun to question in print the assumption that any physically reasonable model must be static.[64] (Although De Sitter immediately criticized Einstein's argument that the universe must be static in 1917, he did not advocate a systematic study of dynamical solutions in print until 1930.) The American cosmologist Richard Tolman had reached a similar conclusion in the previous year, after surveying the problems facing all of the static models (Tolman 1929a).[65] All three suggested that dynamical cosmological models were a suitable topic for further research – only to discover that these models had already been studied during the 1920s. After reading the proceedings of the January 1930 Royal Astronomical Society meeting, Lemaître wrote to remind Eddington of a study of dynamical solutions he had published three years earlier (Lemaître 1927).[66] Eddington promptly advertised Lemaître's "brilliant work" in a letter to *Nature* and published a translation of Lemaître's paper in the *Monthly Notices* (Lemaître 1931), where it would reach a wider audience than the Belgian journal in which it had originally appeared.[67] Lemaître had investigated evolving models without knowledge of an even earlier pair of papers by Friedmann (1922, 1924). Although Friedmann's papers appeared in the prominent German journal *Zeitschrift für Physik*, they also garnered little attention. Einstein's (1922e) initial claim that Friedmann's results rested upon a mathematical mistake can be partially blamed for this lack of interest. To paraphrase (De Sitter 1931, 584), most cosmologists discovered expanding universe models in 1930, several years after they had first appeared in print.

These models can also be derived elegantly by appealing to symmetries, as Robertson and A. G. Walker showed in 1935. As a result they are often called the Friedmann-Lemaître-Robertson-Walker (FLRW) models. Robertson (1929) arrived at expanding models in a mathematical study of all possible solutions to Einstein's field equations with uniform spatial sections. Six years later he considered the relationship between these relativistic models and "kinematical relativity" advocated by E. A. Milne (1935). Milne aimed to derive cosmological models by extending the kinematical principles of *special* relativity. The project was motivated by an operationalist methodology and a conviction that Einstein's *general* relativity was unsound.[68] Robertson (1935, 1936a, 1936b) and Walker (1935, 1936) independently showed that Milne's distinctive approach led, ironically, to the same set of basic models already studied in relativistic cosmology.[69] Walker showed that the FLRW metric follows from Milne's basic principles: the "cosmological principle," which requires observational equivalence among fundamental observers, along

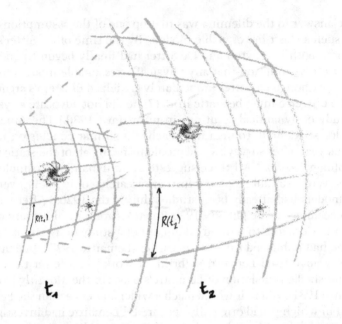

Figure 7.5. *The scale factor* R(t) *measures the rate of change of distance between freely falling particles as a function of time. Although the cosmological expansion has dynamical effects on local systems, these effects are incredibly small – far too small to "stretch" systems along with expansion.* Drawn by Laurent Taudin.

with a further symmetry principle stating that the model is spherically symmetric around each fundamental observer.[70] Robertson and Walker applied group theory to derive the consequences of such strong symmetry principles for the space-time metric. Their work showed how the FLRW metric follows from simple geometrical assumptions.

The discovery of the FLRW models made it clear that the Einstein and De Sitter solutions are special cases of a more general class of solutions. Friedmann retained two of Einstein's starting assumptions: that the space-time can be decomposed into three dimensional spatial sections corresponding to different cosmic times t, and that these sections are uniform, without preferred locations or directions in space.[71] He then showed that these two assumptions could be satisfied in dynamical models, in which the curvature of the spatial sections changes with time. The change in curvature corresponds to expansion or contraction in the sense that the distance measured on any spatial section between freely falling test bodies changes with time; the so-called scale factor $R(t)$ measures the rate of change of this distance (see Figure 7.5).[72] Using the simplifying assumptions stated previously, Friedmann showed that Einstein's field equations – in general a complicated set of ten nonlinear, coupled partial differential equations – reduced to a simple pair of differential equations for $R(t)$.[73]

The set of solutions to these equations describe simple dynamical models that have since become the standard models of relativistic cosmology. There are three classes of solutions to this pair of equations, typically classified according to the curvature of the spatial sections and the corresponding geometry. Friedmann (1922) discovered what are now called the *spherical models*, whose spatial sections have positive constant curvature and finite volume (like a sphere), and Friedmann (1924) introduced hyperbolic models, with infinite spatial sections of constant negative curvature. Robertson (1929) discovered the intermediate flat case between these two, namely, infinite spatial sections with zero curvature. For all three cases, the two dynamical equations relate the scale factor $R(t)$ to the properties of the matter filling the space-time. From the equations for $R(t)$ one can define a critical density that divides the three cases described. In addition, normal matter[74] leads the expansion of the universe to decelerate, that is, $\ddot{R}(t) < 0$. The flat solution corresponds to a universe with exactly the critical density, and the initial velocity of the expansion is delicately balanced with this deceleration due to gravitational attraction. Models with less than the critical density have hyperbolic spatial sections and expand forever, whereas the spherical models have sufficient matter density to stop and reverse the initial expansion. The equations for $R(t)$ also illustrate that the simplest cosmological models are inherently dynamical. Einstein's preference for a static model was satisfied by choosing precisely the value of Λ needed to counteract the attraction of gravitation and yield $\ddot{R}(t) = 0$.

Einstein did not overlook Friedmann and Lemaître's work before 1930, but he rejected expanding models as unphysical due to his persistent preference for a static model. Even after he retracted his initial claim (Einstein 1922e) that Friedmann had made an error, he clearly did not think that the expanding models were of any use in describing the observed universe. Stachel (1986c) discovered that Einstein's draft retraction closed by stating that no physical significance could be attributed to Friedmann's models.[75] Fortunately, this strong statement did not appear in print, but several years later Einstein expressed much the same view in conversation with Lemaître. According to Lemaître's account of their 1927 meeting, Einstein acknowledged that he had to withdraw his criticism of Friedmann's work but still found expanding models "totally abominable" from a physical point of view.[76] Friedmann seems to have shared, to a degree, Einstein's reluctance to attribute physical significance to the expanding models; his papers explored the properties of the hyperbolic and spherical solutions without linking them to the observed universe. Friedmann approached cosmology like a mathematician exploring the space of solutions to a differential equation. By way of contrast, Lemaître (1927) clearly regarded the spherical model (derived independently of Friedmann) as possibly giving a description of the universe's evolution. Unlike Friedmann, Lemaître explicitly derived the redshift effect and argued that his model naturally accounted for observed redshifts, and he also speculated about the physical origins of the expansion.

In the early 1930s theorists began to develop a richer account of the evolution of the universe based on the expanding models. Hubble's results qualitatively agreed with the redshift effect calculated in these models, but the utility of the simple dynamical models depends upon whether the universe is approximately uniform. The status of this assumption was the focus of lively debate, partly due to Milne's advocacy of an alternative theory based on what he called the "cosmological principle." Relativistic cosmologists regarded the idealized uniformity of the FLRW as a simplifying assumption compatible with observations but not justified by them (see, e.g., Tolman 1934, 332), rather than a methodological principle. The unrelenting uniformity built into the FLRW models conflicts with the clear nonuniformities of the stars, star clusters, and galaxies of the local universe, but the models might still serve as a useful approximation if the nonuniformities are negligible at larger scales. In 1926 Hubble initiated an observational program to measure the uniformity of the large-scale distribution of nebulae, but the results of this program were challenged by Shapley.[77] Shapley and Hubble disagreed about the significance of observed small-scale clumping. In any case, the observational debates did not force theorists to abandon the simple expanding models.

There were, however, two widely acknowledged limitations of these simple models. First, extrapolating the models backward in time leads to a genuine singularity (later derisively called the "big bang" by Fred Hoyle) that cannot be blamed on a poor choice of coordinates. Eddington and Lemaître avoided this initial singularity by suggesting that the universe began in an Einstein static state and then decayed into an expanding model. Einstein's cosmological constant could still serve the useful purpose of avoiding this initial singularity, and throughout the 1930s Lemaître and others studied a wide variety of evolving models with nonzero Λ.[78] Lemaître later proposed a speculative theory of the initial state; on this account, the universe began as a "primeval atom" and has a finite age. A more common response, advocated by Einstein as well as Tolman, blamed the existence of the singularity on the strong idealizations built into the simple models. Perhaps a more detailed model describing the lumps and bumps of the real universe would not have evolved from an initial singularity, or so the suggestion went. (The singularity theorems of the 1960s showed that the singularity could not be blamed on the symmetries of the standard models.) The second limitation is that the uniform models provide no insight into the formation of nonuniformities, such as galaxies and stars. The cosmological models did not directly answer the cosmogonic problem that had concerned Jeans and others before the introduction of general relativity. Starting in the early 1930s, Lemaître, McCrea, and others studied the growth of small clumps of matter in a background expanding model with the hope of shedding some light on the formation of galaxies.

Einstein missed the chance to discover the expanding models, and Friedmann and Lemaître's papers did not convince him of their physical importance. Throughout the 1920s he apparently did not waver in his insistence on static models. He did reconsider the status of the cosmological constant but did not abandon it. He acknowledged that his earlier introduction of the cosmological constant was "gravely detrimental to the formal beauty of the theory" (Einstein 1919a), but immediately went on to argue that it could be treated more satisfactorily as a constant of integration.[79] Hubble's observations convinced him to abandon his preference for static models as well as the cosmological constant. During a trip to Pasadena from December 1930 to March 1931, Einstein learned about the latest observations at Mount Wilson firsthand from Hubble and his colleagues. Tolman persuaded Einstein that dynamical models were preferable to static models for describing the observed universe.[80] Shortly after returning to Berlin, Einstein published a paper that discussed the expanding models favorably and noted the shortcomings of his own static model (Einstein 1931b). During another visit to CalTech the following year, Einstein wrote a brief paper with De Sitter describing the properties of a simple expanding model with zero spatial curvature (Einstein and De Sitter 1932). Both of these papers emphasized that Hubble's results completely undermined the original rationale for introducing the cosmological constant, which was to insure the possibility of a static universe with a finite mean matter density. Einstein later reportedly regarded its introduction as his "biggest blunder."[81]

After this decade and a half of sometimes intense work on cosmology, Einstein returned to the subject only occasionally in his later years. His most significant later contribution was a discussion of the impact of cosmological expansion on the gravitational field surrounding a star. The Schwarzschild solution used to describe such a field asymptotically approaches Minkowski space-time as the distance from the star goes to infinity. But this cannot be correct for a star within an expanding universe. What are the consequences of treating the star as part of an expanding model? Einstein and Straus (1945) showed that the Schwarzschild solution could be embedded in an expanding FLRW model, and that despite the time dependence of the background cosmological model the field near the star remains static. This was an important first step in understanding the impact of global cosmological expansion on local physics. Einstein's other research in general relativity periodically touched on cosmology, but it was no longer a major focus of his work.

6. CONCLUSION

Within fifteen years after the introduction of Einstein's theory, theorists had surveyed the properties of a number of simple, idealized solutions of the new gravitational field equations. Einstein's groundbreaking work initiated the study of cosmological models, although several contemporaries were more clear-sighted in understanding the properties of these models, and Einstein's preference for a static model kept him from discovering or initially accepting the expanding models. The exploration of these models touched on a number of the novel features of general relativity, including event horizons and singularities, that have been a focus of further research. The work described has been aptly called "geometrical cosmology," given its very mathematical style. However, Hubble's discovery of the redshift-distance relation provided a link with astronomical observations that encouraged many astronomers and theorists to take these models seriously. The new consensus regarding the expanding models was codified in systematic reviews (such as Robertson 1933) and textbooks (Tolman 1934).[82]

However, this consensus faced a number of objections, ranging from empirical anomalies to questions regarding the legitimacy of taking cosmology to be the study of idealized solutions of Einstein's gravitational field equations. As critics of relativistic cosmology rightly pointed out, the links between the expanding models and observations were remarkably tenuous. The expanding models provided a natural explanation of the observed redshifts of the spiral nebulae. However, the use of these observations to constrain parameters of the models led to a striking problem: for expanding models with a finite age, the "age of the universe" often turned out to be far less than the age estimated for various astronomical objects! The eventual resolution of this problem two decades later came not from a change in the cosmological models but a recalibration of the distance scale based on Cepheids. Prior to this recalibration the opponents of the expanding universe models often cited the age problem as a serious anomaly (among others) for relativistic cosmology.

Debates in the next two decades extended beyond such empirical problems. Critics of relativistic cosmology introduced rival cosmological theories based on allegedly sounder scientific methodology than that followed by Einstein and the other relativists. Although I do not have space to explore these debates in detail, a brief discussion of the status of the cosmological principle will convey some of the issues at stake. Within the standard approach to relativistic cosmology, the cosmological principle was typically treated as a simplifying assumption used to arrive at mathematically tractable models. Although the assumption of uniformity could be precisely characterized mathematically, there was admittedly no physical motivation for demanding such a high degree of symmetry. As a result, it was unclear what, if anything, the failure of the observed universe to satisfy the uniformity assumption indicated regarding the status of the models, and whether the models could be refined to give more realistic descriptions of the observed matter distribution. By way of contrast, the main alternative to relativistic cosmology throughout the 1930s, Milne's kinematic relativity, elevated a version of the cosmological principle to a fundamental axiom. Milne's unabashed rationalism in treating the cosmological principle as an *a priori* axiom led to a series of heated debates regarding method in cosmology, and Milne's distinctive approach influenced work in relativistic cosmology (as noted) even though kinematic relativity never won widespread acceptance. Admiration of Milne's approach also played a role in the discovery of the second major alternative to relativistic cosmology, the steady-state theory introduced in 1948 by Herman Bondi, Thomas Gold, and Fred Hoyle (Bondi and Gold 1948, Hoyle 1948).

Bondi and Gold (1948) proposed the "perfect cosmological principle" as a response to an apparent threat to the extrapolation of local physical laws to cosmological scales. Relativistic cosmology certainly involves an enormous extrapolation of Einstein's theory, from its empirical testing grounds of roughly solar-system scale to cosmological length scales such as the Hubble radius (the length scale of the observable universe), roughly fourteen orders of magnitude larger. The steady-state theorists were not primarily concerned with the threat of new physics arising at greater length scales; instead, they followed Milne in insisting that cosmology faced distinctive methodological problems due to the uniqueness of the universe. In particular, they argued that the distinction between "laws" and "initial conditions" familiar in other areas of physics does not carry over to cosmology. Since this distinction could not be drawn clearly, perhaps there would be some form of "interaction" between local physical laws and the features usually described as initial conditions, such as the global distribution of matter.[83] The perfect cosmological principle was introduced to rule out the possibility that the local physical laws evolve along with the changing universe by simply stipulating that the universe does not change. Bondi and Gold argued for this principle on methodological grounds: if the principle fails, they argued, any extrapolation of local laws to cosmological scales is unjustified, since it might neglect possible interactions. The principle leads to a "steady-state" universe with unchanging global features, and the reconciliation of this picture with the observed expansion forced the steady-state theorists to relinquish conservation of matter. The steady-state theory provoked sharp debates throughout the 1950s, focused in part on whether cosmology should employ such a distinctive methodology. The perfect cosmological principle provided a tight constraint on cosmological theorizing, as the steady-state theorists had hoped, but by the early 1960s the constraint proved too tight – and several advocates of the theory abandoned it in light of new observational evidence.

The status of the cosmological principle changed again in 1965 with the discovery of very low temperature radiation by Arno Penzias and Robert Wilson. The Princeton physicist Robert Dicke immediately interpreted this radiation as the remnant of a hot big bang in the early universe, and it provided a final piece of evidence against the steady-state theory.[84] But more importantly, the discovery of the background radiation provided an empirical touchstone that encouraged a dramatic increase in research effort devoted to cosmology. This effort included the development of detailed accounts of the nuclear reactions taking place in the first three minutes of the universe's history (an idea first introduced by Gamow, Alpher, and Herman a generation earlier, but largely ignored), and theorists also explored competing accounts of the formation of galaxies and other large-scale structure. The uniform temperature of the background radiation indicated that the simple mathematical assumption of uniformity built into the evolving models was far more accurate than cosmologists had expected, and it also provided evidence that these models applied to the universe at very early times.

Cosmologists had reason to be puzzled rather than exhilarated by this result, since they lacked an explanation of why the universe should be in such a highly symmetric state soon after the big bang. The attempt to resolve this final puzzle has pushed cosmological theories to incredibly early times ($t \approx 10^{-35}$ seconds after the big bang!), where exotic features of particle physics theories become relevant. "Particle cosmology" has been an active area of research for the last twenty years, partially because the "poor man's accelerator" (to use the Soviet cosmologist Zel'dovich's apt description of the early universe) provides perhaps the only way of testing aspects of particle physics well beyond the reach of earthbound accelerators. The field has been dominated by inflationary cosmology, introduced in Guth (1981), which explains the puzzling uniformity of the early universe as the consequence of a brief period of exponential ("inflationary") expansion in the early universe. In light of Einstein's early research in cosmology, the theory of inflationary cosmology is doubly ironic, since it reintroduces Einstein's infamous cosmological constant Λ (admittedly for very different reasons) and it makes use of the De Sitter solution, the model Einstein argued so strenuously against.

ACKNOWLEDGMENTS

I have benefited from comments and discussions regarding earlier drafts from Tyler Burge, Sarah Gallagher, David Malament, Charles Midwinter, Sheldon Smith, and especially Michel Janssen. I am grateful to Laurent Taudin, who drew the diagrams for this paper. I also gratefully acknowledge the debt to my two mentors in history and philosophy of physics, John Earman and John Norton, from whom I learned a great deal about the topics discussed herein.

NOTES

1 This brief overview draws on several comprehensive historical accounts, primarily Smith (1982), North (1965), Paul (1993); see also North (1995). Macpherson (1929) and Clerke (1890) offer well-informed contemporary overviews of the field that bracket this time period.

2 Jeans coined the name following Kapteyn's publications in the early 1920s, and Seeliger's research supported a similar model (see Paul 1993, 141–50). Estimates of the scale of the system were subject to a great deal of uncertainty. Eddington (1914) argued that the dimensions are roughly 1,100 light-years (along the shorter axis) by 3,200 light-years (along the galactic plane), where the distance is measured from the center to the region where the density of stars drops to below one-fifth of the maximum density. Six years later Kapteyn and van Rhijn (1920) estimated the size of the galaxy to be roughly 6,100 by 31,000 light-years.

3 Advocates and opponents of the island universe theory advanced a variety of ideas regarding the size, distance, and nature of the spiral nebulae and other astronomical objects; in this sense there was not a single "island universe theory" throughout this period. E.g., one disputed topic was whether globular clusters also lie outside the Milky Way and have some connection with the spirals. For a more fine-grained analysis of these debates that highlights such differences, see Smith (1982).

4 For further discussion of these objections and references to the original literature, see Smith (1982, 1–54) and North (1965).

5 The observational situation was muddied by the failure to distinguish novae from much more luminous supernovae. At the time, the small sample of novae used in the comparison included two supernovae, leaving Shapley and other island universe opponents ample reason to doubt that novae in the spirals were dimmer than those within the galaxy (North 1965, 2–13).

6 There is much less dust along the line of sight to objects near the galactic poles than there is to objects near the galactic plane. Even if the spirals were uniformly spread throughout the sky, they would appear to be clustered around the galactic poles due to this differential absorption.

7 The observed brightness (apparent magnitude) of the star decreases with distance, so if the intrinsic brightness (absolute magnitude) is known the distance can be calculated.

8 The published exchange appeared a year later (Curtis 1921, Shapley 1921). Hoskin (1976) argues that Shapley tailored his performance to secure a position as the director of the Harvard College Observatory, which had recently opened up with the death of E. C. Pickering. See also Smith (1982, 77–90) for a detailed discussion of the Great Debate and its context.

9 In 1917, Shapley accepted the island universe theory but became increasingly critical in 1918 and 1919 (Smith 1982, 85–6).

10 Hubble measured the distance of the spiral nebula Messier 33 to be roughly 930,000 light-years, compared to van Maanen's estimate of 9,800 light-years.

11 Three decades after Hubble's pioneering work, Walter Baade showed that earlier distance measurements had failed to take into account the existence of two *distinct* populations of variable stars, each with a characteristic period-luminosity relation (Baade 1952). Taking this into account dramatically changed the distance scale, literally doubling the earlier distant estimates.

12 Here I draw on Brush's (1996a, 1996b) rich historical account of the nebular hypothesis and the competing ideas introduced in the early twentieth century.

13 The subject of the 1917 Adams Prize at Cambridge was "the course of configurations possible for a rotating and gravitating fluid mass, including the discussion of the stabilities of the various forms" (Jeans 1919, v); Jeans's innovative 1919 book extended his prize-winning essay. See Smith (1982) for an account of Jeans's influence and his interactions with van Maanen, Curtis, and others.

14 E.g., from roughly 1916 to 1924 Jeans thought that van Maanen's measurements of radial velocities in the spiral nebulae fit naturally into his account of the dynamical evolution of spiral nebulae (Smith 1982), an unfortunate association given the later problems with van Maanen's measurements.

15 See Torretti (2000) for an insightful discussion of the nature of cosmological models and what he memorably calls Einstein's "unexpected lunge for totality" in introducing the first cosmological model. I am indebted to Torretti's article and previous work, especially in this section and in Section 5.

16 In this subsection, I draw on Norton's (1999b) comprehensive discussion of the difficulties with Newtonian cosmology and their history; cf. Jaki (1990) and North (1965).

17 Einstein was unaware of Seeliger's papers when he wrote his cosmology paper (Einstein 1917b), but after they came to his attention he duly cited them (see, in particular, Einstein 1919b, 433, footnote 1, and Einstein to Förster, November 16, 1917 [CPAE 8, Doc. 400]). In 1692, Richard Bentley raised the problem in correspondence with Newton while preparing his inaugural Boyle lectures for publication (Newton 1959–77, Vol. III, 233–56).

He pressed Newton to consider the following question. For a given particle of matter in a uniform, infinite mass distribution, the gravitational force in any chosen direction is infinite. But would the pull exerted by the infinity of mass lying in the opposite direction cancel this first infinite force, so that the forces in different directions would be in equilibrium? Bentley balked at comparing infinities, but Newton answered with a definite yes, apparently without recognizing the deeper problem. The force on the particle is found by integrating over the contributions due to all other masses, but this integral fails to converge for a uniform, infinite distribution of mass.

18 Seeliger proposed adding an exponential decay term to Newton's gravitational force law. Einstein's modification is not equivalent to this, although Carl Neumann had proposed a modification equivalent to Einstein's, with different motivations, in 1896 (see Norton 1999b, 293–8).

19 See Norton (1999b, 302–13) for a clear description of the properties of these hierarchic models and an explanation of how these models avoid the divergences.

20 Treating time-varying matter distributions is clearly not sufficient to avoid the difficulty, since even in an evolving model the universe may pass through a uniform state. Milne (1934) and McCrea and Milne (1934) do not address the question directly, and Layzer (1954) argued that the models do not apply consistently to an infinite universe, even though they can be consistently applied to a finite mass distribution. McCrea (1955) and Heckmann and Schücking (1955, 1956a) aptly defended the Neo-Newtonian models along lines similar to those discussed in the text, but without appealing to the geometric formulation (cf. Layzer 1956 and Heckmann and Schücking 1956b).

21 See Malament (1995) and Norton (1995, 2003) for thorough treatments of the geometric formulation and its implications for Newtonian cosmology.

22 Additional solutions to Equation (1) can be obtained by adding a harmonic function (a function ψ such that $\nabla^2 \psi = 0$) to one of these solutions. These additional solutions can be eliminated by requiring that ϕ is isotropic around r_0.

23 One other feature of Einstein's note is worth highlighting: he comments that the latest observations refute the "hypothesis of the equivalence of the spiral nebulae and the Milky Way," although it is not clear which observational results he had in mind.

24 Einstein later added Mach's principle – "The metric field is determined *without residue* by the masses of bodies" – to the list of three foundational principles of general relativity (Einstein 1918e, 242, original emphasis).

25 The editorial note "The Einstein-De Sitter-Weyl-Klein Debate" (CPAE 8, 351–7) gives a thorough discussion of the correspondence between Einstein and De Sitter, and the later participants Weyl and Klein. See also Section 5 of Chapter 6.

26 Einstein (1915h) introduced the boundary conditions of the Schwarzschild solution without commenting on the potential conflict with his Machian ideas. However, there are hints that Einstein had already begun to worry about these issues (see Hoefer 1994, 298–303) in both his comments regarding the status of Minkowski space-time in the context of general relativity and

a letter to Lorentz from the winter of 1915 (Einstein to Lorentz, January 23, 1915 [CPAE 8, Doc. 47]).

27 As Torretti (2000) has emphasized, Birkhoff's (1923) theorem shows that the Schwarzschild solution can be derived from the assumption of spherical symmetry *without* further stipulating boundary conditions at infinity. Thus in the case Einstein had in mind, the boundary conditions are not logically independent from the symmetry requirements.

28 Einstein and De Sitter also studied the transformation behavior of different boundary conditions as a way of testing their consistency with Mach's principle. Einstein apparently thought that degenerate values would be generally covariant, but De Sitter showed that this claim was incorrect. De Sitter pointed out that even the "degenerate values" were not invariant under certain transformations (see De Sitter to Einstein, November 1, 1916 [CPAE 8, Doc. 272] and De Sitter [1916c, 1917a]).

29 De Sitter to Einstein, November 1, 1916 (CPAE 8, Doc. 272).

30 More precisely, Einstein takes the matter distribution and the corresponding metric field to be homogeneous and isotropic. Roughly, homogeneity is satisfied if there are no "preferred locations" on a spatial section, and isotropy holds that at a given location there are no "preferred directions" in space. The modern treatment of these symmetries traces back to the work of Robertson and Walker described briefly in the following text.

31 A few years later Weyl introduced a contrast between "static" and "stationary" solutions that has since become standard: in a stationary solution, the spatial components of the metric are not functions of time but the "mixed" space-time components may be functions of time, whereas for a static solution only the time-time component of the metric is allowed to vary with time. More intuitively, spatial distances remain fixed under time translations in a stationary solution, and this is compatible with a perfectly rigid rotation; there is no such rotation in static solutions.

32 De Sitter to Einstein, April 1, 1917 (CPAE 8, Doc. 321, original emphasis). In a later letter (De Sitter to Einstein, June 20, 1917 [CPAE 8, Doc. 355]) he considered the constraints that low stellar velocities would put on a cosmological solution, and argued that they do not lead to a static solution as Einstein had thought.

33 Einstein to De Sitter, before March 12, 1917 (CPAE 8, Doc. 311).

34 This change is not "perfectly analogous" to the addition of a constant in Poisson's equation, as Einstein claimed. The Newtonian limit of the modified field equations yields $\nabla^2 \phi + \Lambda \phi = 4\pi G\rho$, rather than Equation (3) (Trautman 1965, 230).

35 The Ricci tensor $R_{\mu\nu}$ is defined in terms of the metric field $g_{\mu\nu}$ and its first and second derivatives, and the metric field characterizes the geometrical properties of space-time. The stress-energy tensor $T_{\mu\nu}$ represents the distribution of matter, energy, and stress throughout the space-time.

36 The left-hand side of Equation (7) is the most general symmetric tensor that can be constructed from the metric $g_{\mu\nu}$ and its first and second derivatives, such that the conservation law for $T_{\mu\nu}$ holds. See Renn (2007a, Vol. 2, 493–500) for further discussion of the physical requirements that guided Einstein's derivation of the original field equations.

37 To be more precise, Λ enters into equations governing, e.g., the behavior of nearby test particles with the opposite sign as matter or energy density. The presence of normal matter or energy leads paths of nearby test particles to converge, and this captures the idea that gravitation is a "force of attraction." A positive Λ term leads to divergence of these paths (if it is strong enough to counteract gravitational attraction), and in this sense it is a "repulsive force."

38 De Sitter to Einstein, April 18, 1917 (CPAE 8, Doc. 327).

39 Minkowski space-time apparently constitutes a violation of Mach's principle as formulated, since it is a vacuum solution. Einstein did not explicitly mention that Minkowski space-time is not a solution with Λ ≠ 0, although he undoubtedly appreciated this point (as did Pauli 1921).

40 "It is to be emphasized, however, that our results give a positive curvature of space even if the supplementary term [Λ] is not introduced; that term is necessary only for the purpose of making a quasi-static distribution of matter possible, as required by the fact of the small stellar velocities" (Einstein 1917b, 152). Torretti (2000, 178) suggested that "Einstein could hardly have published that remark" without knowing of a nonstationary solution with closed spatial sections. Torretti's point is plausible, but as far as I know there is no evidence that Einstein had obtained such a solution in 1917.

41 Prior to the introduction of his own model, De Sitter had focused on the status of the time coordinate in Einstein's solution, and he characterized his solution as treating the temporal coordinate as "closed" in the same manner that Einstein closed the spatial dimensions. He gives credit to Ehrenfest for the idea of getting rid of temporal as well as spatial boundary conditions by postulating a "spherical" space-time (De Sitter 1917c).

42 De Sitter to Einstein, March 20, 1917 (CPAE 8, Doc. 313).

43 I am setting aside debates regarding the topology of De Sitter's space-time; Klein, Einstein, and others discussed whether antipodal points should be identified or treated as distinct points. (Antipodal points lie on opposite "sheets" of the hyperboloid and are connected by a line passing through the center of the hyperboloid.)

44 Einstein to De Sitter, June 22, 1917 (CPAE 8, Doc. 356).

45 See, in particular, Chapter 6, Earman and Eisenstaedt (1999), and Goenner (2001) for further discussions of the debates regarding De Sitter's solution.

46 De Sitter to Einstein, June 20, 1917 (CPAE 8, Doc. 355).

47 For discussion of geodesics in the De Sitter solution, see Schrödinger (1956). Geometrically, referring to Figure 6.9, timelike geodesics are defined by the intersection of the hyperboloid and vertical planes through the origin. These are curves of constant spatial coordinates in pseudopolar coordinates, which cover the entire hyperboloid. But the curves of constant spatial coordinates in static coordinates are not geodesics, as can be shown by considering the coordinate transformation from pseudopolar to static coordinates.

48 See Chapter 6 regarding Klein's contribution and his exchange with Einstein, and for further discussion of the properties of the De Sitter solution. See also Schrödinger (1956) for a particularly clear treatment of the De Sitter solution.

49 See "The Einstein-De Sitter-Weyl-Klein Debate" (CPAE 8, 352–4) for a discussion of Einstein's other criticisms and De Sitter's responses.

50 See Earman (1999) for a history of the problems related to singularities, culminating with a discussion of the famous Penrose-Hawking singularity theorems.

51 See Earman and Eisenstaedt (1999, 189–93) for further discussion of Einstein's definition and its relation to an alternative, more stringent definition proposed by Hilbert.

52 Written in static coordinates, the g_{tt} component of the metric is given by $\cos^2(r/R)$ (where r is a coordinate and R is a constant), and this goes to zero at the "mass horizon" $r = \pi R/2$. See also Section 5 of Chapter 6.

53 The construction only appeared in the second edition of *Raum-Zeit-Materie*. Weyl and Einstein exchanged several letters regarding Weyl's results, which Einstein foreshadowed in the closing remarks of (Einstein 1918c). See Goenner (2001) and Earman and Eisenstaedt (1999), as well as the correspondence and the editorial headnote in CPAE 8, for discussions of Weyl's result and its role in Einstein's thinking.

54 Lanczos criticized the mass horizon idea in a second paper (Lanczos 1923); see also Stachel (1994) and Eisenstaedt (1993).

55 Einstein to Ehrenfest, July 6, 1918 (CPAE 8, Doc. 664).

56 Rindler (1956) is the classic paper regarding horizons in cosmology; see Ellis and Rothman (1993) for a more recent and extremely clear discussion.

57 I do not have space to review the disputes regarding redshift during the 1920s. See North (1965), Goenner (2001), and Ellis (1989) for thorough assessments and references to the original papers.

58 See Section 3 of Chapter 6 for a brief explanation of the notion of a geodesic.

59 Following the discussion in Ellis (1989, 374–5), contributions to the overall redshift come from the following six sources: the Doppler shift due to relative motion of the source, gravitational redshift due to inhomogeneity near the source and near the observer, Doppler shift due to the peculiar velocity of the observer, the cosmological expansion or contraction, and the gravitational redshift produced by large-scale inhomogeneities.

60 See Smith (1982), ch. 5, regarding the background of Hubble's work and its reception.

61 De Sitter had voiced some skepticism about this in correspondence with Einstein (De Sitter to Einstein, June 20, 1917 [CPAE 8, Doc. 355]), and Gustav Mie also broached the issue (Mie to Einstein, February 17, 1918 [CPAE 8, Doc. 465], March 21, 1918 [CPAE 8, Doc. 488]), although he argued incorrectly that any inhomogeneities in Einstein's solution would evolve back to a uniform matter distribution.

62 The conclusion follows from Lemaître's equations for the evolution of the scale factor given a perfect fluid with zero pressure: $3\ddot{R}/R = \Lambda - (\kappa\rho/2)$, given that ρ varies with $R(t)$ as $\rho \propto R^{-3}(t)$ whereas Λ remains constant.

63 Tolman (1929b) also criticizes the De Sitter model. Unlike De Sitter, Tolman argued that predicting a correct redshift-distance relation requires ad hoc assumptions regarding the world lines of the spiral nebulae, but like De Sitter he expressed unease at treating galaxies as test bodies within a vacuum solution.

64 At this time the De Sitter solution was regarded as a static solution. Following De Sitter's presentation, Eddington expressed puzzlement regarding why there should be only two solutions, adding that "I suppose the trouble is that people look for static solutions" (Royal Astronomical Society 1930, 39). Similarly, De Sitter (1930) concluded by advocating dynamical solutions in light of the difficulties facing both models; cf. a letter from De Sitter to Shapley quoted in Smith (1982, 187).

65 Tolman remarked that "the investigation of non-static line elements would be interesting" in the final line of a paper presenting the difficulties for static models (Tolman 1929a); in a slightly later paper, Tolman argued that the De Sitter solution does not give a "simple and unmistakably evident explanation" of the distribution and motion of the spiral nebulae, as observed by Hubble (Tolman 1929b).

66 Lemaître had been Eddington's student in 1923–4, and Eddington later remarked that he had probably seen but not appreciated Lemaître's paper before 1930; see Kragh (1996, 31–3) and Eisenstaedt (1993).

67 Lemaître (1931) is not, however, a complete translation of the original paper; Eddington dropped Lemaître's determination of what we now call the Hubble constant, based on the redshift-distance relation for forty-two galaxies.

68 Milne's work and the larger methodological debates it sparked have been the focus of a series of papers by George Gale and various collaborators – see, in particular, Gale and Urani (1993), Gale and Shanks (1996), Gale (2002) for a general overview, and Lepeltier (2006) for a critical response.

69 Despite the significant overlap in mathematical content in these sets of papers, their methodological motivations were quite different. Robertson adopted Milne's operationalist standpoint only to rebut Milne's supposed alternative to general relativity, highlighting its similarities at the level of kinematics and the difficulties facing Milne's kinematical-statistical theory of gravitation. Milne's protegé Walker was, unsurprisingly, much more sympathetic to Milne's approach.

70 In current terminology, the metric is required to be homogeneous and isotropic. Robertson and Walker both set aside Milne's further requirement that the (operationally defined) coordinates assigned by distinct fundamental observers are related by Lorentz transformations, as this singles out the $k = -1$ FLRW model.

71 Torretti (1983, 202–10) describes Friedmann's work in greater detail and compares it to the work of Robertson and Lemaître; cf. Kragh (1996, 22–38).

72 In Einstein's static model the distance between test particles remains constant; De Sitter's solution only appears to be static if one mistakenly focuses on test particles that do not move along geodesics.

73 Friedmann studied solutions of the modified field equations including the cosmological constant Λ, although his results establish the existence of the solutions described with $\Lambda = 0$.

74 Roughly speaking, normal matter is defined as having "positive total stress-energy density." The presence of such matter produces gravitational attraction, other things being equal, in the sense of convergence of the trajectories of freely falling test particles. There are different ways of making the notion

of "normal matter" more precise, leading to various energy conditions in general relativity. Matter which satisfies the so-called strong energy condition contributes to the dynamical equations such that it contributes to deceleration. By way of contrast, a nonzero cosmological constant term does not satisfy this energy condition and does not qualify as normal matter.

75 The last line of the brief correction reads, "It follows that the field equations admit, as well as the static solution, dynamic (that is, varying with the time coordinate) centrally symmetric solutions for the spatial structure, [to which a physical significance can hardly be ascribed]." The phrase in square brackets is canceled in the manuscript (EA 1–206) and does not appear in the printed version (Einstein 1923g).

76 Lemaître described the encounter to Eddington in a letter written in early 1930 accompanying the reprint of his 1927 article. Eisenstaedt (1993, 361) quotes a draft of this letter from Lemaître's papers.

77 See Peebles (1980, 3–11) for an overview of the debate, and Tolman (1934, Secs. 177–85) for a contemporary assessment of the evidence.

78 See Earman (2001) for a discussion of these models and different attitudes to the cosmological constant.

79 Schrödinger (1918) had pointed out another way of treating the cosmological constant: moving it from the left-hand side of Equation (7), where it represents a contribution to space-time curvature, to the right-hand side, where it represents a contribution to the energy-matter distribution. Then it would correspond physically to a kind of cosmic pressure. In his response to Schrödinger, Einstein (1918d) wrote that he had considered this option too but had rejected it as artificial. Einstein (1919a) argues that the cosmological constant term, treated as a constant of integration, also solves a problem in matter theory by contributing a negative pressure term to keep the electrodynamic forces on a small charged particle in equilibrium. For further discussion of the proposal in Einstein (1919a), see Earman (2003).

80 As Stachel (1986c) has pointed out, Einstein's diary from the trip records that he initially had doubts regarding Tolman's argument, but that during this trip he was convinced that Tolman was, in fact, correct (EA 29–134, 27–8). Einstein had a positive impression of the astronomers at Mount Wilson; see, e.g., his letter to Besso written in March 1931, quoted in Stachel (1986c).

81 The canonical source for this story is George Gamow's autobiography: "Einstein's original gravity equation was correct, and changing it was a mistake. Much later, when I was discussing cosmological problems with Einstein, he remarked that the introduction of the cosmological term was the biggest blunder he ever made in his life. But the 'blunder,' rejected by Einstein ... rears its ugly head again and again and again" (Gamow 1970, 149–50).

82 In 1931 the consensus favored a multistage evolutionary model rather than a simple big bang model; see Gale and Shanks (1996) regarding the discussion of this emerging consensus at a 1931 British Association for the Advancement of Science meeting.

83 Mach's principle returned to the limelight as one of the steady-state theorists' main examples of this type of interaction (see, e.g., Sciama 1957). In the case of Mach's principle, the concern was that local inertial structure would change with the overall change in matter distribution in an evolving

cosmos. E.g., Sciama developed a theory according to which Newton's gravitational constant G depended upon the density and expansion velocity of the global matter distribution, such that in an expanding model the current value of G would differ from that at earlier or later times. Bondi and Gold were concerned that other local "laws" of physics may also "evolve" in this sense. Whether the ideas of interaction and evolving laws invoked here make sense is a contentious issue; see Balashov (2002) for a discussion that is sympathetic to the steady-state theorists.

84 Popular accounts of the controversy typically overplay the importance of the background radiation; as Kragh (1996) establishes, the background radiation was the last of several empirical anomalies facing the steady-state theory, and all those who could be convinced to abandon the theory had already done so before 1965.

8   Einstein, Gravitational Waves, and the Theoretician's Regress

Einstein's achievement in formulating the theory of general relativity was so intellectually impressive that the attention of most historians has been focused, justifiably so, on the journey which took him up to that point. However, the development of any new mathematical theory of physics brings with it challenges for physicists, who must interpret this new theory and develop or adapt a new toolbox of mathematical and other techniques, without which the theory cannot make any useful or testable predictions. General relativity seems to have presented an especially daunting face to contemporary physicists, many of whom apparently resented its perceived mathematical complexity, in addition to its conceptual novelty. Perhaps Einstein shared the notion that general relativity was a "difficult" theory, because all of his early calculations of the theory's predictions involved approximate, rather than exact, solutions of the theory. Moreover, these approximations were designed to make the theory appear, as much as possible, similar to existing theories, such as Newtonian gravity and Maxwellian electrodynamics. These sorts of approximation technique present a particular problem in physics, by forcing us to ask how we know that the solution to a set of "approximate" equations is actually numerically close to a genuine solution of the full theory. One can argue that these kinds of theoretical calculations give rise to a problem identified years ago in the experimental sphere by Peter Galison in his book *How Experiments End?* (Galison 1987). How do you know, as a theorist, when to stop your calculation? How do you know at what point your answer is reliably close to the mathematical behavior of the real theory which you are unable to solve exactly? In this chapter we will examine how Einstein struggled with this problem, as well as pointing out certain ways in which his solutions gave rise to further controversy and debate in the decades after his death.

Einstein's earliest calculations in general relativity were based on the theory's Newtonian approximation. This was a natural choice, since it was vital to the theory's acceptability to most physicists that it should be able to reproduce, in some plausible, if approximate, sense, the enormously successful predictions of Newton's gravitational theory. In addition, this approximation scheme could then prove useful in identifying physical scenarios in which higher-order terms in the approximation scheme predicted measurable departures from the Newtonian prediction, the most famous case of this being the Mercury perihelion advance. Einstein's early emphasis on the Newtonian approximation scheme may explain his comment to Karl Schwarzschild, in a letter of February 19, 1916 (CPAE 8, Doc. 194), that gravitational waves did not exist within his new theory. Unfortunately, Einstein's letter does not state the grounds for this belief. Since we know that Einstein had already worked on the problem of Mercury's orbit, it is interesting to note that Mercury's orbit around the Sun should in principle be affected if gravitational waves are emitted by orbiting gravitational systems, but it turns out that radiation-reaction terms of this type do not appear until relatively high order in the post-Newtonian expansion of the problem of motion in general relativity. Thus it is possible that Einstein, not finding any terms of this type in his early Newtonian calculations, was led to conclude that gravitational waves played no role in his new theory.

However, within a few months of this letter, Einstein was to change his mind, and not for the last time, on the existence of waves in his theory. Probably the occasion for this change of heart came about in his investigation of a new method of approximation, comparing his theory to Maxwellian electrodynamics. Interestingly, it could fairly be argued that the analogy with electromagnetism played a more significant role in the development of general relativity than did the analogy with Newtonian gravity (Janssen and Renn 2007, Renn and Sauer 2007). So it was not at all surprising that, soon after the final development of the theory, Einstein should turn to this kind of approximation as he attempted to explore the theory's implications. Following a suggestion of the astronomer Willem de Sitter, Einstein adopted a linearized approximation scheme of his equations, with a coordinate choice (often referred to as De Donder coordinates) which cast the resulting linearized field equations in a form analogous to a familiar version of the Maxwell equations. Once this maneuver had been accomplished, it was trivial to solve the resulting wave equation, using techniques developed in the study of Maxwell's equations. In this way one could demonstrate the existence of a solution of the linearized Einstein equations representing plane gravitational waves. The paper published by Einstein on this linearized approximation in 1916 (Einstein 1916g) thus became the first mathematical theory of gravitational waves, and the approach it pioneered is still fundamental to the design of such existing gravitational wave detectors as the LIGO, VIRGO, and GEO600 detectors operating (respectively) in the United States, Italy, and Germany.

However, we now come across the first problem faced in any such new calculation. While Einstein had made masterly use of a formal analogy between his new theory and the better understood field equations of electromagnetism, analogies are generally not perfect, and spotting the points at which the analogy breaks down is essential to success. It soon turned out that Einstein had made two errors in his 1916 paper, originally a presentation to the Prussian Academy of Sciences. The first mistake he caught before the paper saw print. The second was brought to his attention a year or so later by his young colleague Gunnar Nordström, in a correspondence whose significance was first appreciated by my colleague Michel Janssen, while editing Volume 8 of Einstein's Collected Papers (CPAE 8). Both demonstrated the pitfalls of failing to carefully scrutinize the nature of the analogy which guided his calculation.

The first mistake was made when Einstein announced, during his 1916 presentation, that there were three types of linearized gravitational waves, characterized by different symmetries in the theory. Hermann Weyl, shortly afterward, described these three types of waves as longitudinal-longitudinal, longitudinal-transverse, and transverse-transverse (Weyl 1918b). Einstein was first alerted to something odd about the first two types of waves in this list when he noticed that they did not seem to transport any energy. Ironically, this insight, though ultimately vindicated, was doubly problematic. First, as we will see, he was using an incorrect measure of the energy in the wave field. When corrected, this error did not change the calculation which showed that the first two wave types did not transport energy. Second, as he had already been warned by mathematicians such as Tullio Levi-Civita, and as was realized decades later in the context of gravitational wave studies, the energy in a gravitational wave is a coordinate-dependent quantity and can, in a local sense, be completely transformed away by certain choices of coordinates (Cattani and De Maria 1993). Thus, even if his calculation was correct, it would still have been possible, in principle, to find a coordinate choice in which even the true gravitational wave solution would have appeared, locally, not to carry any energy. Luckily, Einstein was aware that an argument based on energy alone would not suffice. Accordingly, he showed that these first two "types" of wave did not appear when viewed in another coordinate choice. In a later paper (Einstein 1918a) he went further and showed how the metric describing these spurious waves could be transformed, by a change of coordinates, into the metric for flat space. The "waves" he thought he had seen were merely flat space observed through a "wavy" kind of coordinate frame.

The second mistake concerned Einstein's calculation of the energy carried by these waves. In doing so he made use of a somewhat controversial mathematical construct known as a pseudotensor to describe the energy in the gravitational field. He made a mistake in doing so however, which was only discovered when Nordstrom attempted to use the pseudotensor from Einstein's linearized approximation paper to calculate the energy in the gravitational field of an isolated mass. After some to and fro between himself and Nordstrom, Einstein realized the nature of his mistake, which had given rise to an incorrect formula for the energy transmitted in a gravitational wave. He presented a new paper in 1918 which conclusively cleared up these two mistakes, and which derived, for the first time, the famous quadrupole formula for the energy transported by a gravitational wave. Even then he made one further small error, involving a factor of two, which was later corrected by Arthur Stanley Eddington (1922).

It is interesting to note that careful attention to the analogy and where it broke down could have alerted him to the second mistake. Max Abraham, while working on his own field theory of gravitation had noted, some time before, that conservation of momentum demands that there be no dipole emission of gravitational waves, just as conservation of mass prevents the existence of monopole gravitational waves (see Renn 2007d for an account). Essentially, conservation of momentum prohibits the type of motions of masses which would be required to produce waves with a dipole symmetry. This argument was in analogy with an argument known in electromagnetism, in which conservation of charge prevents the emission of monopole electromagnetic waves. Abraham concluded from this that gravitational waves were unlikely to play a role in any future field theory of gravity, thus arguing for a sort of disanalogy with the electromagnetic case. His insight was remarkable, for it is true that the lowest multipole symmetry of gravitational wave emission is the quadrupole. Einstein's formula of 1916 was incorrect in permitting even monopole, and not just dipole, gravitational waves. Even a naive application of the analogy with electromagnetism should have convinced him that there was something wrong with the result of his calculation.

After his presentation of 1918, Einstein let the matter of gravitational waves rest for nearly two decades. When he returned to it, in 1936, it was with a paper, written in collaboration with his young assistant, Nathan Rosen, entitled "Do Gravitational Waves Exist?" (Kennefick 2007). Their answer to this question was no. Einstein and Rosen had set out to find an exact, as opposed to approximate, solution to Einstein's equations representing gravitational waves. They encountered a problem when they observed a singularity, a point in space-time at which one or more quantities in the calculation is not uniquely defined, in their solution. After much effort attempting to find a coordinate system in which the singularity would disappear they concluded that it was endemic, and that its presence was proof that there was no physically acceptable solution, in the full theory, for gravitational waves.

Much later on it was shown that Einstein and Rosen's argument was misguided because the singularity they had noticed is only a coordinate singularity, not a physical singularity (Bondi, Pirani, and Robinson 1959). A good example of a coordinate singularity would be the North Pole of the Earth, as represented in terms of latitude and longitude. No one can say what the longitude of the North Pole is. Since all lines of longitude pass through this point, it could have any value from zero to 360 degrees. But this does not make the Earth, or spheres in general, physically impossible. It's just that their shape is such that it is difficult to define a coordinate system which works everywhere on their surface. But in the period before World War II the mathematical language of singularities had not yet been fully elaborated and mistakes like that made by Einstein and Rosen were not uncommon.

Convinced they had proved that gravitational waves did not exist, Einstein and Rosen sent their paper off to the *Physical Review*, after which Rosen emigrated to the Soviet Union, and Einstein set to work exploring the implications of their result. If the linearized approximation of his theory suggested gravitational waves existed, but the full theory did not actually predict them, what other surprises lay in store for those who had assumed the theory could be reduced, in some sense, to models based on preexisting theories such as electromagnetism? As he said in a letter to his friend Max Born: "This shows that the non-linear general relativistic field equations can tell us more or, rather, limit us more than we have believed up to now" (Born 1971, letter no. 71). Einstein was hard at work on a paper investigating whether something as basic as a solution for a rotationally symmetric stationary gravitational field could exist in the exact theory. The paper, which exists in draft form only, ends abruptly at a point where it seems likely that Einstein realized the whole basis of his argument was wrong (EA 80–974.0). How he came to this point is an interesting story in itself.

When Einstein and Rosen sent their paper to the *Physical Review* it arrived on the desk of its editor, John T. Tate Sr., based at the University of Minnesota. Tate, although not an expert in general relativity, was shrewd enough to realize that there was something troubling about this paper. Although previous submissions by Einstein and Rosen had been published rapidly, after deliberating over this new paper for some time, he sent it on to an acknowledged expert in the theory of relativity, Howard Percy Robertson (see Kennefick 2007 for a more detailed account). Robertson was actually a colleague of Einstein's at Princeton, but in mid-1936 he was enjoying a holiday in the Pacific Northwest after spending time on sabbatical at his alma mater, the

California Institute of Technology. He quickly replied to Tate that he did not think Einstein and Rosen were correct, including a ten-page referee's report on their paper (EA 19–090). Although he suspected that the singularity they were concerned about might prove to be merely a coordinate effect, he also proposed that, even if it was a physical singularity, it could be made less offensive by changing the symmetry of the system from one describing plane gravitational waves to one describing cylindrical waves. In that case the singularity would be present at the source of the wave, a long central axis. By long tradition, physical singularities can be treated as the source of a gravitational field. Since Newton's day it was considered acceptable to model material objects as point masses.

However, when Tate returned the paper to Einstein the great man replied in high dudgeon, rejecting the referee's criticisms out of hand, apparently without even reading them closely, and withdrawing the paper (EA 19–086). He resubmitted it to the *Journal of the Franklin Institute* in Philadelphia. What happened next was reported in a letter of Robertson's to Tate from early the next year.

You neglected to keep me informed on the paper submitted last summer by your most distinguished contributor. But I shall nevertheless let you in on the subsequent history. It was sent (without even the correction of one or two numerical slips pointed out by your referee) to another journal, and when it came back in galley proofs was completely revised because I had been able to convince him in the meantime that it proved the opposite of what he thought. You might be interested in looking up an article in the *Journal of the Franklin Institute*, January 1937, p. 43, and comparing the conclusions reached with your referee's criticisms. (Quoted in Kennefick 2007, 91–2)

Upon returning to Princeton in the fall of 1936, Robertson happened to strike up a friendship with Einstein's new assistant, and successor to Rosen, Leopold Infeld. From discussions with Infeld, Robertson learned that Einstein still believed he had proved that gravitational waves do not exist. As we learn from Infeld's autobiography, Robertson insisted to him that Einstein's argument must be wrong, and went over the matter with him, convincing Infeld of his case (Infeld 1941). Infeld, highly impressed by Robertson's ability to master the argument so quickly, was left unaware that he had, in fact, refereed the paper during the summer! Interestingly, Infeld claims that Einstein realized at about the same time that his argument was not solid. When he did so he adopted Robertson's suggestion, perhaps transmitted by Infeld, that though he had set out to find an exact solution for plane gravitational waves, and then tried to prove that they did not exist, he had in fact stumbled upon a solution for cylindrical gravitational waves. Accordingly, Einstein was obliged to make radical changes in proof to his paper, right down to the title (Einstein and Rosen 1937).

It is noteworthy that in the published version of the paper Einstein still queried whether it would turn out, in the exact theory, that binary star systems would radiate gravitational waves. He was still skeptical whether the approximation scheme suggested by the analogy with electromagnetism would prove robust enough to stand up to more careful scrutiny in the light of the full theory. This skepticism seems to have been passed on to his assistants, for in the postwar years both Rosen and Infeld played a leading role as skeptics of the existence of gravitational waves, or more particularly, of whether they are emitted by binary star systems.

Einstein changed his mind about gravitational waves so many times that it is difficult to be sure what his final position was. According to Infeld, he came to believe what Infeld later espoused, that gravitational waves existed in the theory, but were not emitted by the only systems likely to produce them in any strength, binary stars and other systems moving under the influence of their own internal gravitational interaction. The evolution of Einstein's views on the subject is a case study in the role played by analogy in the development of scientific ideas. The analogy with Maxwell's electromagnetic field theory which suggested the very idea of a field theory of gravitation, naturally also suggested the possibility of a wave phenomenon associated with that field, given the existence of electromagnetic waves. Although Einstein did not do so, several physicists discussed the notion of gravitational waves in the years before he developed general relativity. Yet Einstein seemed skeptical in the immediate aftermath of developing the theory, as he confided to Schwarzschild. This may have been because his work up to that time had focused on the Newtonian approximation of the theory, which is not suggestive of, or particularly well suited to, the study of gravitational waves. Of course, waves play no role in Newton's theory of gravity.

Sure enough, within a few months, once his attention had turned to the linearized approximation and its analogy with electromagnetism, he changed his mind and published a paper whose chief focus was gravitational waves. Even here though, some level of skepticism was warranted. The coordinates chosen to facilitate the analogy with electromagnetism had the unfortunate characteristic that they made flat space-time appear wavy. Thus Einstein discovered three types of gravitational waves, two of which turned out, on closer examination, to be spurious. Here Einstein's skepticism is real but considered appropriate, in hindsight, because we agree with his conclusions. Only one of these "three types" of waves is what we would now consider to be a gravitational wave.

But what seems real in one approximation may, in a more exact calculation, prove to be spurious also. So Einstein and Rosen convinced themselves, twenty years later, that even this "third type" of wave played no role in the exact theory. Even when he convinced himself, or was convinced by others, that this was not true, Einstein seemed to doubt whether the most obvious possible transmitter of gravitational waves, binary stars, might not transmit these waves at all. While the analogy with electromagnetism, and the quadrupole formula, derived in the context of the linearized approximation of general relativity, suggested that binary stars would be good sources of gravitational waves, Einstein, Infeld, and some others advanced objections. The analogy reasoned that since, in electromagnetism, accelerating charges emitted electromagnetic waves, in the case of gravitational waves, one would expect accelerating masses to radiate energy away. But what, said the skeptics, about the case of a body falling freely in a gravitational field, a star orbiting another star for instance, which follows the geodesic of the local space-time and therefore feels no acceleration at all? The linearized approximation fails when applied to such a strongly self-gravitating system, and such a body can, from a relativistic point of view, be likened to a body coasting through empty space, which follows a geodesic of flat space-time, that is a straight line.

From a relativistic point of view, one may ask the following question: Is a body accelerating when it falls, or when it is prevented from falling by the application of a force? Another way of thinking of the problem at issue is to consider two observers watching a falling body. One is falling along with it; the other is standing on the ground. One thinks the falling body is not accelerating, the other says that it is. But a distant observer can surely only agree with one of them as to whether it is emitting gravitational waves. The problem is all the more troubling for devotees of relativity theory since our usual modern answer (that the falling body is emitting gravitational waves which can be detected by the distant observer) suggests that the equivalence principle, the heuristic device which enabled Einstein to build the theory of general relativity, is violated for charged particles. A falling charged particle is seen by a distant observer to be emitting electromagnetic waves as it accelerates downward. The energy carried away by these waves must come from somewhere, presumably from the kinetic energy of the falling charge, which thus falls more slowly than an uncharged particle in the same gravitational field. Perhaps not surprisingly, given the post-Einsteinian prestige of the equivalence principle, in the mid-twentieth century it seems that opinion among relativists was rather evenly divided as to whether falling charges or masses should emit radiation (Kennefick 2007).

What we notice here is that the sort of skeptical argument advanced by Einstein in his paper with Rosen, and taken up by others in the post-war period, is searching for the point at which the analogy breaks down. While arguments from analogy can be very fruitful in science, it can be just as fruitful to seek out the points of disanalogy, where an analogy ceases to hold. Hence Abraham, noting that gravitational waves should not emit monopole radiation, in analogy with the electromagnetic case, spotted that the argument went further in the gravitational case, forbidding also dipole radiation, which is the principle kind of radiation encountered in electromagnetism. Similarly, classical radiation theory would demand that two electric charges in orbit around each other should emit electromagnetic waves. Indeed, the failure of atomic electrons to do so in a continuous way was one of the great mysteries of physics before the development of quantum mechanics. Both Einstein and Infeld argued that here was another possible instance of a disanalogy with electromagnetism. They thought that orbiting masses might not emit gravitational waves.

Einstein might have been influenced here by the line of thought which, as I have argued, went back to his first objection to Schwarzschild. If gravitational waves are emitted by orbiting objects, where is the evidence for this loss of energy in the orbiting systems? The line of work which Einstein engaged in with Infeld and another young assistant, Banesh Hoffmann, was to solve the problem of motion in general relativity within the post-Newtonian approximation. Their work, known as EIH, after the authors' initials, demonstrated that there was no damping of motion, such as might be associated with gravitational wave emission, up to the order of $(v/c)^2$. But Einstein's own quadrupole formula suggested that the lowest order that one would anticipate for this kind of radiation damping would be (v/c).[5] Since Infeld once compared the difficulty of the calculation in EIH to swimming the English Channel, and the difficulty of going to order $(v/c)^4$ as equivalent to swimming the Atlantic, it is not surprising that he and Einstein had no appetite for looking for evidence of gravitational radiation damping in binary stars.

When people did eventually begin to perform this kind of calculation, mostly in the postwar period, it did prove difficult, for reasons discussed elsewhere (Kennefick 2007). Many calculations failed to agree; some results even had binary stars gaining energy as a result of emitting gravitational waves. Because no one could be sure which the right answer was, it was difficult to determine which the right method was. The problem can be described as the Theoreticians' Regress, in analogy with the Experimenters' Regress discussed by Harry Collins in the context of controversies in the attempt to experimentally detect gravitational waves. The regress describes the difficulty, in pure research, of identifying the right method of performing a calculation, when one cannot check the complete calculation against a known result. Ultimately, the controversy over whether Einstein's quadrupole formula applied to binary star systems, which dragged on for decades, was brought to a head by the discovery of the binary pulsar PSR 1913 + 16. Observations of this system by Joe Taylor and collaborators ultimately showed, in late 1978, that the pulsar's orbit about its unseen companion was decaying by an amount in agreement with the prediction of the quadrupole formula. After this it was possible to argue that any calculation which did not recover the quadrupole formula for such systems was simply an incorrect calculation.

Einstein's repeated changes of mind on the existence of gravitational waves reflect the tricky nature of arguing from analogy. Analogy plays a very useful role, as Thomas Kuhn has argued, in permitting scientists to find new roles for their existing toolbox of *exemplars*, Kuhn's word for problems solvable by a familiar mathematical trick which can be widely applied to different areas of physics (Kuhn 2012, 186). The key to the use of exemplars lies in the detection of analogies between new and existing problems, which permit the familiar method of solution to be applied to the unfamiliar problem. But it is worth noting that, for physicists, much of the excitement in such activity lies in the finding of points at which the analogy breaks down, what a physicist would call "new physics." In other words, the ideal kind of problem in physics may be one which is like an existing problem, but which differs from it in interesting ways without departing completely from the analogy. Thus it is interesting that gravitational waves do not exhibit any dipole emission. If gravitational waves did not exist at all, it would be a remarkable departure from the analogy with electromagnetism which underlies the whole concept of a field theory of gravity. It would also be remarkable if binary stars did not emit gravitational waves, although it turns out that they do, as far as we know. Throughout the debate over the nature of gravitational waves one can observe a distinction between those who were inclined to trust the analogy between gravity and electromagnetism, and those who were skeptical of it. What is interesting is that both made use of it, but in different ways. Some tried to follow the analogy as far as it would take them; others were always on the lookout for the point at which it would break down. Often the same behavior of the theory was viewed in a contrasting light by the two groups. For instance, the fact that the quadrupole is the lowest multipole radiation permitted in the gravitational field can be viewed as an instance of the analogy at work (certain conservation laws set limits on multipole radiation in a field), or of an example of the analogy breaking down (the gravitational field, in forbidding dipole radiation, is different from the electromagnetic field). While we cannot know what Einstein came to believe regarding the status of this analogy between gravitational and electromagnetic waves, we do know that he deserves credit, from today's standpoint, for being the one who first fully articulated it in its modern form, in his paper of 1918.

9 Einstein's Unified Field Theory Program

Einstein explicitly used the term "unified field theory" in the title of a publication for the first time in 1925. Ten more papers appeared in which the term is used in the title, but Einstein had dealt with the topic already in half a dozen publications before 1925. In total he wrote more than forty technical papers on the subject. This work represents roughly a fourth of his overall oeuvre of original research articles, and about half of his scientific production published after 1920.

This contribution is an attempt to characterize Einstein's work on a unified field theory from four perspectives, by looking at its conceptual, representational, biographical, and philosophical dimensions. The space spanned by these four dimensions constitutes Einstein's unified field theory program. It is characterized by a conceptual understanding of physics that provides the foundational physical knowledge, general problems, and heuristic expectations. The mathematical representation both opens up and constrains the possibilities of elaborating the framework and exploring its inherent consequences. The biographical pathway, historically contingent to some extent, takes Einstein from the elaboration of one approach to the next. Finally, Einstein's most vivid philosophical concern was at the heart of this enterprise, and his insistence on the possibility and desirability of a unified field theory cannot be understood without fully acknowledging his philosophical outlook. Somewhat paradoxically, the philosophical dimension of his unification endeavors not only reveals him as being deeply rooted in the nineteenth century and its intellectual traditions, but also embodies an intellectual heritage that is perhaps just as important to acknowledge as his more widely celebrated achievements.

Einstein's unification program was a program of reflection. It aimed at understanding a given content of knowledge in physics in a different way, at what Einstein considered an understanding of this content in an emphatic sense. This understanding was to see the whole of physics as an organic entity, where no part can be separated from any other without severe loss of meaning. The underlying motivation for this program of reflection, I want to suggest, was a conception of the task of human reasoning that would be adequate to a holistic understanding of a nature in which human beings live their lives. The explanation of isolated phenomena by subsuming them under general laws was a valid aspect of this task but it concerned, to pick up a terminological distinction of German idealism, our *Verstand* only. Human *Vernunft*, by contrast, aimed at an understanding of nature in toto, with human beings living and acting in it. From this perspective, Einstein's political interventions flow out of the same philosophical worldview as, in later years, a unified field theory would. Moreover, this philosophical outlook was also driving his creative productivity of his early years (Renn 1993, 2006a). For an adequate historical and philosophical understanding of Einstein's intellectual contributions, one must take seriously Einstein's scientific activities that, although puzzling to many, he pursued with steadfast conviction despite many disappointments, in ever-increasing scientific isolation, and without finding a solution.

An account of Einstein's work on a unified field theory that would go into technical detail is beyond the scope of the present contribution. We still lack fine-grained historical investigation of his later work on which we could base such an account, that is, investigations that would discuss his endeavors with technical understanding from a historical point of view and that would take into account his unpublished correspondence and manuscripts.[1] Any discussion of his work in this direction on a technical level also immediately requires a substantial amount of mathematical preliminaries, the exposition of which would either be lengthy or run the risk of being accessible only to a limited audience. Nevertheless, by discussing the dimensions of Einstein's unification program I hope to show that a full appreciation of these different dimensions is essential for a proper *historical* understanding of this substantial aspect of Einstein's intellectual heritage.

1. THE CONCEPTUAL DIMENSION

It has been argued that Einstein's attempts at a unified field theory are to be seen as a continuation of heuristic aspects of his celebrated conceptual breakthroughs to special and general relativity (Bergmann 1979). With respect to special relativity, the reconsideration of the concept of simultaneity in frames of reference that are in relative motion to each other had solved an outstanding conceptual contradiction among foundational aspects of classical physics. Specifically, the redefinition of simultaneity allowed Einstein to resolve the conflict between the universal validity of the principle of relativity of Newtonian mechanics and the constancy of the velocity of light in Maxwellian electrodynamics by providing a conceptual justification for the Lorentz transformations. He thus successfully integrated two major fields of physics in what amounts to a conceptual unification.

But solving the conflict between the principles of relativity and of the constancy of the velocity of light in special relativity had created another problem, to be addressed by a relativistic theory of gravitation. The conceptual foundation of special relativity demanded the existence of an absolute and finite limit to the speed of any signal transmission. This fundamental aspect of special relativity was violated by Newtonian gravitation theory. A gravitational interaction between massive bodies that was an instantaneous action-at-a-distance posed a contradiction to the postulate of the nonexistence of any signal transmission speed exceeding the speed of light. Closely related to this conceptual conflict was another inherent difference between the classical theories of gravitation and of electromagnetism. Newtonian gravitation theory conceptualizes the gravitational interaction as a static interparticle interaction. Maxwellian electromagnetism, however, conceptualizes the electromagnetic interaction in terms of a dynamic field concept. The difference becomes most obvious by considering electromagnetic waves. In Maxwellian field theory, Coulomb's law is a very special case, and all dynamic effects propagate in electromagnetic waves with a finite speed. The Maxwell equations even allow for the existence of electromagnetic waves in vacuum without presupposing the existence of electric sources that would generate the field.

In addressing these issues, Einstein took a decisive turn very early in his investigations by linking these problems to the problem of generalizing the special theory of relativity to noninertial, accelerated frames. By working out the implications of the equivalence hypotheses, he arrived at a general theory of relativity and a relativistic field theory of gravitation. This theory provided a resolution of the conceptual contradictions between classical gravitation theory and the Maxwellian theory of the electromagnetic field. Indeed, the general relativistic theory of gravitation conceptualized the gravitational force as a field, and one of the first elaborations of the new theory, after his breakthrough to the generally covariant field equations of gravitation in late 1915, concerned the existence of gravitational waves.

But, the success of this integration came at a price. The gravitational interaction was conceptualized as a dynamical field by a geometrization. Einstein now believed that his geometrized relativistic theory of the gravitational field demanded again a unification with the concept of the electromagnetic field. The latter had essentially been left unaffected by the reconceptualization of the gravitational interaction brought about by the theory of general relativity.

Einstein thought that the new understanding of gravitation demanded further unification with classical Maxwellian theory of the electromagnetic field. But the conceptual differences between the geometrized gravitational field and the classical Maxwell field did not amount to an open or even only to a hidden inherent contradiction. There was no compelling reason why the electromagnetic field should be reconceptualized following the example of the relativistic gravitational field.

Nevertheless, the successful geometrization of the gravitational field in the general theory of relativity motivated many contemporaries to look for a geometrized theory that would also include the electromagnetic field. Unification at this level pertained to the conceptualization of the two known fundamental fields: the gravitational field and the electromagnetic field. Intrinsic criteria of validation or refutation for attempts at unification were not too sharply defined.[2] They included, first of all, the demand that the known equations for the gravitational and electromagnetic field could be obtained in some limiting case. The extent to which a unified description of the two fields then actually represented a conceptual unification was subject to different criteria. The mathematical representation should assign symmetric roles to the gravitational and the electromagnetic fields in some unspecified sense. In a stricter sense, the representation of the two fields should not merely add them in a way that trivially decomposes into two independent sets of equations governing the gravitational and the electromagnetic field. In other words, the unified description should inherently allow for some kind of nontrivial mixing of the two fields. But this mixing might occur on a purely representational level and thus might be principally unobservable. This latter postulate did not, therefore, also entail necessarily the even stronger condition that the unification must also, at least in principle, predict new physical effects.

The most desirable case of a unified description that would both yield the known laws of gravitation and electromagnetism and would also predict new effects, arising from a combination of the fields inherent in the unified description, and that would also be compatible with known empirical facts, has never been achieved. Nevertheless, many theoreticians would applaud progress toward this goal even if it only achieved a more restricted unification or if it only promised the prospect of achieving a true unification.

But the unified field theory program involved even more than these postulates. A unified conceptualization and mathematical representation of the gravitational and electromagnetic fields in the sense discussed so far was only one aspect of the conceptual dimension of the unification problem in physics at the time. A second aspect, independent of the unification of the fields, concerned the representation of matter.

In Newtonian mechanics as well as in Lorentz's interpretation of Maxwellian electromagnetism, the existence of masses or material points and of charges and currents was independent of the fields. Equations of motion were determined by the interaction of the charged particles with the fields as a set of independent, additional equations of the theory. The coexistence of particles and fields did not represent an inherent contradiction or inconsistency, but was commonly referred to as a dualism,[3] and as such was regarded as a violation of the ideal of unification.

More specifically, it had been Gustav Mie who, in 1912-3, had proposed a way to overcome this dualism of particles and fields that was attractive to many physicists at the time (Mie 1912a, 1912b, 1913). Mie's idea was to look for nonlinear modifications or generalizations of Maxwell's equations that would allow for particlelike solutions of the electromagnetic field equations. We should be able to interpret a solution as a particle if it is spherically symmetric, the field intensity is very high only in a finite region of space, and the field equations imply sensible equations of motion for those localized maxima of the fields. Unfortunately, Mie's only explicit example had not provided a viable model for this program. It implied that two such particlelike solutions would necessarily move toward each other and merge. Hence, no stable configuration of matter was possible for this special case. Nevertheless, as a research program Mie's idea had much appeal for many physicists and mathematicians, among them David Hilbert (Sauer 1999; Corry 2004, ch. 6), Hermann Weyl (Scholz 2004), and Einstein.

The basic properties of matter to be accounted for were the existence of two elementary particles: the electron and the proton. Each carried an elementary charge whose absolute value was the same: the electron as a negative charge, the proton as a positive charge. Although debated for a while in the early 1920s, there was no evidence for the existence of fractions of the elementary charge. Also, the electron has a definite rest mass, as has the proton, and the mass ratio between the proton and the electron is numerically of the order of 2,000.

In the early 1930s, this situation changed with the discovery of additional elementary particles (Kragh 1999, ch. 13). In 1932, both the neutron and the positron were discovered experimentally. The existence of an electrically neutral particle with the same mass as the proton had been discussed as early as 1920 by Rutherford, and the existence of a positively charged electron had been predicted by Dirac in 1931. But since the neutron has roughly the same mass as the proton but does not carry an electric charge, and since the positron has exactly the same rest mass as the electron but has positive elementary charge, the existence of these two elementary particles did not seriously challenge the matter aspect of Einstein's unified field theory program. The only consequence was that the unified theory no longer had to account for the *nonexistence* of neutral particles with proton mass or of a positively charged electron. The postulate and existence of the positron also raised the question of antiparticles to the proton and the neutron. The antiproton and antineutron were only postulated in 1931 and 1935, respectively. They were actually discovered only in 1955 and 1956, after Einstein's death.

More serious from the conceptual point of view of the unified field theory program was the postulate and discovery of mesons. The muon was discovered in 1937. The charged pion was postulated in 1935 and discovered in 1947, and the neutral pion was postulated in 1938 and discovered in 1950. It is with the discovery of these particles that the parameters for a unified field theory actually do change. By the mid-1950s, at the time of Einstein's death, a dozen elementary particles were known. Ten years later the list of experimentally confirmed subatomic particles ran to about a hundred. With the discovery of elementary particles beyond the electron and proton, twentieth-century physics also had to deal with two new kinds of fundamental interactions. The weak interaction which plays its most prominent role in the explanation of radioactive decay, and the strong interaction which came with the discovery of mesons and baryons other than protons and neutrons and their decays.

To the extent that the existence of mesons can be regarded an experimentally well-established fact during Einstein's lifetime, the conceptual dimension of a unified field theory no longer encompasses only two distinct elementary masses and a single elementary charge. To be sure, the history of the discovery of elementary particles is complex and what is identified in hindsight as the discovery of an elementary particle may have appeared less convincingly so to Einstein and his contemporaries. In any case, Einstein's unified field theory program was committed to the "two-particle paradigm" (Kragh 1999, 190) of having to account for the existence of only a proton and an electron, or, we may say, to a generalized two-particle paradigm which would recognize only two distinct elementary masses, that is, proton mass and electron mass, and one elementary charge, if one allows for the existence of neutrons and positrons as well.

The postulates of Einstein's unified field theory programs are still not exhausted with the conceptual unification of the gravitational and electromagnetic field and the accounting for two fundamental matter particles. The existence of an elementary charge as well as of elementary electron and proton masses already represented elements of discontinuity that had to be incorporated and justified in a field theoretical, that is, essentially continuous conceptualization of matter. The development of quantum theory had made it clear that further elements of discreteness had to be represented in the field-theoretical framework. Most prominently, these quantum aspects were evident in processes of emission and absorption of electromagnetic radiation by matter. With the development of relativistic quantum mechanics and quantum electrodynamics, the fields were being quantized, too, and many theoreticians at this point gave up looking for a classical unified field theory of the gravitational and electromagnetic fields.

The third aspect of Einstein's unification program therefore concerned quantum theory. At least on a programmatic level, Einstein did not ignore the theoretical advances in quantum field theory. But for Einstein, the empirical success of nonrelativistic quantum mechanics did not demand that the unified description of the fundamental fields itself should be a quantized one. Rather, the foundational unified conceptualization should somehow provide a conceptual justification for the stochastic aspects of quantum theory and for its violations of classical determinism. Also, whereas others began working on bringing together special relativity and electrodynamics with quantum theory, leaving aside the question of how the gravitational interaction would fit into the scheme, Einstein did not alter his priorities. For him, unification efforts had to start from a theory of the gravitational field and hence be general relativistic. Again, the early conceptual problems of interpreting and dealing with intrinsic problems of quantum field theory (negative energy solutions, etc.) and of quantum electrodynamics (infinities and renormalization procedures) may have given him good reason to do so.

The fact that he considered quantum theory only a preliminary theory, highly successful phenomenologically but unjustified on a foundational level, allowed him to concentrate on attempts of a classical theory without putting himself to the task of explaining quantum theory in the first step. To account for the characteristic consequences of quantum theory remained a goal of his unification endeavors. But in most cases no specifics of the quantum aspects of matter and radiation would or could be accounted for. The explanation of quantum mechanics within a unified field theory remained a programmatic desideratum in Einstein's work.

2. THE REPRESENTATIONAL DIMENSION

The crucial mathematical concept that enabled the original formulation of general relativity is the metric tensor field $g_{\mu\nu}$. In general relativity, it plays a double role. On the one hand, it is at the heart of general relativity as a physical theory, by virtue of its meaning for space-time measurements. On the other hand, it is the mathematical object that is determined by the field equations of gravitation. In four-dimensional space-time, the field of a real, symmetric, second-rank metric tensor is a ten-component tensor, that is, its definition in a given coordinate chart requires specification of ten independent functions, or ten numbers $g_{\mu\nu}(x^\mu)$ at each point x^μ of some open set of the space-time manifold. The tensor character then demands that these ten functions transform in a specific way when we change the coordinates.

The gravitational field equations at the heart of general relativity are ten coupled partial differential equations for the components of the metric tensor. General covariance accounts for four identities among them, but the complexity of the solutions for such a system of coupled partial differential equations is such that the space of solutions cannot be surveyed analytically in its full extension. In certain special situations of high symmetry, or other restrictive conditions, analytical solutions are available. Schwarzschild's solution for spatial spherical symmetry is the most prominent among them. But even today most interesting cases can only be handled by extensive numerical calculations, if at all.

The field equations of general relativity are almost completely determined by the postulate of general covariance, the demand that second derivatives of the metric occur at most linearly, and by some boundary conditions, for example, the demand of regular behavior at infinity. The only freedom that was left within those conditions was the addition of a cosmological term which Einstein introduced in 1917. The framework of the original theory of general relativity, however, did not allow inclusion of the electromagnetic field in any nontrivial sense of unification. In order to achieve unification, the representation therefore unavoidably had to be transcended.

Several ways of extending the representational framework for a unified field theory are possible and have been considered by Einstein and other researchers. Briefly, these extensions either consist in relaxing the conditions imposed on the original formulation, or in introducing other mathematical objects into the theory. Along the first option, one can explore the consequences of giving up the restriction to the requirements of reality or symmetry of the metric tensor components. One may also relax the restriction to four-dimensional space-time by considering higher-dimensional representations. Finally, one may relax the condition of locality of a field theory by introducing nonlocal dependencies into the theory.

Another possibility that opens up new ways to achieve a unification is the introduction of other mathematical objects into the mathematical framework, specifically by using other quantities as the dynamical

Table 9.1. *Einstein's different approaches along the unified field theory program as discussed in the text.*

Year	Einstein's Approach
1919–22	Weyl, Kaluza, Eddington
1923	affine theory (Eddington)
1923	overdetermination
1925	metric-affine approach
1927	trace-free equations
1927	five-dimensional approach
1928–31	distant parallelism
1931	semivector theory
1938–41	five-dimensional approach
1944	bivectors
1945–55	asymmetric theory

variables of the theory. The most prominent example arises from the concept of the affine connection that was introduced into the theory of general relativity shortly after its initial formulation mainly through the work of Tullio Levi-Civita and Hermann Weyl. Another possibility that was used to explore ways of unification was to conceive of the field of tetrads, that is, fields of point-dependent orthonormal bases of the tangent spaces at each point of the space-time manifold, as the fundamental dynamical variables.

Each of the alternatives opens up a horizon of possibilities inherent in the mathematical representation that can be explored for its suitability for the unification program. In principle, any kind of combination of those extensions is possible, too. The problem then arises in each and every case to secure the possibility of a physical interpretation, identify a set of field equations that replaces the gravitational and electromagnetic field equations of conventional general relativity, and get a hold on the manifold of solutions for those field equations. Even before the discovery of further elementary particles the unified field theory program was a research problem that easily transcended the intellectual capacities of a single researcher.

3. THE BIOGRAPHICAL DIMENSION

In this section, we will indicate, in rough strokes, the major approaches and main steps of Einstein's engagement along his unification program.

Einstein's early work on the unification program after the completion of the theory of general relativity was, by and large, a reaction to approaches advanced by others. This is the case for the first geometrization of the electromagnetic field, proposed in 1918 by Weyl; for the first exploration of a five-dimensional theory suggested by Theodor Kaluza in 1919; and for the first attempt to base a unified field theory on the concept of the affine connection, rather than on the metric field, as advanced by Arthur Eddington in 1921.

Weyl's approach was motivated by a more general observation concerning the conceptual foundation of a field theory (Scholz 2001). In Euclidean geometry, we can compare vectors at different points both with respect to their lengths and with respect to their direction. In Riemannian geometry of general relativity, a distant comparison of vector direction is no longer possible. On parallel transport, a vector's direction at a distant point depends on the path along which parallel transport took place, and if transported along a closed loop in a curved manifold, the direction of a vector differs before and after parallel transport along the loop. The length of a vector, by contrast, remains unchanged in Riemannian geometry. In Weyl's understanding, this was an inconsistency in the conceptual foundations of a true field theory, which should be represented by a truly infinitesimal geometry, in the sense that only assertions about neighboring tangent spaces should be meaningful.

Motivated by these considerations, Weyl introduced another geometrical object into the theory that he aptly called a "length connection." It determined the length of a vector on parallel transport just as the Levi-Civita connection determined the direction of a vector on parallel transport. The surprising thing was that on the level of the mathematical representation, Weyl could link the length connection to the electromagnetic vector potential and thus establish a link between the geometrical structure, given by the length connection, and the electromagnetic field.

Einstein's reaction to Weyl's approach was highly ambivalent. Weyl had sent him a manuscript expounding these ideas, asking him to submit it for publication in the Prussian Academy's *Sitzungsberichte*. On receiving the manuscript, Einstein called it "a first-class stroke of genius" (CPAE 8, Doc. 498), "wonderfully self-contained" (ibid., Doc. 499), and "very ingenious" (ibid., Doc. 500). Nevertheless, he had also quickly found a serious objection to Weyl's theory.

The nonintegrability of the vector length on parallel transport implied that the wavelength, for example, of light emitted by a radiating atom would depend on the prehistory of that atom. Empirical evidence, however, suggested that the wavelength of light emitted by atoms is only determined by the constitution of the atom, and not by its prehistory. When Einstein presented Weyl's paper to the Prussian Academy and mentioned his own criticism, fellow academy members vetoed publication. As a compromise solution, the paper was published with an added paragraph in which Einstein put forth his measuring rod objection. As a theory of physical reality, Einstein henceforth considered Weyl's approach "fanciful nonsense" (CPAE 9, Doc. 294). And when Weyl included an exposition of his idea into the third edition of his textbook *Space – Time – Matter*, Einstein judged that Weyl had "messed up" the theory of general relativity (CPAE 9, Doc. 323).

Nevertheless, in March 1921, Einstein picked up on Weyl's theory in print and elaborated on the idea, as a "logical possibility," of giving up the postulate of the existence of parallel-transportable measuring rods as a fundamental assumption of general relativity (Einstein 1921e).

In April 1919, Einstein was confronted with another idea of fundamentally modifying the foundations of general relativity with a view of unifying the gravitational and electromagnetic fields. Kaluza, at the time *Privatdozent* in mathematics at the University of Königsberg, sent him a manuscript in which he introduced the concept of a fifth dimension to the underlying space-time manifold of general relativity (CPAE 9, Goenner and Wünsch 2003).

The basic idea was to represent the electromagnetic vector potential or the electromagnetic field components in terms of the additional components of the metric tensor and the affine connection, respectively, that came with the introduction of the additional fifth dimension.

Although physical meaningfulness suggests that we look only at restricted subgroups of the full diffeomorphism group, the demand of general covariance in five dimensions immediately implies that the components of the electromagnetic four-potential and the gravitational components of the metric tensor are mixed up into each other by means of pure coordinate transformations. In this respect, a unification of the two fields was achieved.

The approach is burdened with a number of difficulties on different levels. From an epistemological point of view, the introduction of another spacelike dimension raises the question of the ontological status of this dimension. A somewhat pragmatic answer to this issue was suggested by Einstein himself in his initial reaction to Kaluza's manuscript. If the trajectories of material particles that follow geodesics in five dimensions are projected down to nongeodesic trajectories in four dimensions, one might be able to interpret the deviations from the four-dimensional geodesic path as a direct effect of the electromagnetic field.

On a representational level the following difficulty arose. The five-dimensional metric tensor has fifteen independent components. Of these, ten are interpreted in the usual sense as components of the gravito-inertial field, while four more are associated with the four components of the electromagnetic vector potential. The problem then arises as to the physical interpretation of the fifteenth independent component, g_{55}. In his original paper, Kaluza tried to ignore this problem by arguing that the metric component g_{55} is not truly independent, but rather implicitly determined through the field equations in conjunction with the equation of motions. In later works along the five-dimensional approach, the problem of finding a meaning of the g_{55} component was interpreted differently. One tried to turn this difficulty into an advantage by assuming that the five-dimensional approach inherently introduced a new scalar field that may be given an independent physical meaning.

An additional difficulty that was pointed out by Einstein after his initial reading of Kaluza's manuscript turned out to be a fatal obstacle for Kaluza's plans to publish his paper. An order of magnitude consideration showed that for the equation of motion of an electron, the influence of the gravitational field turned out to be larger by many orders of magnitude than any reasonable physical interpretation would allow for.

Overall, Einstein's reaction to Kaluza's manuscript was just as ambivalent as was his response to Weyl's geometrization. Initially, he was much impressed by Kaluza's idea and considered it physically much more promising than Weyl's "mathematically deeply probing approach" (CPAE 9, Doc. 26). But on pointing out the difficulty of interpreting the g_{55}-term in the equation of motion for the electron, he asked Kaluza for understanding since "due to my existing substantive reservations" he could not present it himself for publication, notwithstanding his "great respect for the beauty and boldness" of the idea (CPAE 9, Doc. 48).

Thus ended their initial correspondence in May 1919, and Kaluza, unable to rebut Einstein's criticism, apparently did not try to publish his manuscript elsewhere. But Einstein must have continued to think about it and, more than two years later, sent a postcard to Kaluza, expressing second thoughts about his previous rejection of Kaluza's manuscript. On Einstein's invitation, Kaluza now sent a manuscript in which he also expounded the difficulty of finding an interpretation for the different terms in the equation of motion, crediting Einstein with having alerted him to this problem. Einstein in turn now submitted Kaluza's manuscript to the Prussian Academy for publication in its proceedings (Kaluza 1921).

A few weeks later, Einstein and Grommer finished a paper investigating the problem of solutions to Kaluza's five-dimensional theory that are everywhere regular and centrally symmetric (Einstein and Grommer 1923). Their result was that "Kaluza's theory does not contain a solution that only depends on the $g_{\mu\nu}$ and is centrally symmetric and could then be interpreted as a (singularity free) electron" (Einstein and Grommer 1923, vii).

A third approach toward a unified field theory may be called the *affine approach*. It was advanced most notably by Eddington in the early 1920s (Eddington 1921; 1923, ch. 7), and was taken up also by Einstein. The idea was to base the theory on the concept of an affine connection as the fundamental mathematical quantity, rather than on the metric tensor. The starting point is the observation that a manifold that is equipped with a linear affine connection allows the definition of a Riemann curvature tensor and of a Ricci tensor. The latter is, in general, not a symmetric tensor, even if the connection is assumed to be symmetric. Eddington then suggested to identify the antisymmetric part of the Ricci tensor as the electromagnetic field tensor, and to interpret the symmetric part of the Ricci tensor – after some rescaling – as the usual metric tensor field.

The problem that Eddington had left open was to provide field equations that would determine the affine connection. It is this question that Einstein addressed in a series of three brief notes published in 1923 in the *Proceedings of the Prussian Academy* (Einstein 1923a, 1923b, 1923c). He proposed to obtain the field equations from a variational principle, and also suggested a Lagrangian function for the field action. But in the course of trying to touch base with familiar general relativistic gravitation theory and Maxwellian electromagnetism, he also performed some transformation of variables and ended up with a variational formulation that, in fact, was almost equivalent to the variational formulation of the general theory of relativity with a Maxwellian field provided by Hilbert in 1915. This equivalence was highlighted by Hilbert in lectures held in Hamburg and Zurich in the summer and fall of 1923, and prompted his republication of a merged version of his two communications on the "Foundations of Physics" in the *Mathematische Annalen* in 1924 (Hilbert 1924, Majer and Sauer 2005).

A problem in the approach of the affine theory was the proper interpretation of the fundamental variables of the dynamical theory. But also, and more importantly, Einstein found that the theory did not account for the electron-proton mass asymmetry and that no singularity-free electron solution seemed possible.

With Einstein's response to Weyl, Kaluza, and Eddington in the early 1920s we find him reacting to approaches that had been advanced by others. At the end of 1923, after his work on the affine theory, Einstein published a paper that is again more original. It is entitled: "Does the Field Theory Provide Possibilities for a Solution of the Quantum Problem?" (Einstein 1923d).

The point of this paper was to argue that there is, after all, a way to account for quantum phenomena by means of differential equations. Given that partial differential equations had been so successful in classical field theory, Einstein found it "hard to believe that partial differential equations would not be, as a last resort, suitable for an explanation of the facts."[4] The principal solution that he advanced in this paper was to generate discontinuities by overdetermining the classical variables with more differential equations than field variables. Although he gave some more technical details as to how this approach was to be understood, Einstein admitted that he was still unable to show concretely how it would solve the quantum problem. In fact, the idea had been on his mind for some years. Indeed, we find Einstein contemplating this issue as early as 1920:

I do not believe that one must abandon the continuum in order to solve the [problem of] quanta. In analogy, one might have thought it possible to force general relativity from abandoning the coordinate system. In principle, of course, the continuum could be abandoned. But how in the world should one describe the relative motion of n points without a continuum? ... I still believe as before that an overdetermination ought to be sought with differential equations for which the *solutions* no longer have any continuum properties. But how?[5]

The first original approach put forward by Einstein himself was published in a paper of 1925 in which also the term "unified field theory" appeared for the first time in a title (Einstein 1925d). In that paper, he explored a metric-affine approach, that is, he took both a metric tensor field and a linear affine connection at the same time as fundamental variables. Both connection and metric were assumed to be asymmetric. Parallel transport then again defines a Ricci tensor and a Riemann curvature scalar, and Einstein defined tentative field equations in terms of a variational principle, taking the Riemann scalar as a Lagrangian just as in standard general relativity. As regards the interpretation of the mathematical objects, he tried to associate the gravitational and electromagnetic fields with the symmetric and antisymmetric parts of the metric field. In his attempt to recover the known cases, he could show that the metric was symmetric for the purely gravitational case and the usual compatibility condition for the Levi-Civita connection can be recovered. Maxwell's equations could be recovered, in the limit of weak gravitational fields, but only in a slightly different form that is not entirely equivalent to the original equations.

The basic problem of this approach seems to have been that Einstein did not know how to go on from here. Dealing with both an asymmetric metric tensor and an asymmetric connection opened up a vast field of possibilities inherent in the mathematical framework, and many familiar results of the theory of Riemannian geometry no longer held. In particular, verifying the existence of nonsingular, spherically symmetric charge distributions posed a formidable challenge. It was also unclear how to explicitly investigate the nonvacuum case beyond the first-order approximation of weak gravitational fields. Einstein did not pursue this approach any longer in print but he did take it up once more, twenty years later, as his final approach toward a unified field theory, working on it until his death.

In 1927, Einstein published a mathematical note on the geometric interpretation of a modification of the original gravitational field equations of general relativity that he had investigated earlier in 1919 (Einstein 1919a, 1927a). The modification consisted in demanding that the equations be trace-free rather than have vanishing covariant divergence. In 1919, the modification had been motivated by considerations concerning the constitution of matter. In 1927, he argued in reaction to a short note by the mathematician George Yuri Rainich, that this modification could be given a geometric interpretation.

That same year, he also published two papers on Kaluza's five-dimensional theory (Einstein 1927b, 1927c) in which he showed that Kaluza's original results can also be obtained without the restriction to weak gravitational fields and slow velocities. It so happened that Einstein reproduced results that had been published only a year before by Oskar Klein, as he acknowledged in a note to his second paper. But whereas Klein's interest in Kaluza's theory was motivated by the wish to account for quantum phenomena within a unified field theory, Einstein does not mention this concern in his two notes.

The sequence of approaches mentioned here as well as Einstein's publication pace is determined not only by an internal logic, but it is also determined, to some extent, by contingent external factors. Two such external factors that influenced his productivity in the 1920s were his traveling and his health. In spring 1928, while on a visit to Switzerland, Einstein suffered a circulatory collapse. An enlargement of the heart was diagnosed and, back in Berlin, he was ordered strict bed rest. At the end of May he wrote to a friend: "In the tranquility of my sickness, I have laid a wonderful egg in the area of general relativity. Whether the bird that will hatch from it will be healthy and long-lived only the Gods know."[6] A few days later, he presented a short note to the Prussian Academy on a mathematical structure that he called "Riemannian Geometry, Maintaining the Concept of Distant Parallelism" (Einstein 1928a). This first note was of a purely mathematical nature. It was followed a week later by a second note (Einstein 1928b), in which Einstein explored the possibility of formulating a unified field theory within the new geometrical framework.

In this new approach a space-time is characterized by a connection with vanishing Riemann-curvature in conjunction with a metric tensor field (Sauer 2006). The crucial mathematical construct that enabled Einstein to formulate a flat space-time that is nonetheless non-Euclidean was the concept of a tetrad field. Tetrads are orthonormal bases of the tangent spaces. Given a field of tetrads, parallel transport is then defined in a natural way, but this connection differs from the usual Levi-Civita connection. Parallel transport along a smooth tetrad field is curvature-free but the manifold is, in general, still non-Euclidean, since it allows for nonvanishing torsion. Torsion of a manifold is characterized by the absence of parallelograms, that is, if a vector is parallel transported along a closed loop it will coincide with the original vector but if it is parallel transported along four legs of a parallelogram the torsion of the manifold will result in a displacement of the resulting vector from its original position.

Einstein soon was to learn that the mathematical concept of distant parallelism was by no means new and had already been explored by mathematicians, notably by Roland Weitzenböck and Elie Cartan. While immediately acknowledging the priority of others as far as the mathematics was concerned, Einstein nevertheless held high hopes for his idea of formulating a unified field theory within this structure. For him, the critical question was to find a field equation for the components of the dynamical tetrad fields. Each field of tetrads defines a metric tensor field. But the converse is not true, since the metric tensor components can only fix ten of the sixteen components of a tetrad. The additional six degrees of freedom are just what would be needed, so he thought, to accommodate the six degrees of freedom of the Maxwell field in a unified description of gravitation and electromagnetism.

The story of the distant parallelism approach can be told largely as a story of attempts to find and justify a uniquely determined set of field equations for the tetrad components, with the demand that solutions of those field equations be given a sensible physical interpretation. The distant parallelism approach in this respect shows a number of marked similarities with Einstein's search for general relativistic field equations of gravitation in the years 1912–15 (Sauer 2006). In 1912, it had been the introduction of the metric tensor into the theory that had started Einstein's research, and existing mathematical theorems had to be adapted to the theory. In 1928, it was the tetrad fields that allowed the investigation of a non-Euclidean geometry of vanishing curvature and, similarly, Einstein was made aware of existing mathematical results by mathematician colleagues. In both cases, Einstein's research quickly focused on finding a set of field equations for the dynamical variables and, in both cases, it was difficult to satisfy all heuristic requirements. In response to these difficulties, Einstein changed back and forth between two different and complementary strategies, each starting from one particular set of heuristic postulates. In both episodes, Einstein at one point settled on a set of field equations that was justified more by physical considerations rather than by mathematical soundness. In both cases, Einstein continued to work out consequences of the field equations as well as continued to find a satisfactory mathematical justification for these equations. And finally, the demise of both theories came about by a combination of realizing more and more shortcomings of the theory and by discovering that an alternative approach promised to be more successful. However, while in 1915 the more successful theory that Einstein substituted for his earlier so-called *Entwurf* theory was the final version of general relativity (see Chapter 6), the successor approach to the distant parallelism episode turned out to be yet another attempt at a unified field theory.

Einstein abandoned the distant parallelism approach when he realized that the tetrad formalism also allowed a different and new perspective on the Kaluza-Klein five-dimensional approach. Together with Walther Mayer, with whom he had collaborated already during the final stages of the distant parallelism approach, Einstein now explored a variant of the five-dimensional idea that seemed sufficiently new in order to justify again taking up the Kaluza-Klein approach (Einstein and Mayer 1931, 1932a). The novelty of the approach was that it was no longer the space-time manifold which was enlarged by a fifth spacelike dimension. Rather Einstein and Mayer constructed a five-dimensional vector space at each point of four-dimensional space-time. The tetrad formalism allowed for an easy generalization to five dimensions, simply by adding another linearly independent vector to the tetrads. The five-dimensional vector spaces obviously could no longer be identified with tangent spaces of the underlying manifold, but Einstein and Mayer gave a projective mapping from the five-dimensional vector spaces to the four-dimensional tangent spaces.

While Einstein and Mayer succeeded to derive the gravitational and electromagnetic field equations from this new five-dimensional approach, they could not account for the structure of matter. In their first paper, they concluded that the existence of charged particles or currents was incompatible with the field equations. They also remarked that an understanding of quantum theory was not yet conceivable in this approach (Einstein and Mayer 1931). In order to allow for the existence of charged material particles, Einstein and Mayer investigated a generalization of their initial framework. The generalization resulted in a new set of field equations. In a subsequent publication, they investigated mathematical properties of these new equations, specifically the problem of compatibility without, however, commenting on a possible physical interpretation of those equations (Einstein and Mayer 1932a).

Since the five-dimensional vector space approach again ran into difficulties, Einstein and Mayer once more tried another approach (van Dongen 2004). Among all of Einstein's investigations into a unified field theory this would be the one that most directly addressed the problem of quantum theory. This time, the incentive came from Paul Ehrenfest, Einstein's Leyden colleague and one of his closest personal friends. Ehrenfest had closely studied recent investigations of a relativistic quantum theory by Wolfgang Pauli and Paul Dirac, and had introduced the term *spinor* for the two-component complex vector representation of the Lorentz group. Since spinors have somewhat counterintuitive transformation properties, for example, a full rotation of 360° changes the sign of the spinor, Ehrenfest was uncomfortable with the formalism and urged his colleagues to provide a more natural and intuitive mathematical representation. Einstein and Mayer picked up on this problem in four papers, published between November 1932 and January 1934 (Einstein and Mayer 1932b, 1933a, 1933b, 1934).

In essence, what Einstein and Mayer investigated in these papers was the Dirac equation in a different representation. They introduced what they called "semivectors," essentially a four-dimensional real vector representation of the Lorentz group. They argued that semivectors were a more natural concept than the suspicious spinors, most likely because of their similarity to ordinary four-dimensional space-time vectors. As it turned out, however, the field equations for semivectors turned out to be decomposable into equations that were equivalent to field equations using the spinors, which in hindsight is not surprising since semivectors are not an irreducible representation of the Lorentz group, whereas spinors are.

The publication history of the semivector approach also reflects the drastic changes in Einstein's life that took place during these months. The first paper (Einstein and Mayer 1932b) was published in the *Proceedings of the Prussian Academy of Sciences*, as were most of his technical papers on general relativity and unified field theory until this point. But while Einstein spent his third winter as a visiting scientist at the California Institute of Technology in Pasadena in 1932–3, the Nazis came to power and Einstein resigned his membership in the Prussian Academy and never returned to Germany. After his return to Europe in the spring of 1933, Einstein went to Belgium, traveled to Switzerland, and in effect spent most of the year in transit. The second and third notes on the semivector approach (Einstein and Mayer 1933a, 1933b) were published in the *Proceedings of the Amsterdam Academy*, and the fourth (Einstein and Mayer 1934) was published in *Annals of Mathematics*, a journal published at Princeton, where Einstein accepted a permanent position after returning to the United States in late 1933.

Einstein would spend the rest of his life in Princeton, and a number of his later papers on general relativity and unified field theory were published in the *Annals of Mathematics*.

This change in publication record not only reflects Einstein's geographical move but also the fact that his investigations into a unified field theory increasingly took on the character of purely mathematical investigations. The change in publication venues may also be an indication of a change in the audience he was addressing, symptomatic of an increasing isolation from modern physics. However, the latter interpretation is at odds with the fact that, at least in the 1930s, Einstein published a number of investigations in conventional general relativity that were of substantial physical significance: equations of motion, gravitational waves, and gravitational lensing.

During the mid-1930s, Einstein published little on unified field theory. Some of his investigations may simply never have been published, and there is some hope of interesting historical findings in his scientific manuscripts (Sauer 2004). But he also was spending much time and effort on behalf of other scientists and intellectuals who were trying to escape Nazi Germany. In any case, he published again on unified field theory in 1938, in a paper coauthored with Peter Bergmann (Einstein and Bergmann 1938). They reconsidered the ontological status of the fifth dimension. It had been a *scandalon* of Kaluza's idea that the extra spatial dimension was a complete mathematical artifact without any physical meaning. In their paper, Einstein and Bergmann entertained the possibility that the fifth dimension was to be regarded as real. Technically, they investigated consequences of substituting the so-called cylinder condition, which demands that all derivatives with respect to the fifth dimension vanish and makes the physical interpretation of the fifth dimension difficult, with the assumption that space was closed or periodic in the direction of the fifth dimension.

The problem with this new investigation along the Kaluza-Klein approach was that it led to integro-differential equations that were hard to solve. The problem was addressed in a follow-up paper, published three years later, that presented a way of turning those equations into differential equations (Einstein, Bargmann, and Bergmann 1941). But in analyzing these differential field equations, Einstein and his coworkers found it impossible to describe particles by nonsingular solutions. They also found that the gravitational and electromagnetic field equations would be given by the same order of magnitude. This latter characteristic made it impossible to account for the quantitative difference in the strength of the gravitational and electrostatic forces between material particles.

The systematic difficulty of accounting for the existence of matter in a unified field theory was the subject of a little note by Einstein published the same year (Einstein 1941). Einstein proved that the vacuum gravitational field equations do not admit a stationary, singularity-free solution that is embedded in flat space and whose metric tensor would allow a classical limit for large spatial distances from the center that was of the form of a Newtonian gravitational potential for a finite mass.

The proof was reconsidered and generalized two years later in a joint paper with Wolfgang Pauli (Einstein and Pauli 1943), in which the authors prove the nonexistence of regular solutions to the vacuum field equations that would asymptotically behave like the Newtonian gravitational potential, regardless of symmetry conditions for the field in regions of finite field strength. Moreover, the proof showed that this result was valid not only in four dimensions, but also for the Kaluza-Klein five-dimensional theory. In effect, this result indicated that, under very general conditions, any attempt to base a unified theory on the Riemann tensor would necessarily involve singularities in particlelike solutions.[7]

Clearly, the latter result was at odds with core requirements of Einstein's program, and it therefore does not come as a surprise that he was willing to reconsider key assumptions of his earlier efforts. An immediate outcome of such reconsiderations was a new approach that he pursued in two publications in 1943, one of them coauthored with his collaborator Valentin Bargmann (Einstein 1944b, Einstein and Bargmann 1944). The starting point now was to keep the four-dimensionality of the theory and also the requirement of general covariance, but to give up the postulate that a generalized theory of gravitation should necessarily be based on the existence of a Riemannian metric. What Einstein and Bargmann proposed instead was, in effect, an attempt at a nonlocal relativistic theory of gravitation. They investigated the properties of a new kind of mathematical object that they called "bivectors." In contrast to modern usage of the term, these are not asymmetric, second-rank tensors, but rather second-rank tensors that depend on two distinct points of the four-dimensional manifold. The transformation properties of these bivectors depend on the two distinct points of the manifold, each index of the bivector being associated with a different base point.

Although explicitly articulated in the context of the unified field theory program the two papers do not discuss any physical interpretation. Rather, they discuss properties of the mathematical structure that derives from the bivectors. They also discuss field equations for the bivectors, which turn out to be algebraic rather than differential equations. The difficulties of the bivector approach again came with finding and interpreting nontrivial solutions of the fundamental equations. The published papers indicate only preliminary results, partly credited to Einstein colleagues, Bargmann and Pauli, as well as to the Princeton mathematician Carl Ludwig Siegel, and explicitly mention ongoing research with Pauli. It is again possible that a closer scrutiny of Einstein's later research manuscripts might shed further light on Einstein's elaboration of this approach.

Judging by the published record, the bivector episode represents Einstein's penultimate distinct approach in the sequence of attempts to arrive at a unified field theory. Einstein devoted the last ten years of his life to the investigation of a framework that he had already worked on in the mid-1920s, to which he now returned in a first publication of 1945 (Einstein 1945). This last approach of Einstein's work along his unified field theory program was again based on a local Riemannian metric but on an asymmetric one.

Initially, Einstein took the metric tensor to have complex components and demanded Hermitian symmetry. Pauli, however, quickly pointed out to Einstein that the restriction to Hermitian symmetry was not necessary. The subsequent investigations were then following the pattern of earlier ones (Einstein 1946, 1948a, 1950a, 1950b, 1956a, App. II; Einstein and Stauss 1946; Einstein and Kaufmann 1954, 1955). Tentative field equations were tested for their mathematical properties, and it was checked whether the criteria for a physical interpretation could be applied. Along with the mathematical properties, Einstein worried very much about the problem of compatibility as he had with other, earlier approaches. Since the mathematics of a framework based on an asymmetric metric tensor is exceedingly more complex and less well-known than the standard formalism of (semi-)Riemannian differential geometry underlying conventional general relativity, Einstein spent the rest of his life elaborating the asymmetric theory. Einstein's very last considerations in this final approach were presented by his last assistant, Bruria Kaufmann, at the fiftieth anniversary of the relativity theory in Bern in July 1955 a few weeks after Einstein's death (Kaufmann 1956).

4. THE PHILOSOPHICAL DIMENSION

For several reasons, Einstein's unified field theory investigations during the last twenty years of his life, and their broader scientific context, remain largely unexplored. The urgency of his political and social concerns increases dramatically with the Nazis' rise to power, the ever-increasing cruelty of their persecution of Jews, the outbreak of World War II, the Holocaust, and the explosion of the first atomic bombs. His professional activities were more and more dominated by acts of solidarity with fellow emigrés, public statements and interviews, and other activities that he considered necessary to counteract developments that were against any rational organization of the human world. The political turmoil of the years before, during, and after the war forced him to be more and more active in nonscientific matters, and left him less and less occasion to delve into mathematical calculations. This shift in the balance of the theoretical and practical aspects of his worldview is also reflected in our image of the later Einstein.

Nevertheless, as we have seen, Einstein did publish a number of papers on the unified field theory program in the last two decades of his life. These are highly abstract and esoteric theoretical investigations, mostly of a mathematical character, exploring consequences of a generalized mathematics very much like venturing into an uncharted terrain. Many unpublished scientific manuscripts of those years are extant, which give an idea of his struggle with the technical difficulties involved in interpreting a generalized mathematical framework along his program. A closer analysis of these manuscripts still needs to be done and will give a more detailed picture of the rationale of these attempts (Sauer 2004).

The striking contrast between Einstein's perseverance in his scientific program and the detached tone of his scientific publications of the 1940s and 1950s and the urgent and sometimes desperate tone of his political statements need not be seen as a contradiction. Rather, it may be seen as arising from the same source of confidence in the ability and responsibility of mankind for a rational understanding and organization of their world. Despite the political chaos of the world that he lived in during the last two decades of his life, this confidence never failed him in all his political engagement. Despite all new developments in nuclear physics, the discovery of new elementary particles, the successes and paradoxes of early attempts at a quantum field theory, and a quantum electrodynamics, this confidence did not fail him either in his attempt to arrive at a unified theory of gravitation and electromagnetism.

Two aspects of Einstein's later scientific work bear witness to this overarching confidence and its specific appearance. A first comment concerns the sincerity of Einstein's willingness to explain the motivation, content, and problems of his scientific research to a lay audience. In 1952 he gave a characteristic response to the request of a reader of *Popular Science Monthly* as to why the promise of his "great scientific achievement" of finding "one all-embracing formula" that would "solve the secret of the universe" seemed not to be coming along as expected (Einstein 1952). Einstein replied by first passing a shot at the public media and, in particular, at "newspaper correspondents" for the fact that "laymen obtain an exaggerated impression of the significance of my efforts." But he then went on to explain with sober sincerity what he saw as the main problem of his present work. He had been proceeding to generalize relativistic equations of gravitation "by a purely mathematical procedure ... and was hoping that the equations obtained in this measure should hold in the real world." But due to the "complexity of the mathematical problem," so far neither he nor anybody else had been able to find solutions to the equations that would allow the representation of empirically known facts, "so that it is completely impossible to tell whether the theory is 'true.'"

To illustrate this point he drew the analogy to Newton's theory of gravitation. This theory had only been confirmed since Newton had been able to derive Kepler's laws from his general equations, which in turn had only been possible because the mass of the Sun is so much heavier than that of the planets, allowing for approximately valid solutions of Newton's equations. If the bodies of the solar system, by contrast, were of equal mass, such approximate solutions may not be obtained and "we might perhaps never know whether or not Newton's theory holds."

The response is characteristic not only for its sincerity, but also illustrates Einstein confidence in a program that admittedly lacked any empirical support. It is also characteristic in its appeal to the history of physics. The comparison with Newton clearly has a rhetorical dimension but it also points to a significant motivation for Einstein's confidence in the eventual success of his program, notwithstanding agnostic overtones. At other places, when explaining his research into a unified field theory to a wider audience, he explicitly linked it to a progressive continuity in the history of science.

In 1950, Einstein gave such an account of his latest efforts in a contribution to *Scientific American* (Einstein 1950a). The question as to why we "devise theory after theory," or why we "devise theories at all," Einstein (1950a, 13) wrote, has a somewhat trivial answer when we encounter new facts that cannot be explained by known theories. Yet "there is another, more subtle motive of no less importance. This is the striving toward unification and simplification of the premises of the theory as a whole" (ibid.). This latter motive is fed by our inborn "passion for comprehension":

I believe that every true theorist is a kind of tamed metaphysicist, no matter how pure a "positivist" he may fancy himself. The metaphysicist believes that the logically simple is also the real. The tamed metaphysicist believes that not all that is logically simple is embodied in experienced reality, but that the totality of all sensory experience can be "comprehended" on the basis of a conceptual system built on premises of great simplicity. (Einstein 1950a, 13)

And in response to a fictitious objection by a skeptic that this would only be a "miracle creed," Einstein admitted that much "but it is a miracle creed which has been borne out to an amazing extent by the development of science" (ibid.)

Here then we see history of science as an underpinning of Einstein's metaphysical creed. Indeed, Einstein goes back to the atomistic conceptions of Leucippos that were borne out only in the kinetic theory of heat of the nineteenth century, or to Faraday's introduction of the field concept that was theoretically justified in Maxwell's theory and then experimentally proven to be real only after the fact, as it were, by Hertz's discovery of electromagnetic waves. Einstein's "train of thought which can lead to endeavors of such a speculative nature" (ibid., 14) as his own unified field theory maps onto a conceptual history of physics from Newtonian mechanics through Faraday-Maxwell field theory of electromagnetism and Lorentz's theory to the theories of special and general relativity. What emerges from this historical perspective is the significance of the field concept. It is the field concept that embodies the concepts of space and time. And it is the theory of the gravitational field as a special case that gives the "most important clue" (ibid., 16). And this historical perspective also justified Einstein's steadfast conviction that a unified field theory needs to be generally relativistic, that is, in accordance with the principles of general relativity from the outset. He explicitly opposed an approach where "the rest of physics can be dealt with separately on the basis of special relativity, with the hope that later on the whole may be fitted consistently into a general relativistic scheme" (ibid.). Historical continuity placed the endeavor of finding a unified field theory above the theory of gravitation implied by general relativity:

I do not believe that it is justifiable to ask: What would physics look like without gravitation? (Einstein 1950a, 16).

In the course of events, however, it was this conviction that separated Einstein from the majority of his contemporaries.

The Einstein that may emerge from seriously acknowledging the philosophical dimension of his unified field theory program may well be that of an intellectual whose belief in the viability of a unified theory of the gravitational and electromagnetic field was intimately connected to a historically outdated belief in the ability of a single human mind to grasp the mysteries of nature in simple terms. From this belief in the power of the human mind sprang the sincerity of his attempts to explain the motif and rationale of his investigations to a lay audience, and this same belief fueled his political interventions in a barbaric world. The futility of his scientific unification endeavors in the face of developments in theoretical physics, both during and after his life, suggests that the specific form of this belief in the power of the human mind belongs to a tradition of enlightenment whose days are gone. We may recognize the same belief in his early undisputed intellectual achievements, popular writings that we still recommend to students, humanitarian efforts that we still esteem, and explicit contentions against quantum theory that we consider deep and insightful. To the extent that Einstein's work on a unified field theory thus reveals a unity in his intellectual endeavors, we should be aware that we can learn from his intellectual achievements only by transforming his legacy to the more complex parameters of our time.[8]

NOTES

1 For first steps in this direction, see Vizgin (1994), van Dongen (2002a, 2002b, 2004), Goenner (2004), Goldstein and Ritter (2003), Majer and Sauer (2005), and Sauer (2006) upon which much of the following is based. This article was completed in 2007. More recent literature was not incorporated but would not have altered the substance of this article. I nevertheless wish to alert the reader to the book by Jeroen van Dongen (2010) as well as to the latest volumes in the series of the Collected Papers of Albert Einstein.

2 For the following discussion, see also Bergia (1993).

3 See, e.g., Einstein to Kaluza, May 29, 1919 (CPAE 9, Doc. 48).

4 "... schwer zu glauben, daß die partielle Differentialgleichung in letzter Instanz ungeeignet sei, den Tatsachen gerecht zu werden" (Einstein 1923d, 360).

5 "Daran dass man die Quanten lösen müsse durch Aufgeben des Kontinuums glaube ich nicht. Analog hätte man denken können, die allgemeine Relativität durch Aufgeben des Koordinatensystems zu erzwingen. Prinzipiell könnte ja das Kontinuum aufgegeben werden. Wie soll man aber die relative Bewegung von *n* Punkten irgendwie beschreiben ohne Kontinuum? ... Ich glaube nach wie vor, man muss eine solche Überbestimmung durch Differentialgleichungen suchen, dass die *Lösungen* nicht mehr Kontinuumscharakter haben. Aber wie?" Einstein to Hedwig and Max Born, January 27, 1920 (CPAE 9, Doc 284).

6 "Ich habe in der Ruhe der Krankheit ein wundervolles Ei gelegt auf dem Gebiete der allgemeinen Relativität. Ob der daraus schlüpfende Vogel vital und langlebig sein wird, liegt noch im Schosse der Götter." Einstein to Heinrich Zangger, EA 40–069, quoted in Sauer (2006).

7 For a critical discussion of this paper and its claim, see also van Dongen (2002a).

8 I wish to thank Diana Buchwald, Jeroen van Dongen, Larry Gould, Dennis Lehmkuhl, Christoph Lehner, and Tom Ryckman for reading and commenting on earlier versions of this paper.

10 Einstein's Realism and His Critique of Quantum Mechanics

1. INTRODUCTION

A few years before their deaths, Einstein's old confidant Michele Besso must have written a letter to Einstein addressing him in his typically enthusiastic manner as a "giant old friend" (*"gewaltiger alter Freund"*). Einstein responded in his own no less typical sarcasm:

Dear Michele,

An old friend I am, but to the address "giant" I can only add "nebbish" if you are familiar with this telling word of our forefathers. It expresses an alloy of compassion and disdain. All the fifty years of conscious ruminations have not gotten me closer to an answer for the question: "What are light quanta?" These days any rascal may believe that he knows, but he deludes himself. (Albert Einstein to Michele Besso, December 12, 1951; Einstein and Besso 1972, 453)

This often-quoted passage[1] shows Einstein's own rather bleak assessment of his lifelong struggle with the questions posed by his early work on the dual nature of light (see Chapter 4). But more importantly, it illustrates Einstein's conviction that quantum mechanics (and quantum field theory) as formulated after 1925 had done nothing to elucidate these questions, either. Einstein's lack of enthusiasm for these theories was a bitter disappointment for their authors,[2] who mostly admired Einstein as the foremost theoretician of his generation. As quantum mechanics and later quantum field theory became ever more successfully applied as foundational theories (a fact that Einstein readily admitted), Einstein's reservations were increasingly seen as the stubborn metaphysical prejudice of an old man who could not adapt anymore to the demands of modern physics.[3] This assessment, however, has changed with the reopening of the interpretation debate of quantum mechanics since the 1960s. John S. Bell's work on the statistics of measurements on entangled states played a big role in that turn of events. Building on the example that Einstein, Podolsky, and Rosen proposed in their famous thought experiment, Bell gave a clear formal expression to the conflict between quantum mechanics and classical expectations. Bell's work led to the experimental study of entangled states and so the first empirical tests of Einstein's "metaphysical prejudice"; and even though quantum mechanics has passed those tests convincingly, this development led to a renewed interest in the foundational problems of quantum mechanics and a new appreciation for Einstein's critique. Instead of the stubborn old man, Einstein now frequently was portrayed as the valiant rebel against the imposition of Copenhagen orthodoxy.[4]

Before offering an opinion on these starkly different evaluations, I will first of all attempt to understand the basis of Einstein's critique of quantum mechanics. I will argue that the basis of this critique is a position of *realism*, but that this realism is not metaphysical prejudice, but rather a *methodological* principle which Einstein believed he could infer from his most successful work on relativity and statistical mechanics.[5] This principle was supplemented by more specific physical assumptions in Einstein's various arguments considered in this chapter, but it is more fundamental than any of these physical assumptions. To understand these foundations, it will be necessary to spend some time tracing how Einstein's views on the method of physics were formed by this early work, before turning to the topic of quantum physics proper. In Section 2, I will argue that in the years around 1918, under the dominating influence of his experience with general relativity, Einstein's mature views about the nature and method of physics were formed. In the rest of the contribution I will try to show how deeply this view of physics influenced Einstein's position vis-à-vis quantum mechanics and its claim to be the new foundation of physics. In Section 3 I will summarize what Einstein saw as the unsolved riddles of the old quantum theory, especially the wave-particle duality and the lack of a deterministic theory of elementary processes. Einstein's first reactions to the new quantum mechanics will be the subject of Section 4, while Section 5 will treat Einstein's first public critique of quantum mechanics at the 1927 Solvay conference. In Section 6, I will discuss Einstein's most famous response to quantum mechanics, the paper coauthored with Boris Podolsky and Nathan Rosen that first pointed out the most unsettling consequence of quantum mechanics: nonlocal entanglement. Section 7 deals with some less well-known arguments that Einstein made late in his life, trying to show that there is no "macroscopic limit" in which the troubling consequences of quantum mechanics disappear. A short epilogue will discuss the implications of Bell's arguments for Einstein's position and summarize what responses can be given to Einstein's arguments after the quantum mechanical predictions have been confirmed empirically. Finally, I will return in the conclusion to the question regarding whether Einstein's critique of quantum mechanics simply shows his lack of appreciation for the revolutionary novelty of quantum mechanics or whether it is of value for our understanding of its foundations.

2. EINSTEIN'S REALISM

The best-known Einstein quote on quantum mechanics is doubtlessly his dictum "God doesn't play dice," uttered in a letter to Max Born, his colleague and longtime friend who became one of his chief opponents in the dispute about the status of quantum mechanics.[6] Born was one of the creators of matrix mechanics, and is credited with first giving a statistical interpretation of the wave function in his work on the wave mechanical treatment of scattering processes (Born 1926). Like many after him, Born saw indeterminism as the fundamental departure of quantum mechanics from classical physics and took Einstein's criticism of quantum mechanics to be a prejudice against the possibility of a radically indeterministic physics.[7] This reading has become very much the standard view of Einstein's position: the old revolutionary that once abolished absolute space and time, now grown conservative and himself clinging stubbornly to outdated metaphysical assumptions. However, this is a misunderstanding of Einstein's position, maybe understandable from Born's point of view and his pride in having done to determinism what Einstein did to absolute space, but much too simplistic to do justice to Einstein's assessment of quantum mechanics. And although philosophically sensitive historians of science[8] have pointed out that Einstein's views were much more complex (and much better founded), the caricature still looms large outside the small community of Einstein scholars.

I will be pleading the case with a star witness: Wolfgang Pauli, nicknamed the "conscience of physics." Being one of the vigorous defenders of quantum mechanics, Pauli was certainly not soft on Einstein. Just a year before Einstein's death, Einstein and Born once more entered a rather heated debate about quantum mechanics following Einstein's contribution to the Born *Festschrift*, in which he had criticized quantum mechanics with a thought experiment discussed later in this contribution. Pauli, who was at Princeton with Einstein at the time, stepped in to calm the waves. He wrote to Born: "In particular, Einstein does not consider the concept of 'determinism' to be as fundamental as it is held to be (as he told me emphatically many times).... Einstein's point of departure is 'realistic' rather than 'deterministic', which means that his philosophical prejudice is a different one" (Wolfgang Pauli to Max Born, March 31, 1954, in Einstein and Born 1969, 286).

Not surprisingly, the concept of reality is at the heart of the Einstein-Podolsky-Rosen argument (Einstein et al. 1935), and "Quantum

Mechanics and Reality" is the title of a follow-up paper in which Einstein presented a simpler version of the argument (Einstein 1948b). We will find plenty of evidence for Einstein's concern with the reality that quantum mechanics is purported to describe in the course of our discussion. But what is the realism that is Einstein's "philosophical prejudice"? Pauli, who wanted to banish all pictures from the description of atomic phenomena, considered it to be a naive and old-fashioned attempt to continue making intuitive mechanical models in the tradition of nineteenth-century physics, and so, in effect, his evaluation of Einstein's position is not more positive than that of Born. Later, more philosophically oriented commentators concentrated on Einstein's critique of logical positivism and equated his position with a traditional epistemological realism that holds that science is to be a faithful image of "how things really are"; that is, they saw Einstein as making a claim about the relation between an observer-independent truth and our empirical knowledge of it (Holton 1968). Both the logical positivists and the defenders of the Copenhagen interpretation of quantum mechanics liked to label this position "naive realism," implying again that it meant clinging to outdated ideas about the possibility of scientific knowledge. But this dismissal ignored that Einstein, the enthusiastic student of David Hume, Ernst Mach, and Henri Poincaré, was certainly not naive when it came to the epistemology of science.[9] He had long learned the lesson of empiricism and positivism that science is not simply an image of "the world out there"; and he had not forgotten this lesson in old age, as the following quote shows:

The "real" isn't given to us in any way immediately, only the experiences of human beings are given to us.... The positing of the "real" as existing independently of my experience is a totality of intellectual constructs.... Our trust in the belief system about reality rests only on the fact that those concepts and relations [posited as real] stand in a relation of correspondence with our experience; this is the only ground for the "truth" of our statements. (Einstein 1951, 136–8, translation CL)

In this quote from the introduction for a book on realism, Einstein contrasts his view sharply with that of the author, Herbert Samuel, for whom reality is the "out there," which science is supposed to picture. For Einstein, the truth of science does not lie in its being a truthful image of a mind-independent reality, but in its success in accounting for our experiences (here is the empiricism of Hume and Mach); and, most importantly for our purposes, reality is not the mind-independent "out there" of Cartesian epistemology, but an intellectual construct posited by science (here is Poincaré's conventionalism).

If Einstein's realism is not a simple epistemological realism, what is it then? What can it mean to be a realist about an intellectual construct?

Arthur Fine has proposed that Einstein "entheorizes" realism, meaning that Einstein's realism is not an epistemological claim about the relation between science and mind-independent reality, but rather a methodological claim about the correct internal structure of science and the choice of its methodological apparatus. Fine sees Einstein's realism in a bundle of requirements for a satisfactory fundamental theory (Fine 1986, 86–111): the theory should talk about objects independent of observation, represent them in a spatiotemporal framework, and consist of deterministic laws. Fine also adds some secondary requirements that are not as central to Einstein.

Presenting Einstein's realism this way may make it appear less naive, but it still looks rather prejudiced: Einstein clings to a set of *a priori* demands on physics, and he is willing to dismiss the whole thriving field of quantum mechanics because it does not fit with them. Putting Einstein's methodological convictions in a historical context, I will argue that they are not *a priori*, but rather that they sum up the lessons he believed to have learned from his own successes in theoretical physics. This also will sharpen the somewhat vague notion of a bundle of requirements that Fine proposes to one fundamental principle (with a few corollaries) that I propose to call Einstein's *methodological realism*. Lastly, this historical contextualization will make Einstein's demands appear a great deal more plausible as general methodological precepts, even in the light of contemporary physics.

The context to look at is, as I already announced, Einstein's "philosophical" period after the publication of the general theory of relativity, the time in which he tried to understand and justify the principles that had led him to his greatest theoretical success, confronted with skepticism about his radical abolition of the physical reality of space and time. Put in the spotlight of the general public by the spectacular confirmation of his prediction of the bending of light by the Sun, but at the same time criticized for the supposed mathematical abstruseness and unintuitiveness of his theory, Einstein responded with a series of papers defending the physical and epistemological soundness of his theory, both to his colleagues and to the general public. This forced him to think hard about the foundations and methods of physics, and many of the most important of his philosophical papers were written in these years (Einstein 1918e, 1918j, 1919f, 1919g, 1920j, 1921c).

There is no better place to begin but with Einstein's exchange with Moritz Schlick, who had written one of the first philosophical analyses of general relativity, the paper "Raum und Zeit in der gegenwärtigen Physik" (Space and Time in Contemporary Physics). Einstein had praised its "unsurpassable clarity in wording and organization" (Einstein to Schlick, February 6, 1917, CPAE 8, Doc. 297). A few months later Einstein was studying the expanded book version of the paper and had a comment on Schlick's critique of Machian positivism. Schlick had added a new chapter, "Relations to Philosophy," which Einstein also found excellent. There, Schlick argued for a wider concept of physical reality beyond Mach's immediate sensations, and claimed that it is more satisfactory to also call "real" physical events that are not directly perceptible. Einstein's response illustrates his own take on the question of realism beautifully, so I will quote it in full:

> Your conception contrasts to Mach's according to the following scheme: Mach: Real are only sensations. Schlick: Real are sensations and events (of physical nature). It seems to me that the word "real" is understood in two different ways, depending on whether it is said about sensations or about events, i.e. matters of fact in the sense of physics.
>
> If two peoples do physics independently of each other, they will create systems that certainly agree with regard to the sensations ("elements" in the sense of Mach). The intellectual constructs that both think up to connect these "elements" can differ vastly. Both constructs also need not agree in regard to the "events", because they certainly belong among the conceptual constructs. Real in the sense of "irrefutably given in experience" are only the "elements" but not the "events."
>
> However, if we designate as "real" what is arranged by us in the schema of space and time, as you have done in your epistemology, then without doubt it is mainly the "events" that are real.
>
> What we now designate as "real" in physics is doubtlessly the "spatiotemporally arranged", not the "immediately given." The immediately given can be an illusion. Conversely, the spatiotemporally arranged can be a sterile concept that doesn't contribute to the elucidation of the connections between the immediately given. I would like to propose a neat separation of concepts here. (Albert Einstein to Moritz Schlick, May 21, 1917, CPAE 8, Doc. 343)

Einstein is proposing here a clear distinction between a phenomenal reality as the epistemological basis of our knowledge (that which is irrefutably given in experience) and a *physical reality* of spatiotemporally arranged events, which is an intellectual construct (that two peoples – or scientists – do not need to agree on). Why does Einstein propose this distinction? Just for methodological clarification? I would rather assume that Einstein, who famously called himself an "epistemological opportunist" (Einstein 1949b, 684), has a concrete issue in physics in mind. And it is not hard to see what that might be in 1917.

General relativity had introduced an unsettling discrepancy between the space-time structure as postulated by physics and our pretheoretical intuitions about space and time (or maybe just our intuitions schooled by previous theories) embodied in Euclidean geometry. Einstein had to defend this disregard of intuitive convictions, not just against philosophers who protested that the laws of space and time are not a matter for empirical science to decide, but also against many physicists who resented that the foundations of physics became ever more removed to an abstract and unintuitive level of mathematical physics. It was in this situation that he found support in the conventionalism of Henri Poincaré, who postulated that the foundations of science are neither empirical laws nor *a priori* necessities, but free conventions, chosen for the sake of simplicity and coherence of the theoretical structure. As an additional positive argument, Einstein could point out that the separation between the foundations of the theory and the empirical facts was not merely a loss, but also the basis for a new powerful methodological principle: the principle of general covariance. This principle meant that the coordinates were stripped of any physical meaning as a measure of spatial and temporal distance. However, it did not mean, as popular misunderstanding implied, that now "everything is relative." Rather, the particular power of general covariance is the discovery of a new *invariance* structure behind the phenomena.

This discovery was in no way trivial, as Einstein's lengthy struggle with the question of the covariance of the field equations for the gravitational field shows.[10] After the breakthrough in late 1915, however, Einstein could announce that the field equations of general relativity were now generally covariant, that is, valid independently from any specific choice of coordinate system. This fact meant that the evolution of the gravito-inertial structure (the union of gravitational and inertial effects demanded by the principle of equivalence) can be described without referring to a specific choice of coordinate system; the coordinate system itself becomes a mere convention and does not describe any objective features of space-time. This was a conceptual breakthrough in comparison with the previous *Entwurf* theory: in that theory, the field equations were given in certain adapted coordinate systems and it was unclear from this form whether the field equations were covariant in respect to coordinate transformations more general than the Lorentz transformations.

Now, with general covariance, this division is straightforward: *any* coordinate system is equally permissible for the description of space-time. Therefore, only covariant entities like tensors and vectors that are defined independently of a specific choice of a descriptive (coordinate) system can be considered to be defined objectively. The fundamental laws (such as the field equations) can only be defined on such invariant objects. The mathematical invariance property of general covariance therefore becomes a mark of *objectivity* in the sense of observer-independence. By contrast, physical measurements do not reveal those objective entities themselves, but their representations in local coordinate frames, that is, noninvariant quantities. Such measurements are no less real, but they are real in a different sense: their reality is *phenomenal*, dependent on an observer situated in a specific way. If we measure the gravitational force on the surface of the Earth, this measurement has a perfectly good physical sense, but it depends on the specification of a rest frame, that is, the measuring apparatus being fixed relative to the surface of the Earth.

The association of descriptive invariants with objective reality has become a methodological standard in modern physics: it is the conceptual basis not only of general relativity, but of all gauge theories, also in their quantum forms. It also pervades quantum mechanics since its formulation as a *transformation theory* by Fritz London, Paul Dirac, and Pascual Jordan, which again is based on the distinction between physical facts and their representations in different reference systems (here different bases of what soon became known as the Hilbert space). This common feature of relativity and quantum mechanics is observed especially clearly in the preface to Dirac's *Principles of Quantum Mechanics*: "The important things in the world appear as the invariants (or more generally the nearly invariants, or quantities with simple transformation properties) of these transformations. The things we are immediately aware of are the relations of these nearly invariants to a certain frame of reference...." (Dirac 1930, v). The association of invariance with objectivity has also become a well-treated topic in the history and philosophy of modern physics (e.g., Norton 1992a, 1999a; Nozick 2001). However, the association of description-dependent entities with a phenomenal reality (which Dirac clearly picked up in the preceding quote) has not created much of a stir. Therefore, I will look at one striking example of Einstein's own use of this association: the interpretation of the Christoffel symbols of the space-time metric as the theoretical representative of the gravitational force field. The Christoffel symbols describe how the geodesic motion in a given metric deviates from a motion that is "linear" in a given coordinate system. They therefore depend essentially (not merely covariantly) on the choice of coordinate system. For this reason, Einstein's interpretation (and the related claim that the energy of the gravitational field is given by a coordinate-dependent quantity) was criticized even by most adherents of general relativity,[11] and in modern coordinate-free presentations of general relativity, Einstein's point is mostly ignored. Nonetheless, there is good physical sense in Einstein's interpretation. It is exactly the point of the equivalence principle – Einstein's original intuition that led him to general relativity – that the gravitational field is entirely dependent on the choice of reference system: an observer resting on the surface of the Earth feels the Earth's gravitational field; an observer falling freely (like in a satellite circling the Earth) feels "zero gravity." The equivalence principle was a central issue for the critics of general relativity who argued that Einstein muddled the essential difference between the fictitious inertial field and the *real* gravitational field. For us, it is especially interesting how Einstein responded to this criticism:

First of all, I have to point out that the distinction "real" vs. "nonreal" cannot be very helpful here. [Einstein refers to a previous discussion regarding how energy in classical physics is a quantity that depends on the choice of frame of reference.] Instead of distinguishing between "real" and "nonreal" let us distinguish more clearly between quantities that inhere in the physical system by itself (independent of the choice of the coordinates) and such quantities that depend on the coordinate system. It would be immediately plausible to demand that physics should introduce into its laws only quantities of the former kind. However, it has turned out that this path is not realizable practically, as already was shown by the development of classical mechanics.

[Physics] cannot do without the coordinate system, and hence has to use with the coordinates quantities that cannot be understood as the results of definable measurements. According to general relativity, the four coordinates of the spacetime continuum even are completely freely chosen parameters lacking any independent physical meaning. Part of this freedom also affects those quantities that we use to describe the physical reality (the field components).... Therefore, one can neither say that the gravitational field in one point is something "real" nor that it is something "merely fictitious." (Einstein 1918k, 699–700)

The empirical, measurable quantity "gravitational field" cannot be expressed as a purely objective entity, since its measurement depends essentially on the state of the observer. That doesn't make it fictitious. Rather, it is part of what I call the phenomenal reality of a specifically situated observer.

This, then, I claim, is the specific theoretical background for Einstein's distinction of the two meanings of reality in the letter to Schlick. The "Machian elements" are real in this phenomenal sense, and the "physical events" are real in the objective sense (Figure 10.1). But both concepts of reality are quite independent of any epistemological claim of the relation of physics to the "real world." What Einstein addresses is not the Cartesian problem, but the methodology of physics: *within physics*, we need to distinguish between what the theory claims to be objective (description-independent) facts and what it claims to be phenomenal (description-dependent) facts. That implies that Einstein is not concerned with relating them to different sources of knowledge, as Mach had been in the tradition of the classical epistemological debates. Einstein does not need to assume that the objective description is *a priori* in a Kantian sense, that is, free of empirical knowledge, or that the phenomenal description is theory free in the sense of a radical positivism,

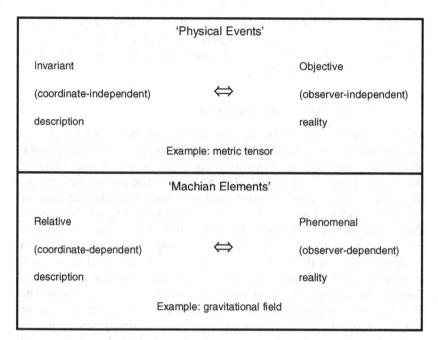

Figure 10.1. *Objective and phenomenal reality.*

that is, limited to pure sense data. Of course, Einstein's distinction is merely analytic and does not impose any restriction on physical theories. But it motivates a substantive principle, namely the methodological requirement: only objective facts (given by the invariants of the description) can enter into the fundamental laws of physics.

One can call this requirement Einstein's *methodological realism*, keeping in mind that it has little to do with the epistemological realism that concerns philosophers.[12] The basis for this methodological principle is not, as naive realism would have it, that a physical theory should provide an "image of reality," but rather that it should be a theory of invariants, that its purpose is to construct a simple and coherent structure behind the dizzying multitude of the phenomena. At the time, Einstein expressed his thoughts on physics and reality most directly in his speech "Motives for Research" (Einstein 1918j), where he stressed the search for the absolute behind the varying phenomena as the most fundamental motivation for doing physics. This methodological principle is similar to Kant's regulative principles, and later in life Einstein would acknowledge this affinity with a philosopher whom he originally disliked as the defender of knowledge *a priori*.[13] The important point is that Einstein's methodological realism does not make any claims about reality beyond its invariance. This is where my reading of Einstein differs most substantially from other proposals such as Arthur Fine's and Don Howard's who maintain that Einstein's realism implies determinism or separability, respectively.

3. EINSTEIN'S DETERMINISM

The early work of Einstein on quantum theory, described by Olivier Darrigol in Chapter 4, was accompanied by his hope to eventually arrive at a constructive theory of quantum phenomena, which the old quantum theory before 1925 described by a rather ad hoc and incomplete set of quantum conditions. Einstein explicitly explained the meaning of "constructive" in a popular article on general relativity for the *London Times* (Einstein 1919f): there he distinguished between "theories of principle" that are founded on general, empirically well-confirmed principles (such as thermodynamics, based on the first and second law of thermodynamics) and "constructive theories," which construct a description of physical reality out of simple parts, such as the atoms of Boltzmannian statistical mechanics, or the electromagnetic field of Maxwell's theory of electromagnetism.[14] In his comparison, Einstein made clear that, although theories of principle had the advantage of being firmly based on empirical knowledge, a truly fundamental physical theory should be constructive since only a constructive theory could be explanatory, that is, make us understand why things happen the way they do. Einstein's ideas about what such a constructive basis for quantum theory might look like are not well documented in his publications; they have to be pieced together from parenthetical remarks and some statements in his correspondence. One characteristic strand in his thinking was that a nonlinear modification of the Maxwell equations might deliver several benefits: an explanation of the quantum nature of electromagnetic radiation and an explanation of the properties of the electron. For a few years around 1909, he vigorously pursued attempts to construct such a theory until his lack of success made him doubt the viability of such an approach (see Chapter 4, Section 8).

After 1915, when Einstein had found the field equations of general relativity after a long and arduous struggle, two changes had taken place that would influence the further course of events. First, in 1913, Niels Bohr had applied quantum theory to the Rutherford model of the atom. This daring combination of two novel theoretical endeavors, quantum theory and atomic physics, propelled quantum theory from a somewhat obscure theoretical specialty to the forefront of research in fundamental physics. It impressed Einstein so much that he wrote a full thirty-five years later in his "Autobiographical Notes":

That this insecure and contradictory foundation [of the old quantum theory] was sufficient to enable a man of Bohr's unique instinct and sensitivity to discover the principle laws of the spectral lines ... appeared to me as a miracle – and appears to me a miracle even today. This is the highest form of musicality in the sphere of thought. (Einstein 1949a, 45–7)

It is plausible to assume that Bohr's success rekindled Einstein's interest in quantum theory. In 1916, he published his paper on emission and absorption of radiation by the Bohr atom, connecting the new model to his old specialty, the light quantum hypothesis, still opposed by most of his colleagues, including Bohr (see Chapter 5). Second, the success of general relativity confirmed Einstein's conviction of the soundness of a constructive program of basing quantum theory on field theory; already in 1919 he proposed a modification of general relativity with the aim to represent electrons as solutions of the field equations (Einstein 1919a). This program would eventually lead to the unified field theory program that Einstein pursued from the 1920s to the end of his life. It is in these years that Einstein's methodological realism discussed in Section 2 found its explicit form. From then on, it would be a criterion by which Einstein judged the success of any attempt to explain the puzzles of the old quantum theory.

First, however, Einstein pointed out a momentous consequence of the success of Bohr's theory. As Darrigol discusses in detail, Einstein was able to show (Einstein 1916j) that Planck's energy distribution for black-body radiation followed from Bohr's atomic model with the help of some simple assumptions about the *probability* of emission and absorption of radiation. He could further show (Einstein 1917c) that the emitted radiation had to have a definite momentum, that is, that it was emitted in a specific direction, not as a spherical wave as demanded by classical electrodynamic theory. This was a rather radical break with the classical picture: there, a radiating body would continuously emit radiation at a certain rate in all directions. In Einstein's model, there were individual acts of emission in specific directions that were only statistically distributed (as Einstein remarked, this resembled the law for radioactive decay). Could it be that the act of emission is fundamentally indeterministic? This is the answer that quantum mechanics eventually would give (at least in its standard interpretation). Einstein, however, was not ready to give up determinism.

What made the assumption of a genuinely statistical law unacceptable for Einstein? The standard account holds that Einstein's commitment to determinism is a deep-seated conservative belief that Einstein would not give up in the face of mounting evidence to the contrary. I want to argue, however, that Einstein's attitude toward determinism was a corollary to his methodological realism discussed in Section 2 and that it was motivated by an application of this methodological realism to statistical mechanics, incorporating conclusions that Einstein had drawn from his previous work on the foundations of statistical mechanics and space-time theory. As Anne Kox discusses in Chapter 3, Einstein saw his central contribution to statistical physics in the years 1902–4 in the clarification of the concept of probability that entered Boltzmann's statistical definition of entropy $S = k \ln W$. Einstein explained the probability of a thermodynamical state as the fraction of time that on average the system spends in that state during its complex dynamical evolution. This implies that a system in equilibrium does not simply stay in its most probable state (the state of highest entropy) but that it will also reach other states with a probability given by their entropy. Hence the system's state will fluctuate around the equilibrium state in a certain well-defined manner. This theory of fluctuations around the equilibrium led Einstein to his most important contributions not only in statistical mechanics (Brownian motion) but also in the early quantum theory.[15] The success supported the still controversial aspects of Boltzmannian statistical mechanics: the second law of thermodynamics is not an exact law, but only a statistical regularity, and the macroscopic thermodynamical quantities do not always give an exact description of the mechanical state of the system. For a system of definite energy, for example, the microstate (the specification of the positions and velocities of all its molecules) has to lie in a specific region of its state space (the energy shell) since the total energy is simply the sum of all the energies of the molecules. A system of a given temperature, however, does not have a definite energy, only its average energy is specified. (Einstein understood such averages as time averages, which are exactly defined only in the limit of infinite time.) Therefore, the microstate of the system is not constrained to a certain region of the state space; one can only say that it is with high probability near the energy shell defined by the average energy. A specification of temperature therefore does not give a description of an objective fact in the sense of Section 2, but only a probabilistic weight over such facts. This confirmed for Einstein a clear division between the status of the microscopic and the macroscopic description of a physical system: the microstates and their dynamics – whose reality was dramatically demonstrated by Einstein's 1905 prediction of Brownian motion – give a complete and objective determination of the state of a physical system. The macrodescription, however, is only an *incomplete* description, specifying our limited knowledge of the objective facts. Such incomplete descriptions cannot be invariant since there always can be many different incomplete descriptions of the same objective fact. Hence incomplete descriptions cannot be objective; rather, they give a phenomenal description in the sense of Section 2 since they describe the objective facts from the perspective of an observer with incomplete knowledge.

This division acquired additional weight with Einstein's development of general relativity. When he announced that "time and space thereby lose the last vestige of physical reality" (Einstein to Schlick, December 14, 1915, CPAE 8, Doc. 165), this had an important implication for statistical mechanics. Since the physically real (in the sense discussed in Section 2) is given by the tenseless four-dimensional structure of space-time events, all events in the past as in the future exist objectively just like present events. Hence, future events are objectively characterized by exactly the same complete microscopic descriptions as past and present events, and there is no possibility for an openness of the future that would allow for an objective indeterminacy of future events.[16] A nondeterministic dynamical law describes the future not in terms of such objective events, but in terms of incomplete statistical descriptions, and therefore cannot be a fundamental law according to Einstein's methodological realism. This, of course, does not mean that there cannot be probabilistic dynamical laws in physics; such laws, however, can only be phenomenal laws describing an observer's ignorance about the objective facts of the future. They do not yield objective descriptions of the future state of the system, but only statistical regularities derived from the law of large numbers.[17] Nothing assures us that objective laws in this sense do exist in nature, but methodological realism commits us to try to find such laws. This is the sense in which Einstein's methodological realism together with conclusions he drew from statistical mechanics and the theory of relativity implied for him a commitment to determinism: just as in the case of his realism, not as a metaphysical belief but as a methodological guideline, since only a deterministic law can give an invariant and objective characterization of the physical facts.

This identification of statistical macrostates with the sense of phenomenal reality discussed in Section 2 would become centrally important for Einstein's understanding of quantum theory. There is, however, an important difference between the two cases of phenomenal states (in relativity and in statistical mechanics) that I have discussed: in statistical mechanics, the description is *incomplete*, it gives only a partial characterization of the objective state; one can say that the relation between description and state is one-to-many (each description refers to many possible objective states). The noninvariant descriptions of relativity are not incomplete descriptions in this sense. Rather, they are "overcomplete," every objective state of affairs can be described in different systems of reference, the relation between description and state is many-to-one. We can say that the description is *not absolute*, but relative to a system of reference. In both cases, this relation is not one-to-one (the linguistic term is *biunique*), but for different reasons.[18] Einstein's principle of methodological realism implies in both cases that the description does not refer to an objective state and therefore cannot enter into fundamental laws.

This all means that the account of the emission and absorption of radiation that Einstein gave in 1916 could for him only be, despite its success, a phenomenological account, not a fundamental dynamical law. Einstein continued to search for this fundamental theory. Hoping to find clues for such a theory, he began to pursue the idea of an *experimentum crucis* in order to decide about the nature of the individual act of light emission from an atom and in the hope of clarifying the most mysterious aspect of Bohr's atomic model: the discontinuous and instantaneous jump between two atomic orbits, which, moreover, according to Bohr, should produce an extended light wave of constant frequency. (Not only was the mechanism of the jump unclear, the fact that it should produce a comparatively gigantic light wave seemed entirely absurd.) Einstein proposed several experiments during the 1920s that should have forced an answer to whether the act of emission was instantaneous or spatially and temporally extended. Darrigol recounts in Chapter 4 (Section 10) one such unsuccessful attempt to use the light emitted from canal rays for an *experimentum crucis*. Einstein's original hopes that it confirmed an instantaneous emission were soon dashed: it turned out that a wave theory would not give a different result. On a later attempt, Einstein collaborated from 1926 with the young experimental physicist Emil Rupp, again designing experiments on canal rays.[19] However, by that time Einstein grew doubtful that an experiment could produce the unambiguous answer that he originally had hoped for. It seems that the failure to devise experiments that furthered his constructive effort made Einstein increasingly skeptical about the possibility of an inductive approach to the desired fundamental theory based on experimental evidence, and led him to rely ever more on the program of constructing a unified field theory based mainly on criteria of mathematical simplicity.

Relatively little is known about Einstein's theoretical speculations in the time between circa 1915 and 1925 since he did not publish his ideas and even seems reluctant to discuss them in his correspondence. One interesting glimpse can be obtained from Hendrik A. Lorentz who, both in a letter to Einstein (H. A. Lorentz to Albert Einstein, November 13, 1921, quoted in CPAE 7, 486) and in a lecture given in Pasadena in 1921 (Lorentz 1927), reports about a theory of light that Einstein tentatively entertained at the time. In this theory, light quanta and waves coexist.[20] While the quanta are point particles carrying the energy packets necessary to explain the light quantum effects, their motion is determined by a guiding wave very much in the way Louis de Broglie would envision a few years later. Einstein never published an account of his ideas. Wigner reports from discussions in Berlin at the time that Einstein gave up the idea of the guiding wave because it could not account for scattering processes: the exact correlation of the scattering angles demanded by energy and momentum conservation could not be caused by waves that emanate in all possible directions of scattering.[21] A similar problem affected the Bohr-Kramers-Slater (BKS) theory proposed soon after (see Section 4). However, this background explains Einstein's early enthusiasm for de Broglie's and Erwin Schrödinger's constructive efforts as well as his keen awareness of their fundamental problems. Einstein was more successful in his work on statistical quantum theory, which, moreover, paved the way for Schrödinger's wave mechanics. As described in Section 11 of Chapter 4, Einstein quickly realized the importance of Satyendranath Bose's derivation of Planck's law, which, different from Max Planck's own, did not rely on a classical electrodynamic account of the interaction of light and matter. Instead, it used a new kind of statistics, which Einstein applied to the theory of the ideal gas, which had not been quantized in a fully satisfactory manner yet. Einstein pointed out that this new statistics led – just as in the case of black-body radiation – to a fluctuation term consisting of a wave and a particle parts,[22] and referred to de Broglie's dissertation proposing the new idea of matter waves as a possible explanation. It was from this hint that Schrödinger, who had been concerned with the quantum theory of the ideal gas, took the idea for wave mechanics.[23]

In 1923 Einstein published a paper with the title "Does Field Theory Offer Possibilities for the Solution of the Quantum Problem" (Einstein 1923d), in which he gave for the first time a detailed and programmatic statement of the hope that he would hold onto for the rest of his life: to derive the properties of elementary particles and their quantum behavior from a unified field theory, which extends general relativity and Maxwellian electrodynamics (see Chapter 9). Einstein proposed in the paper that the quantum conditions should be understood as a consequence of an overdetermination of the field equations of this unified theory. If the theory contained more differential equations than independent field variables, not every set of initial conditions would be allowed. In this way, Einstein hoped to explain both the existence of elementary particles with definite properties and their quantized dynamics. The trick was to find the general field equations, but Einstein hoped he could repeat the success of general relativity: once the right mathematical structure was found (as Riemannian geometry in the case of general relativity), the whole theory could be developed out of very few assumptions of highest generality (as the principle of equivalence). In hindsight, his years of struggle with the physics of the gravitational field[24] seemed to him to be an unnecessary detour from the royal road of mathematics. This confidence is expressed most clearly in his Oxford Spencer lecture of 1933, in which he declared: "Our experience hitherto justifies us in believing that nature is the realization of the simplest conceivable mathematical ideas."[25] Einstein's turn to unified field theory therefore has a double root: the frustration with attempts to build constructive models for quanta and quantum effects, and the hope that the theory that would solve these problems could be found directly by searching for the mathematically most stringent and simple field theory "of everything." The Spencer lecture is the high point of Einstein's declarations of confidence in the heuristic value of mathematics. But even when he doubted in later years whether field theory could ever solve the quantum riddle, he remained convinced that this was the only way open to him. During the 1920s and 1930s, however, he had high hopes for his new research program.

4. THE RISE OF QUANTUM MECHANICS

When attempting to understand Einstein's reaction to the development of quantum mechanics, one has always to keep in mind that he was engaged in an alternative program, his search for a unified field theory. Einstein's skepticism stemming from his failed attempts at finding a theoretical foundation for the old quantum theory is the background for his reactions to the various attempts of others in the mid-1920s: first the BKS theory, then matrix and wave mechanics and their unification into transformation theory, and what we now know as quantum mechanics. He remained an interested outsider, at first even enthusiastic in the case of wave mechanics. But when the developments ran into difficulties or conflicted with what he regarded as well-established general principles, he saw this as an indication of the failure of the "inductive program."

The BKS theory posed serious problems even before it was tested empirically.[26] The basic idea of the theory was to associate with an atom in a certain state an orchestra of virtual oscillators oscillating with all the frequencies the atom could emit in that state. Each emission oscillator emits (virtual) electromagnetic radiation at its oscillation frequency, which propagates classically according to the Maxwell equations. If another atom with a virtual absorption oscillator of the same frequency is exposed to this radiation, there will be a probability determined by the intensity of the radiation for the atom to jump into a higher state. Einstein objected to the theory on various grounds, among them the violation of energy conservation and the violation of determinism. As he wrote to Paul Ehrenfest, "this idea is an old acquaintance of mine but I don't consider it to be the real thing" (Albert Einstein to Paul Ehrenfest, May 31, 1924, quoted in Klein [1970a]). The theory reminded Einstein of his own attempts at a guiding wave theory of light, and so he was well aware of its problems. When the BKS theory was refuted by the Bothe-Geiger and Compton-Simon experiments, Einstein only remarked laconically to Ehrenfest: "We both had no doubts about it" (Albert Einstein to Paul Ehrenfest, August 18, 1925, quoted in Klein [1970a]). Einstein's skepticism would resurface in his later critical questions about quantum mechanics.

When Werner Heisenberg proposed a novel quantum mechanics in 1925, he took from the BKS theory not much more than the idea of virtual oscillators. Heisenberg eliminated electron orbits and described the atomic phenomena entirely in terms of the frequency and amplitude of *transitions* between states. He showed that if one replaced the classical coordinates of the electron by an array of transition amplitudes (formally similar to a Fourier decomposition of a coordinate into harmonic components), one could calculate with these arrays in a way analogous to classical mechanics. Born and Jordan soon gave a mathematically exact formulation to this calculus: it was the calculus of matrices, hitherto only used in linear algebra. Einstein followed the development of matrix mechanics with interest, but also with skepticism. To Ehrenfest, he wrote: "Heisenberg has laid a big quantum egg. In Göttingen they believe in it. (I don't.)"[27] Einstein was certainly not alone with his reservations; matrix mechanics did not only use obscure mathematics, it also was as yet not giving any visualization of the atomic processes it purported to describe. It did not even offer a way to represent a specific state of an atom, only the totality of all possible transitions between states were represented by a matrix. This reservation probably explains Einstein's very cool reaction to Jordan's successful application of matrix mechanics to the question of the dual nature of light. Jordan was able to derive Einstein's 1909 fluctuation formula for black-body radiation, which was the sum of a particle term and a wave term and therefore strikingly expressed the double nature of light, applying matrix mechanics to the harmonic vibrations of the electromagnetic field.[28]

Schrödinger's wave mechanics, presented in early 1926, at first seemed to Einstein to be a much more promising approach to understanding quantum theory. In a letter to Schrödinger, he wrote: "I am convinced that you have made decisive progress with your formulation of the quantum conditions, just as I am convinced that the Heisenberg-Born way is absurd" (Albert Einstein to Erwin Schrödinger, April 26, 1926, quoted after Przibram [1963, 24]). The explanation of atomic orbits as standing matter waves promised the return to a continuous and field-theoretical description of quantum phenomena in line with Einstein's own program. Schrödinger could quickly show that wave mechanics can reproduce the empirical predictions of matrix mechanics. But it held the promise to do much more, since it gave a dynamical account of atomic phenomena, as opposed to the static and abstract character-ization of atomic states in matrix mechanics. However, it became clear quickly that wave mechanics posed serious problems of interpretation. As Schrödinger acknowledged, the wave function of a system of several particles was not a field defined in space-time, but an abstract function in a higher-dimensional space. There seemed to be no way of interpret-ing this function in a space-time framework. Moreover, Schrödinger abandoned de Broglie's idea of a dual representation, which interpreted particles as field singularities guided by the matter wave, and insisted that there was nothing but the matter wave. This caused serious prob-lems to account for the particulate properties of matter. Schrödinger's original hope that he could represent particles as small wave packets ran into the difficulty that such wave packets would smear out over time. Einstein, along with Lorentz, Planck, and Bohr, soon became very skeptical regarding whether such a solution was viable.[29]

Quite a different reading of Schrödinger's wave function was soon established. Based on Born's proposal to understand the outgoing wave after a scattering event as a probability distribution for the particle being scattered in different directions, Pauli, Dirac, and Jordan devel-oped a general probabilistic interpretation of the wave function in the context of transformation theory, which unified matrix and wave mechanics into the first comprehensive systematic account of the new quantum mechanics.[30] It also resolved the problem of the instability of the wave packet. But it thwarted the hope that wave mechanics could give a field-theoretical foundation of quantum theory. Einstein's enthu-siasm waned considerably since he did not believe that the empirical evidence from atomic physics was strong enough to justify changing his fundamental methodological principles as outlined in Sections 2 and 3. If wave mechanics was a statistical theory, he saw it, according to his long-standing convictions about probability, as a phenomenological theory giving an incomplete description of objective reality, just as his account of emission and absorption from 1916. By the end of the year, he wrote to Born: "The quantum mechanics is very awe-inspiring. But an inner voice tells me that it still is not the real McCoy. The theory delivers a lot, but it gets us hardly closer to the secret of the old one. At any rate, I am convinced that *he* doesn't play dice" (Albert Einstein to Max Born, December 4, 1926 [Einstein and Born, 1969, 127]). That this judgment also included wave mechanics can be seen from a letter to Paul Ehrenfest from January 1927 in which he wrote that his heart does not warm to the "*Schrödingerei*" (after Fine 1986, 27). If quantum mechanics is an incomplete theory, then the way to find the underlying objective theory that suggests itself is by adding the particle positions that the wave function does not yield. Accordingly, Einstein attempted to add trajectories of particles to wave mechanics, reviving his earlier idea of a dual theory of particles and a guiding wave. There is an unpublished manuscript[31] from the spring of 1927 about this attempt with a note that Walther Bothe pointed out a difficulty concerning the scheme that Einstein proposed: in the case of a composite system of two noninteracting parts, the wave function of the composite system can be represented as the product of the wave functions of the two parts. If one constructs the trajectory corresponding to the wave function of one part, it is not identical to the trajectory one would obtain if the other part did not exist at all. This conflicts with the assumption that the two parts are noninteracting. Einstein therefore abandoned this attempt at a dual theory of waves and particles, but at the Solvay meeting of 1927 he still considered de Broglie's similar attempts to be promising.

Also, Heisenberg became concerned about the question regarding how to define particle position after the equivalence of matrix and wave mechanics had been established.[32] He realized that he needed to respond to Schrödinger's "wave" interpretation of quantum mechanics with a "particle" interpretation. This meant that he had to abandon the radical positivism of his first paper, in which he had justified the elimination of electron orbits with the argument that such an orbit is unobservable in principle. As he reports in Heisenberg (1969), Einstein protested against this positivism, much to Heisenberg's surprise, since he had accepted the common view that Einstein's theory of relativity was a paradigm of Machian positivism. Heisenberg quotes Einstein as saying "only the theory decides what one can observe" and sees in this exchange the seed for his later idea of the uncertainty relations as a consequence of the quantum mechanical formalism.[33] In the uncertainty paper (Heisenberg 1927), Heisenberg proposed to understand the quantum mechanical matrices as describing physical quantities not as precise numbers but with a certain degree of indeterminacy. This indeterminacy corresponds to our inability to observe the physical quantities with unlimited precision. Heisenberg derived from quantum mechanics an inequality, the Heisenberg uncertainty relation, that limits the definiteness of two complementary properties such as position and momentum of a particle. He then tried to show through a simple thought experiment that this uncertainty relation also describes the limits to our possibilities of measurements. He assumed that one tries to measure the position of a particle through Compton scattering with a light quantum. For a precise position measurement, the light has to be of short wavelength, and therefore the light quantum has to have a large momentum. But this means that the observation disturbs the momentum of the particle. Therefore, the precisions of position and momentum measurements limit each other.

Heisenberg's uncertainty relation should become one of the central issues in Einstein's critique of quantum mechanics. It is easy to see why when we try to interpret Heisenberg in the light of Einstein's distinction between objective and phenomenal states: taken at face value, one could take Heisenberg's argument to imply that the indeterminacy of the quantum mechanical state expresses simply our limited knowledge about the objective state, that is, that it is an incomplete description like a statistical state. However, Heisenberg soon insisted[34] that there could be no more complete description of the objective matters of fact than the quantum mechanical state. But then the quantum mechanical state must be itself the objective state, if we assume with Einstein that there has to be some objective state of the system. Einstein would soon start arguing that the quantum mechanical state could not fulfill this role, and we will see that all his criticisms of quantum mechanics were of this form.

5. THE FIRST DEBATES ABOUT QUANTUM MECHANICS

Bohr, like Einstein, had followed the development of quantum mechanics as an interested spectator. Unlike Einstein, he was not deterred by the difficulties that both matrix and wave mechanics presented to a physical interpretation. As his atomic model had already shown, he was willing to entertain models that contained contradictory elements, and the difficulties of the BKS theory had confirmed his suspicion that the classical spatiotemporal description of atomic processes had to be abandoned. The abstract nature of matrix mechanics appeared to him to be a virtue rather than a vice. While he also welcomed Schrödinger's wave mechanics as an appropriate expression of wave-particle dualism, he disagreed strongly with Schrödinger's hope that it would lead back to a field-theoretical account of atomic phenomena. Instead, Bohr envisioned that quantum mechanics should present a symmetry between a particle and a wave picture of both matter and light. This vision was to develop into Bohr's concept of complementarity, first formulated in his Como lecture in September 1927. This lecture was shaped by Bohr's dispute with Heisenberg about the latter's uncertainty paper, which Heisenberg had sent off for publication without Bohr's consent. Bohr objected to two points in Heisenberg's paper. Heisenberg had committed an error in his thought experiment concerning the position measurement through Compton scattering: the fact that a light quantum imparts a momentum to the observed particle is not in itself an impediment to determining the particle's momentum. One could simply measure the momentum of the light quantum before and after the scattering. More importantly for Bohr, Heisenberg completely relied on the particle picture in his interpretation of quantum mechanics. Bohr showed that a consideration of the double nature of light could solve the problem with Heisenberg's thought experiment. If one tries to determine the position at which the scattering took place, one has to use a lens that captures the light emanating from the point of scattering. This means we now have to consider the wave nature of the light quantum. This imposes a limitation, well-known from classical optics, on the exactness with which this position can be determined. It is inversely proportional to the wavelength of the light, which saves Heisenberg's uncertainty relation. Bohr's Como lecture, which went through many revisions before it was published in the summer of 1928, contains a detailed discussion of this thought experiment. It became the canonical formulation of Bohr's complementarity interpretation of quantum mechanics. It also reflects Bohr's discussions at the Solvay conference in October 1927.

This meeting marks the beginning of Einstein's public critique of quantum mechanics.[35] In the general discussion, Einstein posed a rather demure-sounding question of understanding, which, however, not only confounded Bohr during the discussion, but also has been puzzling to later commentators. But if one interprets Einstein's question in the light of the methodological realism proposed here, it becomes quite obvious how it is to be understood. Einstein is posing a dilemma for any interpretation that takes the wave function as an invariant description of the situation, and concludes that it cannot be objective. Hence, one should search for a more complete description. Notably, Einstein points out a problem that quantum mechanics is still struggling with today: the measurement problem. Einstein proposes the following thought experiment: we are sending particles with a certain momentum (represented in wave mechanics by a parallel train of waves) through a narrow slit. Behind the slit, the wave will be diffracted, emanating in the shape of concentric wave fronts from the slit. We now put a cylindrical screen behind the slit to detect the particles. What we will find are, of course, not waves, but individual detection events (little flashes on a fluorescent screen or little black spots on a photographic plate). Einstein asks how we are to think of the particle *waves* in this case. He considers two possibilities.

The first possibility is that the wave function only describes a totality of many particles.[36] It seems that Einstein's thoughts on light quantum statistics, which he originally intended to present in a talk at the Solvay conference, are the background for this discussion. In the case of light, a viable interpretation of the relation between light wave and light quanta is that light quanta have no individual existence within the wave, but only appear in interaction events. With this picture in mind, Einstein asks whether the wave function has to be seen as the description of a collective of particles rather than applying to a single particle. It does not describe the motion of an individual particle within the collective and thus cannot explain cloud chamber tracks or the correlation between individual particle energies in the Geiger-Bothe experiment. Therefore, it is not an acceptable physical interpretation of the wave function.

The second way to think of the wave function is to understand it as a complete description of an individual particle, which really does not have a definite place. Then the wave function describes probabilities in a different sense: while the particle before the detection did not have a definite place, it will trigger a detection event in a certain spot on the screen with the probability given by the wave function. As Einstein points out, one has to think of one wave function for each individual particle to explain phenomena like the Bothe-Geiger experiment or Wilson chamber tracks. However, in this case, we encounter a problem explaining the detection events on the screen: consider a wave train corresponding to just one particle. Then we know that if there is a flash at one spot on the screen, there will be none anywhere else on the screen. But how does the screen know? After all, the wave impinges homogeneously all over the screen, and if there is, say, a probability of 10 percent that it triggers a flash in one region of the screen, and there is the same probability that it triggers a flash in a second different region of the screen, then there should be a probability of 10 percent × 10 percent = 1 percent that it triggers two flashes in both regions since these two events are statistically independent. But this is not what quantum mechanics predicts; rather it demands that there always be just one flash.

But again, how does the screen know only to flash once? It seems that in the moment that a flash happens in one spot the wave has to be "turned off" everywhere else. Such a process must be, as Einstein puts it: "an entirely peculiar mechanism of action at a distance, which prevents the wave continuously distributed in space from producing an action in *two* places on the screen." Einstein points out a dilemma for our understanding of quantum mechanics: an interpretation as the description of a collective would be straightforward, but it would be in conflict both with the formalism of quantum mechanics and with empirical evidence. An interpretation as the description of an individual particle as a spatially extended object would require an unphysical and nonlocal transition upon its detection. This process is, of course, nothing but the infamous "collapse of the wave function," which von Neumann would formalize a few years later in his treatment of the measurement problem (von Neumann 1932).

The conclusion that Einstein draws from this dilemma is a cautious endorsement of de Broglie's proposal to describe the process with *two* entities: a spatially extended Schrödinger wave and a localized particle, which is guided by the wave but which is responsible for the local interaction with the screen. It is this endorsement that corresponds to Einstein's later claim that quantum mechanics is an incomplete description of individual systems.

The various responses to Einstein's question constitute the first published discussion of the measurement problem of quantum mechanics and certainly merit a more detailed treatment than can be offered here. I will concentrate on Bohr's response, since he already at this point emerges as the speaker for the group of theoreticians defending the finality of quantum mechanics against Einstein's criticism. Bohr responded directly after Einstein's contribution, but the answer was cut from the published proceedings, as were all other contributions by Bohr, to be replaced with a French translation of Bohr's Como lecture. Bohr's original answer is only recorded in rather fragmentary notes from the discussion preserved at the Niels Bohr Archive.[37] Striking in these notes is Bohr's statement that he did not understand Einstein's question. It is not hard to see why: Bohr did not share Einstein's methodological presuppositions and did not understand Einstein's implicit demand for an objective and complete description. Still, one can plausibly assume that Bohr's wish to replace his contributions with a reprint of the Como lecture implies that he saw the lecture (the written version was finished only after the Solvay conference) as a more extensive and coherent expression of his views. Therefore, it seems acceptable to use the Como lecture to extrapolate the meaning of the discussion notes.

Bohr counters Einstein's dilemma with a point he makes explicit in the Como lecture: quantum mechanics is a collection of mathematical methods adequate to describe our observations. But we have to give up hope that any theory can offer a complete description of the processes taking place between these observations. Such a complete theory would require a description that is both spatiotemporal and respects what Bohr calls "causality," that is, the conservation of energy and momentum. Bohr argues that the impossibility of an exhaustive description is due to the impossibility of observing microscopic systems without the measuring instrument interacting with the object. The specific limitations of quantum theory result from the fact that this interaction always involves at least the transfer of Planck's quantum of action. It is not obvious how Bohr's argument about limits of observability justifies his claim about the inapplicability of spatiotemporal descriptions in quantum mechanics. Nothing in Einstein's dilemma depends on an assumption of observability without interference. The only observation that takes place is the single detection event; Einstein's question is rather about the objective description of unobserved processes. Nevertheless, there is a good physical rationale behind Bohr's repudiation of a continuous spatiotemporal representation of atomic processes: it is simply the failure of attempts at finding such a representation, as in the old quantum theory or in Schrödinger's wave mechanics. The epistemological justification of the fundamental nonclassicality that Bohr tries to substitute for a less glamorous acknowledgment of defeat has become a recurring theme in the interpretation of quantum mechanics. However, it did not satisfy Einstein on several grounds.

First, it is based on a sweepingly general statement about the possibilities of empirical observations, which can never be proven in any kind of generality and was quickly contested by Einstein, who tried to give examples of simultaneous measurement of noncommuting observables. The question of the measurability of noncommuting observables became the topic of heated discussions between Bohr and Einstein during the time of the 1927 Solvay conference and in the years after. In these discussions, Einstein tried to construct experimental situations that allowed for the simultaneous measurement of complementary quantities.[38] Bohr countered Einstein's arguments by applying the uncertainty relations also to the measuring apparatus, which introduced an incontrollable influence from the measuring apparatus on the observed object. These discussions played an important role in confirming Bohr's conviction that a careful analysis of measurement uncertainties could serve as a conceptual foundation of quantum physics. Bohr described the exchange later in his contribution to the volume in the Library of Living Philosophers devoted to Einstein (Bohr 1949).

Second, however, Einstein still could maintain that even if one concedes the impossibility of certain measurements of theoretical quantities, this does not generally imply in any way that one cannot form a coherent and empirically successful theory out of these quantities. Therefore, he was not convinced by the impossibility of constructing counterexamples against the uncertainty relations. Einstein could argue that physics is full of theories based on unobservable quantities. This was the case for nineteenth-century statistical mechanics, but it was also true for many theoretical concepts contemporary to quantum mechanics, such as the electromagnetic potential, entropy, or even the all-pervasive concept of energy: in none of these cases is the impossibility of measuring a quantity directly seen as a reason to generally limit its theoretical use.

Einstein's conclusion from the debates was that quantum mechanics is a self-consistent but incomplete description of the objective processes. It is phenomenal in the same way as thermodynamics is phenomenal, operating with description-dependent states. To argue this case, Einstein returned to the kind of conceptual argument like the single-slit example, trying to force the question about the physical reality of quantum mechanical states against Bohr. In this sense, the Einstein-Podolsky-Rosen (commonly called EPR) paper is a direct continuation of Einstein's single-slit argument of 1927.

6. THE ENTANGLEMENT ARGUMENT

Einstein's most famous contribution to the debate over quantum mechanics is the paper "Can Quantum Mechanical Description of Physical Reality Be Considered Complete?" coauthored with Boris Podolsky and Nathan Rosen and published in 1935. After a rebuttal by Bohr, which was rather obscure in content but confident in tone, it was mostly ignored by the physics community. This only changed with Bell's use of EPR's example for an argument that went substantially beyond Einstein's (Bell 1964). Bell's paper was part of a resurgence in the interest in the foundations of quantum mechanics and a reevaluation of the dominant Copenhagen interpretation, which also led to a new appreciation of Einstein's criticism.[39] (I will briefly look at the relevance of Bell's work for Einstein's argument in Section 8.) At the same time, historians and historically sensitive philosophers have uncovered interesting details about the historical background of Einstein's argument.[40] One such new aspect is that it was Boris Podolsky who wrote the paper and that Einstein was unhappy with the way the argument was written up, as he wrote to Schrödinger: "The main point has been buried so to speak beneath learnedness" (Einstein to Schrödinger, June 19, 1935, EA 22–047). In the same letter, he gave a much simpler presentation of the argument, which he repeated soon after in a paper entitled "Physics and Reality" (Einstein 1936). I will look at this presentation first, and then at the full EPR argument. Not only will this help to understand the "learnedness" of the EPR argument, it will also bring out an important difference between the two versions.

In the letter to Schrödinger, Einstein starts with an explanation of his idea of completeness. He takes a simple example from everyday life: consider two boxes with closed lids. Suppose you know that there is one ball in the boxes, which you will always find in one of the boxes when you open the lids. You describe the state of the boxes by saying that the probability for the ball to be in a given box is one half. The question is now: is this a complete description of the physical situation? If you answer no, you suppose that there is an objective matter of fact about the ball being in one or the other box, even if you have not opened a lid. If you answer yes, you have to assume that without the act of opening a lid, there is no matter of fact about the location of the ball. Only once you open a lid and find or don't find a ball, you cause the ball to be in one specific box. Einstein calls the first alternative "Born interpretation," even though Born had long fallen in with the Copenhagen orthodoxy, and the second alternative "Schrödinger interpretation," although Schrödinger points out to him in his response that he is "long past the stage where I thought that you could see the ψ function in any way directly as description of reality" (Schrödinger to Einstein, August 19, 1935, EA 22–051). However, the real enemy is Niels Bohr – unnamed in the letter but referred to as "the Talmudist philosopher who doesn't give a damn about reality, the bugaboo of naive minds" – who claims that both interpretations are only different in wording and that the debate is pointless. Einstein's letter makes quite clear that he is not naively worshiping the bugaboo of reality. Rather, it is an excellent illustration of Einstein's methodological realism (as construed in Section 2): he acknowledges that we have no way of knowing reality independently of its physical description. Hence, we cannot compare the physical description directly with reality. Therefore, it is not immediately clear how we can decide which of the two interpretations is correct. We need some assumption of what happens within the boxes in this situation (i.e., a rudimentary theory), before we can make any meaningful statement about the objectivity of a description of the content of the boxes. This is why Einstein introduces an additional principle that he calls the *principle of separation*: on the assumption that the boxes are spatially separated and that there is no physical interaction between the two, the real state of one box and its contents is independent of what happens to the other box. But then, opening the lid of one box and checking whether there is a ball inside cannot change the real state of the other box. By contrast, after opening the lid of one box, one knows whether the ball is in the other box or not. This fact about the other box must therefore, because of the principle of separation, also have been true before opening the lid of the first box. Hence, a description that only gives probabilities for this situation is incomplete. Einstein stresses that the analogy with the quantum mechanical case is imperfect and that the everyday example should only be seen as an explication of the concept of incompleteness. Discussions of Einstein's principle of separation (Howard 1985, Fine 1986) have often analyzed it as consisting of two logically independent assumptions: *separability*, the existence of independent states of the two subsystems, and *locality*, the assumption that a change of the state of one subsystem does not affect the state of the second. Nobody in Einstein's time doubted locality, and Einstein in his later more explicit statements assumed separability and showed that under this assumption, quantum mechanics will violate locality. Therefore, I generally only talk about separability as the assumption relevant for the discussion.

Einstein's analysis of the quantum mechanical case is based on the representation of states of composite systems in standard quantum mechanics, and especially on the phenomenon for which Schrödinger later coined the term *entanglement* in a paper written in reaction to Einstein's (Schrödinger 1935). Although this representation is implicit in von Neumann's formulation of quantum mechanics, the general question of the description of states in composite systems had not elicited much interest until then. In the standard Hilbert space formalism of quantum mechanics there is, for any physical observable O, a set of values o_i (the *eigenvalues* of the observable) that this observable can assume. There is also a set of *eigenstates* ψ_i corresponding to the eigenvalues o_i, defined by the eigenvalue equation,

$$\hat{O}\psi_i = o_i\psi_i,$$

where $\hat{O}$ is the Hilbert space operator corresponding to the observable O. These eigenstates form a basis of the whole Hilbert space, that is, every state ψ in the Hilbert space can be represented as a linear combination

$$\psi = \sum_i c_i \psi_i.$$

Two different observables O and O' will, in general, have different sets of eigenstates ψ_i and ψ'_i.

Again, Einstein considers two systems, A and B, that are separated from each other and don't physically interact. For any observable O^A on A and O^B on B, we can write any state Ψ^{AB} of the composite system in the form

$$\Psi^{AB} = \sum_{i,j} c_{ij} \psi_i^A \psi_j^B$$

where the ψ_i^A and the ψ_j^B are the bases of eigenstates corresponding to the observables O^A and O^B, respectively. If we choose a different observable O'^A for system A, we can equally represent Ψ^{AB} on the basis of the eigenstates ψ'_i^A of O'^A:

$$\Psi^{AB} = \sum_{i,j} c'_{ij} \psi'_i{}^A \psi_j{}^B.$$

Notice that we still use the same basis of states for system B.

This representation implies a radical break with classical expectations: a state Ψ^{AB} of this form in general cannot be written as a product of a state of system A and a state of system B. This happens in general if the two systems have interacted at some point in the past. One says that the states of the two systems have become *entangled*. This means that quantum mechanics does not give us a definite quantum mechanical state for each individual component system when the composite system is in an entangled state. However, the reduction postulate (i.e., the collapse of the wave function necessary for definite measurement outcomes that Einstein had worried about at the 1927 Solvay conference) implies definite predictions for the states of both of the individual component systems if one performs a measurement on one of the systems. If we measure, for example, observable O^A, the state Ψ^A of system A after the measurement will be one of the eigenstates $\psi_i{}^A$, while the state of the system B will be the corresponding "relative state" $\Psi^B = \sum_j c_{ij} \psi_j{}^B$. If we measure instead O'^A, the state Ψ'^A will be one of the eigenstates $\psi_i{}^A$ and the state Ψ'^B will be $\sum_j c'_{ij} \psi_j{}^B$, which in general is different from Ψ^B.

At this point, Einstein applies the principle of separation: the physical situation of system B cannot be affected by a measurement on system A. Hence, it must be the same regardless of whether we measure O^A or O'^A. This means that the same physical situation is described by two different quantum mechanical states Ψ^B and Ψ'^B. The quantum mechanical states, therefore, are not an invariant description of physical reality. Until here, Einstein's argumentation is unassailable. Moreover, it is a clear reflection of his methodological realism: Einstein does not make any assumptions about objective reality besides that it is the same in both situations. The decisive question is simply whether a theoretical description is invariant in different observational situations.

But Einstein doesn't stop at this point; he immediately concludes that the quantum mechanical description is not biunique because it is incomplete and that therefore quantum mechanics must be a statistical theory. As we have already discussed in Section 3, this conclusion is not warranted logically. A description that is not invariant is not necessarily incomplete: for example, in special relativity one can have different descriptions of the same objective state (corresponding to different frames of reference). This does not imply that the description is incomplete, but rather that it is "overcomplete" or nonabsolute. Moreover,

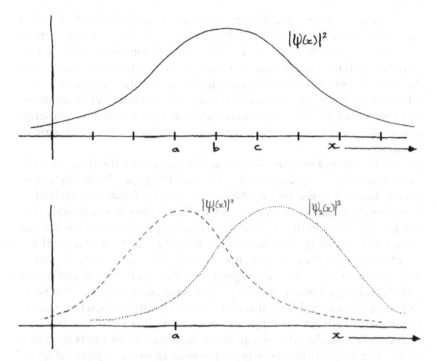

Figure 10.2. *Objective microstates versus probabilistic descriptions.*

the structure of the argument makes it obvious that Einstein did not prove incompleteness but nonabsoluteness: what he showed was that several descriptions apply to one situation and not that one description applies to several situations. The logically unmotivated jump is, however, intuitively plausible if one thinks in terms of statistical mechanics where both forms of nonuniqueness hold; and this was presumably the intuition that Einstein had when he was making the argument.

In statistical mechanics, one microstate can be referred to by various probability distributions, and one probability distribution refers to different microstates. The difference between these two notions of nonuniqueness is illustrated in Figure 10.2 for a one-particle system that can be in various microstates a, b, c, and so forth.

The quantity $|\psi(x)|^2$ plotted in these figures gives the probability of finding the particle in the state $x = a$, and so forth. The top half of the figure illustrates the usual *one-description/many-situations* incompleteness. The same probability function $|\psi(x)|^2$ is compatible with the particle being in many different states a, b, c, and so forth. The bottom half of the figure illustrates the *many-descriptions/one-situation*-type nonuniqueness that we called *nonabsoluteness*: different probability functions – here $|\psi_1(x)|^2$ and $|\psi_2(x)|^2$ – are compatible with the particle being in one and the same state *a*. The description through the statistical macrostate is nonunique in both senses. Einstein had shown that the second form of nonuniqueness holds in quantum mechanics and inferred that quantum mechanics is a statistical theory. While nonabsoluteness would also hold in a statistical theory, a nonabsolute description must not necessarily be statistical, as we have seen in the case of relativity.

This lacuna in Einstein's simple argument may be the reason for the more complicated argumentation in the EPR paper. Podolsky might have noticed the need for an additional step that Einstein perceived as unnecessary learnedness. In any case, the EPR paper shows directly that the quantum mechanical description is incomplete (and not merely not biunique) by proving that the same description can refer to several different states of reality. But for this, it needs the infamous "criterion of reality" that is laid out in the beginning of the paper: "If, without in any way disturbing a system, we can predict with certainty (i.e., with probability equal to unity) the value of a physical quantity, then there exists an element of physical reality corresponding to this physical quantity." This criterion implies that if a quantum mechanical system is described by an eigenstate of the operator corresponding to some physical quantity, there exists an "element of physical reality" corresponding to its eigenvalue. What EPR now proceeds to show is that they can construct a specific type of entangled state of separate systems, the state

$$\Psi_{EPR}(x_A, x_B) = \int e^{(2\pi i/h)(x_A - x_B + x_0)p} dp$$

and show that in this state the decomposition into relative states described previously gives the following specific results: if particle *A* is in a momentum eigenstate with momentum p, then also *B* is in a momentum eigenstate, with momentum $-p$. If particle *A* is in a position eigenstate with position x, then *B* is in a position eigenstate with position $x + x_0$.[41] Thus, if one performs a position measurement on *A*, one can predict with certainty the position of *B*, and the criterion of reality implies that there is an element of reality corresponding to the position of *B*. Equally, if one performs a momentum measurement on *A*, there is an element of reality corresponding to the momentum of *B*. The separation principle, however, implies that what is an element of reality of particle *B* is a matter of fact quite independent from what is done or measured on particle *A*. It follows that both the position and the momentum of particle *B* must be elements of the *same* reality of particle B. However, there is no quantum mechanical state that would allow particle B to have both a definite position and a definite momentum. Hence, the quantum mechanical state is incomplete in the strict sense; it does not give a full characterization of the physical situation.

What allows EPR to arrive at this stronger conclusion is the criterion of reality. This is what allows them to infer from the measurement result on particle A to a statement about the real matters of fact about particle B. However, this implies a stronger sense of realism than Einstein's methodological realism. If one subscribes to traditional realism by assuming that any measurement result reflects an element of reality, then the conclusion is unproblematic. If one, as Einstein says in the letter to Schlick, makes a strict distinction between phenomena and objective physical reality as an intellectual construct, then it is not evident that a measurement and the predictions following from it convey objective reality to the measured quantity: the gravitational field in general relativity can be measured perfectly well; this does not yet show that the gravitational field has objective reality in the fundamental theory. Einstein's discomfort with the EPR paper might also have to do with this additional metaphysical baggage. As Einstein wrote to Schrödinger: "The only thing that is essential to me is that Ψ^B and Ψ'^B are different at all.... Whether the Ψ^B and Ψ'^B can be interpreted as eigenfunctions of different observables *ist mir wurst* [doesn't interest me]." But if they are not, then there are no values of observables to which the criterion of reality could apply and the EPR argument does not go through. Einstein's simple argument suffices to show that quantum mechanics does not offer the invariant description that for Einstein was the mark of objectivity. That certainly was enough for Einstein to justify his program of finding a more fundamental description. However, it does not justify Einstein's confidence that the puzzles of quantum mechanics could be explained as statistical effects within an ensemble interpretation (see Section 5).

It was this confidence that Schrödinger criticized. He sent Einstein a long letter with a rebuttal of his argument that quantum mechanics suffers from incompleteness (Schrödinger to Einstein, August 19, 1935, EA 22–051). He argued that Einstein's ensemble interpretation of the wave function fails if we consider the functional relations between different observables. He considers the set of operators $O_a = (p^2/a + ax^2)/\hbar$ for any value of a. These operators are proportional to the Hamiltonian operator for a harmonic oscillator with mass m and frequency $\omega = a/m$, that is, $O_a = 2H/\hbar\omega$, where $H = p^2/2m + m\omega^2x^2/2$. Since the eigenvalues of H are $(n + 1/2)\hbar\omega$, the eigenvalues of O_a are $(2n + 1)$. The important point for Schrödinger is that O_a is defined for any particle with operators x and p, even if this particle is not a harmonic oscillator. In this case, O_a will not be related to the Hamiltonian of the particle in the way given, but it still has to have the same spectrum of integer eigenvalues. In an ensemble theory, we have to assume that the particle has definite values of p and x. To explain the discrete spectrum of O_a, which in this case is simply a function of p and x, we have to assume that O_a has integer values for all values of a. For some given value of a, we can pick p and x so that $O_a = (p^2/a + ax^2)/\hbar$ is integer. But if we now vary a by a small amount to a', $O_{a'}$ as a function of the chosen p and x and the free variable a', will, in general, not remain integer. Hence not all O_a can have integer values simultaneously, as quantum mechanics demands. This simple example shows that there can be no ensemble interpretation that accounts for the values of functions of observables from the values of those observables.[42] However, Schrödinger's point was lost on Einstein. In his response (Einstein to Schrödinger, September 4, 1935, EA 22–052), he wrote: "If I were to claim that I understood your last letter, that would be a big fat lie." This did not discourage him. Two sentences later on, he introduces his own thoughts on the matter with a cheerful, "Now if I glide over everything I did not understand, I don't see why it is contradictory to assume that the ψ function refers to a statistical ensemble."

Bohr's reply to EPR came in two parts: a short preliminary announcement (Bohr 1935a) and a somewhat longer discussion (Bohr 1935b). The latter again referred to a forthcoming detailed exposition, which, however, never appeared. Bohr had to dispute the EPR criterion of reality in order to avoid EPR's dilemma between nonlocality and incompleteness since he did agree that quantum mechanics should not allow for a mechanical disturbance at a distance. The simplest response would be to deny that there can be a meaningful concept of objective reality in quantum mechanics at all: while the formalism allows us to give successful experimental predictions, we have to accept that there is no consistent picture of reality that can be constructed from it. This would be the radically positivist answer, which was defended, for example, by Pascual Jordan (Jordan 1934). However, Bohr did not want to go that far: he wanted to show that an appropriate sense of realism is compatible with the completeness of quantum mechanics. Therefore, he only argued that EPR's criterion of reality contains an ambiguity that makes it inapplicable in the case considered. While a physical influence from the measurement on one particle to the other particle is excluded, Bohr maintains that "even at this stage, there is essentially the question of *an influence on the very conditions which define the possible types of predictions regarding the future behavior of the system.*" He tries to show that this criterion is not fulfilled in the case of EPR by considering an experimental realization of the EPR state.

Bohr considers two particles, each with definite momentum in relation to some fixed frame of reference, passing simultaneously through a two-slit screen. If we measure the momentum of the screen in relation to the fixed frame of reference before and after the passing of the particles, we also know the total momentum of both particles after the passage. The momentum measurement on the screen prevents us from knowing its position, therefore also the absolute positions of the two particles are unknown, and we also don't know the individual momenta of the particles since they have exchanged an indefinite momentum during their passage through the screen. However, if the slits are narrow enough, we know precisely the difference between the positions of the two particles. Therefore, the setup gives a way to prepare the quantum mechanical state that EPR had described formally. After the passage through the screen, a measurement of the momentum of one particle allows one to deduce the momentum of the other particle since the total momentum is known. Alternatively, measuring the position of the first particle should allow one to deduce the position of the second particle since the distance between the two is known. Of course, carrying out both measurements simultaneously is impossible. This fact would be enough for the radical positivist to deny the cogency of EPR's argument. Bohr, by contrast, is trying to give a physical argument regarding why the measurement on one particle interferes with the "conditions which define the possible types of predictions regarding the future behavior of the system." However, it is not clear how this criterion should be applied to the experimental realization that Bohr lays out. Dugald Murdoch analyzes it as saying that "we cannot meaningfully ascribe a physical property to an object unless the preconditions for the meaningful use of the predicate in question are satisfied, and these preconditions are the presence of an appropriate experimental arrangement that is capable of being used to measure the property in question" (Murdoch 1987, 176).

Unlike Murdoch, I do not believe that there can be an argument from the physical setup of the situation that these preconditions are not fulfilled. The argument would have to be that the measurement of, for example, the position of one particle would preclude the possibility of measuring the momentum of the other particle. Murdoch thinks that measuring the position of the first particle involves a transfer of momentum from the particle to some massive frame that has to be assumed as joining both sides of the experiment and defining the common spatial reference frame; this transfer would interfere with the measurement of the momentum of the second particle. This argument is wrong since the body defining the common reference frame can be made as heavy as required. Therefore, the *velocity* of the frame of reference is affected only by an amount that can be made arbitrarily small (since velocity is momentum divided by mass) and hence the frame can still be considered at rest with an arbitrarily high precision. But this is all that is necessary for it to be used for a momentum measurement on the second particle, which can always be done as a measurement of velocity.[43] Bohr's whole strategy to show the correctness of the quantum mechanical formalism by considerations of the possibilities of measurement would be in jeopardy if it was otherwise, since quantum mechanics demands that the position of the first particle and the momentum of the second particle are commuting variables. Also, Bohr's own words don't clearly support this reading. He writes: "We have by this procedure [measuring position] cut ourselves off from any future possibility of applying the law of conservation of momentum to the system consisting of the diaphragm and the two particles and therefore have lost our only basis for an unambiguous application of the idea of momentum in predictions regarding the behavior of the second particle" (Bohr 1935b, 700). If this means that the transfer of momentum from the first particle to the frame of reference precludes any possibility of applying the law of conservation of momentum to the total system, then it is true but irrelevant. Bohr seems to assume that we can only infer the momentum of the second particle through an application of the law of conservation of momentum. However, we can measure the momentum of the second particle directly, and this is all we need for the application of his own criterion of meaning.[44]

Beller and Fine (1993) propose a more subtle argument on Bohr's side: in Bohr's setup, the correlation between the two particle positions is only valid at the time of the passage through the screen. Immediately afterward, as the particle positions spread, this correlation is destroyed. Therefore, a measurement of the position of the first particle has to take place immediately after the passage through the screen to allow any inferences to the position of the second particle. But in this case, measuring the position of the first particle amounts to a measurement of the position of the screen and therefore interferes with the preparation of the state, which depends on the screen having a definite momentum. Beller and Fine see that as the reason why Bohr says that measuring position on the first particle precludes any possibility of applying the law of conservation of momentum to the total system: the measurement simply interferes with the preparation of the entangled state. On Beller and Fine's reading it is not surprising that Bohr thinks that the measurement of the position of one particle interferes with the conditions for making predictions about this system: it simply interferes with the preparation of the total state. This problem, however, is an artifact of Bohr's realization of the EPR state. The underlying problem is that the correlation of position in the EPR state only exists for one moment in time since position is not an invariant of the motion. However, it is not necessary to assume that the point of time at which the correlation happens is the point at which the preparation of the state was performed. And only in this case is it plausible to assume that the measurement interferes with the preparation.

Bohr's response to EPR still employs and defends his theory of the meaning of quantum mechanical observables from the Como lectures: their meaning is given (and also limited) by the definite experimental setup which can be used to measure them. This implies that once a specific experimental setup is given, a value for the corresponding observable does objectively exist. Against EPR, he argues that their setup is not a well-defined experimental setup: the determination of, say, position on one of the particles interferes with the determination of momentum on the other. However, as we have seen, this argument is not successful. While he did not acknowledge a connection with EPR, it has been commonly observed that Bohr changed his criterion of meaning after 1935, presumably realizing the problematic nature of his reply (Murdoch 1987, Beller and Fine 1993). Bohr demands a much stricter criterion of meaning in (Bohr 1939): a physical quantity can be meaningfully ascribed to a system only once an actual measurement of the quantity has taken place. This criterion, of course, is in conflict with Einstein's methodological realism. It will preclude the possibility to consider different phenomenal descriptions of the same real situation and therefore it does not allow to construct a theory of invariants describing one consistent reality across different experiments. Moreover, it is in conflict with Einstein's intuitions about separability, since the reality of a quantity defined on a system will depend on measurements performed on another, spatially separated system. Bohr's strict criterion was considered a possible objection by EPR: "Indeed, one would not arrive at our conclusion if one insisted that two or more physical quantities can be regarded as simultaneous elements of reality only when they can be simultaneously measured or predicted" (Einstein, Podolsky, and Rosen 1935, 141). They dismissed this view because of its conflict with the assumption of separability: "This makes the reality of P and Q depend upon the process of measurement carried out on the first system which does not disturb the measurement of the second system in any way. No reasonable definition of reality could be expected to permit this" (ibid.).

Einstein's only published reaction to Bohr's response was a brief characterization of Bohr's position that he gave in Einstein (1949b, 681–2, emphasis in original): "... there is no reason why any mutually independent existence (state of reality) should be ascribed to the partial systems

A and B viewed separately, *not even if the partial systems are spatially separated from each other at the particular time under consideration.*" Taking back the charge of unreasonableness, Einstein admitted that such a position is tenable in a contribution to an issue of the philosophical journal *Dialectica* edited by Wolfgang Pauli (Einstein 1948b). In this paper, a detailed exposition of the entanglement argument in the simple form from the letter to Schrödinger, he presents the argument as showing the incompatibility of the assumption that quantum mechanics is complete and the principle of separability. He now makes a detailed argument why the assumption of separability, albeit not logically necessary, is at the foundation of all our physical thinking but especially the concept of a field governed by local field equations. It is a necessary condition for the possibility of formulating and testing laws of physics. Therefore, Einstein concludes that he cannot accept the completeness of quantum mechanics.

7. MACROSCOPIC REALISM

There is a further problem with Bohr's strict criterion of meaning that Einstein pointed out: it is impossible to restrict its use to microscopic phenomena. Adapting Schrödinger's Cat example,[45] Einstein (1949b) considers the following situation: the decay of a radioactive atom is recorded macroscopically by a mark on a paper strip indicating the time of decay. Einstein argues that in this case, the defender of completeness is committed to saying that there is no objective matter of fact about the location of the mark on the paper until an observation has taken place. Again, Einstein admits that such a position is logically tenable, but "there is hardly likely to be anyone who would be inclined to consider it seriously." Although this argument might seem like an appeal to naive realism, it should be seen in the context of the debate with Bohr who did want to maintain realism for the classical physics of macroscopic objects and who tied the nonclassicality of quantum mechanics to the failure of our means of observation in the microworld.

In the vein of this argument, Einstein presented another thought experiment in the *Born Festschrift* (Einstein 1953): he considers a perfectly elastic, macroscopic ball bouncing back and forth between two parallel walls. The energy eigenstates of this ball will be completely delocalized. On a statistical interpretation, this simply means that the probability of finding the ball in a certain location is nearly the same everywhere. But if one wants to maintain that the quantum mechanical description is complete, it is in principle impossible to ascribe a position to the ball. Even if we were to allow only quantum-mechanical states that are approximately localized (wave packets) for the description of the ball, this would not solve the problem since these wave packets would become increasingly delocalized over time. Thus, it is impossible to claim that in the macroscopic limit, the quantum mechanical description will become approximately the same as the classical description of the ball. In the ensuing discussion (Einstein and Born 1969), Born insisted that the spreading wave packet was a manifestation of our incomplete knowledge of the initial conditions of the ball, and that, therefore, the quantum mechanical description exactly mirrors the classical description if one allows for an uncertainty in the initial conditions. Born assumed that Einstein meant to criticize the indeterminism of quantum mechanics expressed in the spreading of the wave packet. He did not understand that his response conceded Einstein's point that the quantum mechanical description can only be understood as the description of a statistical ensemble and not that of an individual case.

As already mentioned in the introduction, Pauli sent two detailed letters about the disagreement to Born (Einstein and Born 1969, 285–91), pointing out that determinism was not the issue that Einstein was interested in. Pauli recognized that Einstein's complaint was that a statistical description could not be a complete description of an objective reality. Pauli had been attacked by Einstein for this: in a letter to Michele Besso, Einstein criticized Pauli's claim that quantum mechanics offers a self-contained conceptual scheme for the description of phenomena: "Did you also notice how illogical Pauli's answer [to Einstein's *Dialectica* paper] was? He denies that this way of description [by a wave function] is incomplete, but he says in the same breath that the ψ function is a *statistical* description of the system, the description of an ensemble of systems. Only that that is just another form of the statement: the description of the (individual) single system is incomplete!" (Einstein to Besso, July 24, 1949, in Einstein and Besso 1972, 403). And neither Pauli nor anyone else could respond to Einstein's challenge regarding how a probability distribution of a single object's states could be objective. Since Pauli could not offer a way of thinking of the statistical description as a description of an objective fact, Einstein maintained that it had to be incomplete.

Pauli's response to Einstein's argument was rather that the idea of observation-independent reality has to be given up in the macroscopic world just as in the microworld. That was the lesson that had to be learned from quantum mechanics: there was no observer-independent, objective reality at all. Einstein's "metaphysical prejudice" was that he kept clinging to this idea. Consequently, Pauli objected to Einstein's talk of an "incomplete description" since it implies that a more complete conceptual scheme could exist. To Einstein, by contrast, giving up the idea of an objective reality behind our subjective perceptions amounted to abandoning a fundamental presupposition of physics. And I hope to have made a convincing argument in the previous sections that what he meant with this was not the need to believe in a transcendent reality behind science, but the metatheoretical observation that physics is a theory of invariants and as such presupposes the distinction between observations and objective facts.

One can still maintain that this distinction is a useful methodological fiction, which does not say anything about how the world *really* is. Such a radical subjectivism can escape Einstein's challenge to quantum mechanics. However, this position is not Bohr's, who maintained a classical realism for macroscopic objects and arguably was also a realist in his understanding of quantum mechanics.[46] Also most contemporary physicists – this is at least my personal, statistically insignificant impression – do not want to give up some sense of empirical realism which should at least support a sense of "the moon is also there when nobody looks"[47] and are therefore willing to accept an explicit nonlocality in quantum mechanics. In the modern popular perception of quantum mechanics, the standard way of thinking consists of a wavering between an ignorance interpretation (there are particles whose positions are not known exactly) and a propensity interpretation (there is a wave function that collapses upon observation). Einstein's thought experiment is first of all an expression of his impatience with this kind of theoretical inconsistence.

In light of Einstein's defense of an interpretation of quantum mechanics as an incomplete theory, it is at first surprising that he was utterly dismissive of David Bohm's attempt to revive and sharpen de Broglie's dual wave-particle interpretation of quantum mechanics (Bohm 1952). When one considers, however, that Einstein had already thought about dual theories at length, both in the early 1920s and in connection with his attempts to understand wave mechanics, his skepticism becomes understandable. Bohm's interpretation suffered from the same problems that made the previous attempts unacceptable to Einstein: the wave function defined in high-dimensional configuration space could not be interpreted as a physical field. Instead it established nonlocal correlations between distant particles, thus giving the least-appealing answer to Einstein's dilemma about entangled states. In the *Born Festschrift* (Einstein 1953), Einstein adds that Bohm's interpretation can also not give a satisfactory account of the macroscopic limit of quantum mechanics: in the Bohm interpretation, the ball that should be bouncing back and forth in a state of definite energy stands absolutely still.

In his later years, after many failed attempts to design a unified field theory, Einstein did acknowledge the possibility that the assumption of an objective and local description of reality that underlies field theory might have to be abandoned and a totally new foundation of physics might have to be found, but that this would mean a total collapse of the foundations of contemporary physics: "However, I consider it entirely possible that physics cannot be founded on the concept of a field, i.e., on continuous constructs. Then, *nothing* will remain of my whole castle in the air including the theory of gravitation but also of the whole rest of contemporary physics."[48] However, he remained convinced that quantum mechanics could not be that foundational theory.

8. EPILOGUE: EINSTEIN'S DREAM SHATTERED?

The EPR argument would come to play an important role in subsequent discussions of the foundations of quantum mechanics. In the 1960s, John Bell showed that it was impossible to recover the statistics predicted by quantum mechanics if one assumes with Einstein that two locally separable states determine the measurement outcomes on either component of the entangled state (Bell 1964). Bell showed that if that was the case, the correlations of the measurement outcomes in EPR-type experiments should satisfy the so-called Bell inequalities. The predictions of quantum mechanics, however, violate these inequalities. Already as a thought experiment, this proof is of great importance because it shows the impossibility of Einstein's idea that quantum mechanics could be understood as an incomplete description of a reality that is objective and locally definite. It remained an empirical question whether Einstein's vision or quantum mechanics would be vindicated.

Many experiments to test the Bell inequalities have been performed since the early 1970s (see Bertlmann and Zeilinger 2002). Their results have increasingly confirmed the quantum-mechanical predictions and the violation of the Bell inequalities, despite some critics who pointed to possible loopholes in the experimental realization (e.g., Fine 1996). Fine also argued that Einstein did not believe in a hidden variable theory that simply adds preexistent value for measurement outcomes to standard quantum mechanics. However, Bell's argument also applies to a local field theory as Einstein had envisioned: also, a locally defined field is a local hidden variable in the sense of Bell, and Einstein would have certainly agreed that the statistical predictions of quantum mechanics should be derivable from his foundational theory, at least approximately. But the Bell inequalities prohibit a local field theory from making predictions that are even approximately identical to those of quantum mechanics.

Even though Einstein's conclusion from EPR has been disproven by experiment, the questions that Einstein asked about the interpretation of quantum mechanics remain valid. The paradoxical effect of Bell's work has been to bring these questions back into the consciousness of more than a tiny minority of physicists and philosophers. Moreover, the conceptual clarity of Einstein's thinking remains useful as a framework for the debates about the interpretation of quantum mechanics. Following this reconstruction of Einstein's criticism of the completeness of quantum mechanics, there are basically three assumptions in the argument that can be modified:

1. *One can give up Einstein's methodological realism.* This means that one renounces the possibility of having a complete and objective state description in quantum mechanics: quantum mechanical states are always phenomenal, but we cannot achieve an invariant description of the facts behind the phenomena. Bohr's hope that this solution could be consistent with a commonsense realism about macroscopic objects has become less and less plausible. Since the quantum world is not nearly as impenetrable experimentally as it was in Bohr's days, it has become hard to believe that it should be epistemologically fundamentally different from the macroworld. But then, one has to accept Pauli's position and give up in general even a minimal methodological realism. As Einstein stressed repeatedly, this is a logically tenable position. It is not only held by the defenders of the Copenhagen orthodoxy, but also by proponents of an information-theoretic approach and probably by most adherents of pragmatist or instrumentalist readings of quantum mechanics. However, this position is at odds with the practice of physics, today as much as in Einstein's days: the idea of descriptive invariants has become one of the fundamental theoretical tools in all of contemporary physics, especially in quantum mechanics and quantum field theory. Without a methodological realism in Einstein's sense, this idea is completely ad hoc.

2. *One can accept Einstein's methodological realism (quantum theory needs some kind of complete description of an objective state), and accept Einstein's contention that the standard quantum mechanical description is incomplete.* If one follows Einstein's argument for the need of a methodological realism and accepts his contention that the standard quantum mechanical description (at least in a case like EPR) is incomplete, then one has to construct a more complete description, either by adding additional physical variables or by modifying the dynamics to allow for a "physical collapse," some mechanism that gives physical reality to measurement outcomes before their observation. But as the Bell inequalities have shown, this requires an explicit nonlocality in

the dynamics. This applies to the various types of hidden-variable interpretations such as the de Broglie-Bohm interpretation or the modal interpretation, and also to physical collapse theories like nonlinear quantum mechanics or the Ghirardi-Rimini-Weber interpretation. There are at least two reasons that make this solution unappealing on physical grounds: as Einstein already pointed out, accepting the possibility of nonlocal interactions defeats the theoretical justification of any field theory, including quantum field theory. There is an even weightier argument that has, to my knowledge, not been considered in this context. Quantum mechanics confirms the locality of all observable physical processes, as is demonstrated by the various no-signaling theorems, which show that entanglement cannot be used for the transfer of information. However, this means that the nonlocality of the interpretations mentioned resides entirely in the theoretical structure and not in the phenomena.

3. *One can maintain both Einstein's methodological realism and his principle of separation* by rejecting his conclusion that the nonuniqueness of quantum mechanics implies its incompleteness. As pointed out in Section 6, the EPR proof of incompleteness of quantum mechanics relies on a stronger sense of realism than what Einstein was generally willing to admit. But Einstein's favored entanglement argument can only show the nonuniqueness of quantum mechanical descriptions, not their incompleteness. Is it possible to construct an interpretation of quantum mechanics that understands quantum mechanical states as nonabsolute descriptions in the same sense as in relativity? Hugh Everett has explored this possibility in his "relative state interpretation" of quantum mechanics by applying Einstein's distinction between phenomenal reality and objective reality in a radically new way (Everett 1973). Everett proposed that measurement results do not give us direct information about the objective state, which never collapses. Measurement results are only reflected in a noninvariant relative state, which, like the coordinate-dependent descriptions in relativity, describes a nonunique phenomenal perspective. This allows for a natural explanation of Einstein's simple entanglement argument: the two quantum-mechanical states for the second particle that one arrives at depending on the measurement on the first particle are simply two such phenomenal perspectives. Unlike what is assumed in EPR and by common sense, they do not reflect elements of objective reality. Therefore, no violation of Einstein's principle of separation is necessary to

explain the quantum mechanical predictions. Everett's interpretation, however, deviates radically from Einstein's, and not only Einstein's, intuitions about what the world is like. It is committed to assuming the physical existence of an infinite manifold of possible worlds, inhabited by an infinite manifold of observers almost like ourselves. Nevertheless, as contrary as it is to our traditional metaphysical assumptions, it is, as Everett pointed out, the most faithful to Einstein's methodological principles.

All these modifications require giving up well-established principles of our scientific practice. At this point, we do not have a commonly agreed upon criterion that tells us which of these principles we have to give up. However, as I have tried to show in the case of Einstein's methodological realism, this is not a question of metaphysical speculation, but rather a question of which methodology one sees as the most promising for the further progress of physics.

9. WHY EINSTEIN DID NOT LIKE QUANTUM MECHANICS

After having spent a good amount of time trying to understand Einstein's arguments criticizing quantum mechanics, I want to return to the question of their relevance. This question is even more acute given that Einstein's "gut feeling," which he quoted often against quantum mechanics, was demonstrated to be wrong by the confirmation of the violation of the Bell inequalities. Were the younger physicists that developed quantum mechanics right when they thought that Einstein simply had become too old and conservative to be able to accept the revolutionary novelty of quantum mechanics? Einstein himself sometimes made ironic comments of this kind in his correspondence. There is certainly some truth to this assessment, as the exchange with Schrödinger discussed in Section 6 shows – Einstein was not willing to try to understand even this relatively basic argument. Even less did he show interest to keep up with the rapid technical development of the theory in his later years. This may have to do with Einstein's frequent complaint regarding how difficult it was for him to "calculate." One has to see, however, that he did not shy away from the equally demanding mathematics of unified field theory. Just as he spent years struggling with its difficulties and eagerly recruiting junior collaborators to deal with its technicalities, he could have done the same in quantum theory.

The more plausible answer is that by the time quantum mechanics arrived on the scene, Einstein was already firmly committed to a very different type of foundational theory. His experiences in finding general relativity, the greatest triumph of his scientific career, gave Einstein a strong sense of confidence that a theory along the lines of general relativity unifying gravity and electromagnetism would in the end also explain all quantum phenomena. He was set on his own research program, and he cared very little that hardly any of his colleagues followed him on this program. This had been very much part of Einstein's *modus operandi* all along: just as he had pursued the light quantum in opposition to most of his colleagues for years and as he had developed general relativity against the skepticism of even his most sympathetic followers in splendid isolation, so he also was not impressed by the turn of practically the whole physics community to quantum mechanics. He had to admit that the program of quantum mechanics had much more to show for itself than his own efforts in unified field theory. But quantum mechanics was someone else's program; Einstein decided to stick to his own. This occupation with unified field theory led to a perhaps overly glib disregard of arguments against his ensemble interpretation of quantum mechanics. Einstein's hope that quantum theory could be resolved into a unified field theory does not seem tenable anymore today.

However, unlike his contemporaries who resented Einstein's refusal to join their effort, we can, in hindsight, appreciate his role as a critical outsider. Unlike the quantum physicists caught up in the dramatic development of their theory, Einstein had the distance to see the problematic nature of the foundations of quantum mechanics. As one of a very small group of critics like de Broglie, Schrödinger, or David Bohm, Einstein was not satisfied with the fact that quantum mechanics worked, and exactly for that reason he probed deeper into what it meant than his technically better-versed colleagues. Einstein made a very telling comment in this respect in a letter to Ehrenfest in 1925. Ehrenfest had become increasingly frustrated at not being able to follow the mathematically sophisticated arguments in quantum theory. Einstein offered the following consolation: "There are principle jugglers [Prinzipienfuchser] and virtuosos. All three of us [Bohr, Ehrenfest, and Einstein] belong to the first kind and have little virtuoso talent (certainly at least the two of us). Therefore the effect on meeting distinct virtuosos (Born or Debye): discouragement. Works the other way around, too, by the way" (Einstein to Ehrenfest, September 18, 1925, EA 10–111). There is nothing vain or self-serving about this assessment. Einstein saw his weaknesses as well as his strengths with great and almost detached clarity. Principle juggling certainly was his strength. It had led him, in his *annus mirabilis*, to see connections between problems that everybody else thought were entirely disconnected, it had led him to general relativity from questions that nobody else thought worthy of asking, and it had led him to question quantum mechanics when everyone else was struck by its successes. If anything, Einstein's brilliantly clear thinking about the fundamental principles underlying physical theories became better with age.

The problems in the understanding of quantum mechanics that Einstein pointed out and its conflict with the basis of field theory are still very much with us: we still have not understood what quantum field theory really is about, and we know even less as to how to combine it with what we have learned from general relativity. To have pointed that out is the most lasting value of Einstein's critique of quantum mechanics.

ACKNOWLEDGMENTS

Work on this essay has been supported by the German Israeli Foundation and the Innovation Fund of the Max Planck Society. I am indebted to Jürgen Renn for his continued intellectual and institutional support. My thanks go to Laurent Taudin for drawing the figures in this chapter, and to Lindy Divarci for meticulous language editing. I also want to thank the many colleagues who have helped me with their comments on this essay: first of all, Michel Janssen, who has been involved beginning with the first draft; second, my colleagues at the Max Planck Institute for the History of Science, Jürgen Renn, Christian Joas, Alexander Blum, and Martin Jähnert; last, but not least, the many colleagues all over the world who have read and discussed the chapter with me: among them Christian Camilleri, Olivier Darrigol, Jeroen van Dongen, George Musser, John Norton, and Chris Smeenk. I apologize to anybody whose remarks I don't have a written record of and whom I forgot to mention here.

NOTES

1 Such as in Klein (1970a, 38–9), Stachel (1986b, 379–80), and Duncan and Janssen (2008, 649). Stachel (1986b) presents a rich selection of quotes from Einstein on quantum mechanics.

2 E.g., Duncan and Janssen (2008) point out that Einstein was rather unimpressed by Pascual Jordan's derivation (Born et al. 1926) of Einstein's dual wave-particle fluctuation formula for black-body radiation from matrix mechanics (see Section 4). Some other exchanges with Niels Bohr, Max Born, Werner Heisenberg, and Wolfgang Pauli will also be discussed.

3 Even Abraham Pais, in his otherwise very thoughtful biography of Einstein, is still completely dismissive of Einstein's critique of quantum mechanics (Pais 1982, 456).

4 The reopening of the interpretation debate thus triggered a set of new historical works challenging the previous rather whiggish accounts of the ascendancy of the Copenhagen interpretation (such as Fine 1986, Cushing 1994, and Beller 1999).

5 One might argue that an analysis attempting to find a simple basic principle for Einstein's dispute with the various defenders of an orthodox reading of quantum mechanics will necessarily oversimplify the historical complexities and disregard a host of nontheoretical factors that contributed to this dispute. While it is certainly true that the full history involved more than the abstract debate between two theoretical positions, it is equally certain that one cannot even begin to tell such a full story without understanding the theoretical issues at stake. Conversely, I will claim that the philosophical issues cannot be understood without paying attention to the concrete historical context of the debate.

6 Albert Einstein to Max Born, December 4, 1926 (Einstein and Born 1969, 127).

7 This view is expressed in Born's commentaries in Einstein and Born (1969).

8 Classical discussions of Einstein's philosophical views have been given by Jammer (1974), Howard (1984, 1985, 1993), and Fine (1986).

9 See Chapter 11 and the discussion in Howard (1993).

10 Chapter 6 treats the complicated development of Einstein's ideas during this period. See also Janssen and Renn (2007).

11 See Einstein (1918f) on energy and momentum conservation in general relativity, as well as the discussion and references in CPAE 8, Introduction, p. li.

12 Of course, both forms of realism do not necessarily contradict each other, and one can argue that Einstein at times also defended a stronger and more traditional epistemological realism. However, as the quotes in this section show, there is sufficient evidence that at least in his more philosophically oriented writing he was rather critical of it. More importantly, as I will argue in the following sections, for his critique of quantum mechanics his methodological realism is sufficient.

13 See Chapter 12, which discusses Einstein's assessment of Kant in detail.

14 This distinction is discussed in the introduction to this volume. See also Stachel (1986b, Sec. 3) and Howard (2004a) for a more detailed analysis.

15 See the discussion of Einstein's fluctuation arguments to derive consequences from the law of black-body radiation in Chapter 4.

16 See Balashov and Janssen (2003) for a detailed exposition of the relation between relativity and a tenseless theory of time. While it is logically possible to maintain a tensed theory in the face of relativity, Einstein certainly did not hold such a position, as Balashov and Janssen show with his well-known remark: "To us believing physicists the distinction between past, present, and future has only the significance of a stubborn illusion" (Einstein to Vero and Bice Besso, March 21, 1955; Einstein and Besso 1972, 538).

17 Note that Einstein's distinction removes the tension between the directionality of thermodynamics and the missing direction of time in mechanics that had worried the defenders of statistical mechanics in the nineteenth century. Both the temporal directionality and the indeterminacy of macroscopic processes are part of phenomenal reality and hence not in conflict with the timelessness of objective reality.

18 The important role of uniqueness of description in Einstein's thinking about general relativity was first pointed out by Howard (1992).

19 This episode is quite remarkable since it turned out later that Rupp made up experimental results. Van Dongen (2007a, 2007b) shows that there were obvious inconsistencies also in the results that Rupp produced in the collaboration with Einstein, who ignored these problems, however.

20 See Stachel (1986b, Sec. 5), who points out that Born stated repeatedly that he took the idea for the statistical interpretation of the wave function from this theory of Einstein's.

21 As reported by Wigner (1980, 461), quoted in Howard (1990b).

22 See Chapter 4, Section 7, for Einstein's dual fluctuation formula.

23 See also Klein (1979) and Darrigol (1986, 1991) for a detailed account.

24 See Chapter 6 for an account of the role of physical arguments in the development of general relativity.

25 Einstein (1933b), quoted after Einstein (1954, 274). On the Spencer lecture, see Norton (2000) and van Dongen (2002b).

26 See Duncan and Janssen (2007, Sec. 4) for an account of BKS and its problems, and Klein (1970a) for Einstein's critique.

27 Einstein to Ehrenfest, September 30, 1925, EA 10–113.

28 Born et al. 1926. See Chapter 4, Section 7, for Einstein's dual fluctuation formula, and Duncan and Janssen (2007) for an account of Jordan's derivation and its reception.

29 See Przibram (1963) for the correspondence of Schrödinger with other physicists about these questions. See Perovic (2006) for an account of the exchange with Bohr.

30 See Lacki (2004) and Duncan and Janssen (2009) for Jordan's transformation theory.

31 Einstein Archive Nr. 2–100. The manuscript is discussed in Belousek (1996).

32 Camilleri (2009) studies the development of Heisenberg's philosophical views.

33 See Stachel (1986b, Sec. 6) for a more detailed account of the exchange between Heisenberg and Einstein.

34 In Heisenberg (1927), Heisenberg still professed agnosticism about the possibility of a more complete description. However, one has to see this statement in the context of his positivistic conviction that a description is meaningless if it is not operationally defined, and that it is impossible to perform measurements more precise than the uncertainty relation allows.

35 An English translation of the proceedings with commentary can be found in Bacciagaluppi and Valentini (2009).

36 Commonly, this position has been equated with Einstein's later ensemble interpretation of quantum mechanics (while individual particles have well-defined trajectories, the wave function gives only an incomplete description of these trajectories). See, e.g., the discussions in Fine (1986), Howard (1990), Home and Whittaker (2007), and Bacciagaluppi and Valentini (2009). That this cannot be the case is shown by the fact that Einstein goes on to say that the first interpretation cannot account for the validity of conservation laws, the result of the Geiger-Bothe experiment, or the reality of cloud chamber tracks. The alternative reading proposed here explains why Einstein did not endorse the first interpretation as all commentators seem to assume.

37 Reprinted in part in Bohr (1985, 99–106) and in Bacciagaluppi and Valentini (2009, Part III).

38 The discussions between Einstein and Bohr are analyzed in Jammer (1974, 126–58).

39 See Jammer (1974) for an account of the earlier reactions to EPR and see Auletta (2001) for an extensive account of more recent developments.

40 Well-known discussions are Howard (1985), Fine (1996), and Beller and Fine (1993). Further references can be found in Home and Whitaker (2007, 138–40). See Jammer (1974, 166–84) for the prehistory of the argument in Einstein's thinking, especially an earlier paper by Einstein, Podolsky, and Richard Tolman (Einstein et al. 1931) that treats a similar situation of two correlated particles.

41 This specific example is somewhat unfortunate since momentum eigenstates are completely delocalized (a plane wave that extends over all space). Therefore, it is problematic to assume that the particles can be spatially separated. This problem was avoided by David Bohm, who proposed for the first time (Bohm 1951) to consider a state where particle spin is entangled. Such states can be well localized, so there is no problem to think of a state of two widely separated noninteracting particles, which nevertheless exhibit spin entanglement. Modern discussions of the EPR argument usually operate with this state. Also, spin-entangled states of particles and photons have been the usual experimental realization of the EPR situation and have become an object of extensive experimental study. See Bertlmann and Zeilinger (2002) for various papers about the history of experimental realizations of the EPR-Bohm experiment.

42 This is a more specific version of von Neumann's argument against hidden variables in Neumann (1932), which has been unjustly criticized as being physically unfounded. In Schrödinger (1935), Schrödinger chooses an example that is even closer to physical observation: instead of the O_a he uses the set of angular momenta L_a with respect to any origin of coordinates a. Again, $L_a/\hbar$ has to be an integer at all times, even though we can move a around in a continuous manner.

43 Similar cases are considered in Bohr and Rosenfeld (1933) where Bohr correctly recognizes the possibility of such measurements.

44 If this is the correct reading, Bohr's mistake might stem from equating the EPR example with the case of a delayed-choice measurement that he discusses in Bohr (1935b) and in a later discussion of EPR (Bohr 1939). In the case of a delayed-choice measurement, the conservation of momentum plays an important role since it allows the determination of the initial particle momentum from the momentum of the screen after the measurement and the final particle momentum.

45 As Fine (1986) has pointed out, Schrödinger's (1935) cat paradox may well have been inspired by an example Einstein used in a letter to Schrödinger (Einstein to Schrödinger, August 8, 1935, EA 22–049). Einstein used a similar example to the one in the letter (Einstein 1949b).

46 The issue of Bohr's realism has been much debated. For different positions on this issue, see Faye and Folse (1994).

47 Abraham Pais relates that Einstein once asked him this question in a discussion about quantum mechanics (Pais 1979, 907).

48 Einstein to Besso, August 10, 1954 (Einstein and Besso 1972, 572, translation from Stachel 2002a, 395).

11 Einstein and the Development of Twentieth-Century Philosophy of Science

1. INTRODUCTION

What is Albert Einstein's place in the history of twentieth-century philosophy of science? Were one to consult the histories produced at mid-century from within the Vienna Circle and allied movements (e.g., von Mises 1938, 1939; Kraft 1950; Reichenbach 1951), then one would find, for the most part, two points of emphasis. First, Einstein was rightly remembered as the developer of the special and general theories of relativity, theories which, through their challenge to both scientific and philosophical orthodoxy made vivid the need for a new kind of empiricism (Schlick 1921) whereby one could defend the empirical integrity of the theory of relativity against challenges coming mainly from the defenders of Kant.[1] Second, the special and general theories of relativity were wrongly cited as straightforwardly validating central tenets of the logical empiricist program, such as verificationism, and Einstein was wrongly represented as having explicitly endorsed those same philosophical principles.

As we now know, logical empiricism was not the monolithic philosophical movement it was once taken to have been. Those associated with the movement disagreed deeply about fundamental issues concerning the structure and interpretation of scientific theories, as in the protocol sentence debate, and about the overall aims of the movement, as in the debate between the left and right wings of the Vienna Circle over the role of politics in science and philosophy.[2] Along with such differences went subtle differences in the assessment of Einstein's legacy to logical empiricism. Philipp Frank – a dissenter from central points of right-wing Vienna Circle doctrine – deserves particular mention for his more accurate reading of Einstein's position on such issues as the place of convention in scientific theory (Frank 1949b, 1949c). Still, on the whole, the Einstein legacy is rendered with an unhappy mixture of hagiography and persuasive misreading in the service of appropriating the mantle of his authority to legitimate logical empiricism as it sought at mid-century to establish itself as the authoritative philosophical voice on the nature and status of scientific knowledge.

354

A more accurate picture of Einstein's relationship to logical empiricism began to emerge at about the time in the mid-1960s when challenges to logical empiricist orthodoxy from the likes of Norwood Russell Hanson, Stephen Toulmin, Thomas Kuhn, and Paul Feyerabend began to create a space in which dissent was possible. The single most important early contribution to this new picture of Einstein's connection with logical empiricism was Gerald Holton's 1968 study, "Mach, Einstein, and the Search for Reality" (Holton 1968). Though one can argue with Holton's uninflected interpretation of Einstein's post-1915 view as a version of what later came to be termed "scientific realism" and with his missing the important theme of Einstein's sympathy for Pierre Duhem's brand of conventionalism (see Howard 1993), we nevertheless owe Holton a major debt for making clear the extent to which, already by the late 1920s, Einstein had explicitly forsworn any sympathy for verificationism and the antimetaphysical agenda of logical empiricism, this at the very time when Vienna Circle pamphleteers were trumpeting his support for the movement (see, e.g., Neurath, Hahn, and Carnap 1929).

Lacking, however, in both the original logical empiricist hagiography and in Holton's revisionist picture is a clear sense of what might well be Einstein's major legacy to twentieth-century philosophy of science, which is the fact that Einstein was one of the most important, active, constructive contributors to the development of a new empiricism in the 1910s and 1920s. Einstein did not merely provide theoretical grist for the philosophical mill of logical empiricism, that and possibly a benediction. On the contrary, virtually all of the major figures involved in the debates out of which logical empiricism grew worked out their ideas through an engagement not just with general relativity but with the author of the theory. This was true of central founding figures in the logical empiricist camp, such as Moritz Schlick, Hans Reichenbach, Philipp Frank, and Rudolf Carnap. It was true, as well, of figures associated with other traditions, such as the neo-Kantian Ernst Cassirer and Hermann Weyl, whose chief philosophical debt was to the phenomenology of Edmund Husserl. In conversation, correspondence, and a wealth of published essays and reviews, Einstein tutored, criticized, and learned, his interventions concerning large questions of doctrine and fine points of detail.

Moreover, while Einstein produced no philosophical magnum opus to rival Schlick's *Allgemeine Erkenntnislehre* (1918), Weyl's *Philosophie der Mathematik und der Naturwissenchaft* (1927), or Reichenbach's *Philosophie der Raum-Zeit-Lehre* (1928), the position that he defended from early to late – as mentioned, a variant of Duhemian conventionalism – has much to recommend it as, in its own right, a cogent, comprehensive philosophy of science, from which there is still much to be learned that is of relevance to contemporary philosophical controversies (see, e.g., Howard 2010). A full history of twentieth-century philosophy of science should, therefore, devote a whole chapter to Einstein as a figure at least as important and influential as those already canonized.

2. BACKGROUND AND EARLY INFLUENCES

Einstein was like many scientists of his generation in his early and regular engagement with philosophy. He is reported to have read Kant's first *Critique* already at the age of thirteen (Beller 2000). As a physics student at the *Eidgenössische Technische Hochschule* (ETH) in Zurich in the late 1890s, Einstein was required to attend lectures on "Theorie des wissenschaftlichen Denkens" given by August Stadler, a neo-Kantian philosopher who was Hermann Cohen's first doctoral student at Marburg, and Einstein elected to enroll also in Stadler's lectures on Kant (Beller 2000). On his own while a student and, a few years later with his friends in the discussion group they constituted in Bern around 1903, the so-called Olympia Academy, Einstein read carefully many of the most important works of figures like Ernst Mach, Henri Poincaré, John Stuart Mill, Richard Avenarius, Karl Pearson, Richard Dedekind, and David Hume.[3] He read Friedrich Albert Lange's *Geschichte des Materialismus* (Lange 1873–5), Eugen Dühring's *Kritische Geschichte der Principien der Mechanik* (Dühring 1873), and Ferdinand Rosenberger's *Isaac Newton und seine physikalischen Prinzipien* (Rosenberger 1895).[4] During his university years and repeatedly thereafter, Einstein read widely in the works of Arthur Schopenhauer (Howard 1997).

From the beginning, Einstein understood such reading in the history and philosophy of science as making an important difference to the way in which one does physics. Thus, in a 1916 memorial notice for Mach, Einstein wrote:

How does it happen that a properly endowed natural scientist comes to concern himself with epistemology? Is there no more valuable work in his specialty? I hear many of my colleagues saying, and I sense it from many more, that they feel this way. I cannot share this sentiment. When I think about the ablest students whom I have encountered in my teaching, that is, those who distinguish themselves by their independence of judgment and not merely their quickwittedness, I can affirm that they had a vigorous interest in epistemology. They happily began discussions about the goals and methods of science, and they showed unequivocally, through their tenacity in defending their views, that the subject seemed important to them. Indeed, one should not be surprised at this. (Einstein 1916c, 101)

And in a November 28, 1944, letter to Robert Thornton he echoed those words of nearly thirty years earlier:

I fully agree with you about the significance and educational value of methodology as well as history and philosophy of science. So many people today – and even professional scientists – seem to me like somebody who has seen thousands of trees but has never seen a forest. A knowledge of the historic and philosophical background gives that kind of independence from prejudices of his generation from which most scientists are suffering. This independence created by philosophical insight is – in my opinion – the mark of distinction between a mere artisan or specialist and a real seeker after truth. (Einstein to Thornton, December 7, 1944, EA 61–574)

Many of the mentioned authors had an important influence on the development of Einstein's thinking, foremost among them Kant, Poincaré, Mach, and Hume (Beller 2000, Howard 2004a, Norton 2010). Much has been made of the influence of Mach, in particular (see, e.g., Frank 1949c). But the nature of that influence is complicated.

Einstein acknowledged an important debt to Mach in the genesis of both special relativity (Norton 2010 provides helpful detail; see also Holton 1968) and general relativity (see Hoefer 1995, Norton 1995, and Howard 1999). But on various later occasions Einstein stressed his dissent from Mach's doctrine of the elements of sensation as the basis of all scientific knowledge, as in a letter to Besso of January 6, 1948:

I see his great service as residing in the fact that he dispelled the dogmatism that reigned in the foundations of physics in the 18th and 19th centuries. Especially in the Mechanik and the Wärmelehre, he sought to show how concepts grow up out of experience. He convincingly defended the view that these concepts, even the most fundamental ones, obtain their justification only from experience.... I see his weakness as residing in the fact that he more or less believed that science consists in the mere "ordering" of empirical materials; i.e., he misunderstood the free, constructive element in the formation of concepts. In a sense, he believed that scientific theories arise through discovery and not through invention. (Einstein and Besso 1972, 390–1)

More than anything else, the positive lesson that Einstein drew from Mach concerned the historicity of scientific concepts and theories. Both the Mechanik (Mach 1883) and the Wärmelehre (Mach 1896) advertise in their subtitles that they are essays in an "historical-critical" method, and, as in the contemporary biblical hermeneutics literature that employed the same characterization of its critical method (see, e.g., Strauss 1835–6), the point was to exploit the historicity of texts, concepts, or theories – their having been authored in specific historical circumstances by specific individuals working in specific problematic settings – to deprive those texts, concepts, and theories of their authority as received wisdom. Precisely this aspect of Mach's legacy was stressed by Einstein in the just-quoted letter to Besso but also, more than thirty years earlier, in the cited obituary notice for Mach:

The fact is that Mach exercised a great influence upon our generation through his historical-critical writings.... Concepts that have proven useful in ordering things easily achieve such an authority over us that we forget their earthly origins and accept them as unalterable givens. Thus they come to be stamped as "necessities of thought," "a priori givens," etc. The path of scientific advance is often made impassable for a long time through such errors. For that reason, it is by no means an idle game if we become practiced in analyzing the long commonplace concepts and exhibiting those circumstances upon which their justification and usefulness depend, how they have grown up, individually, out of the givens of experience. By this means, their all-too-great authority will be broken. They will be removed if they cannot be properly legitimated, corrected if their correlation with given things be far too superfluous, replaced by others if a new system can be established that we prefer for whatever reason. (Einstein 1916c, 102)

Mach's emphasis on the historicity of concepts and theories is continuous with his preferred description of his point of view as a "biological-economical" one, whereby he meant that the capacity for scientific knowledge, with its "economical" representations of experience through concepts and theories, should be viewed as an adaptive trait of the human species.

As great as was the influence of Mach on Einstein, there is another author whom Einstein seems also to have encountered at an early date, an author whose influence was, arguably, still more important, namely, Pierre Duhem. The story of Einstein's first acquaintance with Duhem is instructive not only as regards the content of Einstein's philosophy of science but also from the point of view of appreciating the manner in which Einstein and his contemporaries regarded the intermingling of physics and philosophy.[5]

When, in the fall of 1909, Einstein left his job at the Patent Office in Bern and returned to Zurich to take up his first formal academic position at the University of Zurich, he happened to rent an apartment at Moussonstraße 12, directly upstairs from his old friend and fellow Zurich physics student, Friedrich Adler. As chance would have it, Adler had been the other finalist for the very job Einstein was now assuming and – coincidence added to coincidence – just one year earlier Adler published his German translation of Duhem's La Théorie physique: son objet et sa structure (Duhem 1906, 1908). We know from Adler's letters to his father, Viktor Adler, cofounder of the Austrian Social-Democratic Party (see Ardelt 1984), that he and Einstein set up for themselves in the attic of the apartment building a shared study where they could work and converse, away from the noise of the children playing in the court-yard below. How did Adler recall their interactions?

We stand on very good terms with Einstein, who lives above us, and indeed as it happens, among all of the academics, we are on the most intimate terms precisely with him. They have a bohemian household similar to ours, one boy of Assinka's age, who is very often at our place.... The more I speak with Einstein – and that happens fairly often – the more I see that my favorable opinion of him was justified. Among contemporary physicists he is not only one of the clearest, but also one of the most independent minds, and we are of one mind about questions whose place is generally not understood by the majority of other physicists. (Ardelt 1984, 166)

Adler later gained notoriety for assassinating the Austrian Prime Minister, Graf Stürgkh, in 1916 as a protest against Austrian policies in World War I. In 1909, however, he was merely a *Privatdozent* at the University of Zurich, where he held the *venia legendi* for "experimental and theoret-ical physics, as well as their history and epistemological foundations" (Ardelt 1984, 157–66) and was well-known in both philosophical and political circles for his writings on the philosophy of science, most of them written from a broadly Machian perspective, as with his interven-tion on Mach's side in the just then raging debate between Mach and Max Planck (see Adler 1909; Planck 1909, 1910; and Mach 1910). Adler's praise of Einstein's "independent" mind anticipates Einstein's own later stress on the independence of judgment that he thought the most valu-able consequence of studying the history and philosophy of science.

The young Adler and the young Einstein were not at all unusual for the time in thus viewing the history and philosophy of science as an integral, critical part of the larger project of physics. Other examples are everywhere to be found, from Einstein's good friends Michele Besso (Einstein and Besso 1972), Maurice Solovine (Einstein 1956b), and the Habicht brothers, Conrad and Paul (Conrad and Solovine part of the Olympia Academy), to future colleagues, like Philipp Frank (Frank 1949a, Howard 2004c), Erwin Schrödinger (Moore 1994), Wolfgang Pauli (Enz 2002), Paul Ehrenfest (Klein 1970b), and Max Born (Born 1969 [CPAE 8, Doc. 575]). Many of this generation's teachers likewise evinced the same synthetic understanding of a philosophical physics, whether it be Mach (Banks 2003) or Planck (Heilbron 1986), Hermann von Helmholtz (Krüger 1994) or Ludwig Boltzmann (Broda 1983). Major journals promoted the integration of philosophy and physics, from Wilhelm Ostwald's *Annalen der Naturphilosophie* and *Zeitschrift für den physikalischen und che-mischen Unterricht* to Arnold Berliner's *Die Naturwissenschaften*. The philosopher-physicist, if perhaps not quite the norm, was also far from being the exception.[6]

The main philosophical lesson that Einstein and Adler took from their reading of Duhem concerned the empirical content and warrant for physical theories. Duhem argued that it is not individual propositions but only whole theories that possess empirical content and that, in consequence of this holism, theory choice is underdetermined by empirical evidence, since the fit between theory and evidence can be maintained by making adjustments in various different places within the total body of theory. For Duhem, the choice among these options is a conventional one.

Duhem's position stands opposed to a stern empiricism, sometimes wrongly associated with Mach,[7] but rightly associated with the later verificationism of the Vienna Circle, according to which each proposition or even each scientific concept must have its own determinate empirical content. Duhem's position stands equally opposed to Poincaré's view that some scientific laws are not at risk of empirical refutation as a result of their playing a uniquely definitional role in theory, as with the law of free fall (alleged to define the notion of an inertial trajectory), for Duhem argues that there is no principled basis upon which to parse a theory into propositions that function as definitions and propositions that function as synthetic, empirical assertions.

That Einstein learned this lesson from Duhem is evident from the way in which he explained to his students in a course on electricity and magnetism in the winter semester of 1910–11 (CPAE 3, Doc. 11) how one can make good empirical sense of the notion of electrical charge within a solid charged body even though one cannot introduce a test particle whereby to gauge the electrostatic field in the interior of that solid:

We have seen how experience led to the introduction of the concept of electrical charge. It was defined with the help of forces that electrified bodies exert on each other. But now we extend the application of the concept to cases in which the definition finds no direct application as soon as we conceive electrical forces as forces that are exerted not on material particles but *on electricity*. We establish a conceptual system whose individual parts do not correspond immediately to experiential facts. Only a certain totality of theoretical materials corresponds again to a certain totality of experimental facts.

We find that such an el[ectrical] continuum is always applicable only for representing relations inside ponderable bodies. Here again we define the vector o[f] el[ectrical] field strength as the vector of the mech[anical] force that is exerted on a unit of pos[itive] electr[ical] charge inside a ponderable body. But the force thus defined is no longer immediately accessible to exp[eriment]. It is a part of a theoretical construction that is true or false, i.e., corresponding or not corresponding to experience, only *as a whole.* (CPAE 3, Doc 11, [p. 12])

One could not ask for a more succinct statement of the central epistemological lesson of Duhem's *La Théorie Physique.*

3. A NEW EMPIRICISM AS AN ANSWER TO THE NEO-KANTIANS

Logical empiricism was the philosophical movement with which Einstein later came to be most closely associated; this in no small measure because of Einstein's personal connections to several of its chief architects. For example, as early as 1907, Einstein began a correspondence with Philipp Frank, then a central member of what is now known as the "first Vienna Circle" (Frank 1949a, Haller 1985) and to the end of his life prominently associated with the Vienna Circle and its descendent institutions (Stadler 1997, 681–7). Einstein had read Frank's 1907 paper on causality (Frank 1907) and wrote to offer some suggestions about the manner in which causality might be seen as having a conventional and an empirical character (Frank 1949a, 22). In 1912 Frank succeeded Einstein to the chair in theoretical physics at the German university in Prague, this upon a recommendation from Einstein that included mention of Frank's epistemological writings, and in later years he became Einstein's biographer (Frank 1947) and an important commentator on Einstein's philosophy of science (Frank 1949b, 1949c; see also Howard 2004c).

Moritz Schlick became, after 1922, the guiding spirit of the Vienna Circle. Schlick and Einstein were in correspondence by late 1915 when Schlick sent Einstein a copy of his paper on the philosophical significance of the theory of relativity (Schlick 1915), a paper that Einstein commended in part because of Schlick's having appreciated the importance of Mach and Hume for the development of special relativity.[8] Theirs was an especially close relationship through the early 1920s, Einstein working to promote Schlick's career and the dissemination of his writings, as with his helping to arrange an English translation of Schlick's 1917 monograph, *Raum und Zeit in der gegenwärtigen Physik* (Schlick 1917), the main aim of which, according to Schlick, was to explain the chief philosophical implications of the general theory of relativity.[9]

Einstein first met Hans Reichenbach in the late 1910s, when, as a student, Reichenbach audited Einstein's lectures on relativity at the University of Berlin. Einstein did not wholeheartedly commend Reichenbach's early efforts to develop an interpretation of the theory of relativity as demonstrating a central role in physics for a contingent, constitutive *a priori* (Reichenbach 1920), a view that Reichenbach quickly modified in the direction of the metric conventionalism for which he later was famous. Nevertheless, in the mid-1920s Einstein's esteem for Reichenbach was such that he collaborated with Planck to create for Reichenbach a chair in the philosophy of science in the physics department at Berlin (Hecht and Hoffmann 1982), from which position Reichenbach went on to build the Berlin outpost of the Vienna Circle.

Schlick and Reichenbach played the leading role in the early to mid-1920s in shaping the philosophy of science that we remember as logical empiricism, much of the fine structure of which reflects their efforts, together with Einstein, to craft a form of empiricism adequate to the task of defending the empirical integrity of general relativity against various challenges, including most prominently that issuing from neo-Kantians such as Cassirer (Cassirer 1921). The Kantian challenge to general relativity came in many forms, ranging from those who simply rejected as impossible general relativity's assertion of a non-Euclidean metrical structure for space-time, and those who took refuge in an arguably quite un-Kantian distinction between physical space and psychological space, to those technically more able philosophers who, like Cassirer, sought to save Kant by retreating from ascribing *a priori* status to a specific metrical geometry of space-time, locating some weaker *a priori* structure in, say, local topological structure.[10] As mentioned, in his first book on relativity Reichenbach sought to defend Kant by tearing apart the apodictic and constitutive aspects of the *a priori*, rejecting the former and retaining the latter – now regarded as a contingent *a priori* – as that which effects the needed coordination between theory and world and thereby endows a theory with empirical content (Reichenbach 1920).

Schlick was not enthusiastic about any of the Kantian responses to general relativity, Reichenbach's included. Schlick's own earlier exposition of the central philosophical implications of general relativity in his *Raum und Zeit in der gegenwärtigen Physik* (1917) emphasized the theory's positing a real space-time event ontology and, thereby, its incompatibility with a philosophy of science that would accord reality only to observables, such as Mach's elements of sensation. But Schlick also stressed the ineluctable moment of convention in scientific theorizing, inspired both by his reading of Poincaré and, even more so, by his own still earlier and then quite influential theory of truth as involving only a univocal, many-to-one coordination of theory to world (Schlick 1910), which entailed an underdetermination of theory choice by evidence not unlike that central to Duhem's philosophy of science. Einstein, who first read the book in manuscript, wholeheartedly commended both the critique of Mach and Schlick's views on the empirical underdetermination of theory choice. In the first edition of his *Allgemeine Erkenntnislehre* (1918), Schlick again gave center stage to the role of convention in scientific cognition, basing the argument once more on his view of truth as univocal coordination but employing also David Hilbert's idea of a theory's implicitly defining its primitive terms through the systematic role they play in the theory's axioms – as opposed to each term's being given an explicit empirical definition – for the purpose of stressing that the coordination relates theory to world only as a whole and not one concept or one proposition at a time.

When, by the early 1920s, the gathering neo-Kantian reaction to relativity finally elicited a focused and thoughtful reply from Schlick, it was the notion of convention that provided Schlick an alternative to the *a priori*. In private correspondence with Reichenbach and in a published review of Cassirer's book on Einstein's theory of relativity (Cassirer 1921, Schlick 1921), Schlick suggested that the "constitutive principles" whereby experience is ordered and interpreted are more helpfully characterized as conventions than as elements of either a contingent or an apodictic *a priori* component of scientific cognition.[11] Einstein – with whom Schlick was then in close and regular contact – had been suggesting much the same thing for several years. Thus, in a letter to Max Born of July 1918, Einstein wrote:

I am reading Kant's *Prolegomena* here, among other things, and am beginning to comprehend the enormous suggestive power that emanated from the fellow and still does. Once you concede to him merely the existence of synthetic a priori judgments, you are trapped. I have to water down the "a priori" to "conventional," so as not to have to contradict him, but even then the details do not fit. Anyway it is very nice to read, even if it is not as good as his predecessor Hume's work. Hume also had a far sounder instinct. (Born 1969, 25–6)

Einstein repeated the same point – that what Kant regards as *a priori* is more properly seen as conventional – in virtually every one of the many comments he penned on the subject through the mid-1920s (see Howard 1994).

As the discussion involving Einstein, Schlick, Reichenbach, and Cassirer progressed in the early 1920s, the effort to craft an empiricist analysis of general relativity that would force the point with the neo-Kantians led Schlick and Reichenbach to an historically important refinement in their understanding of where the conventional moment in science was to be located, and it led to a parting of the ways, ironically, with Einstein. Schlick and Reichenbach wanted to be able to say to the neo-Kantian that the attribution of a specific metrical structure to space-time was, in the end, an empirical matter, this in spite of there being a conventional aspect to ascriptions of metrical structure. They took their cue from Poincaré's well-known view of conventions as disguised definitions (Poincaré 1902b).

As first clearly presented in Reichenbach's 1924 book on the axiomatization of space-time theory (Reichenbach 1924) and then more famously in Reichenbach's 1928 *Philosophie der Raum-Zeit Lehre* (Reichenbach

1928) or Schlick's 1935 "Sind die Naturgesetze Konventionen?" (Schlick 1935), the idea was to restrict the moment of convention to the coordinating definitions that are assumed to effect links between a theory's primitive empirical concepts and relevant aspects of the world that the theory seeks to describe. Paradigmatic coordinating definitions would be the associations of basic notions like "spatial interval" and "temporal interval" with, respectively, a specific kind of practically rigid measuring rod and a specific kind of regular clock. Fix the empirical meaning of a theory's primitive terms by such conventional coordinating definitions and each of the synthetic empirical propositions in a theory thereby acquires a determinate empirical content such that its truth or falsity is, in principle, univocally determined by the corresponding experience. Differences do arise through our choosing different coordinating definitions. But such differences are inconsequential, for they depend, at root, only on arbitrary, conventional definitions, and so can be no more significant than the difference between choosing English or metric units. Moreover, for such empiricists, empirical content is the only content, which implies that, since empirically equivalent theories have the same empirical content, they can only be alternative ways of expressing that same content, just as "il pleut" and "es regnet" are merely alternative ways of expressing the same meteorological proposition, "it is raining."

If this were the right way to think about how theories acquire empirical content, it would have represented a persuasive reply to the neo-Kantian critics of general relativity, for it would have made the choice of a metrical space-time geometry a straightforwardly empirical matter. But the maneuver rests upon one crucial assumption, namely, that one can in a principled manner distinguish within a theory among those propositions that, by their very nature, function as mere conventional definitions and those that function as synthetic, empirical claims. In another guise, this is the analytic-synthetic distinction.

Schlick and Reichenbach might have thought that they were simply elaborating a view about empirical content already put forward by Einstein in a widely read lecture from 1921, *Geometrie und Erfahrung* (Einstein 1921c). But when that lecture's discussion of the empirical interpretation of general relativity is read against the background of Einstein's then at least decade-old sympathy for Duhemian holism, it becomes clear that Einstein had in mind a very different picture of the place of convention in scientific theories.

In *Geometrie und Erfahrung*, Einstein distinguished "pure" and "practical" geometry. Pure geometry is an uninterpreted mathematical formalism the primitive terms of which – for example, "point" and "line" – are defined implicitly, – à la Hilbert, through the systematic role they play in the formalism's axioms. Practical geometry differs by associating or coordinating those primitive terms with some empirical or physical objects whereby the geometry becomes, in effect, a physical theory, a theory about the behavior of those objects. The crucial question concerns the manner in which the primitive terms are thus associated with empirical or physical objects.

Consider the primitive term "segment of a straight line." Our intuition tells us to associate this with a physical structure – say the path of a ray of light or the edge of a rigid measuring rod – of such kind that information gleaned from the behavior of those structures concerns exclusively the geometrical properties of the space or space-time in which the objects live and not physical facts about the objects. We all know, however, that real measuring rods are not perfectly rigid, suffering thermal deformations and mechanical stresses. Should measurements conducted with such rods appear to show deviations from Euclidean geometry, how much of that "deviant" behavior should be attributed to the geometry of space-time and how much to the physics of the rod? In "L'expérience et la géométrie" (1902) Poincaré had, in effect, argued that the defender of Euclidean geometry can always exploit this ambiguity by ascribing the deviance to the physics and that one would want to do so, in order to save Euclidean geometry, because Euclidean geometry is simpler than its non-Euclidean rivals. Of course, Einstein disagreed, arguing that we should choose not what yields the simpler geometry, but what maximizes the simplicity of geometry plus physics, as does, by his lights, general relativity.

There is, however, a serious obstacle to our maximizing the simplicity of geometry plus physics, for while we know, at one level, that various aspects of the behavior of the rod are physical effects, as with thermal deformations, we lack at present a complete, fundamental, physical theory of structures like measuring rods and clocks and so cannot, in fact, actually pose the simplicity question on the physical side. Under such circumstances, the best that we can do is, by stipulation and with a modicum of scientific good sense, to pick some practically rigid body as the physical or empirical interpretation of "segment of a straight line," likewise for some practically regular clock and the primitive term "temporal interval," and then to let measurements carried out with these rods and clocks settle the question of the metrical structure of space-time.

The resulting picture looks much like Reichenbach's. Geometrical primitives are assigned empirical interpretations using stipulative, hence conventional, coordinating definitions, whereafter the truth or falsity of synthetic, empirical propositions such as that concerning the space-time metric is univocally determined on the basis of the relevant

experience. But what for Reichenbach and Schlick was a first principles story about how all theories get their empirical content was, for
Einstein, but a provisional, stop-gap procedure forced upon us by the
lack of a sufficiently complete fundamental theory:

The idea of the measuring rod and the idea of the clock coordinated with it in
the theory of relativity do not find their exact correspondence in the real world.
It is also clear that the solid body and the clock do not in the conceptual edifice
of physics play the part of irreducible elements, but that of composite structures,
which must not play any independent part in theoretical physics. But it is my
conviction that in the present stage of development of theoretical physics these
concepts must still be employed as independent concepts; for we are still far
from possessing such certain knowledge of the theoretical principles of atomic
structure as to be able to construct solid bodies and clocks theoretically from
elementary concepts. (Einstein 1921c, 8)

For Einstein, the correct first principles story was very different. It was
essentially Duhem's holism.

This was not the first time that Einstein had contrasted an in-
principles story akin to Duhem's with an in-practice story portraying a
tighter link between theory and experience. In a 1918 address celebrat-
ing Planck's sixtieth birthday, "Motive des Forschens," Einstein compared the underdetermination of theory choice in principle with the
well nigh universal impression of physicists that, in practice, a single
theory always stands out as superior:

The supreme task of the physicist is ... the search for those most general,
elementary laws from which the world picture is to be obtained through pure
deduction. No logical path leads to these elementary laws; it is instead just the
intuition that rests on an empathic understanding of experience. In this state of
methodological uncertainty one can think that arbitrarily many, in themselves
equally justified systems of theoretical principles were possible; and this opinion
is, *in principle*, certainly correct. But the development of physics has shown
that of all the conceivable theoretical constructions a single one has, at any
given time, proved itself unconditionally superior to all others. No one who
has really gone deeply into the subject will deny that, in practice, the world
of perceptions determines the theoretical system unambiguously, even though
no logical path leads from the perceptions to the basic principles of the theory.
(Einstein 1918j, 31)

But between 1918 and 1921, the question of the extent to which and
manner in which straightforward empirical warrant could be claimed
for a space-time theory had become, for Einstein, a most acute one,
thanks mainly to the challenge posed by Weyl's attempt at a unified
field theory (Weyl 1918a, 1918b).

Weyl's attempt to unify gravitation with electricity and magnetism
employed a geometrical framework that was more strictly "local" than

that of Einstein's general relativity in the sense that it did not assume the cogency of direct distant comparisons of length. In Weyl's geometry, length was path dependent: two bodies congruent at one space-time point need not be congruent at another if they followed different paths (world-lines) to the latter. Einstein objected that this feature of Weyl's space-time geometry left it without determinate empirical content, since in a Weyl space the geometrical concept of the space-time interval could, therefore, not be given a univocal empirical interpretation (Einstein 1918g, 1918h; see also Einstein to Walter Dällenbach, after June 15, 1918 [CPAE 8, Doc. 565]). In effect, Einstein was arguing that, in a Weyl space, there could be no univocal definition of the interval using a practically rigid rod.[12] At the time, Einstein seemed to regard the impossibility of our giving such a direct empirical interpretation to the interval to be a nearly fatal methodological flaw, though he recognized that Weyl would not grant the point, since Weyl would insist on viewing rods and clocks as structures derived from within one's fundamental theory, as opposed to their being specified from without by a coordinating definition. By 1921, however, Einstein seems to have come to regard Weyl's point of view as, in principle, the correct one. Had Einstein really changed his mind?

The clearest statement of Einstein's view on the empirical interpretation of general relativity is found in a 1924 review of a book on Kant and relativity by Alfred Elsbach (Elsbach 1924). Elsbach had commended several arguments originally adduced in Paul Natorp's *Die logischen Grundlagen der exakten Wissenschaften* (1910), among them the claims (1) that if deviations from Euclidean geometry were discovered they could be accounted for by making changes in physical laws and (2) that the metrical structure of space cannot be determined experimentally because space is not real but ideal. Einstein comments that:

The position that one takes on these claims depends on whether one grants reality to the practically-rigid body. If yes, then the concept of the interval corresponds to something experiential. Geometry then contains assertions about possible experiments; it is a physical science that is directly underpinned by experimental testing (standpoint A). If the practically-rigid measuring body is accorded no reality, then geometry alone contains no assertions about experiences (experiments), but instead only geometry with physical sciences taken together (standpoint B). Until now physics has always availed itself of the simpler standpoint A and, for the most part, is indebted to it for its fruitfulness; physics employs it in all of its measurements. Viewed from this standpoint, all of Natorp's assertions are incorrect.... But if one adopts standpoint B, which seems overly cautious at the present stage of the development of physics, then geometry alone is not experimentally testable. There are then no geometrical measurements whatsoever. But one must not, for that reason, speak of the "ideality of space." "Ideality" pertains to all concepts, those referring to space and time no less and no more than all others. Only a complete scientific conceptual system comes to be univocally coordinated with sensory experience. On my view, Kant has influenced the development of our thinking in an unfavorable way, in that he has ascribed a special status to spatio-temporal concepts and their relations in contrast to other concepts.

Viewed from standpoint B, the choice of geometrical concepts and relations is, indeed, determined only on the grounds of simplicity and instrumental utility.... Concerning the metrical determination of space, nothing can then be made out empirically, but not "because it is not real," but because, on this choice of a standpoint, geometry is not a *complete* physical conceptual system, but only a part of one such. (Einstein 1924b, 1690–1)

When Einstein speaks here of granting "reality" to the practically rigid rod, he means simply our taking it as the putative referent of the geo- metrical primitive, "segment of a straight line." According no "reality" to the practically rigid rod means simply that there is none such that can stand on the other side of a conventional coordinating definition of "segment of a straight line," for the reason previously mentioned, namely, that real rods (and clocks) are not perfectly rigid (or regular). If there are no such rigid rods and regular clocks, then the alternative is to regard such structures as being defined implicitly within the theory (or some future, complete, fundamental theory), not stipulated from with- out, which, in turn, implies, as Einstein notes, that it is only the whole theory, "the complete scientific conceptual system" that has determi- nate empirical content.[13]

Reichenbach and Schlick argue that empirical content attaches to the- ories one proposition at a time. Einstein, following Duhem, argues that, in principle, it attaches only to whole theories, even if, for practical rea- sons, we proceed as if the Reichenbach-Schlick picture were the right one. What, then, of the empiricist reply to the neo-Kantians that Reichenbach and Schlick were seeking? Einstein, no less than they, wanted a reply to the neo-Kantians, but he found it precisely in Duhemian holism. Earlier in his review of Elsbach, after asserting that, if one thought relativity a reasonable theory, then one must reject the Kantian doctrine of the *a priori* character of space and time, Einstein added:

This does not, at first, preclude one's holding at least to the Kantian *problematic*, as, e.g., Cassirer has done. I am even of the opinion that this standpoint can be rigorously refuted by no development of natural science. For one will always be able to say that critical philosophers have until now erred in the establishment of the a priori elements, and one will always be able to establish a system of a priori elements that does not contradict a given physical system. Let me briefly indicate why I do not find this standpoint natural. A physical theory consists of the parts (elements) A, B, C, D, that together constitute a logical whole which correctly connects the pertinent experiments (sense experiences). Then it tends to be the case that the aggregate of fewer than all four elements, e.g., A, B, D, *without* C, no longer says anything about these experiences, and just as well A, B, C without D. One is then free to regard the aggregate of three of these elements, e.g., A, B, C as a priori, and only D as empirically conditioned. But what remains unsatisfactory in this is always the *arbitrariness in the choice* of those elements that one designates as a priori, entirely apart from the fact that the theory could one day be replaced by another that replaces certain of these elements (or all four) by others. (Einstein 1924b, 1688–9)

Einstein's named target here is Kant. But Einstein's argument bears just as well against the kind of conventionalism defended by Reichenbach and Schlick, which, recall, arose out of Schlick's telling the Reichenbach of 1921 that his *a priori* principles of coordination were more helpfully regarded as conventional coordinating definitions. Call them *a priori* principles or conventional definitions. The fact remains that one requires a principled basis upon which to distinguish propositions of that kind from the synthetic propositions thereby allegedly endowed with empirical meaning. Einstein's point is that there is no principled basis for thus distinguishing the components of a theory, any such parsing of theory being, from a principled point of view, purely arbitrary.

We see here in germ an argument better known to most philosophers in the form famously given it by Quine in "Two Dogmas of Empiricism" (Quine 1951). The point of view concerning empirical content being developed by Schlick and Reichenbach, with its assumption of some kin to the analytic-synthetic distinction as the basis upon which to distinguish the nonempirical from the empirical components of a theory, is the progenitor of the verificationism mainly targeted by Quine ("reductionism" in Quine's vocabulary),[14] and Einstein's deploying Duhemian holism against Schlick and Reichenbach anticipates Quine's principal critical arguments against both verificationism and the analytic-synthetic distinction.

Just how close Einstein comes to anticipating Quine's more famous argument is evident from Einstein's "Reply" to Reichenbach's contribution to the Einstein volume in the Library of Living Philosophers. In his essay, Reichenbach had defended a first-principles view of the empirical character of geometry much like the provisional, stop-gap view that Einstein presented in *Geometrie und Erfahrung*, with the exception that Reichenbach invoked explicitly his distinction between coordinating definitions and empirical hypotheses, portraying Einstein's identification of the geometer's "rigid body" with the physicist's "practically rigid rod" as an instance of the former (this is what is meant by a definition of "congruence"):

The choice of a geometry is arbitrary only so long as no definition of congruence is specified. Once this definition is set up, it becomes an empirical question which geometry holds for physical space.... The conventionalist overlooks the fact that only the incomplete statement of a geometry, in which a reference to the definition of congruence is omitted, is arbitrary. (Reichenbach 1949, 297)

In his reply, Einstein constructs an imaginary dialogue between "Reichenbach" and "Poincaré," about which Ernest Nagel remarked in his review, "the spokesman for Poincaré does not obviously have the worst of the argument" (Nagel 1950, 293–4).[15]

The dialogue begins with "Poincaré" asserting that geometrical theorems are not, by themselves, testable because there are, in fact, no rigid bodies by which to interpret them. "Reichenbach" replies, à la the Einstein of 1921, that we can do well enough with the almost rigid bodies of our experience, as long as we make obvious corrections for such factors as changing temperature. Einstein's "Poincaré" notes that in making these corrections we must use physical laws that presuppose Euclidean geometry, and concludes that what is at stake in an experiment is, thus, the entire body of law consisting of physics and geometry. Here Einstein interrupts with the following parenthetical remark: "(The conversation cannot be continued in this fashion because the respect of the writer for Poincaré's superiority as thinker and author does not permit it; in what follows therefore, an anonymous non-positivist is substituted for Poincaré)" (Einstein 1949b, 677).

As the dialogue resumes, "Reichenbach" grants a certain attractiveness to "Poincaré's" point of view, while insisting that in fact, if not in theory, it would have been impossible for Einstein to have developed general relativity "if he had not adhered to the objective meaning of length," that is, to the concept of the practically rigid body regarded as an unanalyzed primitive. "Reichenbach" goes on to argue that, if physics and geometry *are* tested together, then our aim should be the development of the simplest possible total system, not the simplest geometry alone.

Thus far the dialogue has proceeded in a predictable fashion. But now it takes an unexpected turn. Having wrung from "Reichenbach" the grudging admission that "Poincaré" might be correct, in theory, about physics and geometry being tested together, though with the stipulation that it is then the simplicity of this whole – physics plus geometry – that must be judged, the "Nonpositivist" points out that "Reichenbach" has thereby contravened one of his own fundamental postulates – the equation of meaning with verification:

Nonpositivist: If, under the stated circumstances, you hold distance to be a legitimate concept, how then is it with your basic principle (meaning = verifiability)? Do you not have to reach the point where you must deny the meaning of geometrical concepts and theorems and to acknowledge meaning only within the completely developed theory of relativity (which, however, does not yet exist at all as a finished product)? Do you not have to admit that, in your sense of the word, no "meaning" can be attributed to the individual concepts and assertions of a physical theory at all, and to the entire system only insofar as it makes what is given in experience "intelligible?" Why do the individual concepts that occur in a theory require any specific justification anyway, if they are only indispensable within the framework of the logical structure of the theory, and the theory only in its entirety validates itself? (Einstein 1949b, 678)

Notice that Einstein here asserts not just the epistemological holism of Duhem but also the semantic holism fundamental to Quine's critique of logical empiricism.[16]

4. NOT A "MACHIAN," BUT NOT A "REALIST" EITHER

By the late 1920s Einstein, once Schlick and Reichenbach's close philosophical friend, had become an open critic of logical empiricist orthodoxy, notwithstanding repeated efforts by members of the Vienna Circle to claim him as an ally (see, e.g., Neurath, Hahn, and Carnap et al. 1929).[17] There were many points of disagreement, logical empiricism's antimetaphysical bias prominent among them. Thus, in a letter to Schlick from November 1930, commenting on a manuscript in which Schlick opposed his own neo-Humean view of causality to more robustly metaphysical notions of causality, Einstein wrote:

From a general point of view, your presentation does not correspond to my way of viewing things, inasmuch as I find your whole conception, so to speak, too positivistic. Indeed, physics *supplies* relations between sense experiences, but only indirectly. For me *its essence* is by no means exhaustively characterized by this assertion. I put it to you bluntly: Physics is an attempt to construct conceptually a model of the *real world* as well as of its law-governed structure. To be sure, it must represent exactly the empirical relations between those sense experiences accessible to us; but *only* thus is it chained to the latter.... You will be surprised at the "metaphysician" Einstein. But every four- and two-legged animal is de facto in this sense a metaphysician. (EA 21–603; as quoted in Howard 1994)

Einstein might well have been puzzled over how his old friend, Schlick, could have strayed so far from the realism central to the argument of *Raum und Zeit in der gegenwärtigen Physik* (Schlick 1917). Einstein had repeatedly stressed his preference for realism, as when he wrote in the cited review of Elsbach:

Is there not an experiential reality [Erlebnis-Realität] that one encounters directly and that is also, indirectly, the source of that which science designates as real [wirklich]? Moreover, are the realists and, with them, all natural scientists (those who do not philosophize from the very start) not right if they allow themselves to be led by the startling possibility of ordering all experience in a (spatio-temporal-causal) conceptual system to postulate something real that exists independently of their own thought and being? (Einstein 1924b, 1685)

A disagreement over metaphysics or realism was not, however, the main issue, or at least surely not the only issue, that led to Einstein's estrangement from logical empiricism. More fundamental, in my view, was the disagreement with Schlick and Reichenbach over the role of convention in scientific theory. They held that the moment of convention is confined to conventional choices of coordinating definitions, the stipulation of which thereby endows each empirical proposition with its own, determinate, empirical content. Einstein, the student of Duhem, held that only whole theories have empirical content and so regarded every component of a scientific theory to be equally conventional and empirical. But can a disagreement over such a recondite question about the semantics of scientific theories really have been the main difference separating Einstein from mainstream logical empiricism?

Recall Einstein's major complaint about Mach. It was not his denying the reality of unobservables, however much that had been an issue in Schlick and Einstein's discussion of the philosophical implications of general relativity when Schlick was writing *Raum und Zeit in der gegenwärtigen Physik* (Schlick 1917). As Einstein recalled in his "Autobiographical Notes" (Einstein 1949a, 48), Mach turned out not to be dogmatic on this point, readily granting the reality of atoms after experimental confirmation circa 1909 of Einstein's own earlier theoretical work on Brownian motion. No, what chiefly bothered Einstein was Mach's neglect of the moment of invention or free creation in the development of scientific theories.

That theories are the "free creations of the human intellect" (Einstein 1921c, 5), was a recurring theme in Einstein's writings (see also, e.g., Einstein 1936, 314). It was always opposed to overly strict empiricist accounts of the relation between concepts or theories and experience: "In general, the weakness of the positivists appears to be that, on their view, the logical independence of concepts with respect to sense experiences does not stand out clearly" (Einstein 1924b, 1687). Einstein was not alone in seeing this tension. Philipp Frank saw it as well, contrasting Mach's emphasis upon the empirical legitimation of theories with what he took to be Poincaré's stress on their character as free creations, and credited both Duhem (Frank 1949a, 26) and Einstein (1949b) with having shown us how to reconcile the two by asserting that, since it is only whole theories that possess empirical content, not individual concepts or propositions, the theorist is, therefore, free to make changes anywhere in the total body of theory, as long as the whole theory conforms with empirical fact.

For neither Duhem nor Einstein was the freedom such as to be an obstacle to progress in science. Duhem emphasized that all creative scientific activity takes place in an historical context that constrains, because of contingent historical circumstance, the freedom that the theorist enjoys as a matter of logic alone (Duhem 1906, ch. 7). Einstein contrasted the fact that, in principle, there is no univocal, logically determined path from experience to theory with the nearly universal recognition among scientists that, in practice, theory choice seems univocally determined (see Einstein 1918g, 1933b). Yet both Einstein and Duhem saw in the fact that the theorist is, thus, in principle, a free epistemic agent as that which secures for theory, as such, its central role in science. Were theoretical concepts eliminable using definitions in terms of phenomenalist protocol terms, or were there, in some other form, an empirical theory choice algorithm, then science could be left to the experimentalist.

That realism was not the main issue in Einstein's dispute with mainstream logical empiricism is evident also from the fact that Einstein was, in fact, rather skeptical about blanket claims of the metaphysical realist kind. Surely the most amusing example of this skepticism is found in a letter of September 25, 1918, from Einstein to the Bonn mathematician, Eduard Study, commenting on Study's book *Die realistische Weltansicht und die Lehre vom Raume* (1914). About Study's defense of realism against both positivists and conventionalists, Einstein wrote:

"The physical world is real." That is supposed to be the fundamental hypothesis. What does "hypothesis" mean here? For me, a hypothesis is a statement, whose *truth* must be assumed for the moment, *but whose meaning must be raised above all ambiguity.* The above statement appears to me, however, to be, in itself, meaningless, as if one said: "The physical world is cock-a-doodle-doo." It appears to me that the "real" is an intrinsically empty, meaningless category (pigeon hole), whose monstrous importance lies only in the fact that I can do certain things in it and not certain others. This division is, to be sure, not an *arbitrary* one, but instead....

I concede that the natural sciences concern the "real," but I am still not a realist. (CPAE 8, Doc. 624)

In what sense might science, nonetheless, concern the real on Einstein's view? Arthur Fine has made the insightful suggestion that it was typical of Einstein to "entheorize" a methodological notion like realism, turning it into a physical thesis, more specifically, the thesis of determinism (Fine 1986). But whereas Fine sees Einstein entheorizing realism as the thesis of determinism, I see him entheorizing it as the thesis of spatial or spatiotemporal separability. Consider what he wrote to Max Born on March 18, 1948:

I just want to explain what I mean when I say that we should try to hold on to physical reality. We are, to be sure, all of us aware of the situation regarding what will turn out to be the basic foundational concepts in physics: the point-mass or the particle is surely not among them; the field, in the Faraday-Maxwell sense, might be, but not with certainty. But that which we conceive as existing ("real") should somehow be localized in time and space. That is, the real in one part of space, A, should (in theory) somehow "exist" independently of that which is thought of as real in another part of space, B. If a physical system stretches over the parts of space A and B, then what is present in B should somehow have an existence independent of what is present in A. What is actually present in B should thus not depend upon the type of measurement carried out in the part of space, A; it should also be independent of whether or not, after all, a measurement is made in A.

If one adheres to this program, then one can hardly view the quantum-theoretical description as a *complete* representation of the physically real. If one attempts, nevertheless, so to view it, then one must assume that the physically real in B undergoes a sudden change because of a measurement in A. My physical instincts bristle at that suggestion.

However, if one renounces the assumption that what is present in different parts of space has an independent, real existence, then I do not at all see what physics is supposed to describe. For what is thought to be a "system" is, after all, just conventional, and I do not see how one is supposed to divide up the world objectively so that one can make statements about the parts. (Born 1969, 223–4)

That separability was central to Einstein's conception of physical reality makes sense when it is understood that it was Einstein's commitment to separability that underlay his decades-long opposition to quantum mechanics as a preferred framework for fundamental physics. This was because quantum entanglement seems hard to reconcile with the assumption that spatial separation is a sufficient condition for the individuation of physical systems.[18]

5. CONCLUSION

The development of Einstein's philosophy and the development of logical empiricism were both driven in crucial ways by the quest for an empiricism that could defend the empirical integrity of general relativity in the face of neo-Kantian critiques. But logical empiricism was more than a philosophy of relativity theory, and Einstein's philosophy of science was more than an answer to Kant.

A fuller account of Einstein's philosophy of science would have to include discussion of his belief in simplicity as a guide to truth, especially in areas of physics comparatively far removed from extensive and direct contact with experiment, as in his own long search for a unified field theory (see Howard 1998, Norton 2000, van Dongen 2002b, and Janssen and Renn 2007). A fuller account would also investigate Einstein's largely original and, I think, quite profound distinction between "principle theories" and "constructive theories," the former constituted of mid-level, empirically well-grounded generalizations like the light principle and the relativity principle, which, by constraining the search for constructive models, often facilitate progress in science, as Einstein thought was the case in his discovery of special relativity (Einstein 1919f; see also Howard 2004a). And a fuller account would examine Einstein's appropriation of what Joseph Petzoldt dubbed "the law of univocalness" (Petzoldt 1895), in effect the requirement that theories determine for themselves unique models of the phenomena they aim to describe, for this idea was central to Einstein's thinking about a permissible space-time event ontology, his solution of the "hole argument" using the "point-coincidence argument" in the genesis of general relativity, and his more general attitude toward physical reality and objectivity (see Howard 1992 and 1999). And partly through its influence on Einstein, this idea of Petzoldt's also played a significant role in the history of logical empiricism, especially in the development of Carnap's thinking (see Howard 1996).

Still, the struggle to craft a compelling response to neo-Kantian critiques of general relativity was, in my view, the single most important factor shaping the development of the story about empirical content that defined mainstream logical empiricism, and Einstein's central role in this development through his personal and intellectual relationships with logical empiricism's chief architects has yet to receive the attention that it is due in the historical literature. Moreover, understanding Einstein's role in this history helps us to put into context Einstein's late and often-quoted characterization of himself as an "epistemological opportunist" (Einstein 1949b, 684). Yes, Einstein's philosophy of science borrowed from realism, positivism, idealism, and even Platonism. It might appear to the "systematic epistemologist" to be mere opportunism. But when viewed in its proper historical setting, it emerges as an original synthesis of a profound and coherent philosophy of science that is of continuing relevance today, the unifying thread of which is, from early to late, the assimilation of Duhem's holistic version of conventionalism.

NOTES

1 For a helpful survey of philosophical reactions to relativity, see Hentschel (1990).
2 Stadler (1997) is a compendious source of information on all aspects of the history of logical empiricism. On the protocol sentence debate, see Zhai (1990). Howard (2003) provides background on the role of politics in the development of logical empiricism.

3 For a complete list of the readings of the Olympia Academy, see CPAE 2, xxiv–xxv.
4 Lange (1873–5), Dühring (1873), and Rosenberger (1895) are among the many titles preserved in Einstein's personal library, now housed with the Albert Einstein Archives at the Hebrew University and National Library in Jerusalem.
5 For a fuller discussion of Einstein and Duhem, see Howard (1990a).
6 For helpful discussions of the late-nineteenth- and early-twentieth-century phenomenon of the "philosopher-physicist," see Stöltzner (2003) and Howard (2004b).
7 That such could not have been Mach's view is evident from his praise for Duhem both in his preface to Adler's translation of Duhem (1906) and in the laudatory comments he added to the 1906 second edition of his *Erkenntnis und Irrtum* (Mach 1906).
8 See Einstein to Schlick, December 14, 1915, EA 21–611, CPAE 8, Doc. 165.
9 For more on Einstein's relationship with Schlick, see Howard (1984, 1994).
10 Hentschel (1990) surveys the variety of Kantian reactions to relativity.
11 See Schlick to Einstein, October 9, 1920, EA 21–580, and Schlick to Reichenbach, November 26, 1920, HR 015–63–22 (Hans Reichenbach Papers, Archive of Scientific Philosophy, University of Pittsburgh).
12 See Ryckman (2005), esp. ch. 3, for further helpful discussion of Einstein's objection to Weyl's unified field theory.
13 More or less the same argument, using almost exactly the same formulations, is repeated in Einstein (1925c).
14 Michael Friedman gives a very different reading of the history of verificationism, locating its birth in Einstein's point-coincidence argument (Friedman 1983, 24). For a critical evaluation of Friedman's argument, see Howard (1999).
15 An enduring puzzle is why Einstein so frequently attaches the name of Poincaré to the holistic version of conventionalism that was famously defended by Duhem in explicit opposition to the views of Poincaré, another well-known example of his doing so being Einstein (1921c). For further discussion, see Howard (1990a).
16 Quine reports that he had not known of Einstein's quoted remarks from Schilpp (1949) when he wrote "Two Dogmas of Empiricism" (private communication). But given Quine's close connections with Philipp Frank at Harvard in the late 1940s and the latter's deep immersion in Einstein's philosophy of science at that very time (see, e.g., Frank 1947, 1949b, 1949c), one suspects that there is more to the story than Quine remembered.
17 While Einstein disagreed, as indicated, with what came to be regarded as mainstream logical empiricism, his philosophy of science actually had much in common with the views prevalent among the members of what is now dubbed the "left wing" of the Vienna Circle, including Neurath, Hahn, and Frank, all of whom shared with Einstein the experience of being deeply influenced by Duhem. For more on the Vienna Circle's left wing, see Howard (2003).
18 For more on the role of separability in Einstein's critique of quantum mechanics, see Howard (1985, 1990b, 1997).

12 "A Believing Rationalist"

Einstein and "the Truly Valuable" in Kant

I did not grow up in the Kantian tradition, but came to understand quite late the truly valuable which is to be found in his doctrine, alongside of errors that today are quite obvious. It is contained in the sentence: "The real is not given to us, but rather put to us (*nicht gegeben sondern aufgegeben*) (by way of a riddle)." This obviously means: there is such a thing as a conceptual construction for the grasping of the interpersonal, the authority of which lies purely in its validation. This conceptual construction refers precisely to the "real" (by definition), and every further question concerning the "nature of the real" appears empty.

Einstein 1949b, 680

1. INTRODUCTION

It is long observed that a significant shift occurred in Einstein's philosophical views as a result of his work on the general theory of relativity. In Gerald Holton's influential account, Einstein's "philosophical pilgrimage ... start(ed) on the historic ground of positivism" (Holton 1968, 244), heavily under the influence of Mach.[1] Yet following the completion of general relativity in late 1915, his "apostasy" from Mach became more and more apparent. The end point of Einstein's philosophical odyssey lay in his conversion to what Holton termed a "rationalistic realism," namely the conviction "that there exists an external, objective, physical reality which we may hope to grasp – not directly, empirically, or logically or with fullest certainty, but at least by an intuitive leap, one that is only guided by experience of the totality of sensible 'facts'" (Holton 1968, 263). In support of this assessment, Holton cited Einstein's letter of January 24, 1938, to the mathematical physicist Cornelius Lanczos. Here Einstein confessed: "Coming from skeptical empiricism of somewhat the kind of Mach's, I was made, by the problem of gravitation, into a believing rationalist, that is, one who seeks the only trustworthy source of truth in mathematical simplicity" (Holton 1968, 259). Lanczos, Einstein's Berlin collaborator on unified field theory in 1928–9, was not only a friend but also a philosophical soul mate. For he was, as Einstein observed in a letter four years later (EA 15–294), the only other physicist Einstein knew who held "the same attitude towards physics," namely "belief in the comprehensibility of reality through something logically simple and unified."[2] This "credo" (Holton's term) of mathematical simplicity notably found public expression in Einstein's (1933b) Herbert Spencer lecture, "On the Methods of Theoretical Physics," delivered at Oxford on June 10, 1933 (e.g., Holton 1968, Miller 2000, Norton 2000). Lecturing for the first time in English, Einstein in no uncertain terms expressed his faith that the "right method" of theoretical physics consisted in seeking laws of nature with the simplest mathematical formulation.

Our experience hitherto justifies us in believing that nature is the realization of the simplest conceivable mathematical ideas. I am convinced that we can discover by means of purely mathematical constructions, the concepts and the laws connecting them with each other, which furnish the key to the understanding of natural phenomena. Experience may suggest the appropriate mathematical concepts, but they most certainly cannot be deduced from it. Experience remains, of course, the sole criterion of the physical utility of a mathematical construction. But the creative principle resides in mathematics. In a certain sense, therefore, I hold it true that pure thought can grasp reality, as the ancients dreamed. (Einstein 1954, 274)

If one considers that as recently as his 1923 "Nobel Lecture" at Gothenburg, Einstein had insisted in near-textbook positivist fashion that "concepts and distinctions are only admissible to the extent that observable facts can be assigned to them without ambiguity" (Einstein 1923f, 482; see also the encomium to Mach, Einstein 1916c), these elevated declarations in Oxford could hardly be less than revelatory, an apparent endorsement of a severe kind of Platonism, essentially *identifying* fundamental physical reality with the abstract reality of ideal mathematical structures. Yet recent scholarship (van Dongen 2010) gives pause, suggesting that such expression of an extreme rationalism scarcely voiced elsewhere, hardly consistent with Einstein's many subsequent and more nuanced philosophical avowals, merits a highly context sensitive interpretation. In point of fact, the Herbert Spencer lecture occurred toward the end of a two-year period of intense cultivation of a new physical proposal for which Einstein had held high hopes. Formulated in collaboration with his assistant Walter Mayer, the theory was based on a "new sort of field" from which Dirac's spinor fields for electrons could be regarded as a special case. The so-called semivector fields purportedly "described, in a natural way, certain essential properties" of oppositely charged particles of unlike mass (Einstein 1954, 275). Returning to the theme of mathematical simplicity, Einstein hailed semivectors as "after ordinary vectors, the simplest mathematical fields possible in a metrical continuum of four dimensions" (ibid.). But even as the lecture prominently lists "semivector fields" in confirmation of the thesis that "nature is the realization of the simplest conceivable mathematical ideas," giving an unmistakable impression of Einstein's self-assurance in this new theory, his correspondence at precisely the same time reveals that confidence in the indicated physical interpretation of semivectors "was already waning" (van Dongen 2010, 121). As van Dongen further documents, early in 1934 Valentin Bargmann, then a doctoral student under Pauli in Zurich but later to become a research assistant to Einstein and professor of mathematics in Princeton, pointed out that the Einstein-Meyer semivector concept possessed an inert index under Lorentz transformation (hence, was inelegant) while showing their analysis to be "redundant" since it "might as well have been carried out for spinors" (ibid., 126). So perhaps the vainglorious proclamation should not be taken at face value but rather *cum grano salis*. In any case, it certainly falls into Einstein's oft-noted pattern of issuing highly enthusiastic statements regarding the bright prospects of each newly minted candidate unified field theory, venting an optimism that invariably proved to be short-lived. In this regard, it is not without interest that Lanczos entertained the suspicion that Einstein's claims regarding the extent of Mach's influence on his earlier self "may have been self deception" (Lanczos 1974, 42).

As we shall argue, Einstein's "believing rationalist" is neither a Platonist, nor really a rationalist at all, at least as that term is understood within the tradition stemming from Descartes, Spinoza, and Leibniz. Notwithstanding his well-known railings against Kant's unsustainable doctrine of synthetic *a priori* concepts and judgments, the later Einstein affirmed that while experience alone remained the ultimate arbiter of any theory, a "belief in the comprehensibility of reality through something logically simple and unified" (see note 2) was a methodologically legitimate heuristic, whose general form Kant had thoroughly probed in the Transcendental Dialectic of the *Critique of Pure Reason*. There Kant had given a tempered account of the cognitive role of *reason* in science, outlining its necessary but "hypothetical employment," issuing in an injunction to seek systematic unity in nature. So it is that Einstein came to understand "the truly valuable in Kant" lay in the doctrine of the regulative principles of reason, according to which the very *concept* of an order of nature presupposes a *decision* to seek unity through systematic connections.

As Einstein would observe, it was Kant who first pointed out that the assumption that nature *is* comprehensible is a necessary precondition for the postulation of the real external world that physics attempts to systematically investigate.[3] This means that the idea of the systematic unity of nature *cannot* be an empirical hypothesis, but is posed *only* as a problem or a task: namely, fundamental physical theory does not find but rather projects systematic interconnection into the world. The question whether such an order *really exists* is in fact meaningless, for the *only justification* any particular theory of systematic interconnection *can ever possess* is that it organizes the totality of physical observations ("sense experience") into a comprehensively ordered and empirically adequate arrangement. To be sure, Einstein's occasionally expressed, but apparently not univocally held, opinion that at any given time such an arrangement may be unique, as in an oft-quoted remark that "the world of phenomena practically determines the theoretical system uniquely" (Einstein 1918j, 31), has been cited as evidence of his Platonic tendencies.[4] But, as the opening quotation affirms, for the later Einstein systematic interconnection formulated in any such a theory is, and can be, only a *projected* unity, it is "the real" *by definition*. As we shall see, what is ultimately Einstein's nonconventionalist attempt to justify the use of concepts and hypotheses of nonempirical origin in theoretical construction closely parallels Kant's doctrine of the hypothetical employment of reason in the Transcendental Dialectic. Similarly, although Einstein's endorsement of a heuristic methodology of mathematical simplicity may be reminiscent of Poincaré's geometric and physical conventionalism, a closer look reveals that, for Einstein, the choice of simplicity was not motivated on grounds of convenience, but rather by the ever-growing distance between fundamental field theoretical concepts and principles and possible confirming or disconfirming observations. As Hilbert expressly did also, with Einstein's gravitational theory specifically in mind (Hilbert 1992, 66–70, 90–101; Brading and Ryckman 2008a), Einstein – unless we are to read him as a Platonist – appears to have recognized that legitimacy for use of a methodology of mathematical simplicity in physics lay within Kant's doctrine of the hypothetical employment of reason.

Claims of Kantian influences on Einstein, thought outlandish in the era when philosophy of science lay under the domination of logical empiricism, have more recently been considered on their merits (e.g., Scheibe 1992, Paty 1993, Beller 2000, Friedman 2008, Torretti 2008). Moreover, the still-lurking suspicion that the theory of general relativity was a fatal disconfirmation of *any* form of transcendental idealist understanding of physical theory flies in the face of recent studies (Ryckman 2005, Brading and Ryckman 2008) documenting that a highly significant cohort of the initial leading contributors to the articulation and further development of general relativity – David Hilbert, Hermann Weyl, Arthur Eddington, and Max von Laue – all affirmed a broadly "transcendental idealist" understanding of the theory, the spirit of which is captured in the "Copernican reversal" invoked by Kant's statement: "Thus the order and regularity in appearances that we call Nature, we introduce ourselves, and indeed we could not find them there had not we, or the nature of our mind, put them there originally" (A 125).[5]

This is just to say that there can only be as much knowledge of the formal and conceptual aspects of experience as is projected through the spontaneous activity of the cognizing subject.[6] Certainly, as general relativity would show, at least a major component of Kant's own conception of transcendental philosophy in the Transcendental Analytic, that regarding supposed necessary *a priori* conditions (intuitions, concepts, and judgments) constitutive of objects of possible experience, required clarification or revision. But from a more ecumenical perspective, Kant's error testifies merely to an excusable historical blindness regarding the impermanence of Newton's magnificent achievement, having no more enduring significance for transcendental idealism than the Pythagorean restriction of number to positive rational fractions has for modern analysis.

2. WHAT DID "THE PROBLEM OF GRAVITATION" TEACH?

Einstein began his Oxford lecture with a provocative challenge: "If you want to find out anything from the theoretical physicists about the methods they use, I advise you to stick closely to one principle: don't listen to their words, fix your attention on their deeds" (Einstein 1954, 270).

Puzzling advice from a theoretical physicist lecturing on the methods of theoretical physics! But the sage counsel to examine deeds, not words, has been closely followed by a group of recent philosophers and historians of general relativity who have closely scrutinized Einstein's notebooks and papers recording his struggle with the problem of gravitation between 1912 and 1915 (Norton 2000, Janssen and Renn 2007, Renn and Sauer 2007). These researchers have determined that during this crucial period Einstein pursued a "double strategy" in trying to find the field equations of gravitation, oscillating between two distinct and not necessarily complementary tracks, one physical and the other mathematical. Working the physical side, Einstein took the limiting case of Newtonian gravity as the starting point, attempting to recover the limit cases of Newtonian gravity ("weak field approximation") and static gravitational fields in a relativistic theory of gravity. He then sought to determine (for details, see Renn and Sauer 2007) conservation laws for energy and momentum, and only if finding success there, attempted to make the theory covariant according to the largest possible class of coordinate transformations.

By contrast, Einstein's so-called mathematical strategy rests on the postulate of a generalized principle of relativity of motion, formally expressed as the requirement that the theory's equations be generally covariant, that is, that they remain valid under the largest possible ("arbitrary") class of differentiable coordinate transformations. To be sure, while accelerated reference frames of a generalized principle of relativity are naturally represented by the arbitrary (smooth) curvilinear coordinate systems permitted under the purely formal condition of general covariance, the connection between the two principles is not so straightforward as Einstein originally imagined.[7] However, through his friend Grossmann in Zurich, Einstein had discovered the existence of an already formulated mathematical language, the "absolute differential calculus," that seemed ideally suited to a generalized principle of relativity, since the objects of such a calculus – tensors – are invariant under transformation between arbitrary coordinate systems. This ensured that the fundamental mathematical relations between physical quantities represented as tensors, that is, the field equations, would be by default generally covariant, having the same form in all coordinate systems. So, following the mathematical strategy, Einstein started from a generalized principle of relativity of motion. Then, under the supposition that the Riemannian metric tensor $g_{\mu\nu}$ represented the "potential" of the gravitational field, he sought (for details, see again Renn and Sauer 2007) a generally covariant differential operator on $g_{\mu\nu}$ coupling the ensuing tensorial gravitational field strengths to the energy-momentum sources of the field. The latter must also be represented by a tensor if the principle of conservation of energy is to be satisfied. Hence the gravitational field equations will be generally covariant tensor equations in accordance with the generalized principle of relativity. As Einstein already recognized in the so-called Zurich Notebook dating from 1912–13 (CPAE 4, Doc. 10),[8] the starting point for seeking such an operator is naturally found in the Riemann curvature tensor $R_{\mu\nu\sigma\tau}$ since this quantity implicitly contains the complete system of generally covariant differential operators possible in the underlying (pseudo-)Riemannian manifold. However, stymied in part by the "Hole Argument," in part by difficulties with energy-momentum conservation and his failure to recover the Newtonian limit (see Section 3 of Chapter 6), two long years passed before Einstein found the right answer, formulating the so-called Einstein tensor on the left-hand side of his field equations from an algebraic combination of differential invariants constructed from the Riemann curvature tensor.

For our purposes here it is only necessary to observe that Einstein's "mathematical strategy" – the speculative route based upon a generalized principle of relativity and the requirement of general covariance – was *essentially* philosophically motivated by a desire to eliminate from physics *a priori* privileged structures of space-time, in particular global inertial frames of reference. It is well-known that Einstein's commitment to a generalized principle of relativity stemmed from a Machian-inspired wish to entirely eliminate privileged frames of reference – inertial systems – from physics. As Einstein later but characteristically observed, such a system, "acts upon all processes but undergoes no reaction," and in this regard the concept of an inertial system "is in principle no better than that of the center of the universe in Aristotelian physics."[9] Of course, the concept of a global inertial system, the last vestige of Newtonian absolute space, still survived in the special theory of relativity, a defect of that theory Einstein had pointed out already in 1907. Philosophical repugnance of absolute space thus lay behind Einstein's attraction to a generalized principle of relativity, however imperfectly expressed by the principle of general covariance. In turn, his efforts to implement a philosophically satisfactory remedy led to a gravitational theory according to which there is no coordinate-independent way to distinguish between gravitational and inertial "forces," and so in which there are no global inertial frames, no prior background geometry of space-time. The result, in the theory of general relativity, is that the very metric of space-time becomes a dynamical field quantity, meaning that its properties depend in part on its specific matter (stress-energy) content, just like electric and magnetic fields.

While a consensus exists among the previously mentioned scholars regarding the presence of twin physical and mathematical strategies in Einstein's notebooks and papers prior to November 1915, disagreement emerges regarding which strategy ultimately proved successful, leading to general relativity. Norton (2000) has argued that it was the mathematical strategy; Janssen and Renn (2007) that it was the physical strategy.[10] Einstein provided conflicting retrospective assessments (ibid., 911–17). In the immediate aftermath of the completion of his theory, he sought to defend it by emphasizing its mathematical pedigree in the "absolute differential calculus." All the same, against the early and purely mathematically driven speculative attempts of David Hilbert, Hermann Weyl, Arthur Eddington, and Theodor Kaluza to generalize his theory, Einstein insisted on the necessity of physical foundations and tended to downplay the role of mathematical speculation, perhaps in rhetorical response (cf. Ryckman 2005). But he did not allow this tempered caution to hamper his own research; by 1923 he was explicit about following a speculative methodology based upon mathematical simplicity in pursuit of unified field theory.[11] The result is that during the decade of the 1920s, a cognitive fissure opened between his own pronouncements against the speculative mathematical approach of Weyl and others, and the methodology he had adopted in constructing one unified field theory after another. Reflecting back on this decade in 1933, perhaps the opening lines of the Oxford lecture cited previously were intentionally self-referential. However that may be, by the end of the 1920s Einstein had become quite forthcoming about his conviction that a methodology coupling mathematical speculation with a belief in "the uniformity of natural laws and their accessibility to the speculative intellect" had not only yielded general relativity, but might well do the same in the project of unified field theory (Einstein 1929, 114).[12] Post-1930, ever more deeply immersed in a paradigm that viewed unified field theory as the necessary completion of the problem of gravitation as well as the fundamental basis of a theory of matter, from which the empirical successes of quantum mechanics might be seen as having merely statistical validity, Einstein repeatedly stressed that pure mathematics, not physical insight or physical requirements, had been the winning ticket back in 1912–15. The message that mathematical simplicity had provided the needed key to unlock the door shielding general relativity was reiterated many times in his last years, including to de Broglie, on February 15, 1954, just a little more than a year before his death: "The equations of gravitation were *only* found on the basis of a purely formal principle (general covariance), that is to say, on the basis of trust in the greatest logical simplicity of laws of nature thinkable" (de Broglie 1979, 56). Scholars who, on examination of the extant materials, conclude that it was, to the contrary, the physical strategy that opened the door to general relativity have suggested that these later pronouncements indicate that Einstein had a selective memory (who hasn't?) (Janssen and Renn 2007, 841). Other factors help explain why Einstein played the mathematical card as trump. The fruitfulness of the mathematical strategy was illustrated by the example of Hilbert, who certainly pursued a purely mathematical path ("the axiomatic method") that quickly brought him into the race for the generally covariant gravitational field equations in November 1915 (Sauer 1999, Brading and Ryckman 2008). Norton (2000, 138) points to this example in arguing that Einstein switched to the mathematical strategy in late 1915. Whether or not he did adopt this strategy at that time, it was only natural for Einstein to justify (perhaps even to himself) the speculative mathematical approach guiding his unified field theory program by a claim that such a methodology had already yielded success (cf. van Dongen, 2010). It is easily documented that the further Einstein went down the road of unified field theory, the more he stressed his "belief in the comprehensibility of reality through something logically simple and unified" (see note 2), legitimation for which, I argue next, he found belatedly in Kant.

3. "THE TRULY VALUABLE" IN KANT

Transcendental idealism is perhaps most familiar as the thesis that we can know only things as they appear (in Kant's technical sense, and so in accordance with the subjective conditions of the human faculties of understanding and sensibility), that is to say as mere representations, not as they are in themselves. It will be simply assumed here that, according to Kant, the *Dinge an sich* are not a realm of unknowable *objects* (a *noumenal world*) but rather limit *concepts*, correlates to our cognitive representations of objects.[13] Of course, rather straightforwardly, relativity theory, both special and general, called for distinct modifications of central sections of the *Critique of Pure Reason*.

Let us recall that the first and by far major portion of *The Critique of Pure Reason*, "The Transcendental Doctrine of Elements," is divided into two parts, a Transcendental Aesthetic and a Transcendental Logic, where the latter is partitioned into two subdivisions: Transcendental Analytic and Transcendental Dialectic. Undoubtedly, the Transcendental Aesthetic and the Transcendental Analytic are the most widely read and best-known parts of the *Critique;* the Transcendental Aesthetic contains the doctrine that space and time are *a priori* forms of sensibility, while the Transcendental Analytic presents Kant's answer to the question of how synthetic *a priori* judgments are possible: namely, as judgments structured by the categories that, as schematized by the forms of intuition, prescribe precise conditions within which all human cognition of objects occurs. Obviously, it is the doctrine of space and time in the Transcendental Aesthetic that most egregiously requires modification in the light of relativity theory; in particular, Kant's account of the *necessary structure* or *form* of the subjective conditions of human sensibility, understood as mandating a globally Euclidean structure of space and an absolute Newtonian character of time, cannot be sustained.

However, in arguing that synthetic *a priori* judgments can only be established for cognitions within the domain of sensible experience, the Transcendental Analytic already initiated a critique of traditional "dogmatic" metaphysics, wherein *a priori* judgments transcend the boundaries of possible experience. The goal in the Transcendental Dialectic is similarly critical though somewhat more indirect. No mere skeptic of metaphysics, Kant wished to show that, although the questions that preoccupy metaphysical inquiry are inevitable, as *inherent in the nature of human reason itself,* they are nonetheless deceptive, and so always must be understood in the right (i.e., "critical") manner, on pain of falling into metaphysical dogmatism. Though the matter cannot be argued here, recent work provides grounds for viewing the Transcendental Dialectic as indispensable to Kant's theory of science and toward understanding the account of cognition given in the *Critique* as a whole.[14]

From this perspective, the central task of the Transcendental Dialectic is to complete the account of cognition presented in the Transcendental Analytic (pertaining to sensibility and the categories of the understanding) by including the cognitive role of "theoretical reason" in its "critical" employment. In general, Kant viewed reason (both "theoretical" and "practical") as rooted in the human mental capacity to project beyond given experience in order to seek the totality of possible experience, the "totality of all conditions" or the "unconditioned" presupposed by any series of conditions. Since such a totality can never be an object of possible experience, it cannot be legitimately considered an object of cognition. Nonetheless, the ideas of theoretical reason that do project such a totality are essential to natural science where they have only a regulative, not a constitutive, sense: they give expression to reason's capacity to surpass the confines of experience through the *hypothetical* adoption of maxims of systematic unity or unity of nature. Kant is quite careful in drawing boundaries around such a use of reason; in particular, the rationality it prescribes is merely projected: "systematic unity (as mere idea) is, however, only a *projected* unity, which one must regard not as given in itself, but only as a problem" (A647/B675). The ideal concepts or principles of this projected unity are the product of the "hypothetical use of reason, on the basis of ideas as problematic concepts," and are "really not *constitutive*" (for "constitutive" in Kant's sense pertains only to objects of experience). I submit that this, in Einstein's assessment, is just the core of "the truly valuable in Kant."

Thus in the Transcendental Dialectic, emphasis is not on the *constitutive rules* of the understanding but on *regulative principles*, transcendental ideas or concepts of the faculty of reason. Its most celebrated chapter, on the Antinomy of Pure Reason, shows that the antinomies stem from an *uncritically* accepted directive of reason to find, for the cognitions provided by the understanding (whose a priori conditions are inventoried in the Transcendental Analytic), the unconditioned totality or unity of all such cognitions. However, such an unconditioned totality can, according to the Transcendental Analytic, never be given as an object of possible experience, in contrast to the assumptions of a dogmatic metaphysics that treats appearances as things-in-themselves. The resolution of the Antinomy of Pure Reason comes through critical reflection upon *the demand of reason* to bring all cognition under a principle of systematic unity; while the idea of such unity *is necessary* to the guidance, and so, proper functioning, of the understanding, it is nonetheless *not constitutive of an actual object of experience*. Rather it is *an indispensable regulative idea*, in Kant's projective geometric terms, a *focus imaginarius* toward which all cognition is directed, providing an ideal goal and direction of inquiry that mark out the way knowledge is to be sought and organized. The assumption that nature embodies such a unity cannot be disconfirmed by recalcitrant experience, for it is a pre-supposition of seeking physical laws at all: "These concepts of reason [i.e., ideas] are not derived from nature; on the contrary, we interrogate nature in accordance with these ideas, and consider our knowledge as defective so long as it is not adequate to them" (A 645–6/B 673–4).

From the standpoint of epistemology of science, Kant's purpose in invoking such a regulative use of reason is to demonstrate that empirical knowledge presupposes a general framework of unity within which specific empirical claims can be situated. The regulative use of reason, by specifying the *ideal structure of a completed system of scientific knowledge*, provides the context within which specific scientific theories are located. In this way, scientific theorizing requires a *transcendental*, not merely a logical (methodological, instrumental) use of ideas, articulating the ideal of a completely unified system of explanation to which, however, any current state of knowledge of the world only very imperfectly approximates and, in fact, can never be attained. But then the regulative use of reason involves a fundamentally different use (and meaning) of *a priori* knowledge than that in the Transcendental Analytic. Empirical science requires the assumption that nature accord with reason's interest in unity; nonetheless, the way, or degree to which, this demand may be satisfied cannot be specified *a priori*. By paying attention to this neglected aspect of Kant's account of the nature of empirical knowledge, Einstein arguably came to see that, despite "errors that today are quite obvious" in Kant's doctrine of necessary *a priori* structures of the Transcendental Analytic, Kant had shown how ideas transcending experience have an independent and essential role to play in theoretical cognition.

4. EINSTEIN'S KANT

In the course of a celebrated discussion with French luminaries at the Collège de France in April, 1922, Einstein pointedly replied to philosopher Leon Brunschvicg's question concerning the bearing of Kant's philosophy upon the theories of relativity: "As concerns Kant's philosophy, in my opinion, every philosopher has his own Kant" (Einstein 1922d, 101). After noting that he was unsure of Brunschvicg's own interpretation of Kant as manifested in the latter's lengthy intervention, Einstein went on to say,

It seems to me that the most important matter in Kant's philosophy is that one speaks of a priori concepts in the construction of science. But here there are two opposing viewpoints: the apriorism of Kant, in which certain concepts preexist in our mind, and the conventionalism of Poincaré. These two points of view agree on the point that science requires, for its construction, arbitrary concepts; with regard to whether these concepts are given a priori or are arbitrary conventions, I cannot say. (ibid.)

It is somewhat surprising that in such an august public venue Einstein chose to sit on the fence, refusing to clearly voice the opinion already stated in correspondence with Schlick and Born among others,[15] that such arbitrary concepts are conventions, and so to state his opposition to any Kantian or Kantian-like doctrine holding that "certain concepts preexist in our mind." As is widely known, this dissent emphasized that the concepts of theoretical science ("categories") are not "unalterable"; they are by no means necessary ("conditioned by the nature of the understanding"), but are "(in the logical sense) free conventions" (Einstein 1949b, 674). Accordingly, what might be termed the *negative moment* of Einstein's Kant targets the orthodoxy of the Transcendental Analytic, and the Transcendental Aesthetic that it presupposes. To the extent that these doctrines are regarded as comprising the core of Kant's theory of cognition, as logical empiricists such as Schlick (1920) and Reichenbach (1922) affirmed, there could be but one attitude to Kant in the aftermath of general relativity: *utter rejection*.

   Still, the real significance of Einstein's remarks on Kant in Paris lies in their implied dissent to pure empiricism or positivism, a criticism contained in the statement that "science requires, for its construction, arbitrary concepts"; here "arbitrary" is an ellipsis for "logically arbitrary" which, in Einstein's somewhat peculiar epistemological vocabulary, has the meaning "not derived from sense experience." That physical theory requires freely posited concepts or "free conceptual construction" (Einstein 1949a, 49) is a long-recognized tenet of Einstein's epistemology, one that *prima facie* appears to be rooted in conventionalism, in the sense of Poincaré. And, though not in Paris, it is just here, in an insistence that "all concepts ... are from the point of view of logic free conventions" (Einstein 1949a, 13; translation modified slightly), that Einstein locates his departure from Kant – that is to say, from the doctrine of *a priori* categories in the Transcendental Analytic. But as noted in Section 3, it is to the Transcendental Dialectic that we must turn if we are to discern Einstein's positive appreciation of Kant. In this *positive moment*, two nonconventionalist components stand out quite distinctly. First, there is an abiding concern with the "justification" (*Berechtigung*) of the use of theoretical concepts in physics neither analytic nor derived from sense experience. Second, is the supreme importance of the idea of systematic unity in Einstein's conception of physical theory. The relationship Einstein affirmed between these two aspects can be brought out in view of our previous discussion of the Transcendental Dialectic.

A prominent feature of Einstein's later writings on the epistemology of science points, no doubt in response to the new vogue of positivism among philosophers and quantum physicists beginning in the late 1920s, to the insufficiency of the empiricist thesis that all knowledge rests solely on the deliverances of the senses. Einstein counters empiricism's shortcoming with an emphasis that "reason" or "pure thought" is not an eliminable, albeit "'metaphysical'" (scare quotes always added by Einstein), factor in cognition. But as Einstein recognized, this presents an epistemological challenge: with what "justification" ("*Berechtigung*") does the physicist employ concepts that are neither analytic nor derived from sense experience? Kant, in the opinion of the late Einstein, at least provided a partly correct statement of the problem: "The following, however, appears to me to be correct in Kant's statement of the problem: in thinking we use, with a certain 'justification' such concepts, to which there is no access from the material of sense experience, if one considers the matter from the logical standpoint" (Einstein 1944a, 285–7; translation modified slightly). As Einstein went on to say, physico-mathematical concepts are "free creations of thought which cannot be inductively derived from sense experience" (ibid.). Now the mantra of concepts as "free creations of the human mind" is one of the most familiar components of Einstein's epistemology of science, one he repeatedly stressed against empiricists, from Russell to Reichenbach. But then if concepts (certain ones, surely, more than others) do lack empirical justification, just what other kind of justification might Einstein have had in mind?

His answer, triangulating between realism and conventionalism, is perhaps most fully stated in the course of an extended discussion covering several pages in the 1949 "Reply to Criticisms" (Einstein 1949b). Ostensibly engaging once more in battle against the positivist epistemology of many quantum theorists, Einstein articulated a viewpoint decidedly nonpositivist, but also nonrealist and nonconventionalist, by focusing on the justification of one such concept, that of "the real" or "being," whose extension includes the "not observable." Derided by positivists as a metaphysical excrescence on the fabric of science, Einstein proceeded to give reasons for considering the concept not merely convenient, but essential.

"Being" is always something which is mentally constructed by us, that is, something which we freely posit (in the logical sense). The justification of such constructs does not lie in their derivation from what is given by the senses....

The justification of the constructs which represent "reality" for us, lies alone in their quality of making intelligible what is sensorily given (the vague character of this expression is here forced upon me by my striving for brevity). (Einstein 1949b, 669)

Einstein's striving for brevity should not occlude two important points stated here. First, "being" – in this context a clear reference to the portrayal of physical reality in theoretical physics – is *always* a "mental construction" (i.e., "conceptual construction for grasping the interpersonal," in the quotation beginning this chapter), freely posited by the theorist. Of course, this admission distances Einstein from realism, whether a "rationalistic realism" or scientific realism more generally, since the core of realism far outstrips an avowal of a mind-independent external reality – a commandment of sanity violated only by Berkeley and some seemingly like-minded quantum theorists. Rather, any realism about physical theory worthy of the title must in addition hold firm to a conception that our best physical theories are at least *approximately* true, that only *à la façon de parler* might they be considered "mental constructions," in the rather trivial sense that it takes a theorist to write down a theory, just as it takes a photographer to snap a portrait. But for the realist, the theorist/photographer is irrelevant except for "stylistic" details, while the ensuing portrait is not a "mental construction" at all for it corresponds to, maps, or represents in literal fashion, a mind-independent reality.

Second, the passage affirms that the *only* justification conceptual constructs representing "being" or "reality" *can have* is that they serve to make our sense experience "intelligible." It cannot be, as conventionalism prototypically maintains, that there is "no fact of the matter" about which concepts are correct, or that justification is the subjective matter of choosing concepts that pragmatically prove *most convenient* or *commodious* in managing experience. It is "intelligibility" that matters, and for Einstein, intelligibility required that conceptual constructs of fundamental physical theory manifest to the greatest extent possible the unity of systematic interconnection of an ideal order of nature, an ideal projected by the hypothetical employment of reason. It is just here that Einstein points to "the truly valuable in Kant," that the real is not "*given*" but "*aufgegeben*" – posed as a problem (for theoretical construction). As noted in the epigram to this chapter, he paraphrases this to mean: "There is such a thing as a conceptual construction for the grasping of the interpersonal, the authority of which lies purely in its validation. This conceptual construction refers precisely to the 'real' (by definition), and every further question concerning the 'nature of the real' appears empty" (Einstein 1949b, 680). To the extent that this conceptual construction enables us "to grasp experiences intellectually," it

can be regarded as "'knowledge of the real'": "Insofar as physical thinking justifies itself, in the more than once indicated sense, by its ability
to grasp experiences intellectually, we regard it as 'knowledge of the
real'" (Einstein 1949b, 673–4).

Einstein then comments that this sense of the "real" in physics "is
to be taken as a type of program, to which we are, however, not forced
to cling *a priori.*" And he concludes:

The theoretical attitude here advocated is distinct from that of Kant only by the
fact that we do not conceive of the "categories" as unalterable (conditioned by
the nature of the understanding) but as (in the logical sense) free conventions.
They appear to be *a priori* only insofar as thinking without the posit of categories
and concepts in general would be as impossible as is breathing in a vacuum.
(Einstein 1949b, 674)

There is little doubt that these late philosophical pronouncements, and
others (e.g., Einstein 1936) stressing the role of reason in creating the
conceptual structure within which empirical phenomena find representation but which structure in turn *can* only be justified by the possibility of such a representation, are not merely *ex cathedra* but reflect
Einstein's own mature appreciation that progress in unifying physical
theory rested not only on new physical and mathematical ideas but also
on formulating an epistemology appropriate to an ever-growing distance
between observation and the fundamental mathematical concepts and
relations.

5. PROJECTING THE ORDER OF NATURE

As he freely and repeatedly admitted (e.g., Einstein 1936, 1944a, 1949a,
1949b), Einstein's final philosophical vantage point developed as the
product of a long preoccupation with what he delicately termed "the
present difficulties" (Einstein 1944a, 279) of physics. These "difficulties" are presumably a reference to the two principal concerns of his last
three decades: (1) his critique of the quantum theory as incomplete, and
his related attempt to counter the positivism of the quantum theorists;
and (2) his struggles with unified field theory, which epistemologically
speaking, presented new concerns on account of the tenuous and indirect connection between fundamental concepts and possible observable
evidence. Dovetailing philosophical commitments emerged from these
twin battles. As seen previously, Einstein stressed the legitimacy of a
concept of "the real" that necessarily contains the presumption of comprehensibility by the human mind: that nature is intelligible. Arguing
for the incompleteness of the quantum theory, he stressed locality (or,
as we shall see, "separability") and determinism as core aspects of the

intelligible real in the projected order of nature. But a considerably more filled out picture emerges from his quest for unified field theory. Here, above all, we see that Einstein's hypothetical projection of the intelligible real is that of a nonlinear, continuum-based field theory whose laws are generally covariant. Einstein's repugnance to a back-ground space-time was not a passing fancy; in his later life Einstein still regarded the generalized principle of relativity as not only the "focal point" (*Schwerpunkt*) of the theory of general relativity, but also as an indispensable cornerstone of any projected unified field theory.[16] This has some reasonably well-known consequences in restricting the choice among possible theories and it has certain desirable implications, such as the fact that the existence and form of the equations of motion are a direct consequence of the covariant character of the field equations.[17] But in addition, there are some far-reaching ramifications for the ensuing conception of physical reality.

Now, as was argued in Section 2, for Einstein the fundamental point about the requirement of general covariance, understood in its broadest sense, may be concisely expressed: since there can be no principled distinction between space-time structure and the "contents" of space-time, any *a priori* reasonable dynamical theory must be formulated without reference to a background space-time. This holds *a fortiori* for any theory of "the total field" (Einstein 1949a, 89). However, in such a clipped formulation, it is easy to overlook the striking transformation such an injunction entails in the usual conception of the individuation and identity of physical objects. For example, in a "consistent field theory" implementing the unified field theory program of "representing reality by everywhere continuous, indeed even analytic functions," the very notion of "'particle' does not exist in the strict sense of the term" (Einstein to Besso, April 15, 1950, in Einstein and Besso 1972, 257). For Einstein, this new situation called for some kind of principle of individuation in lieu of the usual identification of physical systems by reference to a fixed background of space and time (e.g., "the particle located at such-and-such coordinates").

Perhaps following a suggestion Schlick had made fifteen years earlier, in a June 19, 1935, letter to Schrödinger (quoted in Fine 1996, 36), Einstein invoked "a principle of separation" (*Trennungsprinzip*) to this end. Schlick's concern in 1920 had been to respond to period objections that Einstein's general relativity was not compatible with the requirement of causality – that is, the existence of causal laws acting contiguously from point to point.[18] He observed that causality in this sense presupposes a principle of separation (*ein Prinzip der Trennung*) according to which "occurrences can be *alike* without being *identical*." Such a principle of separateness in nature must keep similar things apart from one another without in any way influencing them materially (Schlick 1920, 467). As Schlick noted, the same point had essentially already been affirmed by Maxwell, that differences between events depend not on their space or time coordinates, but only on "differences in the nature, configuration, or motion of the bodies concerned" (Schlick 1920, 468; quoting Maxwell 1876). With reference to the principle of general covariance, Schlick concluded that a principle of separation, whereby like things or events in the world can be individuated (thus objectified) without relying on physically meaningless coordinates, is a presupposition of causal laws of nature. This same concern over individuation of physical systems by causal separation will resurface in Einstein's criticism of the quantum theory, an objection that centers on the quantum theory's violation of the separation principle's criterion of objectification, according to which we may speak of spatially separated physical systems as distinct.

As documented by Fine (1996), Einstein's correspondence with Schrödinger in the period immediately following publication of the famous EPR paper (Einstein, Podolsky and Rosen 1935) explicitly brings out an assumption on which the EPR argument relies: a "principle of separation" (*Trennungsprinzip*) affirming the independent existence of what is spatially separated. In the EPR context, separation also embodies the idea of action-by-contact, that is, "the claim that whether a physical property holds for one of the particles does not depend on measurements (or other interactions) made on the other particle when the pair is widely separated in space" (Fine 1996, 37). Although the language here is that of measurement, or rather of what has been called *outcome-independence* of measurements, the underlying idea is clear enough: outcome-independence in the two wings of the EPR situation presupposes a conception of physical reality structured by a principle of separation, such that a measurement made on one of two *separated* (in space) physical systems can in no way have an immediate influence on the other physical system. For Einstein, a principle of separation serves as a principle of physical objectification, by which one can distinguish whether there are *two* physical systems and not just one. Returning to the programmatic context of a "consistent field theory," the separation principle may be interpreted to mean: field magnitudes "at A" (whose values individuate the region A) comprise an individual physical system, the properties of which exist (have physical reality) whether or not a simultaneous measurement is made of field magnitudes "at B," distantly separated (spacelike separation) from A. Individuation through spatial separation no doubt implies a principle of locality, a physical injunction against superluminal causal "disturbances," but it *is* a criterion of objectification – for example, of the individual physical system "at A" (i.e., the physical reality of A). One might call this the *ethos* of any field theory, and it is the core conception behind the infamous "EPR criterion of reality": "If, without in any way disturbing a system, we can predict with certainty (i.e., with probability equal to unity) the value of a physical quantity, then there exists an element of physical reality corresponding to this physical quantity" (Einstein, Podolsky, and Rosen 1935, 777). Thus the principle of separation implicit in the EPR paper, brought out explicitly in Einstein's discussions related to it, attempts to formulate, in informal "particle" terms, the field theoretic presupposition regarding "what the external world must be like" – if the general theory of relativity (and more particularly its generalization in unified field theory) is on, or near, the right track.[19] What made the development from the general theory of relativity seem natural to Einstein is the cluster of notions associated with what he broadly called "general covariance" for this includes a criterion of objectivity or physical reality which shows that the latter is not the expression of a simple-minded realism ("whether something one cannot know anything about exists all the same") but rather a presupposition for the application of causal laws in the physical description of the world. Of course, quantum theory is a nonlocal theory and in this regard nature appears not to be bound to the projected order of nature of Einstein's "consistent field theory." Yet it must be remembered that this projected order can only be a *hypothetical order*, one that guides and instructs but cannot compel. Einstein recognized this in many places, including in what Stachel has described as his last published words:

One can give good reasons why reality cannot be represented by a continuous field. From quantum phenomena it appears to follow with certainty that a finite system of finite energy can be completely described by a finite set of numbers (quantum numbers). This does not seem to be in accordance with a continuum theory, and must lead to an attempt to find a purely algebraic theory for the description of reality. But nobody knows how to obtain the basis of such a theory.[20]

6. CONCLUSION

We have argued that the contesting currents of Einstein's "believing rationalist" can best be understood through the prism of what he had come to regard as "the truly valuable in Kant." In choosing to identify this with the cryptic catechism *nicht gegeben, sondern aufgegeben*, Einstein cannot be thought unaware of its import in the context of the regulative use of reason as characterized in the Transcendental Dialectic. Nor was Einstein alone in making this identification; philosophers of widely differing neo-Kantian tendencies such as Heinrich Rickert, Ernst Cassirer, and Josef Winternitz also found the very kernel of transcendental idealism in the stenographic phrase. Other Kant-inspired thinkers, of whom the most notable was Hermann Weyl, saw in these words an expression of the concession that theoretical physics must make to idealism.[21] What conceivable relevance might this have today? Here is but one suggestion. Unifiers in contemporary physics can and do have quite divergent assessments of the lasting value of Einstein's own quest for a unified field theory. But insofar as a legacy of unification in physics still linked to Einstein survives, the epistemology associated with his own attempts at unification might be better, and more productively, understood.

NOTES

1 Page references to the reprint in Holton (1988). To be sure, Holton also identifies nonpositivist aspects in the philosophy of the early Einstein but considers these influences subordinate.

2 Einstein to Lanczos, March 21, 1942, as cited and translated in Lanczos (1998, 2–1526, note 9).

3 "One may say 'the eternal mystery of the world is its comprehensibility.' It is one of the great realizations of Immanuel Kant that the postulation of a real external world would be senseless without this comprehensibility" (Einstein 1936, 292). The page number refers to the reprint in Einstein (1983).

4 Miller (2000, 205) cites this as evidence of "Einstein's Platonic tendencies." Janssen (2002a, 510) cites a passage from another 1918 text in which Einstein expressed the diametrically opposite viewpoint, arguing that "the intuitive preference of the scientist" provides the only criterion for choice between two theories "compatible with all the data."

5 The standard convention of referring to the 1781 edition of The Critique of Pure Reason as "A," the 1787 edition as "B," is followed.

6 Cf. Buchdahl (1969, 508). In Eddington's formulation, the sense is almost identical to Kant's: "In the end, what we comprehend about the universe is precisely that which we put into the universe to make it comprehensible" (1936, 328).

7 Einstein admitted as much (1918e, CPAE 7, Doc. 4). The principles of general relativity and of general covariance are not coextensive; see Section 3 of Chapter 6, Norton (2000, 145, note 17), and Renn and Sauer (2007, 301–2). Dieks (2006) provides a careful assessment of the significance of the differences.

8 For a facsimile reproduction of the notebook, with transcription and commentary, see Renn (2007a, Vols. 1–2, 314–714).

9 Einstein to George Jaffe of January 19, 1954, quoted by Stachel (2002a, 143).

10 See also Section 6 of Chapter 6.

11 Einstein (1923e, 448): "The search for the mathematical laws which shall correspond to the laws of nature thus resolves itself into the answer to

the question: What are the formally most natural conditions that can be imposed upon an affine relation?"

12 The full passage reads: "The characteristics which especially distinguish the General Theory of Relativity and even more the new third stage of the theory, the Unitary Field Theory, from other physical theories, are the degree of formal speculation, the slender empirical basis, the boldness in theoretical construction, and finally the fundamental reliance on the uniformity of natural laws and their accessibility to the speculative intellect" (Einstein 1929, 114). This is one of several remarks on the epistemological novelty posed by unified field theory, made in the context of the theory based on the concept of "distant parallelism."

13 Arguments for this reading of Kant are found in Buchdahl (1969), Allison (2004), and Bird (2006).

14 Buchdahl (1969), Neiman (1994), and Allison (2004).

15 See, e.g., the letter to Moritz Schlick dated December 14, 1915 (CPAE 8, Doc. 165) or the undated letter to Max Born from the summer of 1918 (CPAE 8, Doc. 575).

16 Letter of January 17, 1955, of Einstein to Leopold Infeld, quoted in Infeld (1978, 152, 201). The message is prominently announced in the first sentence of Einstein (1945, 578): "Every attempt to establish a unified field theory must start, in my opinion, from a group of transformations which is no less general than that of the continuous transformations of the four coordinates."

17 Einstein (1949b, 675–6): "The theory of gravitation showed me that the non-linearity of these equations results in the fact that this theory yields interactions among structures (localized things) at all. But the theoretical search for non-linear equations is hopeless (because of too great variety of possibilities) if one does not use the general principle of relativity (invariance under general continuous co-ordinate transformations). In the meantime, however, it does not seem possible to formulate this principle, if one seeks to deviate from the above program. Herein lies a coercion which I cannot evade. This for my justification." See also Bergmann (1949) and, on the restrictive character of the requirement of general covariance, see Brown and Brading (2002).

18 In particular, Schlick (1920) analyzed Einstein's (1916e, Sec. 2) explanation of the cause of the difference in shape of the two relatively rotating bodies. Drawing upon what Schlick (erroneously) considered a general relativity of motions established by Einstein's theory of gravitation, Schlick inferred that in this theory the kinematic and dynamical concepts of motion coincided, and moreover, the theory demonstrated that an ontology of events and processes replaced classical physics' ontology of substances endowed with qualities. Since the concepts of cause and effect were applicable, Schlick argued, only to processes, he concluded that Einstein's theory was not only consistent with the causal principle but also extended the domain of causal explanation.

19 Einstein (1948b, 321) notes the origin in "everyday thought" of the notion that things existing at a given time but lying "in different parts of space" are independent of one another, observing that "the field theory has carried through this principle to the furthest extent (*Die Feldtheorie hat dieses*

Prinzip zum Extrem durchgeführt) by localizing the elementary objects on which it is based and which exist independently of each other, as well as the elementary laws which have been postulated for it in the infinitely small (four-dimensional) elements of space." This expresses what we have termed the "ethos" of field theory.

20 Einstein (1956a, 166, the closing paragraph of an appendix rewritten for this edition of the book); also cited by Stachel (1993, 288).

21 Cassirer (1918, 320), termed this distinction "the transcendental insight": "that the 'absolute' is not so much 'given' as 'set as a task.'" Heinrich Rickert, the leading philosopher of the so-called southwest school of neo-Kantianism, defined the object of knowledge in transcendental idealism in just these terms as well: "For the transcendental idealist, the object of knowledge is ... neither immanent nor transcendently 'given' (*gegeben*), but rather 'set as a task' (*aufgegeben*)" (Rickert 1921, 316). Josef Winternitz, son of a friend and colleague of Einstein's from his Prague days and a young associate of Einstein in Berlin in the 1920s (see, e.g., Infeld 1941, 96–100), was the author of a neo-Kantian book on general relativity. Winternitz (1923, 201–2) deemed the contrast to be "the essential in the Kantian view" and the "most general, highest principle of transcendental philosophy." This has particular resonance as Einstein (1924d) wrote a review of Winternitz's book appearing the next year. Finally, in a well-known passage, Weyl (1927, 117) observed: "Science concedes to idealism that its objective reality is not given but set as a task (*nicht gegeben, sondern aufgegeben*), and that it cannot be constructed absolutely but only in relation to an arbitrarily assumed coordinate system, and in mere symbols."

13 Space, Time, and Geometry

1. INTRODUCTION

Einstein's theories of relativity – especially the general theory – exerted a profound influence on twentieth-century philosophy of geometry, and this story begins (as do so many episodes in twentieth-century thought) with the refutation of Kant. One of the pillars of the Kantian system was the idea that the science of Euclidean geometry, in particular, is paradigmatic of synthetic *a priori* truth: truth that is certain, necessary, and known independently of any particular experience but, nonetheless, genuinely descriptive of and applicable to physical or empirical reality. Kant concluded, accordingly, that Euclidean geometry is not a merely logical or analytic science arising in the pure understanding or pure intellect, but is rather an essentially nonlogical or synthetic science articulating the structure of what he calls the pure sensible form of our spatial intuition or perception. This pure form of sensibility – what Kant calls the pure intuition of space – is independent of any particular spatial sensation or empirical physical object that may be given to us, yet, at the same time, it is still necessarily applicable to any perceivable spatial object. As Kant puts it in the *Critique of Pure Reason* (Kant 1781/1787, A165/B206): "Empirical intuition is only possible through pure intuition (of space and time); hence what geometry says of the latter is unquestionably valid of the former."[1] All physical objects are in space; but space, first and foremost, is the object of the mathematical science of Euclidean geometry; therefore, Euclidean geometry (which, for Kant and virtually everyone else in the seventeenth and eighteenth centuries, is obviously certain and *a priori*) is necessarily descriptive of and applicable to the behavior of all physical objects.

Kant's conception of the relationship between mathematical geometry and its application to objects of experience made particularly good sense in the context of the science of the seventeenth and eighteenth centuries – the age of the triumph of Newtonianism. Euclidean geometry was the only mathematical science of space then in existence, and Newton had unquestioningly applied this geometry in his theory of universal gravitation throughout the solar system and beyond. The general theory of relativity, however, takes the revolutionary step of using a non-Euclidean geometry of variable curvature (depending on the distribution of mass and energy) to represent the phenomenon of gravitation, where idealized bodies or "test particles" ("freely falling" under no other influence than gravity) are not affected by an externally impressed gravitational force but rather follow geodesics or straightest possible paths in a variably curved four-dimensional space-time manifold. And this four-dimensional space-time geometry then has implications for three-dimensional, purely spatial geometry as well. One can visualize the purely spatial geometry of the solar system in two dimensions, for example, by imagining that the Sun is a massive ball placed on a rubber sheet – the resulting "Schwarzschild geometry" is spherically symmetric with increasing negative curvature as one radially approaches the central Sun. Gravity is not an external force acting on the planets (now viewed as comparatively much less massive test particles), but is rather a direct manifestation of the strongly non-Euclidean variable curvature increasing in the neighborhood of the Sun: only at great distances from the Sun does the geometry (the rubber sheet) approach a state of Euclidean flatness or zero curvature.

Early twentieth-century logical empiricism, as represented by Moritz Schlick, Hans Reichenbach, and Rudolf Carnap, took this revolutionary application of non-Euclidean geometry to physics (the very first such application) to be a decisive refutation of the Kantian philosophy of geometry – and, by implication, a decisive refutation of the Kantian philosophy of the synthetic a priori more generally.[2] Not only are non-Euclidean mathematical geometries perfectly possible or consistent from a logical point of view, but Einstein has shown that one such geometry is applicable to nature. Euclidean geometry is not true of the actual empirical spatial world in which we live, much less necessarily or a priori true, and so Kant's conception of the synthetic a priori status of geometry is obviously false. Indeed, this fundamental anti-Kantian move was in fact the starting point of the philosophy of logical empiricism. Schlick published a very well-known exposition and defense of Einstein's new theory, Space and Time in Contemporary Physics, which went through four editions between 1917 and 1922. Largely on the basis of this work, which stressed the radical implications of the new perspective on the philosophy of geometry opened up by general relativity, Schlick was then appointed (apparently with some help from Einstein) to the Chair for the Philosophy of the Inductive Sciences at the University of Vienna in 1922.[3] A philosophical circle quickly formed around Schlick, and what we now know as the Vienna Circle of logical positivism or logical empiricism was born.

Logical empiricism did not rest content, however, with showing, by appeal to Einstein's creation of the general theory of relativity, that Kant's doctrine of the synthetic *a priori* character of geometry is false, it also developed a new philosophy of geometry intended, at least in part, to explain why Kant had mistakenly believed that geometry is, at the same time, both *a priori* and applicable to physical or empirical reality. We must sharply distinguish, on this new conception, between pure or mathematical geometry, which is an essentially uninterpreted axiomatic or deductive system making no reference whatsoever to spatial intuition or any other kind of extra-axiomatic or extraformal content, and applied or physical geometry, which then attempts to coordinate such an uninterpreted formal system with some domain of physical facts given by experience: we may coordinate the uninterpreted terms of pure mathematical geometry with light rays, stretched strings, or rigid bodies, for example. Given such a particular interpretation or "coordinative definition" (in Reichenbach's terminology), a system of axiomatic geometry (Euclidean or non-Euclidean) is then empirically true or false of the objects to which it is coordinated, but this kind of empirical or physical truth cannot possibly be *a priori*. By contrast, purely mathematical or deductive geometry is certain and *a priori*, but this pertains only to an uninterpreted, purely formal system where all we have really established is that *if* the axioms are true (under a certain interpretation) *then* the theorems are true (under this same interpretation) as well. That the axioms are in fact true of the physical or empirical world (under a certain interpretation or coordinative definition) can only be established empirically. Thus, mathematical geometry is certain and *a priori*, but it is also purely logical or analytic; physical geometry is synthetic, but it is in no sense *a priori*. Kant's fundamental mistake was precisely to confound the two.

Reichenbach articulates this new philosophy of geometry especially clearly and explicitly in a popular exposition of the "results" of logical empiricism, *The Rise of Scientific Philosophy*, published in 1951:

This consideration shows that we have to distinguish between mathematical and physical geometry. Mathematically speaking, there exist many geometrical systems. Each of them is logically consistent, and that is all a mathematician can ask. He is interested not in the truth of the axioms, but in the implications between axioms and theorems: "if the axioms are true, then the theorems are true" – of this form are the geometrical statements made by the mathematician. But these implications are analytic; they are validated by deductive logic. The geometry of the mathematician is therefore of an analytic nature. Only when the implications are broken up, and axioms and theorems are asserted separately, does geometry lead to synthetic statements. The axioms then require an interpretation through coördinative definitions and thus become statements about physical objects; and geometry is thus made into a system which is descriptive of the physical world. In that meaning, however, it is not a priori but of an empirical nature. There is no synthetic a priori of geometry: either geometry is a priori, and then it is mathematical geometry and analytic – or geometry is synthetic, and then it is physical geometry and empirical. The evolution of geometry culminates in the disintegration of the synthetic a priori. (Reichenbach 1951, 139–40)

Reichenbach's penultimate sentence is a self-conscious allusion to a famous address by Einstein on the relationship between general relativity and the philosophy of geometry, *Geometry and Experience*, published in 1921.

In particular, in perhaps the most well-known passage of this famous paper Einstein puts the point this way:

In so far as the propositions of mathematics refer to reality they are not certain; and in so far as they are certain they do not refer to reality. Full clarity about the situation appears to me to have been first obtained in general by that tendency in mathematics known under the name of "axiomatics." The advance achieved by axiomatics consists in having cleanly separated the formal-logical element from the material or intuitive content. According to axiomatics only the formal-logical element constitutes the object of mathematics, but not the intuitive or other content connected with the formal-logical element. (Einstein, 1921c, 3–4)

And it is clear, in context, that Einstein's opening sentence is implicitly directed against the Kantian synthetic *a priori*, in precisely the same sense in which logical empiricist philosophy of geometry (as represented by Reichenbach) is also fundamentally opposed to Kant. The connection between the philosophy of geometry characteristic of twentieth-century logical empiricism and Einstein's creation of (and reflections upon) the general theory of relativity therefore goes very deep. As we shall see, however, this connection is by no means as simple and straightforward as it first appears.[4]

2. EINSTEIN, SCHLICK, AND POINCARÉ

We noted that Schlick was appointed to the University of Vienna in 1922, largely on the strength of his work as an expositor and apologist for Einstein's new theories of relativity, and that Einstein's own approval of this work probably played a significant role here. Indeed, Einstein had worked closely with Schlick in the composition of *Space and Time in Contemporary Physics*, having read and commented upon several different drafts (see Howard 1984). One of the most important philosophical lessons Schlick infers from Einstein's discoveries is that the variably curved space-time of general relativity, in particular, must be viewed as an entirely abstract, entirely nonintuitive "conceptual construction," which can only be related to experience and the physical world by a similarly abstract, entirely nonintuitive relation of "designation" or "coordination."[5] In his *General Theory of Knowledge*, written virtually simultaneously with *Space and Time in Contemporary Physics*, and first appearing in 1918, Schlick extends this epistemological conception to all of empirical science. We must sharply distinguish, on this view, between abstract formal *concepts* [*Begriffe*], which receive their meaning only through their logical roles within a given axiomatic or deductive system, and concrete intuitive *images* [*Vorstellungen*], which are directly linked to sensible or perceptible contents. The rigor and objectivity of scientific concepts quite generally depend precisely on their purely formal, nonintuitive character – on the logical meaning bestowed on them through the "implicit definitions" set up by the axioms of a given formal system. Such a system of implicit definitions then receives empirical content by a formal correspondence relation Schlick terms designation or coordination, as the uninterpreted terms of physical geometry, for example, receive their interpretation in terms of physical structures or process (such as light rays, stretched strings, or rigid bodies). And the model for this notion of implicit definition, as Schlick makes amply clear, is David Hilbert's modern axiomatic formulation of Euclidean geometry, first appearing in 1899.[6]

It is no wonder, then, that Einstein, at the beginning of *Geometry and Experience*, articulates a virtually identical conception of axiomatic geometry and explicitly refers, in particular, to Schlick's work in *General Theory of Knowledge*:

Geometry treats of objects that are designated with the words line, point, etc. No kind of acquaintance or intuition of these objects is presupposed, but only the validity of those axioms which are likewise to be conceived as purely formal, i.e., as separated from every content of intuition and experience.... These axioms are free creations of the human spirit. All other geometrical propositions are logical consequences of the (only nominalistically conceived) axioms. The axioms first define the objects of which geometry treats. Schlick therefore designated the axioms very appropriately as "implicit definitions" in his book on theory of knowledge.

The conception represented by modern axiomatics purifies mathematics from all elements not belonging to it, and thus removes the mystical obscurity that previously clung to the foundations of mathematics. But such a purified presentation makes it also evident that mathematics as such may assert nothing about either objects of intuitive representation or objects of reality. In axiomatic geometry we understand by "point", "line", etc. only contentless conceptual schemata. What gives them content does not belong to mathematics. (Einstein 1921c, 4–5)

So it is clear, in context, that "modern axiomatics" refers to that tendency of thought brought to fruition by Hilbert and also, as we have already suggested, that one of the most important elements of "mystical obscurity" thereby purified from mathematics is the Kantian conception of the synthetic *a priori* character of geometry based on pure intuition.

Einstein's paper also presents a particular conception of physical as opposed to purely mathematical geometry – a particular view of what a "physical interpretation" of mathematical geometry amounts to. We obtain such an interpretation, according to Einstein, when we relate the axiomatic structure of pure geometry to "practically rigid bodies" and their "situational possibilities [*Lagerungsmöglichkeiten*]." More specifically, by bringing intervals marked on such bodies (e.g., in the form of rigid measuring rods) into coincidence with one another, we can empirically determine relations of distance and thereby obtain a physical interpretation of "congruence." In this way, we obtain what Einstein calls "practical" as distinct from "pure axiomatic geometry":

It is clear that the conceptual system of axiomatic geometry alone can supply no assertions about the behavior of those objects of reality that we wish to designate as practically rigid bodies. In order to be able to furnish such assertions geometry must go beyond its solely formal-logical character, so that experienceable objects of reality (experiences) are coordinated [*zugeordnet*] to the empty conceptual schemata of axiomatic geometry. In order to accomplish this one needs only to add the proposition:

Solid bodies relate to one another with respect to their situational possibilities as do bodies in three-dimensional Euclidean geometry; then the propositions of Euclidean geometry contain assertions about the behavior of practically rigid bodies.

The thus expanded geometry is obviously a natural science; we can actually consider it as the oldest branch of physics. Its assertions rest essentially on inductions from experience, and not only on logical inferences. We wish to call the thus expanded geometry "practical geometry" and to distinguish it in what follows from "pure axiomatic geometry." The question whether the practical geometry of the world is Euclidean or not has a clear sense, and its answer can only be supplied by experience. All length measurement in physics is practical geometry in this sense – geodetic and also astronomical, if one uses the empirical proposition that light propagates in a straight line, precisely a straight line in the sense of practical geometry. (Einstein, 1921c, 5–6)

Practical geometry coordinates the "empty conceptual schemata" of pure geometry with real physical bodies (practically rigid bodies) and thereby transforms geometry into an interpreted empirical science. And, when this interpretation is actually put into effect, we know, according to precisely the general theory of relativity, that Euclidean geometry turns out to be empirically false.

So far, then, Einstein and Schlick appear to be entirely in agreement, a circumstance that is further underscored by Einstein's use of Schlick's idea that geometry receives a physical interpretation insofar as empirical objects are "*coordinated* to the empty conceptual schemata of axiomatic geometry" (my emphasis). Just here, however, we also encounter a fundamental divergence between the two. For Schlick maintains that there is always an essentially arbitrary or conventional element in this process of coordination, and, in particular, that what Reichenbach had called *coordinative definitions* should be conceived as arbitrary stipulations or conventions in the sense of Henri Poincaré. Poincaré had argued that the question whether Euclidean or non-Euclidean geometry should be applied to nature is not a matter of fact but rather a matter of convention, like the choice of a system of units or coordinate system.[7] Accordingly, since Euclidean geometry is mathematically the simplest, it is always to be preferred; and experimental phenomena appearing to indicate that space might be non-Euclidean are always to be explained by physical rather than geometrical hypotheses. Non-Euclidean behavior of apparently rigid bodies, for example, should rather be explained by a disturbing field of force in virtue of which such bodies are not truly rigid after all. Indeed, what can we possibly mean by "rigid" if we do not have a geometry (and therefore a notion of same size or congruence) already in place? Although Schlick does not agree that Euclidean geometry is always to be preferred, he does agree with Poincaré that the choice between Euclidean and non-Euclidean geometry is a matter of convention not a matter of fact – and he does agree, in particular, with Poincaré's critique of the concept of rigid body. The reason non-Euclidean geometry is preferable in the case of Einstein's general theory of relativity is not that rigid bodies have empirically been found to behave in this way, but rather that our total system of geometry plus physics (including our theories of the forces affecting the internal constitution of what we take to be rigid bodies) thereby becomes mathematically simpler – just as Poincaré had previously argued in the case of (Euclidean) geometry alone.[8]

Einstein's own position in *Geometry and Experience*, by contrast, is that the notion of "practically rigid body" must be taken (at least provisionally) as fundamental and primitive in the physical interpretation of mathematical geometry. And, even more strikingly, he says that he "attach[es] particular importance to this conception of geometry, because without it I would have found it impossible to establish the theory of relativity." Finally, Einstein takes the only live *alternative* to his point of view to be precisely Poincaré's conventionalism:

If one rejects the relation between the bodies of axiomatic Euclidean geometry and the practically rigid bodies of reality, then one easily arrives at the following conception, which the perceptive and deep thinker H. Poincaré has, in particular, embraced. Euclidean geometry is picked out from all other thinkable axiomatic geometries by simplicity. And since axiomatic geometry *alone* contains no assertions about experienceable reality, but only axiomatic geometry in connection with propositions of physics, then it may be possible and rational – no matter how reality may be constituted – to hold fast to Euclidean geometry. For one will prefer to opt for an alteration of the laws of physics rather than alter the laws of geometry in case there are contradictions between theory and experience. If one rejects the relation between the practically rigid body and geometry one will in fact not easily free oneself from the convention according to which Euclidean geometry is to be held fast as the simplest. (Einstein 1921c, 7)

Thus, Poincaré's opposing conventionalist conception of geometry must be rejected, for Einstein, because it then becomes impossible to see how one could empirically discover that space in fact has a non-Euclidean geometry.

Einstein also admits, however, that there is considerable justice in Poincaré's critique of the concept of rigid body: "the actual rigid bodies of nature, on closer consideration, are not rigid, because their geometrical behavior – i.e., their relative situational possibilities – depend on temperature, external forces, etc.," as a result of which "[t]he original, immediate relation between geometry and physical actuality appears thereby to be destroyed." Nevertheless, although Poincaré, from this point of view, is strictly speaking correct, it is still necessary, at least provisionally, to take the concept of rigid body as primitive in the current state of physics:

Sub specie aeterni Poincaré, in my opinion, is correct in this conception. The concept of measuring rod, and also the concept of clock that is coordinated to it in relativity theory, has no exactly corresponding object in the actual world. It is also clear that the rigid body and clock do not play the role of irreducible elements in the conceptual framework of physics, but rather the role of composite structures that cannot have any independent status in the construction of theoretical physics. But it is my conviction that these concepts must still be called upon as independent elements in the present stage of theoretical physics; for we are still far from such a secure knowledge of the theoretical foundations that would enable us to give exact theoretical constructions of such structures. (Einstein 1921c, 7–8)

For Einstein, therefore, Poincaré's critique of the concept of rigid body, in the context of Einstein's own construction of the general theory of relativity, must be (at least at present) ignored; and it is for precisely this reason that Einstein here adopts a naively empiricist conception of the application of geometry to physics diametrically opposed to the sophisticated conventionalist conception defended by Poincaré.

Now Schlick, as we have seen, attempts to harmonize Einstein and Poincaré by extending Poincaré's conventionalist understanding of the choice of Euclidean geometry in particular to our total system of geometry plus physics. Just as Poincaré had argued that the choice of Euclidean geometry is conventionally chosen (like a given coordinate system) on the basis purely of its greater mathematical simplicity, Schlick now argues that general relativity is similarly conventionally chosen on the basis of its greater mathematical simplicity as a total system of geometry plus physics – as opposed, for example, to a perfectly possible Euclidean alternative in which the apparently non-Euclidean behavior of rigid bodies is explained by a distorting field of force.[9] But this position, too, is completely at odds with what Einstein says in *Geometry and Experience*. Immediately preceding the passage last quoted, Einstein explains the view he now wants to *reject* as follows:

The original, immediate relation between geometry and physical actuality appears thereby to be destroyed [by Poincaré's critique of rigidity], and we feel ourselves forced to the following more general conception, characteristic of Poincaré's standpoint. Geometry (G) asserts nothing about the behavior of actual things, but only geometry together with the totality (P) of physical laws. We can say, symbolically, that only the sum (G) + (P) is subject to the control of experience. So (G) can be chosen arbitrarily, and also parts of (P); all of these laws are conventions. In order to avoid contradictions it is only necessary to choose the remainder of (P) in such a way that (G) and the total (P) together do justice to experience. On this conception axiomatic geometry and the part of the laws of nature that are elevated to conventions appear as epistemologically of equal status. (Einstein 1921c, 7)

Thus, whereas for Schlick our conception of rigid body is to be absorbed into a conventionally chosen part of our total system of geometry plus physics, for Einstein the concept of "practically rigid body" remains primitive and "naive" – whereby it can then yield an entirely nonconventional determination of the unique mathematical geometry that empirically fits our actual experience.

3. EINSTEIN'S INTRODUCTION OF NON-EUCLIDEAN GEOMETRY AND THE NINETEENTH-CENTURY TRADITION

In order to make sense of Einstein's argument in *Geometry and Experience*, and its relationship, in particular, to Poincaré's geometrical conventionalism, we need to consider this argument against the background of a preceding conception of geometry – one that was dominant in the late nineteenth century but has now, largely through the influence of "modern" views like Einstein's, receded far into the background in contemporary philosophical discussion. This earlier tradition had its home in projective geometry and group theory, and it found its canonical expression in the famous Erlanger program of Felix Klein.[10] Here, as Klein puts it, pure or mathematical geometry is by no means conceived as an "empty conceptual schema," but is rather understood as necessarily connected to our "spatial intuition" (see Klein 1909, 383–4 [translation, 187]). Mathematical geometry, on this view, describes the most general and abstract features of our perception of space – namely, the "perspectival" features of space as we move in and through it and perceive spatial objects from different points of view. These features are not precise and specific enough to yield Euclidean geometry in particular, however, but only that structure common to the three classical geometries of constant curvature (Euclidean or zero curvature, hyperbolic or negative curvature, and elliptic or positive curvature), which, in terms of the Erlanger program, emerge naturally within the more general framework of projective geometry and group theory. So only considerations of convenience and expediency (especially simplicity), not deliverances of our spatial intuition, can then explain our choice of specifically Euclidean geometry. This view of geometry has obvious roots in the conception articulated and defended by Kant at the end of the eighteenth century, but it aims to *generalize* the Kantian picture to take account of nineteenth-century discoveries in projective and non-Euclidean geometry.[11]

For our present purposes, the most important result of this tradition is what we now call the Helmholtz-Lie theorem, which was first articulated by Hermann von Helmholtz in connection with his psycho-physiological researches into space perception and then rigorously proved by Sophus Lie within Lie's theory of continuous groups.[12] Helmholtz (1868) was inspired by Bernhard Riemann's celebrated habilitation on "n-fold extended manifolds" (Riemann 1867) to attempt to derive Riemann's fundamental assumption, that the line element or metric is Pythagorean or infinitesimally Euclidean, from what Helmholtz took to be the fundamental "facts" generating our perceptual intuition of space. Helmholtz's starting point was that our idea of space is in no way immediately given or "innate" but instead arises by a process of perceptual accommodation or learning based on our experience of bodily motion. Since our idea of space arises kinematically, as it were, from our experience of moving up to, away from, and around the objects that "occupy" space, the space thereby constructed must satisfy a condition of "free mobility" that permits arbitrary continuous motions of rigid bodies.[13] And from this latter condition one can then derive the Pythagorean form of the line element.[14] Since, however, the

Riemannian metric thereby constructed has a group of isometries or rigid motions mapping any point onto any other, it must have constant curvature as well.[15] So the scope of the Helmholtz-Lie theorem (and the entire Kleinian tradition) is much less general than the full Riemannian theory of metrical manifolds, which also includes manifolds of *variable* curvature.[16]

The Helmholtz-Lie theorem fixes the geometry of space – and, according to Helmholtz, thereby expresses the "necessary form of our outer intuition" – as one of the three classical geometries of constant curvature: Euclidean, hyperbolic, or elliptic. But how do we know which of these three geometries actually holds? At this point, on Helmholtz's view, we investigate the actual behavior of rigid bodies (e.g., of rigid measuring rods) as we move them around in accordance with the condition of free mobility. That physical space is Euclidean (which Helmholtz assumed) means that physical measurements carried out in this way are empirically found to satisfy the laws of this particular geometry to a very high degree of exactness. Thus Helmholtz's view is Kantian insofar as space has a "necessary form" expressed in the condition of free mobility, but it is empiricist insofar as which of the three possible geometries of constant curvature actually holds is then determined by experience.[17]

Now it was precisely this Helmholtzian view of physical geometry which set the stage, in turn, for the contrasting conventionalist conception articulated by Poincaré. Indeed, Poincaré developed his philosophical conception immediately against the background of the Helmholtz-Lie theorem, and in the context of his own mathematical work on group theory and models of hyperbolic geometry.[18] Following Helmholtz and Lie, Poincaré viewed geometry as the abstract study of the group of motions associated with our initially crude experience of bodily "displacements." So we know, according to the Helmholtz-Lie theorem, that the space thereby constructed has one and only one of the three classical geometries of constant curvature. Poincaré disagreed with Helmholtz, however, that we can then empirically determine the particular geometry of space simply by observing the behavior of rigid bodies. No real physical bodies exactly satisfy the condition of geometrical rigidity, and, what is more important, knowledge of physical rigidity presupposes knowledge of the forces acting on the material constitution of bodies. But how can one say anything about such forces without first having a geometry in place in which to describe them? We have no option, therefore, but to *stipulate* one of the three classical geometries of constant curvature, by convention, as a framework within which we can then do empirical physics.[19] Moreover, since Euclidean geometry is mathematically the simplest, Poincaré had no doubt at all, as we have said, that this particular stipulation would always be preferred.

We know that Einstein was intensively reading Poincaré when he was creating the special theory of relativity in 1905, and it seems very plausible, accordingly, that Poincaré's conventionalism played a significant role in philosophically motivating this theory.[20] More specifically, whereas Poincaré had argued, against both Kant and Helmholtz, that the particular geometry of space is not dictated by either reason or experience but rather requires a fundamental decision or convention of our own, Einstein now argues, similarly, that simultaneity between distant events is not dictated by either reason or experience but requires a new fundamental *definition* based on the behavior of light.[21] Moreover, Einstein proceeds here, in perfect conformity with Poincaré's underlying philosophy, by "elevating" (in Poincaré's terminology) an already established empirical fact – the invariant character of the velocity of light in different inertial reference frames – into the radically new status of what Poincaré calls a convention or "definition in disguise" (here a definition of simultaneity).

As we also know, however, Einstein tells us in *Geometry and Experience* that he needed to *reject* Poincaré's geometrical conventionalism in order to arrive at the general theory of relativity. In particular, Einstein here adopts a Helmholtzian conception of (applied or physical) geometry as a straightforward empirical theory of the actual physical behavior of "practically rigid bodies," and he claims, as we have seen, that "without [this conception] I would have found it impossible to establish the [general] theory of relativity." Here, as Einstein explains in the same passage, what he has in mind specifically is the following line of thought (see Einstein 1921c, 6–7): according to the principle of equivalence (based on the equality of gravitational and inertial mass) gravitation and inertia are essentially the same phenomenon. So, in particular, we can model gravitational fields by "inertial fields" (e.g., involving centrifugal and Coriolis forces) arising in noninertial frames of reference (accelerating and rotating frames of reference).[22] If we now consider a uniformly rotating frame of reference in the context of special relativity, we then find that the Lorentz contraction differentially affects measuring rods laid off along concentric circles around the origin in the plane of rotation (due to the variation in tangential linear velocity at different distances along a radius), whereas no Lorentz contraction is experienced by rods laid off along a radius. Therefore, the geometry in such a rotating system will be found to be non-Euclidean (the ratio of the circumference to the diameter of concentric circles around the origin in the plane of rotation will differ from π and depend on the circular radius).

The importance of this line of thought for Einstein is evident in virtually all of his expositions of the general theory of relativity, where it is always used as the primary motivation for introducing non-Euclidean geometry into the theory of gravitation.[23] Moreover, as Stachel (1989a) has shown, this particular thought experiment in fact constituted the crucial breakthrough to what we now know as the mathematical and conceptual framework of the general theory. For, generalizing from this example, Einstein quickly saw that what he really needed for a relativistic theory of gravitation is a *four-dimensional* version of non-Euclidean geometry (comprising both space and time). He quickly saw that a variably curved generalization of what we now call the flat Minkowski metric of special relativity should serve as the representative of the gravitational field, and, turning to the mathematician Marcel Grossmann for help, he then discovered the Riemannian theory of manifolds. Einstein's repeated appeal to the example of the uniformly rotating frame of reference in his official expositions of the theory therefore appears to reflect the actual historical process of discovery very accurately, and to explain, in particular, how the idea of a variably curved four-dimensional space-time geometry was actually introduced into physics in the first place.[24]

Einstein's introduction of non-Euclidean geometry, therefore, followed a remarkably circuitous route, and it involved, in particular, a process by which Einstein delicately positioned himself within the debate on the foundations of geometry between Helmholtz and Poincaré – a debate which was framed, as we have seen, by a nineteenth-century mathematical tradition in group theory and projective geometry. In creating the special theory of relativity in 1905 Einstein took inspiration from Poincaré's conventionalist scientific epistemology – not, however, as applied to spatial geometry but rather to what we now call the (four-dimensional) geometry of Minkowski space-time. Once the special theory was in place, Einstein then faced a radically new situation in the theory of gravitation, for the Newtonian theory of universal gravitation – based, as it was, on an instantaneous action at a distance and thus on absolute simultaneity – was incompatible with the new conceptual structure of the special theory based on a relativized conception of simultaneity. Einstein was therefore faced with the problem of adjusting the theory of gravitation to this new relativistic conceptual structure, and he addressed this problem, in the first instance, by appealing to the already well-established empirical fact that gravitational and inertial mass are equal. This fact led him to his principle of equivalence – the idea that gravitation and inertia are the very same physical phenomenon – which he then applied, as we have seen, to noninertial frames of reference within the conceptual structure of the special theory (within what we now call Minkowski space-time).[25] This led him, in turn, by the example of the uniformly rotating frame, to a non-Euclidean spatial geometry (now linked to the action of a gravitational field), which he was then able, finally, to generalize to the non-Euclidean *space-time* geometry of general relativity.

Moreover, the crucial thought experiment of the uniformly rotating frame of reference essentially involved, as Einstein tells us in *Geometry and Experience*, a naively Helmholtzian rather than a sophisticated Poincaré-inspired perceptive on the relationship between the behavior of rigid bodies and physical geometry. Indeed, it was necessary for Einstein to have already rejected the more sophisticated perspective on rigid bodies suggested by Poincaré in creating the special theory of relativity. For, as is well-known, Poincaré was actually the first person to discover what we now know as the Lorentz group governing inertial reference frames in special relativity, and Poincaré had already formulated, accordingly, a Lorentzian version of the mathematics of special relativity still in some sense committed to a classical aether.[26] For Poincaré, we might say, the Lorentz group thus operated at the level of electrodynamics – governing the microscopic electromagnetic structure ultimately responsible for physical rigidity – but not, as in Einstein, at the more fundamental kinematical level governing the basic concepts of space, time, and motion formulated prior to and independently of any particular dynamical theory. Just as, in the special theory, Einstein takes the Lorentz contraction as a direct indication of fundamental *kinematical* structure, independently of all dynamical questions about the microphysical forces actually responsible for physical rigidity, here, in the example of the uniformly rotating reference frame, Einstein similarly takes the Lorentz contraction as a direct indication of fundamental *geometrical* structure. And without this remarkably circuitous procedure of delicately situating himself, as it were, between Helmholtz and Poincaré, it is hard to imagine how Einstein could have ever discovered the idea of a variably curved four-dimensional space-time geometry in the first place.[27]

4. MORALS FOR TWENTIETH-CENTURY PHILOSOPHY OF GEOMETRY: FROM SPACE TO SPACE-TIME

There is an important sense, however, in which this same idea of a variably curved space-time geometry, once discovered, renders the entire preceding debate between Helmholtz and Poincaré irrelevant. For, as we have suggested, this debate is framed by the Helmholtz-Lie theorem and is therefore limited – along with the entire Kleinian tradition in group theory and projective geometry within which it is articulated – to spaces of constant curvature. But the general theory of relativity, of course, employs a space-time of variable curvature (depending on the distribution of mass and energy) where Helmholtz's principle of free mobility therefore fails (see note 16). As a result, the characteristically late-nineteenth-century conception of pure geometry as describing the "perspectival" features of our spatial intuition also fails, and we are left with the characteristically twentieth-century conception, originally derived from the work of Hilbert, of pure mathematical geometry as an abstract deductive system having no intrinsic relation at all to our spatial perception or any other kind of experience.[28] So it is no wonder, as we saw in Section 2, that Einstein appeals in *Geometry and Experience* to precisely this Hilbertian conception.

Indeed, as we have also seen, Einstein bases this appeal, more generally, on Schlick's elaboration of the notion of "implicit definition" in *General Theory of Knowledge* – which is based, in the present context, on Schlick's virtually simultaneous work on the philosophical significance of the general theory of relativity. Here, in particular, Schlick portrays the variably curved space-time of general relativity as an entirely abstract, entirely nonintuitive "conceptual construction," which can only be related to experience and the physical world by a similarly abstract, entirely nonintuitive correspondence relation (designation or coordination), in virtue of which the purely mathematical "conceptual construction" represented by Einstein's formulation of the general theory of relativity can then receive empirical content in terms of physical measurement.[29] And it is precisely here, in this procedure of physical coordination, that Schlick finds that Poincaré's conventionalist philosophy of geometry still holds. In particular, we still have a choice regarding whether to use a non-Euclidean or a Euclidean physical geometry even within the context of Einstein's new theory: it is just that in the latter case we would have to introduce further complications into the simple and "natural" physical coordination effected by Einstein's principle of equivalence, and we would thereby introduce onerous complications into our total system of geometry plus physics.[30] But Poincaré had maintained that only mathematical simplicity explains our preference for Euclidean *spatial* geometry, and all we are now doing, in the context of Einstein's new theory, is extending Poincaré's viewpoint to *space-time* geometry.[31]

This attempt to link Poincaré's conventionalist philosophy of geometry and the general theory of relativity is certainly very plausible, and, accordingly, it has been extraordinarily influential in twentieth-century scientific thought. Nevertheless, in light of the historical background we sketched in Section 3, we can now see that it is subject to very significant difficulties. In the first place, the problem of coordination as Schlick understands it – the problem of relating an abstract "conceptual construction" to concrete empirical reality – did not exist for Poincaré. For Poincaré's own work on geometry – mathematical and philosophical – is entirely framed, as we have seen, within the late-nineteenth-century tradition in group theory and projective geometry associated with Klein's Erlanger program. Pure geometry, in this tradition, is by no means an uninterpreted axiomatic system but rather an expression of the abstract "perspectival" features of our spatial intuition or perception. Geometry, on this conception, therefore has *space* as its object, the very space in which we live, move, and perceive, and so the problem of coordination or designation as Schlick understands it simply does not arise. Indeed, this problem, as we have seen, is in an important sense a product of Schlick's assimilation of the radically new conception of space and geometry embodied in Einstein's theory of relativity, not something that was present and available all along.

In the second place, and even more importantly, Poincaré's conception of space and geometry is also entirely based, in accordance with this very same late-nineteenth-century tradition, on the principle of free mobility first formulated by Helmholtz and later brought to precise mathematical fruition in the Helmholtz-Lie theorem. For it was precisely this principle that formed the indispensable link between pure or mathematical and applied or physical geometry within the nineteenth-century tradition in question. Here the relationship between pure and applied geometry was not understood as that between an uninterpreted axiomatic system and a possible (empirical) interpretation of that system. It was rather understood as a relationship between the otherwise empty space in which material bodies are contained and these material bodies – as a relationship, in the original Kantian sense, between a form of intuition and the physical objects or material content contained within that form (see note 17). Unlike in the original Kantian conception, however, this tradition operates within a generalization of the notion of form of intuition including all spaces of constant curvature; and, in this context, the principle of free mobility then serves as our crucial coordinating principle – our crucial link between pure and applied geometry – by coordinating purely geometrical notions, like that of geometrical equality or congruence, to the idealized behavior of physical rigid bodies (see DiSalle 1995 §2).

And it is precisely here, in the context of the Helmholtz-Lie theorem, that a remarkable conceptual situation then arises. For it now turns out that there are three and only three possible geometries compatible with the principle of free mobility – and therefore compatible, as we have seen, with our fundamental coordinating principle linking pure and applied geometry. Our fundamental coordinating principle leaves the choice of Euclidean or non-Euclidean geometry entirely open, and it thus makes perfectly good sense, in this very special conceptual situation, for Poincaré to maintain that the choice of Euclidean geometry is then determined by a convention or stipulation based on its greater mathematical simplicity. In the radically new conceptual situation created by the general theory of relativity, however, this particular view no longer makes good sense at all. Not only is the space-time structure of general relativity incompatible with the fundamental presupposition of the Helmholtz-Lie theorem, the principle of free mobility, but, in the general theory, there is only one empirically meaningful way to effect the required coordination between our purely mathematical formulation of the theory (now conceived as an entirely formal "conceptual construction") and concrete physical reality – namely, the principle of equivalence, which results in a direct coordination of the purely mathematical notion of a (semi-)Riemannian space-time geodesic with the behavior of freely falling idealized bodies or test particles affected only by gravitation.[32] Here, unlike in the very special situation addressed by Poincaré, we are not faced with what we might call a common or generic coordinating principle, which leaves the more specific geometrical structure for physical space still open, but rather with a singular or unique coordinating principle (the principle of equivalence) compatible with one and only one geometrical structure: the geometrical structure for physical space-time described in Einstein's formulation of general relativity.[33] Here the idea of an arbitrary or conventional *choice* of physical geometry has lost all real meaning and application – from a mathematical and from an empirical point of view.

By contrast, Einstein's own engagement with the problematic of conventionalism, and, in particular, with the debate on the foundations of geometry between Helmholtz and Poincaré, was an especially timely and fruitful one. It allowed him, as we have seen, to take the critical step, through the principle of equivalence, from the interpretation of three-dimensional, non-Euclidean spatial geometry to that of four-dimensional, non-Euclidean space-time geometry. And it is in precisely this sense, we might say, that Einstein made the crucial transition from late-nineteenth- to twentieth-century philosophy of physical geometry. One unforeseen consequence, however, was that the fundamentally new perspective on the foundations of geometry actually created by Einstein in this way has proved much more difficult to grasp than it otherwise might. In particular, in the philosophy of geometry bequeathed to us by logical empiricism we remained preoccupied with the problematic of conventionalism and the behavior of rigid bodies long after these had lost all specific relevance to physical theory – where a new concern for *space-time* geometry and the (essentially four-dimensional) problem of *motion* can now be seen as the true successor to the late-nineteenth-century tradition in the mathematical and philosophical foundations of geometry that was subject to a far-reaching and radical transformation in the work of Einstein.[34]

What we are now in a position to see, finally, is that the crucial step in this transformation is strikingly and accurately depicted in the passage on the uniformly rotating frame of reference from *Geometry and Experience*. For it is precisely here that Einstein joins two previously independent problems involving the physical interpretation of space, time, and geometry in a quite remarkable and unexpected fashion. Since Newton's original creation of a coherent mathematical-physical theory of space and time in the seventeenth century, the problem of the relativity of motion had persistently afflicted its conceptual foundations. The discovery of the concept of inertial reference frame in the late nineteenth century introduced considerable clarity into this situation (see Torretti 1983 §1.5; DiSalle 1991), and Einstein had already put this concept to fruitful new use in his creation of the special theory of relativity. In view of the principle of equivalence, however, Einstein now sees that an extension of the special principle of relativity is called for and, in particular, that noninertial frames of reference, in the context of the mathematical structure of special relativity (what we now call the structure of Minkowski space-time), can be used to model the gravitational field. Here Einstein encounters the decisive example of a uniformly rotating frame of reference, whose three-dimensional, purely spatial geometry turns out to be non-Euclidean. And Einstein is able to take this example as a direct physical indication of fundamental geometrical structure (and not as a mere field of force causing apparently rigid bodies to be distorted) by delicately positioning himself, as we have said, within the late-nineteenth-century debate on the foundations of physical geometry between Helmholtz and Poincaré.[35] But the upshot of Einstein's radical transformation (and eventual transcendence) of this debate is that the conceptual foundations of geometry can no longer be fruitfully pursued in independence of the problem of motion. The new problem of physical geometry is precisely the problem of space-time.

NOTES

1 Page references are to the first (1781, "A") and second (1787, "B") editions respectively. All translations from German originals are my own.

2 See Schlick (1915, 1917), Reichenbach (1920, 1928), and Carnap (1922). For further discussion of their (rather different) relations to Kant see Friedman (1999).

3 For discussion of the circumstances of Schlick's appointment see Stadler (1982, 117–20). This chair had previously been occupied by Ernst Mach and Ludwig Boltzmann.

4 This chapter presents an account – in a much more detailed and developed (and I hope also clearer and more accurate) form – of the relationship between the problem of geometry in general relativity and the philosophy of logical empiricism which I have told twice in print before (Friedman 2002b, 2008a). I am indebted to Michel Janssen, in particular, for very helpful comments on the penultimate draft of the present version.

5 See the final chapter of Schlick (1917), which, in turn, is closely related to the "method of coincidences" discussed in Schlick (1918 §30 [§31 of 2nd ed., Schlick 1925]). For further discussion, see Friedman (1997, 2002a).

6 See Schlick (1918 §7). The reference is to Hilbert (1899).

7 See Poincaré (1902a). For further discussion, see Friedman (1999, 2000).

8 The influence of Poincaré's conventionalism is already evident in Schlick (1915), which deals primarily with the special theory of relativity. Schlick (1917) then extends this conventionalism to the general theory. Compare also Carnap's later exposition of this viewpoint in Carnap (1966).

9 This kind of Euclidean alternative to general relativity is explicitly developed for the case of the three-dimensional Schwarzschild geometry in the neighborhood of the Sun in Carnap (1922, 1966).

10 For discussion of Klein's Erlanger program, see Torretti (1978 §2.3). Torretti (1983) depicts the development of relativity theory as the outcome of two quite different nineteenth-century programs in the foundations of geometry – Klein's Erlanger program and Bernhard Riemann's general theory of manifolds. The special theory of relativity can be seen as based on the former program, whereas the general theory can only be based on the latter.

11 See Torretti (1978 §2.3.10) for a discussion of Klein's view of spatial intuition. For a somewhat more sympathetic discussion of the relationship between (closely related) nineteenth-century views of spatial intuition and the original Kantian conception, see Friedman (2000).

12 I am grateful to my colleague Dagfinn Føllesdal for helpful discussions of the relationship between Lie and Klein.

13 For a discussion of Helmholtz's mathematical results in the context of his theory of space perception, see Richards (1977) and Friedman (1997).

14 For the work of Helmholtz and Lie, see Torretti (1978 §3.1). For a philosophically and mathematically sophisticated discussion of Helmholtz and Riemann, see Stein (1977).

15 In two dimensions, e.g., such a group of isometries or rigid motions consists of all transformations under which a given figure can be transported throughout the space while remaining congruent to itself. In the Euclidean plane such transformations consist precisely of translations and rotations. In a non-Euclidean surface of constant curvature – such as the surface of a sphere, which has constant *positive* curvature – the isometries consist of suitably generalized versions of translations and rotations. By contrast, in a surface of *variable* curvature, such as the surface of an egg, figures cannot be isometrically transported throughout the space – a triangle at the fat end of the surface of an egg, e.g., necessarily has a different angle-sum from (and therefore cannot be congruent to) a triangle at the pointy end.

16 In the variable negative curvature of the Schwarzschild geometry in the neighborhood of the Sun, e.g., a body can be isometrically transported on a spherical surface at a given radial distance from the Sun (due to spherical

symmetry), but it cannot be isometrically transported away from or off of the given surface. This is the fundamental reason Klein's Erlanger program (which assumes a full group of symmetries mapping every point onto any other) is not applicable to the space-time of general relativity (see note 10).

17 Helmholtz characterizes space as a "subjective *form* of intuition" in the sense of Kant and as the *"necessary* form of our outer intuition" in his famous address, "The Facts in Perception," delivered in 1878. See Helmholtz (1921, 117) and Cohen and Elkana (1977, 124). Helmholtz viewed the condition of free mobility, in particular, as a necessary condition of the possibility of spatial measurement and thus of the application of geometry. For further discussion, see the works cited in notes 13 and 14.

18 For a discussion of the Poincaré models, see Torretti (1978 §2.3.7).

19 For further discussion see Friedman (1999, ch. 4).

20 See Miller (1981, ch. 2) for a discussion of Einstein and Poincaré. There is no doubt, in any case, that this view of the relationship between Einstein and Poincaré played an absolutely crucial role in the reception of special relativity by the logical empiricists (see esp. Schlick 1915).

21 Indeed, Poincaré (1898) had already argued that distant simultaneity requires a convention or definition essentially involving the behavior of light.

22 For a detailed discussion of the principle of equivalence, based on Einstein's work in the years 1907–12, see Norton (1985).

23 See, e.g., Einstein (1916e, 776; 1917a §§23–8; 1922c, 537–9).

24 As is well-known, Einstein (and Grossmann) required several years of struggle between 1912 and 1915 to find the final field equations of general relativity – during which, in particular, they were sidetracked by the now notorious "hole argument." See Stachel (1989b) and Norton (1984). The present point, however, is that the essential geometrical idea underlying general relativity – the idea of representing gravitation by a variably curved four-dimensional space-time metric – had already been clearly articulated by 1912. This is not to say, of course, that the further story of how Einstein actually arrived at his field equations between 1912 and 1915 is not important: on the contrary, it has been the subject of intensive and very fruitful investigations by a number of scholars, especially those in the group associated with Jürgen Renn at the Max Planck Institute for the History of Science in Berlin (Renn 2007a). But I am here only attempting to depict the influence of previous mathematical and philosophical work on the foundations of geometry (by Helmholtz and Poincaré, in particular) on Einstein's initial formulation of the basic geometrical idea in the period 1907–12.

25 Again, Norton (1985) provides a detailed analysis of the principle of equivalence as Einstein developed it in the years 1907–12. Norton shows, in particular, how well-defined three-dimensional geometries are induced in noninertial frames of reference in Minkowski space-time so long as they are *uniformly* accelerating or rotating.

26 See Miller (1981 §1.14.) In the Lorentz-Poincaré theory of the (deformable) electron, the Lorentz contraction is a genuine physical or dynamical phenomenon (judged from the point of the privileged aether frame), ultimately due to the microphysical effects of electromagnetic forces.

27 From our present, post-general-relativistic point of view, the most natural procedure is to begin with the flat four-dimensional geometry of

Minkowski space-time and then use the principle of equivalence to motivate the idea that freely falling trajectories can be conceived as geodesics in a variably curved "perturbation" of an initially flat Minkowski geometry. This line of thought was definitely not available in the actual historical context within which general relativity was created, however. For, on the one hand, Einstein did not appreciate the importance of Minkowski's four-dimensional reformulation of special relativity until he was already well on his way to creating the general theory. And, on the other hand, no one but Einstein had the pivotal idea of exploiting the well-known equivalence between gravitational and inertial mass so as to create a relativistic field theory of gravitation modeled on the behavior of noninertial frames of reference. In particular, Minkowski was engaged in an attempt to formulate a relativistic theory of gravitation by a generalized action-at-a-distance theory (see Corry 1997, 286–92). Thus the only effective line of thought available at the time was the one Einstein actually followed: beginning from a three-dimensional formulation of special relativity we apply the principle of equivalence to (three-dimensional) noninertial reference frames, and we are then led using the example of the uniformly rotating frame of reference to a (three-dimensional) non-Euclidean spatial geometry – which we are only at this point in a position to generalize to a (four-dimensional) non-Euclidean *space-time* geometry.

28 Riemann's theory of manifolds can be seen as an intermediate stage in this development, where geometry is characterized nonintuitively, and in this sense purely conceptually, within analysis in terms of n-tuples of real or complex numbers. It was Hilbert, however, who created the first purely conceptual (abstract) *synthetic* geometry, and who then clarifies the relationship between geometry in this (axiomatic) sense and analysis using a representation theorem. For discussion of Hilbert's achievement, see Torretti (1978 §3.2.8).

29 See again the references cited in note 5. Schlick models his "method of coincidences," in particular, on Einstein's principle of general covariance – and thus on the use of generalized coordinates in the analytic treatment of n-fold extended manifolds. Schlick then relates such a coordination to the Hilbertian notion of implicit definition along the lines of note 28.

30 See notes 8 and 9, together with the passages to which they are appended. Schlick understands the principle of equivalence as asserting that the laws of general relativity (including its "metrical determinations") approach those of special relativity in infinitely small space-time regions. And this implies, for Schlick, both that freely falling trajectories follow geodesics of the new variably curved space-time geometry and that the three-dimensional spatial geometry, as determined by the behavior of (infinitesimal) rigid bodies, is infinitesimally Euclidean in the sense of Riemann's theory of manifolds (see note 14, together with the paragraph to which it is appended). From a modern point of view, however, we can entirely dispense with rigid bodies in the foundations of general relativity and conceive the principle of equivalence as simply motivating the idea that freely falling trajectories are four-dimensional geodesics of our space-time manifold (compare note 32). The crucial point is that in a relativistic space-time manifold (but not in a Newtonian space-time manifold – see following text) the geodesics or

"affine structure" entirely determine (together with the trajectories of light rays or "null geodesics") the spatial and temporal metrical structures. For further discussion and references, see Friedman (2002a).

31 Schlick became fully clear about this extension of Poincaré's conventionalism to general relativity only in the fourth (1922) edition of *Space and Time in Contemporary Physics* and the second (1925) edition of *General Theory of Knowledge*. For discussion, see again Friedman (2002a). The key idea is that Euclidean geometry can be retained by introducing what Reichenbach (1928) calls "universal forces." And we can understand this most clearly and precisely, from a modern point of view, in the context of the Cartan-Trautmann reformulation of the original Newtonian theory of gravitation – which is based, like general relativity, on the principle of equivalence and which, accordingly, employs a variably curved four-dimensional space-time structure (affine structure) to represent the action of gravity. When we then recover the traditional formulation – which is based, from a modern point of view, on a flat four-dimensional space-time structure – gravity in its traditional (Newtonian) form appears as precisely a "universal force" in the sense of Reichenbach. Following Reichenbach's methodological prescription to "set universal forces equal to zero," in this context, therefore amounts to adopting the principle of equivalence and rejecting the traditional flat space-time structure. For discussion of the Cartan-Trautmann formulation in this regard, see Friedman (1983 §§III.4, III.8). Note that here, because the Newtonian affine structure does *not* determine the spatial and temporal metrical structures (compare note 30), gravity viewed as a "universal force" in this context has nothing to do with any distortion of apparently rigid bodies: the purely *spatial* geometry is Euclidean in both versions of Newtonian gravitation.

32 More precisely, the relationship between the principle of equivalence (gravitational and inertial phenomena are identical) and the geodesic principle (freely falling trajectories are represented by geodesics in a nonflat space-time manifold) is as follows. The idea that gravitational and inertial phenomena are identical entails, from a modern point of view, that one cannot locally distinguish the traditional inertial trajectories of Newtonian gravitation theory (geodesics in a flat Newtonian space-time) from freely falling trajectories in a gravitational field: one cannot determine a unique decomposition of the nonflat affine connection in the Cartan-Trautmann formulation of Newtonian gravitation theory into a flat affine connection plus gravitational potential by the empirically given local motions. We are motivated, therefore, to drop the traditional flat affine connection (along with the traditional inertial frames) and retain only the nonflat connection determined by the freely falling trajectories – which are now *directly* coordinated to the new affine structure entirely independently of a presumed flat affine structure plus gravitational potential. For this understanding of the principles of equivalence, in relation to some other standard interpretations, see Friedman (1983 §V.4). (I am here especially indebted to Michel Janssen for prompting me to make my understanding of the relationship between the principle of equivalence and the geodesic principle more explicit.)

33 For the principle of equivalence as the fundamental coordinating principle for the space-time structure of general relativity (and as the successor, in

this sense, of the nineteenth-century principle of free mobility), see DiSalle (1995). Note that to abandon the principle of equivalence and allow non-zero "universal forces" (note 31) is to introduce empirically meaningless elements into one's formulation admitting no univocal coordination with empirical phenomena. From the point of view of the Cartan-Trautmann formulation, e.g., traditional Newtonian gravitation theory involves an arbitrary choice of a flat affine structure (plus gravitational potential) which is not uniquely determined by the empirical local motions. By contrast, the principle of equivalence avoids all such arbitrariness by enforcing a direct coordination between the nonflat affine structure and the empirical local behavior of freely falling bodies (with no need for a gravitational potential here figuring as a "universal force"). Compare note 32, and, for further discussion, see Friedman (1983 §§V.4, VII.2).

34 In particular, using certain privileged *motions* to define our space-time structure (as in the principle of equivalence) can now render the traditional concern with the behavior of (three-dimensional) rigid bodies irrelevant, because the (four-dimensional) *relativistic* space-time structure already entirely determines the (three-dimensional) purely spatial structure (compare notes 30 and 31).

35 Once again, it is precisely because we are working in the context of a relativistic (infinitesimally Minkowskian) space-time manifold that a non-Euclidean *spatial* geometry appears here: in the context of a four-dimensional Newtonian space-time, by contrast, no purely spatial curvature can appear either in a noninertial frame of reference within the flat four-dimensional formulation or due to the four-dimensional *space-time* curvature arising in the nonflat (Cartan-Trautmann) formulation (see again notes 30, 31, and 34). Indeed, it is precisely because we are now applying the principle of equivalence to a relativistic space-time that the characteristic new predictions of general relativity (such as the anomaly in the perihelion of Mercury) then follow (whereas the Cartan-Trautmann theory is empirically equivalent to traditional flat Newtonian gravitation theory). Thus Einstein's appeal to the example of the uniformly rotating frame of reference, in the context in the debate on the foundations of three-dimensional, purely spatial geometry between Helmholtz and Poincaré, made perfect sense, even though, as we have also seen, the new four-dimensional geometry eventually arising thereby in fact renders this earlier debate irrelevant. The crucial point is that there was no rational and effective way then available to arrive at the new geometry in the first place except the remarkably circuitous route Einstein actually took (see note 27, together with the paragraph to which it is appended).

14   Einstein's Politics

> In a long life I have devoted all my faculties to reach somewhat deeper
> insight into the structure of physical reality. Never have I made any
> systematic effort to ameliorate the lot of men, to fight injustice and
> suppression, and to improve the traditional forms of human relations. The
> only thing I did was this: in long intervals I have expressed an opinion on
> public issues whenever they appeared to me so bad and unfortunate that
> silence would have made me feel guilty of complicity.[1]

> Einstein was a unique phenomenon: a theoretical scientist whose area
> of expertise was far removed from the everyday concerns of his fellow
> mortals, a man without interest or training in the workings of politics
> who ... came to play a critical role in the public life of his epoch as
> preeminent moral figure of the Western world.

> Sayen 1985, 4

1. INTRODUCTION

Universally recognized as a physicist of the first rank, Albert Einstein
as a political figure is far more difficult to assess. He never engaged
systematically in the activities of any political party and remained
throughout his life above the political fray. The idiosyncratic cast of
his political thinking further complicates the issue. And yet for many
there is almost a mystical identification of his person and name with
integrity in the political sphere. The key to understanding this identification lies in recognizing its moral roots, a fact which explains both
the powerful hold on the public his political pronouncements continue
to exert to this day and the source of the myth of Einstein's naiveté
in politics. Idealistic views unchecked by a realistic assessment of the
everyday world to which they must apply may fairly be called naive.
This charge does not apply to Einstein. The same muscular pragmatism which marked the early stages of his personal and professional
life came to shape his political views, once these emerged after World
War I, above all a willingness to evaluate ideas with respect to their

421

consequences in everyday life. Unlike a professional politician, however, he was not accountable to any constituency other than himself.
Drawing on his international fame as the second Newton, he enjoyed
the luxury of reflecting on politics and directing his energies to social
ends without making the usual political compromises. Events were to
alter his views but never to force concessions of principle.

The primacy of Einstein's achievements in physics is indisputable.
It also has two important consequences for the formation of his political sensibility: it defines and explains the emergence of his political
commitment at the relatively late age of forty. These consequences also
dictate the structure of this chapter.

Einstein's political interests only crystalized in 1919 after his most
startling discoveries in physics lay behind him. In the critical years
leading up to his greatest scientific achievements his political and
social interests lay fallow, their moral roots unarticulated. After an initial period of fiercely independent religious piety, attributable in part
to youthful rebellion against his parents' rejection of religious Judaism,
Einstein was drawn by his readings in the natural sciences during his
early teens into the web of natural phenomena and a lifelong attempt to
comprehend them. A self-conscious distancing from emotional involvement followed, as well as contempt for everyday concerns and a reveling
in his outsider status. Merely personal concerns were relegated to the
margins. In the process of single-mindedly defining his path, Einstein
fashioned for himself a set of friendships, affiliations, and commitments
which were at some remove from the world at large. He created for himself a solitary world, choosing a circle of friends and finding mentors
that shared his disdain for convention and pursued similar unorthodox
career paths. The lack of a well-established academic physics discipline
at the end of the nineteenth century also served to reinforce Einstein's
sense of isolation. This coupled with the young man's strong antipathy
toward authority and ruthless definition of priorities, led to an indifference to engaging the external world, an indifference that persisted
through his early academic career and was only modified decisively
in 1919.

Thrust into the limelight after the solar eclipse expeditions of 1919,
Einstein made his first overt political statement at the end of that year
after the announcement of verification of his theory of general relativity
by the British Royal Society. This dramatic announcement by Britain's
most prestigious scientific body conferred upon him an unparalleled
international legitimacy, reinforced by a public fascination with the idea
of finding certainties in the stars which had proved elusive in the turmoil of war. Thus was born the Einstein myth of the all-knowing sage,

equally conversant with the laws of physics as with those of human affairs. Thus were also born the unrealistic expectations of those that sought from him oracular truths, as well as the unreasonable charges of naiveté which surround the legend of the political Einstein to this day. Most importantly, this watershed marked the emergence of an Einstein who committed himself actively to the pursuit and implementation of his social ideals.

In this chapter, we explore the origins and development of Einstein's political views, seeking to uncover unresolved trace elements in his early life which prefigured the course of later political reflections and actions. This will be necessary to understand the specific form that those reflections and actions took. Next we provide some context for the critical year 1919 when Einstein's newfound fame provided the platform for his pronouncements. Lastly, we present a thematic survey of those pronouncements and the events which elicited them from 1920 on. Emphasis is given to the critical decade and a half until his emigration to the United States (1919–33). A brief sketch of the American period follows. Because of its centrality to Einstein's concerns and because it distills the social goals he valued most, a special focus is reserved for Einstein's Zionism.

2. BACKGROUND AND YOUTH: MAKING A VIRTUE OF NECESSITY

Culturally assimilated and religiously indifferent as were most German Jews born around the middle of the nineteenth century, Einstein's parents were eager to obtain for themselves the civil liberties and economic opportunities only recently and still grudgingly conceded to their Jewish brethren. The confluence of an economic upsurge generated by the Industrial Revolution, the rising tide of democratic liberalism, and the resulting hard-won emancipation of Jews proved most favorable to their generation. Secular education (*Bildung*) and an enterprising spirit became the major avenues of social advancement into the middle class.

Once the old barriers began to crumble, numerous Jews became successful in the rapidly expanding new technologies to which the Industrial Revolution gave rise. More importantly, they also sought to acquire the trappings of cultured success, a universal humanism that transcended the categories of religion, ancestry, and tradition.[2] Hermann and Pauline Einstein, née Koch, came from the southern German region of Württemberg with its long tradition of small-scale entrepreneurial firms, an area where socioeconomic advancement was matched by geographic mobility. Availing themselves of new opportunities, many Jews abandoned traditional trades and rural backgrounds to try their hand in the new economic landscape. After completing an apprenticeship as a merchant, Hermann Einstein, born in the village of Buchau am Federsee, moved to Ulm and joined two cousins as a partner in a bedfeather venture (Specker 1979, 55). He married Pauline Koch in 1876 and with capital from his in-laws set up a small electrical business that soon folded. One year after son Albert's birth, he moved on to Munich, the capital of Bavaria and regional center for the rapidly expanding electrical power industry. Together with his brother Jakob, an engineer trained at the Polytechnic in Stuttgart, he founded a company, Jakob Einstein & Co., that initially installed water and gas lines and then evolved into a factory for the production of electrical equipment including dynamos and telephones (Pyenson 1985, 35–57).

Pauline, also from Württemberg, came from a family that had already enjoyed prosperity in the preindustrial era as cattle merchants, beneficiaries of the status accorded to purveyors to the royal court. Toward the end of the nineteenth century, this family too made the leap into the new age, as Pauline's brothers became international grain traders with seats in Genoa, Brussels, and Buenos Aires. While Pauline did not pursue a career, as was typical of women of her generation, she instilled in her only son a fierce determination to succeed. Proudly it was recounted in the family that Albert, even when in grade school, braved the streets of Munich alone on the way to school, where he was one of a handful of Jewish pupils.[3]

The trajectory of Hermann's career was less promising than anticipated. After some initial successes, Jakob Einstein and Co. retreated from the German market, increasingly dominated by emerging giants such as Siemens and Schuckert. In 1894, the brothers moved to northern Italy, there once again to fail in a number of endeavors. Albert stayed behind, already exhibiting the streak of willful independence that marked much of his youth.

As he was later to recount in his *Autobiographical Notes*, until the age of twelve Einstein observed the dietary laws and other Jewish religious traditions with almost fanatical zeal. In a rather sudden change of heart, perhaps brought on in part by an unhappy anticipation of his bar mitzvah the following year, he turned his interests at the age of thirteen, in 1892, to the intellectual mastery of nature (Einstein 1979, 2). His most important intellectual guides were his Uncle Jakob and family acquaintance Max Talmey. The latter introduced him to Aaron Bernstein's twelve-volume compendium on the natural sciences, a work that reflected the rational optimism of the Enlightenment and earnestly imparted to its readership how the natural world might be comprehended through the power of scientific materialism (Bernstein 1853–7).

Enthusiasm for readings at home was not matched by attentiveness in school. According to legend, he was repeatedly reproached by one of his instructors because of his lack of respect, an attitude which was destructive of order in the classroom. Repugnance at the discipline in the schools; distrust of the authorities that wielded it; willfulness and independence of judgment; and a conscious sense of being an outsider, were the emotional reserves from which his later political views were drawn. These factors also fueled his resolve to flee what he considered a bland diet of rote learning in Munich. His sister Maja gives us corroboration of the brashness that allowed him to remain in the Bavarian capital after his parents' departure for Italy when he was fifteen. This is further underlined by Maja's claim that he defined his career path on his own and decided to apply to the *Eidgenössische Technische Hochschule* (ETH) without even bothering to consult his parents. Moving from Munich to Italy to Switzerland, all before the age of sixteen and all on his own initiative without the security of emotional and significant financial support from his immediate family gives some indication of the steely resolve in his character and the sense of being rootless that marked his passage. Describing himself "as a person rooted nowhere," he proclaimed not only his marginality but a genuine pride in being footloose (Einstein to Max Born, March 3, 1920 [CPAE 9, Doc. 337]). He was everywhere an outsider – a German Jew in Switzerland, metaphorically orphaned by his decision to strike out on his own – grounding this isolation in a willed act of self-reliance, living the "gypsy life" (*Zigeunerleben*) of the perennial outsider. This and his strength of character were to contribute decisively to his political awakening.

Failing his entrance exam to the ETH in 1895, due probably to deficiencies in knowledge of Swiss literary and political history, Einstein at age sixteen agreed to finish a final year of secondary school education at the Aargau cantonal school in Aarau, a small town west of Zurich (Einstein 1955, 145–6). He and his cousin were the only Jewish pupils. The historian Jost Winteler, with whose family Einstein boarded during the school year, was a vehement opponent of great-power pretensions. During Winteler's university training in Germany in the 1870s, his political view had been conditioned by his negative reaction to the great-power triumphalism that accompanied the founding of the German Reich. Coupled with the natural aversion of many Swiss to the autocratic German political system of the day was Winteler's freethinking contempt for religious authority. To what degree Winteler influenced, even if only tangentially, the latent political musings of the young Einstein is unclear. A hint of it is, however, preserved in a contemporary letter to Winteler, in which he delivers one of his most quoted epigrams: "Slavish obedience to authority is the greatest enemy of the truth" (Einstein to Jost Winteler, July 8, 1901 [CPAE 1, Doc. 115]). His affinity extended to other members of the Winteler clan. He referred to Jost's wife Pauline as his "second mother" ("Mamerl Nr. 2"). With daughter Marie he developed a more complex relationship. It illustrates his growing ability to sublimate his emotions with theatrical justifications. In 1897, when breaking off his relationship with Marie, he wrote to mother Pauline with an indirection and hyperbole that revealed an individual willfully distancing the self from engagement with the world around him. This psychological device, reinforced by his readings in Schopenhauer, was decisive. He must, he wrote, find his destiny in the stars not in human companionship. "Strenuous intellectual work and the study of God's Nature are the reconciling, fortifying, yet relentlessly stern angels that will lead me through all of life's troubles" (Einstein to Pauline Winteler, May 1897 [CPAE 1, Doc. 34]).

Accompanying this partially self-imposed isolation was a self-confidence, even presumptuousness, on display for instance in an essay he wrote for his Aarau leaving examination in French at age sixteen ("My Future Plans," September 18, 1896 [CPAE 1, Doc. 22]). He planned to pursue the study of theoretical physics, a daring proposition as this discipline was still in its infancy at the end of the nineteenth century. One of its main attractions for Einstein was the independence that such study would afford him. Reveling in this freedom from attachment was underscored when he declared on a Swiss citizenship application in October 1900 that he was without religious affiliation ("Questionnaire for Municipal Citizenship Applicants," October 11–26, 1900 [CPAE 1, Doc. 82]).

Einstein dealt with the contradiction between the great promise for socially mobile Jews at the end of the nineteenth century and the failed expectations of his parents for themselves by a heightened sensitivity on his part to social failure and a fierce empathy with the underdog, both reinforced by a native skepticism of the norms imposed with empty authority by others. These sentiments remained latent during this period. The philosophy of Schopenhauer, references to which are occasionally seeded in the Einstein letters of this phase in his life,[4] provided him with a vocabulary that enabled him to deal with conflicting loyalties and his ambivalent reactions to them – the frustration at what he perceived to be the failed and nonintellectual life of his parents, and yet his dependence on them; his enforced isolation, and yet his feeling of claustrophobia in the small world that he had created for himself;

and his love for the higher values of science, and yet the sometimes humiliating need of having to struggle to make a scientific career.

Priding himself on his outsider status allowed him to subsume his emotional needs to the ideal of an objective search for truth. This consisted of sustained bursts of scientific creativity in which he was to engage for more than twenty years – until virtual confirmation of general relativity in late 1919. In effect, he spun himself a psychological cocoon. Much as he would embrace the autodidactic path in his university learning, paralleling the official coursework offered at the ETH, so he also forged for himself in his youth a parallel social sphere in which his closest companions rejected or were considered marginal by the mainstream of society. Driven also by a need to meet the expectations of his family, this created in him a lack of interest in the world beyond his study, classroom, and intimate circle of friends. He assumed the protective coloration of a self-imposed isolation which matched the isolation imposed on him by circumstance, a case of making a virtue of necessity. In the same letter to Pauline Winteler mentioned previously, he put it this way: "... occasionally in a lucid moment I appear to myself as an ostrich, which puts its head in the desert sands.... One creates a small little world for oneself, and as lamentably insignificant as it may be in comparison with the ever changing magnitude of real existence, one feels miraculously great and important, like a mole in his self-dug hole" (Einstein to Pauline Winteler, May 1897 [CPAE 1, Doc. 34]). After graduating from the ETH he described his life in Schaffhausen as a physics tutor: "I live here as if I were totally alone, i.e., I have no personal contact with anyone. Almost every day I take a short walk ..., the rest of my free time I study Voigt's compendium of theoretical physics, from which I have already learned quite a bit" (Einstein to Mileva Marić, November 28, 1901 [CPAE 1, Doc. 126]). Three weeks later: "It's really a very droll life that I lead here, completely in the sense of Schopenhauer. With the exception of my private student, I speak with no one the entire day.... I find that alone I am in the best company, except when I am with you" (Einstein to Mileva Marić, December 17, 1901 [CPAE 1, Doc. 128]). Out of this isolation Einstein distilled a resolute sense of ambition and opportunistic pragmatism, conditioned also by a determination to succeed where his father had failed. His early career path, to which we now turn, provides ample evidence.

3. CROSSCURRENTS OF THE EARLY CAREER

A mixture of bravado and practicality marked Einstein's university years and after. His isolation remained a constant. Disappointed by the traditional physics coursework offered at the ETH, he cut classes with abandon in order to study physics texts of his own choosing. Yet he was careful to pore over a friend's class notes in order to pass his examinations. On writing his dissertation for the University of Zurich he abandoned an earlier topic on the electrodynamics of moving bodies, which was probably considered overly speculative by his advisor, and submitted in its stead a thesis which gave free play to theory within the constraints imposed by an experimentally oriented academic physics environment. In obtaining a post as technical expert at the Swiss Federal Patent Office in Bern he secured a well-paying temporary position that allowed him to bide his time productively while an academic position was created for him at the University of Zurich.[5] The Patent Office was not a backwater to which Einstein was relegated; he actively sought it as a marshaling ground for writing scientific papers that would keep alive his eligibility for the Zurich position (Schulmann 1993). His Bern posting also afforded him creative isolation and shielded him from excessive bureaucratic demands. An unorthodox schedule allowed him to pursue his own interests half the day after examining the practicability of electrical devices submitted for patent approval the other half.

Matching the unconventionality of his professional life at the Patent Office, Einstein founded and participated in the Olympia Academy, a casual group of bohemians that mirrored more formal societies dedicated to inquiry into the natural sciences and philosophy. Much as he was to hew a solitary path to academic respectability, so he carved out an iconoclastic circle of friends; his fellow Olympians, Maurice Solovine and Conrad Habicht.[6] A similar sense of kinship bound him to the two outsiders he had met at the ETH: his future wife, Mileva Marić, and his best friend, Michele Besso. Mileva, a Serbian classmate at the ETH and fellow bohemian was a privileged comrade in his closed world, the only woman in the mathematical-physics division of the ETH. She shared, at least at this early stage, Einstein's pride in pitting himself against the conventions of bourgeois society.[7] Besso, the "eternal student" and perfect foil, exasperated Einstein with his impracticality and was loved by him for his lack of guile and indifference to worldly ambition.

Einstein described the contours of his closed world this way: "Persons [such as myself] lead an existence in which external experiences, including contacts with other people, generally have only secondary importance.... Professional and personal bonds often inhibit and distract rather than having a positive effect on such lives, while the decisive experience takes place within as if nourished by an invisible source" (unpublished foreword to Frank 1947, EA 28–581).

In 1907 Einstein submitted seventeen published articles in theoretical physics for the habilitation, a postdoctoral degree that confers the right to teach at the university level. Included in this number were the groundbreaking papers of the *annus mirabilis*, 1905. But the University of Bern insisted on one customized article; Einstein cut his losses. He resubmitted the following year with "a specialized investigation of a scientific topic" and was awarded the degree (Albert Gobat to Einstein, February 28, 1908 [CPAE 5, Doc. 89]). His single-minded, if unorthodox, pursuit of an academic career began to bear fruit. The new position that he had anticipated and coveted at the University of Zurich was finally funded. Over the next five years he received appointments there, at the Charles University of Prague, and at the ETH. The move from Zurich to Prague in 1911 was painful for Einstein and his family; Mileva and he felt most comfortable in Switzerland but advancing his career came first, and the Charles University afforded him a full professorship. Any attempt to attribute Einstein's later affiliation with the Zionist movement to involvement with it in Prague is belied by later testimony.[8] One of his acquaintances from that period categorically denied that the newly minted professor had shown any interest, and Einstein expressed contempt some years later for the "Zionist-contaminated circle … of outsiders with a medieval cast" that were loosely grouped around philosophers at the university in Prague.[9] The choice of terms is interesting: "outsider" could fairly be applied to Einstein, but at this point Zionism was not the focus of his own "marginal" pursuits. Concentrated research work into the quantum and the beginnings of his general theory of relativity held pride of place. There is no question that he felt uncomfortable with the stuffy imperial traditions in Bohemia, both academic and otherwise. But pragmatism prevailed. When confronted with the imperial dictate that all civil servants declare a religious affiliation, Einstein, who had asserted his lack of the same in Switzerland, now declared with equanimity that "returning to the bosom of Abraham meant nothing to me. A piece of paper that had to be signed," though he drew the line at conversion (cited in Paul Ehrenfest to Tatiana Ehrenfest, February 25, 1912, Museum Boerhaave, Leyden, Ehrenfest Archive, EPC:3, Section 6).

Though there were harbingers of political awakening during his early career, they appear more as a slow wading into the untested waters of a sensibility that lay outside the scientific realm. In spring 1909, a newly minted associate professor at the University of Zurich, he vented a smoldering resentment of the "tail-wagging" ("*schweifwedeln*") of young German Jews, who abased themselves in order to obtain civil-service posts (Einstein to Jakob Laub, May 19, 1909 [CPAE 5, Doc. 161]). Mistakenly informed that he would not receive a call to Prague in 1910, he attributed it to "my Semitic ancestry" (Einstein to Jakob Laub, August 27, 1910 [CPAE 5, Doc. 224]). In Prague he was fully aware of the animosity between Czechs and Germans (Einstein to Heinrich Zangger, April 7, 1911 [CPAE 5, Doc. 263]) and his sympathy for the underdog was expressed in 1912 in solidarity with the Serbs in their confrontation with Austro-Hungary (Einstein to Helene Savić, after December 17, 1912 [CPAE 5, Doc. 424]). All evidence, however, points to continued indifference: Einstein's refusal to go to anti-Semitic Russia (Einstein to Petr Petrovich Lazarev, May 16, 1914 [CPAE 8, Doc. 7]), another faint indication of his awareness of a pressing social issue, is outweighed by an observation made by Lise Meitner during the war, in late 1916: "Last night I was at Plancks'. They played two marvelous trios, Schubert and Beethoven. Einstein played violin and occasionally made amazingly naïve and really quite peculiar comments on political and military prospects. That there exists an educated person in these times who does not so much as pick up a newspaper, is really a curiosity" (Archive for the History of the Max Planck Society, Berlin, Otto Hahn collection).

In fact, even during the war years Einstein exhibited no real political commitment. His moral outrage at the German violation of Belgian neutrality at the beginning of the war derived directly from his instinctive internationalism, as well as, closer to home, from his sense of betrayal at the blind solidarity expressed by his colleagues toward depredations of the German Army in Belgium. The document that is most closely associated with Einstein's awakening political consciousness – the Manifesto to the Europeans, written in response to the Manifesto of the ninety-three that defended Germany's invasion in the West in autumn 1914 – was composed in a first draft by a colleague, Georg Nicolai. Subsequently, he, with Einstein and two others, "collectively decided in mid-October 1914 on the ... final version" (Nicolai 1917, 12). In late March 1915, almost eight months after the outbreak of the war, Einstein joined a pacifist organization, the Association for a New Fatherland. The immediate spur for this seems to have been his wish to serve on a committee drafting an appeal to intellectuals that urged international solidarity among individuals of goodwill. His cousin (and future second wife) Elsa Einstein accompanied him. Though this was scarcely a politically informed action, it does show that he was beginning to think about his social obligations to further causes he believed in. In an essay written later in the year, however, he still was able to write: "Should everyone perhaps devote a considerable portion of his abilities to politics? I really believe that the intellectually more experienced people in Europe have sinned in their neglect of general political questions; yet I do not regard the pursuit of politics as the path to an individual's greatest effectiveness ..."

("My Opinion on the War," October 23–November 11, 1915 [CPAE 6, Doc. 20]).

Einstein's sporadic participation in the Association for the Like-minded, a group of German intellectual cosmopolitans attests similarly to his uncertainty. His deafening silence during vehement exchanges in the Prussian Academy of Sciences on whether foreign corresponding members from nations at war with Germany should be stripped of their membership was certainly not the behavior of an individual with firmly formed views. A police report on Einstein in the middle of World War I provides us perhaps with the most objective and sharply etched portrait. It states that while the subject subscribes to a liberal newspaper and is an adherent of the pacifist movement, he has not developed a political profile (Gülzow 1969, 234). A self-portrait that Einstein draws of himself during the period makes a similar point. He refers to himself as a staff member in an insane asylum and characterizes a prominent and active pacifist as an "optimist" for trying to free inmates from the grip of madness (Einstein to Heinrich Zangger, circa April 10, 1915 [CPAE 8, Doc. 73]). As the empire is collapsing in late 1918, Einstein once more pithily summarizes his position in response to a request to sign yet another appeal to German intellectuals: "Keep your trap shut" (Einstein to Leo Arons, November 12, 1918 [CPAE 8, Doc. 653]).

What does emerge in the wartime period, however, is Einstein's obvious disdain for the jingoistic breast-beating of a nation at war. As he boasts to his close friend Paul Ehrenfest, even as he signs Nicolai's clarion call to the Europeans, he is dozing off in his peaceful musings, feeling only a mixture of pity and disgust with the world (Einstein to Paul Ehrenfest, August 19, 1914 [CPAE 8, Doc. 34]). At this juncture, anything but his obsessive search for the field equations of general relativity is a mere distraction. His apparent indifference to military applications of his occasional work on gyroscopes and on airplane design tends also to dispel the persistent myth that he was an avowed pacifist during the war.[10]

A fitting capstone to this chapter of Einstein's life is provided by his speech on the occasion of Max Planck's sixtieth birthday, April 1918, an address in which the major themes of his early social isolation are sounded with great eloquence. It also serves as a haunting evocation of the letter to his "second mother" more than twenty years earlier. What has guided those who search for the temple of wisdom? "With Schopenhauer I believe that one of the most powerful motives that lead to art and science is the 'escape from daily life' ("Flucht aus dem Alltagsleben") with its painful coarseness and pitiful desolation, from the chains of one's own constantly shifting desires" (Einstein 1918j).

4. CONVERGENCE WITH THE WORLD: DISTILLING
POLITICAL POSITIONS OUT OF MORAL CONCERNS

The end of imperial Germany in November 1918 altered much in the political and social life of Germany, though the precise degree of continuity versus change in the transition to a republican form of government is still hotly debated. While the old order had certainly not been swept away, Einstein like many other intellectuals experienced a sense of liberation. Opinions no longer needed to be confined to private correspondence. An increasingly aggressive press provided a mass audience for political pronouncements of all stripes. The injunction, "Keep your trap shut," no longer obtained. Einstein's political initiation was facilitated by a number of factors, most obviously the authority conferred on him by the verification of his theory of general relativity and his virtual canonization by the British Royal Society in late 1919.[11] Important in this respect also was an attempt to escape private travails, including a protracted and unpleasant battle to divorce Mileva and to postpone marriage with his cousin Elsa.

New ideologies and the parties that espoused them vied with one another for influence in Weimar Germany. All were beneficiaries of newly won freedoms but extremists on the left and right frequently availed themselves of violent means to enlist support and gain their ends. Virulent anti-Semitism, projected by many right-wing nationalists and the radical right, increasingly reared its ugly head. Though Einstein's personal encounters with this virus initially were few, he witnessed public humiliations of friends and especially of the large East European Jewish community in Berlin. The blind racial hatred directed toward these Jews shocked him. In 1921 he wrote: "Until seven years ago, I lived in Switzerland. In the whole time I lived there, I was not aware of being Jewish. Nothing in my life that might have touched and aroused my Jewish sensibility was evident. This changed as soon as I moved to Berlin. There I encountered the plight of many young Jews. I witnessed how anti-Semitic surroundings robbed them of the opportunity to pursue regular studies and gain a secure livelihood. This is especially the case with the East European Jews, who are subject to constant harassment.... This and similar experiences have awakened in me Jewish national feeling" (Einstein 1921h).

After decades of lone searching for physical truths, Einstein yearned for a sense of authentic community and found one in the young underprivileged students of Berlin. He empathized with a group, whose situation he identified both with his own financial plight as a student and with the marginal existence of his small band of comrades-in-arms in Zurich and Bern. Though Einstein's transformation did not occur

overnight, it was an epiphany of sorts. By abandoning the assimilating "mimicry of the butterfly" one could begin to develop pride in
self and in the community of which one is a part (*Vossische Zeitung*,
March 27, 1921). The new impetus to enlist his fame in the defense
of the underprivileged did not mean that he would abandon his lifelong flight from the merely personal. His emergence from the "self-dug
hole" and venturing into the greater world of politics would be offset
by strictly maintaining an emotional distance from it. His isolation
had ended, but within the framework of his own idiosyncratic compromise. He summed up the dissonance between commitment and isolation as follows:

My passionate interest in social justice and social responsibility has always stood
in curious contrast to a marked lack of desire for direct association with other
human beings. I am a horse for single harness, not cut out for tandem or team
work. I have never belonged wholeheartedly to country or state, to my circle of
friends, or even to my own family. These ties have always been accompanied by
a vague aloofness....[12]

What were the nodes of engagement that Einstein developed in his
political phase? In what follows we look at three areas – Zionism, socialism, and pacifism – in which elements from his formative years shaped
his interpretations of these three "isms." In spite of his early social
isolation, Einstein had absorbed a certain degree of academic elitism
during his budding career. Lacking any exposure to the working-class
movement in the waning decades of the German Empire, he felt an
inevitable discomfort with the mass organizations which sprang up or
developed a new lease on life in the recently declared German republic.
Rather than the emotional harangues and frequent violence advocated
by mass parties on the right and left, he placed his faith in appeals to
reason by a liberal intelligentsia, which, in availing itself of the deco-
rous and principled use of manifestos, might best guide the fortunes
of republican Germany. Arriving at a considered opinion on an array
of public questions was not a simple matter, however. The first challenge for his developing political sensibility was to reconcile his genuine affection for German culture with his abhorrence of its prevailing
militarism. Another was to find a balance between empathy for specific constituencies such as his fellow scientists in Central Europe or
his Jewish brethren and his commitment to internationalism, firmly
rooted in his belief in the transnational character of science. It is the
dynamic contradiction between national allegiance and international
sympathies that defined Einstein's search for a Jewish identity and his
growing commitment to Zionism. Much as the assimilated Jew had lost
his way by denying solidarity with his brethren, so Einstein recognized

that he had, in his self-absorbed world, lost a sense of kinship with the larger world, especially his fellow Jews.

5. RECOVERING JUDAISM: EINSTEIN'S VERSION OF ZIONISM

Einstein viewed German wartime jingoism as a moral affront. After the war, this amorphous sensibility took on a specifically political character as Einstein confronted the growing hostility toward Jews in German society. In characterizing this swelling tide of anti-Semitism, he revealed a particular sensitivity to its psychological component: "I see how schools, the satirical press, and countless other cultural institutions of the non-Jewish majority undermined the confidence of even the best of my fellow-Jews and felt that it should not be allowed to continue in this fashion" (Einstein to Willy Hellpach, October 8, 1929 [EA 46–657]). This resolve, which stemmed from something "purely instinctive," was the spur to recovering "the Jewish soul" (Einstein to Kurt Blumenfeld, March 25, 1955 [EA 59–274]).

Anti-Semitism harbored another, even uglier lie, one that particularly outraged Einstein. Gentile contempt for the Jew was one thing, but even more despicable he found the scorn which his fellow German Jews heaped on their Russian and Polish brethren, the *Ostjuden*. These socially and economically disadvantaged members of the Jewish community, numbering about thirty thousand in Berlin, represented somewhat more than one-quarter of the total Jewish population of Berlin. Various measures contemplated by the Prussian government to control and even deport East European Jews after the war exacerbated the fear of many German Jews that they were the real targets of official displeasure.[13] Einstein dismissed this fear with the acerbic observation that it was "a Jewish weakness ... always and anxiously try to keep the Gentiles in good humor" (Einstein to Felix Frankfurter, May 28, 1921 [EA 36–210]). Jewish anti-Semitism, he thought, represented nothing less than a degrading ritual of redirecting onto the most vulnerable members of the community the Gentile anti-Semitism intended for all Jews. In assimilating to the demands of the larger society and internalizing its expectations, Jews were abasing themselves. "Let us leave anti-Semitism to the Aryan and save love for our own kind" (Einstein 1920h). Given the depth of his feeling on the matter, it comes as little surprise then that Einstein's first public stance on a political matter was a protest of the official and unofficial discrimination practiced against *Ostjuden*, which appeared as an article in a liberal Berlin daily at the end of 1919 (Einstein 1919h).

Acutely Einstein had come to realize "the excessive price at which the blessings of assimilation are bought by Jewish communities ... – a loss of solidarity, of moral independence and of self-respect" (Einstein 1931a, 25–6). The solution, he thought, was straightforward: "More dignity and more independence in our own ranks! Only when we have the courage to regard ourselves as a nation, only when we respect ourselves, can we win the respect of others, or put another way, the respect of others will then follow of itself" (Einstein 1920h; Rowe and Schulmann 2007, 146). By expressing solidarity with an East European Jewish community that had maintained authenticity through loyalty to its roots, their assimilated German brethren too might recapture a sense of communal worth. Einstein's conclusion: "Zionism [is] the only effort which leads us closer to [the] goal" of healing divisiveness and uniting all Jews (Einstein to Heinrich York-Steiner, November 19, 1929 [EA 48–835]).

Though he was aware of the contradiction with his internationalist sentiments, Einstein argued that an embrace of Zionist nationalism was a necessary step in the elevation of Jewish self-esteem. He couched his explanation for resolving the contradiction in the form of a parable:

I am against nationalism, but I am in favor of the Zionist cause. The reason for this has become apparent to me.... If a man has both arms and incessantly boasts of [the strength of] his right arm, then he is a chauvinist. If a man be missing his right arm, however, he must do all in his power to compensate for the missing limb. For this reason I am, in my personal attitude, an opponent of nationalism. As a Jew, however, I support ... the Jewish-national efforts of the Zionists. (Seelig 1956, 76)

The fatal illusions of assimilation could best be avoided by undertaking tasks in common. Of these the most important was the restoration of Jewish national life in Palestine, the very heart of the Zionist enterprise. Two visions of this enterprise dominated. One stressed the spiritual and cultural character of Zionism, the other, its political goals. Cultural Zionists directed their energies toward a Palestine which would rekindle the spiritual life of the Diaspora, and in which Palestine and the Diaspora would stand in a "constant reciprocal spiritual relationship" (Poppel 1977, 97–8). This group sought relatively modest funding for individual enterprise committed to the social and economic development of Palestine as a cultural homeland of world Jewry. The political Zionists, by contrast, believed that the political struggle had only just begun and stressed the importance of mass colonization and ambitious supporting budgets to advance the cause of a Jewish state there.

Einstein's adherence was clear from the beginning. His preference for Zionism's cultural aspect always outweighed purely political considerations. Shortly after his first and only visit to Palestine, in 1923, he wrote a friend: "On the whole, the country [Palestine] is not very fertile. It will become a moral center, but will not be able to absorb a large proportion of the Jewish people" (Einstein to Maurice Solovine, May 20, 1923 [EA 21–189]). In her unpublished biography, Einstein's sister states unequivocally that his impulse to support Zionism sprang out of "a sense of responsibility for providing an independent domain for those kinfolk, who as Jews were denied a place where their intellectual prowess in the sciences might enjoy creative play" (CPAE 1, Introduction, p. lx). Creating such a place was the centerpiece of the cultural Zionist vision: an institution of higher learning, the Hebrew University, which it was hoped might instill pride and provide a livelihood for countless young students and their teachers. Einstein argued that its establishment was "not a question of taste but of necessity" (article on his speech to Manchester University Jewish Students' Society, in *Manchester Guardian*, June 10, 1921).

Prospects for the university initially looked poor. The European leadership of the World Zionist Organization in London was strapped for funds and embroiled in political bickering with its largest affiliate, the Zionist Organization of America. The Europeans favored the formation of a political entity in Palestine while the American Zionists supported a cultural settlement. In spite of Einstein's natural inclination to the "American position," when the London organization requested that he join its delegation on a fundraising mission to the United States in 1921, he agreed "without giving it more than five minutes of thought" (Einstein to Fritz Haber, March 9, 1921 [EA 12–332]). Much as he had made a principled compromise in accepting Zionist nationalism in the face of a bedrock faith in internationalism, so Einstein, ever the pragmatist, accepted in the name of solidarity the political agenda of the World Zionist Organization though his sympathies lay elsewhere. The trip to America, though modest in its fundraising success, took on the character of a triumphal procession for Einstein, who was singled out for adulation by large crowds of recent East European Jewish immigrants. Equally triumphant was his only visit to Palestine in early 1923 on his return from a lengthy stay in Japan. The chairman of the event in which Einstein delivered the Hebrew University's first-ever scientific lecture concluded his introductory remarks with these momentous words: "Mount the platform which has been waiting for you for 2,000 years" (Samuel 1945, 253).

There was a faulty premise in the Zionist calculation shared by both wings of the movement that would come to haunt Einstein at the end of the 1920s. He and many others had taken at face value the Zionist slogan that Palestine was a land without people for a people without land. By the end of the decade, the acquisition of land through purchase from absentee Arab landowners by private capital as well as by the Jewish National Fund had created acute tension between Jew and Arab. The

Palestine riots of August 1929 undermined the beginnings of coexistence between the two parties and set nationalist passions ablaze. Einstein's shock was palpable. In compromising his internationalist principles to embrace Zionist nationalism, he had not counted on such deep-seated Arab hostility toward Jewish settlement nor the bitter accusations of Jewish chauvinism and exclusivity. The domination of Jew over Arab in Palestine threatened to transform the one-armed man of Einstein's parable into a two-armed braggart. The very basis for Zionist legitimacy lay in the preservation at all costs of what Einstein called its "spiritualized" character. During a reception given in his honor in the city hall of New York City at the end of the following year, he reiterated his view that the "unity of Jews the world over is in no way a political unity and should never become such. It rests exclusively on a moral tradition. Out of this alone can the Jewish people maintain its creative powers, and on this alone it claims its basis for existence" ("The Jewish Mission in Palestine," December 13, 1930 [EA 28–121]). The moral basis had been reinforced by a subtle messianic tone: "Revitalized through the mysticism of Zionism," Jews might "without being ridiculously arrogant about it regain an awareness of that mission to mankind, which they embody" ("Mission," *Jüdische Rundschau* 30 [February 17, 1925], no. 14, 129). At the same time this very same moral tradition demanded a just solution of the conflict between Arab and Jew, an end that seemed less and less attainable as the political situation in Palestine continued to deteriorate.

A solution, Einstein believed, might be found in a binational formula that prescribed a common state in historic Palestine shared between Jewish and Arab populations. Events on the ground, however, were dictating a different reality. The quota system imposed on immigration to the United States in 1924 diverted East European emigration to Palestine, a pressure counterbalanced by forceful appeals on the part of the Arabs to halt the population inflow and land transfer. The appeals were made to Great Britain, which had been granted the postwar mandate over Palestine by the League of Nations. Mutual intransigence exposed the overly optimistic character of the binational formulation, supported on the Jewish side by a minority consisting for the most part of cultural Zionists. The need to provide security and the material means of advancement to a swelling Jewish population ensured the triumph of adherents of the political Zionist position. For the Arabs, the presumed economic advantages conferred by Jewish colonization were overwhelmed by political fears that they would cease being masters in their own house. In 1918 there were fewer than sixty thousand Jews in Palestine; by the end of the 1930s, the number had grown to just less than five hundred thousand. Einstein had underestimated the determination of the Jewish settlers as well as the intensity of Arab opposition, writing off the differences in their views as "more psychological than real" (Einstein 1954, 172).

The British response served only to infuriate both parties. Fearful of compromising its access to the Suez Canal; the oil fields of Iraq and the Arabian Peninsula; and India beyond, Great Britain initiated a number of measures to curtail immigration of Jews and mollify the Arabs. None went far enough to satisfy the latter and did everything but reassure the former. Einstein, too, reserved his harshest criticism for what he considered the incompetent and mean-spirited, if not explicitly anti-Jewish, administration of the British. As early as 1929, soon after the Hebron riots, he advised against "leaning too much on the English. If we fail to reach real cooperation with the leading Arabs, we will be dropped by the English, not perhaps formally but *de facto*" (Einstein to Chaim Weizmann, November 25, 1929 [EA 33–411]). Following the Arab general strike of 1936 and the ensuing British restriction of Jewish immigration by 80 percent, the Peel Commission issued a report the following year, providing a sober account of the hopelessness of the situation. The report concluded that there was no common ground between the parties, that the British mandate was in shambles and that binationalism was an illusory goal. The commission recommended that Palestine be partitioned, a position that was accepted with major qualifications by the World Zionist Organization and rejected outright by the Arabs. What followed was a tragic inversion of Einstein's advice to avoid British mediation and deal directly with the other party. The immediacy of the relationship between Arab and Jew was defined increasingly by mutual terrorizing, not cooperation.

Undaunted, Einstein continued to lay out his principles of an understanding between the two peoples. In an address in New York in April 1938, entitled "Our Debt to Zionism," he first praised the role of Zionism in returning a sense of community and self-respect to the Jews, but then warned darkly against the "narrow nationalism" with which the movement was countering Arab violence. In doing so, he once again decisively threw in his lot with the cultural rather than the political Zionists:

I should much rather see reasonable agreement with the Arabs on the basis of living together in peace than the creation of a Jewish state. Apart from the practical consideration, my awareness of the essential nature of Judaism resists the idea of a Jewish state with borders, an army, and a measure of temporal power no matter how modest.... A return to a nation in the political sense of the word would be equivalent to turning away from the spiritualization of our community which we owe to the genius of our prophets. If external necessity should after all compel us to assume this burden, let us bear it with tact and patience. (Einstein 1954, 190; Rowe and Schulmann 2007, 301)

Though Hitler had come to power in January 1933, his regime only began to unleash a systematic fury on the German Jewish population at Kristallnacht, in early November 1938. British timing could not have been worse, for it was at this point that the British Foreign Office withdrew its partition plan from the table. Two weeks after the Night of Broken Glass, ten thousand Jewish children from Germany were denied entrance to Palestine. Einstein was stung by this betrayal, which signaled a British retreat from the Balfour Declaration of two decades earlier. Overnight, German Jews were granted only limited access to Palestine even as immigration there became increasingly a matter of life and death. In addition, no serious effort was undertaken by His Majesty's government to find alternative sanctuaries for Jews remaining in Europe. The final straw was embodied in the 1939 White Paper, which declared a limit of fifteen thousand a year for five years on Jewish immigration and none after that, unless the Arabs relented. Much as Einstein had predicted, British betrayal of Jewish and Arab interests was evident. Zionist rejection of the White Paper was matched by Arab demands that all Jewish immigration be halted at once. Terrorism on both sides began to dictate the policy of both parties. On the Jewish side, any pretense of accepting binational parity was abandoned as creation of a Jewish state became the rallying cry of the day. For the Arabs, the hope of a permanent Arab majority was slipping away.

Einstein's reaction to these events was an almost feverish attention to securing affidavits for European refugees seeking entry into the United States. In fact, the sheer number of the affidavits that he signed soon diluted their impact on the authorities. To his closest friend, Einstein wrote in the autumn of 1938 that he could not "give any more affidavits and would endanger those that are still pending if I were to give more.... The pressure on us from the poor people over there is such that one almost despairs, faced as one is with the depth of misery and the few possibilities of helping" (Einstein to Michele Besso, October 10, 1938 [EA 7–376.1]). Appalled at new measures being taken by the American State Department to exclude refugees from the United States, he was able after lengthy entreaties and with the help of Rabbi Stephen Wise to prevail upon President Roosevelt to create the War Refugee Board, which afforded European Jews more generous terms of entry (January 1944). The effort came far too late, but it served in small part to relieve the suffering of those who had escaped the ravages of Hitler's regime.

The Zionist leadership's decision in 1942 to embrace officially the goal of a Jewish state was determined by Nazi depredations in Europe. David Ben-Gurion's resolve and that of his colleagues was reinforced by the persistence of British exclusivist policy toward immigration, and the intransigence of His Majesty's government only grew in the postwar

period as fears of Soviet influence in the Middle East and Arab alien-
ation assumed greater importance. In 1944 Einstein summarized the
British position:

The promise held out to the Jews in the Balfour Declaration after the First World
War has been whittled down bit by bit in the course of the British appeasement
policy yielding to interests partly British, partly Arabian.... Palestine is a link
in the lifeline of the British Empire between the Near East and India; and
the Jewish people, by necessity a dependable ally of the British, have been
sacrificed to the Arabs who, by their numerical and political strength and the
trump of the Islamic portion of the Indian population, were in a position to
sell even their neutrality dearly in the present conflict. ("Palestine, Setting
of Sacred History of the Jewish Race," with Erich Kahler, *Princeton Herald*,
April 14, 1944)

In an effort to embroil the Americans in their Palestinian nightmare,
Great Britain formed an Anglo-American Committee of Inquiry on
Jewish problems in Palestine and Europe. In mid-January 1946, Einstein
was called before the committee to testify, heaping a veritable torrent
of abuse on British policy. He located the root of the problem in a policy of divide and conquer – fueled by British indifference to the incite-
ment of the Palestinian fellahin by Arab landowners and exacerbated by
active British discouragement of attempts for Arab-Jewish cooperation.
Nor were the political Zionists spared. Arguing that a Jewish majority
in Palestine was not important, Einstein dismissed the goal of a Jewish
state: "The state idea is not according to my heart. I cannot understand
why it is needed. It is connected with many difficulties and a narrow-
mindedness. I believe it is bad." That Arab and Jew live in harmony was
a far greater guarantor of a just political arrangement (Einstein testimony to the Hearing Before the Anglo-American Committee of Inquiry
on Jewish Problems in Palestine and Europe, Vol. 5, January 11, 1946,
118–35). Nevertheless, he remained a strong proponent of unlimited
Jewish immigration into Palestine. Many of Einstein's recommenda-
tions were accepted by the committee, which produced a final report
some months later calling for a binational solution and increased possibilities of immigration, while urging a campaign to improve the Arab
standard of living.

Good intentions dissolved into chaos after the leveling of the King
David Hotel by the Jewish terrorist organization, the Irgun Zwai Leumi,
in which almost one hundred Jews, Arabs, and Britons were killed.
Einstein summarized the result: "one can say what one thinks, but the
facts ... will be determined mainly by the cerebellum – that is, by the
'men of action'" (Einstein to Hans Mühsam, April 3, 1946 [EA 38–351]).
A state of undeclared war now existed between the parties. Some of his

harshest barbs were saved for his fellow Jews: "With respect to Palestine we have advocated unreasonable and unjust demands under the influence of demagogues and other loudmouths. Our impotence is bad. If we had power it might be worse still. We imitate the stupid nationalism and racial nonsense of the *goyim* even after having gone through a school of suffering without equal" (Einstein to Hans Mühsam, January 22, 1947 [EA 38–360]).

The British decided to cut their losses, and in late November 1947 the UN General Assembly accepted a plan to partition Palestine. Until their departure in mid-May 1948 the British adopted a passive attitude in the ongoing civil war, though their sale of arms to the Arabs and support of the Arab Legion placed the Jews at a distinct disadvantage. One month before the declaration of the Jewish state, Einstein and the eminent German-Jewish rabbi, Leo Baeck, issued a statement in which they called for the establishment of a homeland for the Jews "on a peaceful and democratic basis ... in accordance with the fundamental spiritual and moral principles inherent in the Jewish tradition and essential for Jewish hope" (*The New York Times*, April 18, 1948). Einstein had, however, by this time accepted with serious misgivings the fact that his hopes for a binational entity were irreparably dashed. With Arab armies and Jewish fighters still locked in combat in autumn 1948, he admitted that "there is no going back, and one has to fight it out. At the same time, we must realize that the 'big ones' [the United States and the Soviet Union] are simply playing cat and mouse with us and can ruin us any time they seriously want to" (Einstein to Hans and Minna Mühsam, September 24, 1948 [EA 38–379]).

On the death of President Chaim Weizmann of Israel in November 1952, negotiations were begun to woo Einstein as Weizmann's successor to this largely ceremonious office. Einstein's response to these overtures reveals the remarkable consistency with which he addressed the question of his relationship to Jewry since his first flirtation with Zionism at the end of 1919. Pleading advancing age and a special affinity for dealing with objective matters, Einstein expressed regret at having to decline "because my relationship to the Jewish people has become my strongest human bond, ever since I became fully aware of our precarious situation among the nations of the world." A practical consideration also intruded: "I also gave thought to the difficult situation that could arise if the government or the parliament made decisions which might create a conflict with my conscience; for the fact that one has no actual influence on the course of events does not relieve one of moral responsibility" (Nathan and Norden 1960, 572–3). Ambivalence may also have clouded the Israeli offer. "Aware of Einstein's unpredictable political independence," the Israeli premier David Ben-Gurion is reputed to have asked his secretary what to do if Einstein accepted, predicting trouble if the greatest living Jew agreed to shoulder the burden (Sayen 1985, 347).

Neither by ideology nor by his ideals was Einstein blinded to the complexities of the situation in Palestine. Nevertheless, he became disoriented by his lack of familiarity with its realities. In this he was not alone. Politically astute politicians such as Weizmann also suffered their disappointments. Even before the destruction of European Jewry gave a desperate urgency to finding a homeland for the displaced and dispossessed, Weizmann and moderates such as Einstein failed to take the true measure of nationalist fervor on both sides of the divide in Palestine, though his point of departure, from which he never wavered, was the fundamental moral necessity of Arab and Jew to live in mutual respect with the other.[14] Einstein's conciliatory appeals, with their characteristic mixture of hardheaded pragmatism and refusal to make concessions of principle, were ignored. And still we struggle with the formidable task of reconciling security for Israel with the legitimate interests of Palestinian Arabs.

6. ACHIEVING SOCIAL JUSTICE: THE SOCIAL DEMOCRATIC SOLUTION

Until the end of his life Einstein playfully referred to himself as a subversive and revolutionary.[15] When making his debut on the public stage in 1919, however, he was far more sensitive to charges that he was "a Communist and anarchist.... Nothing is farther from my mind than anarchist ideas. I do advocate a planned economy, which cannot, however be carried out in all workplaces (Betriebe), and in this sense I am a socialist" (interview of December 18, 1919, Neues Wiener Journal, December 25, 1919). How seriously are we to take a self-characterization drawn at the time of Einstein's political debut? Paramount for him always and consistently was the struggle for social justice, a core value best achieved within a democratic framework and dependent on the political and moral qualities of the citizenry.

The advance of this ideal, he argued, could best be guaranteed through the efforts of an intellectual elite rather than through direct democracy. This view was in part a direct consequence of Einstein's lack of familiarity with the politics of the postwar era and a reaction to the ever-more prevalent subsuming of the individual in mass organizations. After all, Einstein had his social sensibilities honed during an ancien régime of elitist politics and widespread suspicion of popular movements. Nothing demonstrates this more clearly than a statement expounding his credo: "What is truly valuable in our bustle of life is not the nation, I should say, but the creative and impressionable individuality, the personality – he who produces the noble and sublime while the common herd remains dull in thought and insensible in feeling" ("What I Believe," 193–4). He expanded on this theme in an article, which never appeared in print: "The state is certainly necessary, in order to give the individual the security he needs for his development. But when the state becomes the main thing and the individual becomes its weak-willed tool, then all finer values are lost" (article, 1934, intended for the journal *Tolerance* [EA 49–94], published in Dukas and Hoffmann 1979, 88–90).

Einstein's politicization after World War I, directly stimulated if not precipitated by his newfound Jewish sensibility, was expressed above all in terms of anti-Semitism's devastating effects not on the individual's material or economic conditions, but as a psychological phenomenon. What particularly distressed him was how it stripped the individual of self-respect and confidence. Sympathies exhibited by Einstein while still tentatively exploring the political sphere during World War I are echoed very forcefully fifteen years later in a letter of early 1932 to Sigmund Freud:

In our time, the intellectual elite does not exercise any direct influence on the history of the world; the very fact that of its division into many factions makes it impossible for its members to co-operate in the solution of today's problems. Do you not share the feeling that a change could be brought about by a free association of men whose previous work and achievement offer a guarantee of their ability and integrity? Such a group of international scope, whose members would have to keep contact with each other through constant interchange of opinions, might give a significant and wholesome moral influence on the solution of political problems if its own attitudes, backed by the signatures of its concurring members, were made public through the press...." (Einstein to Sigmund Freud, 1932 [EA 32–557])

The appeal to an intellectual elite of proven ability and integrity is a vestige of Einstein discomfort with mass democracy and the demands of a working-class movement. It also runs counter to the Marxist view that human nature is shaped by economic interests and that war and other conflicts reflect class interests. Einstein's serious political activity was directed foremost toward the "practical" problem of preventing war and aggression. Einstein's socialism was a "social-ethical" philosophy, whose purpose is "to overcome and advance beyond the predatory phase of human development," which has been characterized by conquest, subjugation, and exploitation ("Why Socialism?," *Monthly Review, an Independent Socialist Magazine* 1 [May 1949], no. 1, 9–15).

Initially, Einstein had been fascinated by the suppleness of market-driven economies, though even in 1918 he approved of a qualified planned economy. At the same time, he stoutly protested any programmatic loyalty to the Social Democratic Party or to any political party, including the German Democratic Party of which he was a founder (interview of December 18, 1919, *Neues Wiener Journal*, December 25, 1919). Einstein's contradictory positions on "socialism" at this early stage are the first tentative steps of someone who is feeling his way in unfamiliar territory. Twenty years later, he regretted his initial faith in the market (Einstein to Yisrael Doryon, April 18, 1939 [EA 32–765]), placing his appreciation of a planned economy in an expectation that it would curb capitalism's interference with the private pursuit of the public good. In the early 1930s, as capitalist societies teetered on the brink of collapse, a belief in a planned economy was almost second nature. Einstein's familiarity with the German tradition of state-guaranteed social services was to make him even more uneasy in the face of a market-driven ideology that he encountered on taking up residence in the United States.

Socialism seemed also to provide the solution to a problem that had long rankled Einstein. Since 1917, when he took the educational system in Germany to task for an excessive reliance on rote learning and exaggerated competition (Einstein 1917h), Einstein had decried the deadening effect of school curricula on the creative individual. Only the establishment of a socialist economy could, he thought, lead to coursework that was oriented toward social goals, thereby eliminating capitalism's emphasis on training students "to worship acquisitive success as a preparation for a future career." Caution was necessary though to preserve the paramount interest of society – the guarantee of an individual's rights:

It is necessary to remember that a planned economy is not yet socialism. A planned economy as such may be accompanied by the complete enslavement of the individual. The achievement of socialism requires the solution of some extremely difficult socio-political problems: how is it possible, in view of the far-reaching centralization of political and economic power, to prevent bureaucracy from becoming all-powerful and overweening? How can the rights of the individual be protected and therewith a democratic counterweight to the power of bureaucracy be assured? (Einstein 1954, 151–8; Rowe and Schulmann 2007, 446)

Einstein's socialism was distinctly democratic. His primary critique of capitalism was its potential for the "crippling of individuals," a feature which, in his view, it shared with bureaucratic socialism. By directing the individual's energies away from the full development of her creative potential and thus denying innovative contributions to the public good, capitalism bore some resemblance to its polar opposite, the antidemocratic, bureaucratic Soviet state.

When asked in March 1931 by the leader of the American Socialist party, Norman Thomas, whether a permanent and lasting peace was to be constructed on lines of cooperation expressed in socialism, Einstein's responded that "pacific solutions" must precede economic and political ones (Einstein to War Resisters International, Lyons, France, March 4, 1931, *The New York Times*, August 2, 1931). Implicit in this judgment lay an interesting insight into Einstein's political views. He was more at home with the liberal organizing principle of mobilizing individuals to "practice the ideal of pacifism every week through social gatherings with cultural programs in which peace is especially emphasized. Only in this way can the ideal of world peace get under the skin of the common people. The aim of pacifism, therefore, must be to function in the daily life of the people" ("Statement on a Kellogg League," July 1931, Einstein 1933a, 42). Missing is the language of working-class solidarity. Einstein appeals to the defiance of voluntary associations of concerned citizens, not to that of a professionally disciplined trade-union movement. "Liberty is the prerequisite of everything of real value" (Einstein to Isaac Don Levine, March 15, 1932 [EA 50–922]).

At the heart of Einstein's political analysis lay his faith in the ability of well-intentioned elites to solve intractable problems. In detailing his remedy for the impasse which Arab and Jew had arrived at in 1930, he urged the formation of a privy council to which Jews and Arabs would each send four representatives, independent of all political parties, each composed of "a doctor, elected by the Medical Association, a lawyer, elected by the lawyers; a working men's representative, elected by the trade unions; an ecclesiastic, elected by the ecclesiastics." These eight were to meet once a week and refrain from espousing the sectional interests of their profession or nation "but conscientiously and to the best of their power to aim at the welfare of the whole population of the country." Deliberations were to be secret. If three members on each side concurred in a decision on any subject, it should be published, but only in the name of the whole council (Einstein to the Editor of *Falastin*, March 15, 1930 [EA 46–154]).

In spite of this elitist bias, he clearly recognized that liberalism had lost much of its credibility through disregard of the basic truth that freedom cannot be guaranteed without first satisfying fundamental material needs. For this reason, as he stated in 1945, "the conditions for eking out individual freedom for the majority are more favorable in a socialist state than in an economic system founded on private property" ("Is There Room for Individual Freedom in a Socialist State?" after June 27, 1945 [EA 28–661]).

The elitist prejudices persisted nevertheless. In response to a question posed to him on his sixtieth birthday, he thought that a body of twenty wise men with the proper intellectual and moral qualities might act as the conscience of humanity. "If such a corporation could be formed, which by virtue of its intellectual and moral qualities represented a kind of conscience for humanity, such a body could exert through its resolutions a beneficial and, in the course of time, even decisive influence on the shape of social and economic relationships in the world." He did not, however, minimize the difficulties that stood in the way of such an endeavor. His foremost concern after the successful formation of such a body was how vacancies were to be filled in such a way "that the degeneration of quality is prevented to the degree that this is possible." The only, not completely satisfactory solution, he felt was "an election by the members of the corporation itself, as otherwise external influences would very quickly lead to a leveling effect." Not surprisingly, he bases this concept on the Academy of Plato ("Thoughts on Forming a Council of the Wise; Reflections on his 60th Birthday," March 14, 1939 [EA 28–473]).

7. THE FLEXIBLE PACIFIST

Much as he trimmed Zionist sails to conform to his individual vision, so, too, with his interpretation of pacifism. As in all his politics, Einstein's pacifism should be understood as a function of a moral stance rather than of a particular political commitment and also as an attitude which assumed a number of guises over time. He stated that his "pacifism is an instinctive feeling, a feeling that possesses me because the murder of men is disgusting. My attitude is not derived from any intellectual theory but is based on my deepest antipathy to every kind of cruelty and hatred" (Einstein to Paul Hutchinson [Editor of the *Christian Century*], July 1929, reprinted in Nathan and Norden 1960, 98). He was committed to solving the practical problem of preventing war by curbing aggression, and at the heart of this endeavor, even beyond morality, lay Einstein's concern with biology and with "the sexual singularity of the human male, which from time to time results in such wild explosions" (Einstein to Heinrich Zangger, after December 27, 1914 [Vol. 8, Doc. 41a in CPAE 10]). Both in his "Opinion on the War" from 1915 and in exchanges with Sigmund Freud, he continued to depict war as a collective manifestation of latent human tendencies toward aggression: "The psychological roots of war are – in my opinion – biologically founded in the aggressive characteristics of the male creature. We 'jewels of creation' are not the only ones who can boast this distinction; some animals outdo us on this point, e.g., the bull and the rooster...." ("My Opinion on the War," October 23–November 11, 1915 [CPAE 6, Doc. 20]).

Nowhere are the motives for his initial embrace of pacifism more apparent than during his first visit to Germany's western neighbor after the war, in April 1922. In revulsion at destruction on the battlefields of eastern France, he urged that "we ought to bring all the students of Germany to this place – all the students of the world – so they can see how ugly war is." Knowledge learned of others in books was partly to blame: the German view of the French was "far too literary," not practical (Charles Nordmann, "Avec Einstein dans les régions devastées," *L'Illustration*, April 15, 1922). The older generation, such as the professors who had poisoned the minds of their students during the war, was incorrigible. But the young might heed such an appeal, an appeal to an "internationalism that implies a rational relationship between countries, a sane union and understanding among the peoples, mutual cooperation and advancement.... Should rationality fail and war recur, it would ruin our civilization completely" (interview with Elias Tobenkin, *New York Evening Post*, March 6, 1921).

In order to guarantee this international agreement, Einstein was clear about the need for an international political organization:

Through the development of new technical devices, distances between men and between their institutions have suddenly shrunk to one-tenth their former size. In consequence mankind has organized a production system whose components are spread over the globe. It would not only be reasonable but it is actually necessary that the enlargement of the territories utilized in human production should be followed by the development of an appropriate political organization. (Address to the German Peace Congress, June 11, 1922, Einstein 1933, 11–13)

Einstein joined the Committee on Intellectual Cooperation of the League of Nations in May 1922, only to resign from it the following March. In withdrawing from the activities of the league he expressed his disappointment to the main German pacifist journal that it "not only does not embody the ideal of an international organization but actually discredits such an ideal" ("Statement on Resigning from the Committee on Intellectual Cooperation," *Die Friedens-Warte* 23 [June 1923], 186). The following year when Germany joined the league, he had once again shifted his opinion on the committee's efficacy, praising concrete steps that were being taken to address national animosities:

Rather than entertaining utopian schemes, the committee has initiated several modest but fruitful projects on a small scale, such as the international organization of scientific reporting, the exchange of publications, the protection of literary property, the exchange of professors and students among various countries, etc. Thus far, the greatest progress has been achieved in the sphere of international reporting. ("Report on Meeting of the Committee on Intellectual Cooperation," *Frankfurter Zeitung*, August 29, 1924)

Such shifts in opinion were dictated by a healthy sense of realism, as when in May 1925 he stated that "without the stabilizing influence of the United States this and other international organizations cannot exist" (interview with Herman Bernstein, *The New York Times*, May 17, 1925).

As pointed out earlier, Einstein's realism was of the muscular sort. On addressing a group of pacifists in late 1930, he warned them to avoid incestuous preaching:

When those who are bound together by pacifist ideals hold a meeting they are always consorting with their own kind only. They are like sheep huddled together while the wolves wait outside. I think pacifist speakers have this difficulty: they usually reach their own crowd, who are pacifists already. The sheep's voice does not get beyond this circle and therefore is ineffective.... Real pacifists, those who are not up in the clouds but who think and count realities, must fearlessly try to do things of practical value to the cause and not merely speak about pacifism. Deeds are needed. Mere words do not get pacifists anywhere.... (Address to the New History Society, "The Two Percent Speech," December 14, 1930, Einstein 1933, 34–7; Rowe and Schulmann 2007, 240)

One tack was his advocacy of uncompromising war resistance where conscription was established; where it did not obtain, Einstein floated his "two-percent solution." "For the timid who say, 'What is the use? We might be shut up in prison,' I add: even if only two percent of those supposed to perform military service should declare themselves war resisters and assert, 'We are not going to fight ...' the governments would be powerless – they could not put such masses into jail." The following year he established the Einstein War Resisters' International Fund to implement an appeal to a broad public to declare its refusal to give any further assistance to war or the preparation for war. "I ask them to tell their governments this in writing and to register their decision by informing me that they have done so" (Einstein to War Resisters' International meeting, Lyon, France, August 1, 1931). Einstein's idealism, as always, was tinged with a hard realism. Though statesmen and many individuals were really bent on creating a lasting peace, "the unremitting increase of armaments shows too clearly that they are no match for the hostile powers that press for preparedness for war" (Address to Flemish Peace Demonstration, August 23, 1931, EA 48–30).

Morality, not ideological purity, is the deciding factor. This was all the more important in the age of organization and standardization, which threatened to overwhelm the individual.

I regard it as the chief duty of the state to protect the individual and give him the opportunity to develop into a creative personality.... The state transgresses this commandment when it compels us by force to engage in military and war service, the more so since the object and the effect of this slavish service is to kill people belonging to other countries or interfere with their freedom of development. ("The Disarmament Conference of 1932," September 4, 1931, *The Nation* 133 [September 23, 1931], no. 3455, 300)

Only sacrifices that "contribute to the free development of the human individual" should be countenanced.

The failure of the Geneva Disarmament Conference was a severe blow. Nothing was left but to accept unconditional pacifism's failure "to overcome and advance beyond the predatory phase of human development," as he put it in the "Why Socialism" essay (reprinted in Einstein 1954, 151–8). Einstein's subtlety and pragmatism are clearly on display in an essay that appeared in 1935. Though vilified by his former pacifist allies, Einstein made no bones about the need at least in the short run to abandon his former position in the face of Nazi Germany's menace: "Under present circumstances, I do not believe that passive resistance, even if carried out in the most heroic manner, is a constructive policy. Other times require other means, although the final goal remains unchanged" ("A Re-examination of Pacifism," *Polity* 3.1 [January 1935]: 4–5). As Hitler began to raise his hand against the Jews of Germany, but well before he initiated the destruction of East European Jewry, Einstein saw the need to change course.

8. EINSTEIN IN AMERICA: THE NONPARTISAN CHAMPION OF THE UNDERDOG

Throughout his life Einstein recognized the inherent conflict between the freedom of the individual and the growing power of modern states and social organizations. Thus it is no accident that those whom he admired most in political life were men like Gandhi who risked everything to challenge the authority of the repressive political system under which they lived.[16] His transition from refugee and "bird of passage" to stalwart partisan of the "American way of life" came quite naturally given the sense of urgency he felt not only about the fate of his Jewish brethren but of European civilization in general. In 1943, in an effort to contribute to the war effort, he participated on a part-time basis on research into the nature of explosions for the Bureau of Ordnance of the U.S. Navy. Only after the war did he begin to voice his views, always in a diplomatic tone, regarding anti-Semitism and racism in the United States, as well as speak out on the madness of nuclear confrontation.

Einstein's reputation as the "father of the bomb" was firmly implanted in the public's imagination after the war. In Japan, the issue resurfaced in 1952 when the government finally released photographs of the devastation wrought by the bombs that destroyed Hiroshima and Nagasaki. Thirty years after his triumphant visit, Einstein was contacted once again by a Japanese journal requesting his reaction to the photographs. He vigorously rejected the paternity suit: "My participation in the production of the atomic bomb consisted of one single act: I signed a letter to President Roosevelt" (Einstein to F. D. Roosevelt, August 2, 1939 [EA 33–088]). In addition, his reply revealed a reappraisal of pacifism: "Only the radical abolition of war and the threat of war can remedy the situation. This is what we should work toward, remaining steadfast in order to avoid any actions that would divert us from this goal. This is a harsh demand for the individual who is conscious of his dependence on society; but it is not an impossible demand" (Einstein to the Editor of *Kaizo*, September 20, 1952).

In the wake of the successful campaign to defeat fascism on two fronts, gratitude to his adopted country initially seemed almost boundless. In 1945, he explained the reasons for his happiness: "America is a democratic land not only through its laws and institutions but also through the spirit of its inhabitants. Here no one degrades himself before others and each is aware that they should feel and show respect toward all other beings. No one takes themselves to be terribly important, and a person would look foolish if they were to come forward in some kind of pompous manner." But he did not neglect to point out at the same time that the "dark shadow of racial prejudices, particularly against blacks" was a "shameful evil" that cried out for eradication ("On Political Freedom in the USA," 1945, EA 28–627). He picked up this theme the following year, arguing that there was

A somber point in the social outlook of Americans. Their sense of equality and human dignity is mainly limited to men of white skins. Even among these there are prejudices of which I as a Jew am clearly conscious; but they are unimportant in comparison with the attitude of the "Whites" toward their fellow-citizens of darker complexion, particularly toward Negroes. The more I feel an American, the more this situation pains me. I can escape the feeling of complicity in it only by speaking out. ("The Negro Question," *Pageant*, New York, January 1946)

As the Cold War loomed, Einstein's confidence in American foreign policy quickly came to an end. His support for Henry Wallace, the estranged New Dealer who ran against Truman in 1948, led to accusations that he was a fellow traveler, one of many "Moscow dupes." A heated exchange with the political philosopher Sidney Hook, a staunch anti-Communist, reveals how Einstein attempted to balance his condemnation of Soviet bureaucratic socialism with his growing dismay at the policies of the Truman administration for exacerbating the conflict between the world's two superpowers (EA 58–299 to 58–301). Even more strident was his opposition to the newly emerging national-security state. He remained convinced that only through the creation of a supernational regime with far more authority than the League of Nations could humanity hope to survive in the long run. This survival, in turn, would have to be based on the suppression of man's instinctive aggressiveness. The radical step that Einstein envisioned went far beyond merely banning nuclear weapons. Instead he sought support for the internationalization of atomic energy and the advocacy of world government, motivated "by the belief that the concept and practice of unlimited national sovereignty in the age of atomic energy could result in the catastrophe of war – an evil of such magnitude that it dwarfed all differences over political and economic matters" (Sidney Hook, "My Running Debate with Einstein," Commentary [July 1982], 43).

One of Einstein's political platforms in this period was his chairmanship of the Emergency Committee of Atomic Scientists from 1946 until 1948. Though only the committee's figurehead, he attempted to educate the public about nuclear energy. The evisceration of the Acheson-Lilienthal plan, which sought to place the military use of atomic energy under the jurisdiction of an international agency, proved deeply disappointing to him. His disappointment was expressed vividly in a letter at the end of his tenure with the committee: "If you ask yourself who, since the termination of the war, has threatened his opponent to a higher degree by direct action: the Russians the Americans or the Americans the Russians?" The answer for him was clear and based on the fact that "the military strength of the USA is at present much greater than that of Soviet Russia. It would therefore be sheer madness if the Russians would seek war. On the other hand I have heard influential people in this country pleading for 'preventive war,' even before the last war was finished" (Einstein to Sidney Hook, April 3, 1948 [EA 58–300]).

By the early 1950s, J. Edgar Hoover, Senator Joseph McCarthy, and the right wing of the Republican Party had seized center stage in American politics. Their frenzied search for closet Communists and their sympathizers became a national hysteria. Among the numerous charges made against Einstein was one by which he stood accused of sponsoring the principal Communist causes in the United States, serving as either chairman, member, sponsor, or patron of "10 organizations which have been cited by the Attorney General ... as being Communistic groups." Hoover's attention was drawn to Einstein from the very beginning of the exile's stay in the United States by the watchful Daughters of the American Revolution (DAR). When Einstein applied for a visa application in winter 1932/1933, the DAR protested, which elicited more amusement than annoyance from the object of scrutiny:

Never yet have I experienced from the fair sex such energetic rejection of all advances; or if I have, never from so many at once. But are they not quite right, these watchful citizenesses? Why should one open one's doors to a person who devours hardboiled capitalists with as much appetite and gusto as the Cretan Minotaur in days gone by devoured luscious Greek maidens, and on top of that is low-down enough to reject every sort of war, except the unavoidable war with one's own wife? Therefore give heed to your clever and patriotic womenfolk and remember that the Capitol of mighty Rome was once saved by the cackling of its faithful geese. (Einstein to Daughters of the American Revolution, December 3, 1932 [EA 28–213])

What had appeared to him as ridiculous in the early 1930s assumed a new gravity in the McCarthy period. Inevitably, he viewed this postwar destruction of civil liberties through the prism of his post–World War I experience. Once again, the state, he felt, was seeking to interfere with the free communication of opinion and scientific inquiry. It had created a huge police apparatus which engendered widespread political distrust and instilled fear in individuals. Many avoided anything that might be construed as suspicious and threaten their livelihood. Yet these were only symptoms.

The real ailment, however, seems to me to lie in the attitude which was created by the World War and which dominates all our actions; namely, the belief that we must in peacetime so organize our whole life and work that in the event of war we would be sure of victory. This attitude gives rise to the belief that one's freedom and indeed one's existence are threatened by powerful enemies. ("Symptoms of Cultural Decay," *Bulletin of the Atomic Scientists* 8.7 [October 1952]: 217–18)

The only way for intellectuals to confront the McCarthyite hysteria was to refuse to testify if called before a congressional committee. The solution was a harsh one, however, almost ruthless in its simplicity. One must be prepared for jail and economic ruin, "in short, for the sacrifice of his personal welfare in the interest of the cultural welfare of his country.... If enough people are ready to take this grave step they will be successful. If not, then the intellectuals of this country deserve nothing better than the slavery which is intended for them" (Open Letter to William Frauenglass of May 16, 1953, *The New York Times*, June 12, 1953). A year later he underlined the point with a comment that seems impish at first glance but actually belies a deep sense of consternation. On the occasion of J. Robert Oppenheimer's summons before a committee of the Atomic Energy Commission, Einstein noted that "[i]f I were a young man again and had to decide how to make a living, I would not try to become a scientist or a scholar or teacher. I would rather choose to be a plumber or a peddler, in the hope of finding that modest degree of independence still available under present circumstances" ("On Intellectual Freedom," *The Reporter* 11.9 [November 18, 1954]: 8).

In the last month of his life, he summed up the lessons of his political journey. He dwelled on what he called "the big problem in our time," the power conflict between East and West. Yet, as always, he tried to wrest hidden truths from behind appearances. He viewed the struggle as an "old-style struggle for power, once again presented to mankind in semi-religious trappings. The difference is that, this time, the development of atomic power has imbued the struggle with a nightmarish character; for both parties know and admit that, should the quarrel deteriorate into actual war, mankind is doomed." The only means of overcoming the nightmare, of holding out the promise of peace, was "the course of supranational security" a position that he had supported since the infancy of the League of Nations some thirty-five years earlier. If no politician could afford to advocate such a course, one that would be tantamount to committing political suicide, Einstein, the nonpolitician, dared. He offered his opinions on public issues "whenever they appeared to me so bad and unfortunate that silence would have made me feel guilty of complicity."[17]

NOTES

1 Transcript of Einstein's recorded address "Human Rights," on accepting award of the Chicago Decalogue Society of Lawyers, February 20, 1954 (EA 28–1016); published in *The New York Times* of the following day, and in Einstein (1954, 34–6).

2 See Cahnmann (1989) and Bach (1984, 112). Both are cited in Stachel (2002a, 76, notes 5 and 6), in his discussion of the emergence of a Jewish middle class in nineteenth-century Germany.

3 For a partially firsthand account of Einstein's youth, see the excerpt of Maja Einstein-Winteler, "Albert Einstein – Beitrag für sein Lebensbild" (CPAE 1, [xlviii]–lxvi).

4 E.g., Einstein to Marcel Grossmann, September 6, 1901 (CPAE 1, Doc. 122).

5 For an extensively documented account of the Bern period, see Flückiger (1974).

6 Einstein (1956b), which presents Einstein's letters to Solovine, also includes an introduction by Solovine on the Olympia Academy.

7 On the further development of his relationship with Mileva, see Trbuhović-Gjurić (1993), Renn and Schulmann (1992), Overbye (2000), and Stachel (2002a); on his lifelong correspondence with Besso, see Einstein and Besso (1972).

8 Lewis Feuer, e.g., makes such a claim in the preface to the second edition of his book, Feuer (1982).

9 Bergman (1974, 390); Einstein to Hedwig Born, September 8, 1916 (CPAE 8, Doc. 257); and Einstein to Heinrich York-Steiner, November 19, 1929 (EA 48–835).

10 For a discussion of the contradiction between Einstein's pacifist stance and his lack of concern about military applications of his work on a gyrocompass and on airplane design, see Fölsing (1993, 446–9).

11 "Joint Eclipse Meeting of the Royal Society and the Royal Astronomical Society," in *The Observatory* 42 (1919): 389–98.

12 "What I Believe," in *Forum and Century* 84.4 (October 1930): 193–4; reprinted in Einstein (1954, 8–11).

13 Antagonism between the two communities was heightened by German Jewish fears that Eastern European Jews endangered their goal of full assimilation. See Schulmann (2000, 576).

14 He remained true to this sentiment until the end of his life: "The attitude we adopt toward the Arab minority will provide the real test of our moral standards as a people." Einstein to Zvi Lurie, January 4, 1955 (EA 60–388).

15 See numerous such instances in Johanna Fantova's diary account of telephone conversations with Einstein in the 1950s, Princeton University Library, Fantova Collection.

16 See, e.g., an interview with UN Radio, June 16, 1950; reprinted in *The New York Times*, June 19, 1950.

17 Transcript of Einstein's recorded address "Human Rights," on accepting award of the Chicago Decalogue Society of Lawyers, February 20, 1954 (EA 28–1016); published in *The New York Times* of the following day, and in Einstein (1954, 34–6).

Appendix
Special Relativity

1. THE RELATIVITY POSTULATE, THE LIGHT POSTULATE, AND THEIR STRANGE CONSEQUENCES

1.1. The Two Postulates of Special Relativity and the Tension between Them

When Einstein first presented what came to be known as special relativity, he based the theory on two postulates or principles, called the "relativity postulate" or "relativity principle" and the "light postulate." Both postulates are supported by a wealth of experimental evidence. The combination of the two, however, appears to lead to contradictions. To avoid such contradictions, Einstein argued, we need to change some of our fundamental ideas about space and time.

Einstein formulated the relativity postulate as follows: "The same laws of electrodynamics and optics will be valid for all frames of reference for which the equations of mechanics hold good" (Einstein 1905r, 891). Such frames of reference are called inertial frames and an observer at rest in one of them is called an *inertial observer*. A few examples will suffice to understand both the concept of an inertial frame and the meaning of the relativity postulate. First consider a plane which starts out sitting on the tarmac, proceeds to fly through clear skies, and eventually hits turbulence. All the while a passenger is nursing a cup of coffee. Sipping coffee without spilling is easy during the smooth portion of the flight. This is because the laws governing the behavior of the coffee in the frame of reference of the plane flying at constant velocity are the same as in the frame of reference of the airport.[1] These laws hold in any frame moving uniformly (i.e., with constant velocity) with respect to the frame of the airport. Drinking coffee without spilling when the plane ride gets bumpy is much harder. The laws for the coffee in noninertial frames, such as the frame of a plane encountering turbulence, are more complicated than in inertial frames. As a second example, consider a cruise ship that sets out from its port of origin, sails smoothly on a calm sea, and eventually is caught in a storm. All the while two passengers engage in a drawn-out tennis match on the ship's upper deck. No matter whether the ship lies anchored in the harbor or is sailing on calm seas, the balls will bounce and spin according to the same physical laws. Playing tennis on the deck of a cruise ship is no different from playing tennis on any other court. Once the ship hits bad weather, however, the laws governing the motion of the balls will get more complicated and the players will have to start adjusting their strokes accordingly.[2]

For purely mechanical phenomena such as the bounce of balls or the flow of fluids, the relativity principle had been known since the days of Galileo. For reasons that will become clear, nineteenth-century physicists expected the principle to break down for electromagnetic and optical phenomena. But no violations were ever found. Einstein could thus extend the principle to all of physics, in particular to electromagnetism and optics.

To the relativity postulate Einstein added the light postulate: "light propagates through empty space with a definite velocity [c, about 186,000 miles per second or 669,600,000 miles per hour] which is independent of the state of motion of the emitting body" (Einstein 1905r, 891). That the velocity of light is independent of the velocity of its source is one of the key features of the electromagnetic theory developed in the second half of the nineteenth century by Maxwell, Lorentz, and others. The light postulate is thus indirectly supported by all the evidence amassed during the nineteenth century for this powerful theory.

When Einstein introduced his second postulate, he immediately warned his readers that it is "apparently irreconcilable" (ibid.) with the first. That the two postulates would seem to be incompatible with one another is not difficult to see; the hard part is to see that this incompatibility is only apparent.

Consider the situation in Figure A.1, showing two SUVs and a sedan driving down the highway. The sedan is moving at 60 mph in one direction, the SUVs are moving in the other direction, one at 50 mph, the other at 70 mph. Common sense tells us that the drivers of the SUVs will give different answers when asked how fast the sedan is approaching them. The first driver, going 50 mph, will say 110 mph; the second one, going 70 mph, 130 mph. Now ask both drivers how fast they think the light from the headlights of the sedan is coming toward them. The answers will depend on whether the drivers think of light as waves in some medium, like sound in air, or as particles, emitted like bullets from a gun. If light consists of waves in a medium, the velocity of light is the velocity with which these waves propagate through the medium. If light consists of particles, the velocity of light is the velocity with which these particles are emitted from their source.

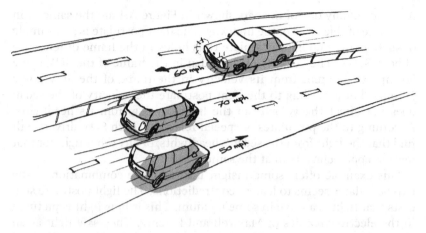

Figure A.1. *The postulates and their strange consequences for the behavior of light.*
Drawn by Laurent Taudin.

If the drivers of the SUVs think of light as particles, they have to take
into account both the velocity of their SUVs and the velocity of the
sedan. The velocity of the sedan, 60 mph, will be a component of the
velocity of the light particles emitted by its headlights. For the drivers of
the SUVs approaching the light, their own velocity will be another component of the light's velocity. Hence, the driver going 50 mph will say
that the light is approaching him with $c + 110$ mph, while the one going
70 mph will say that the light is approaching him with $c + 130$ mph.

If the drivers of the SUVs think of light as a wave in medium, they
have to take into account the velocity of the medium, the velocity of the
SUVs, but not the velocity of the sedan. It is a general property of waves
in a medium that their velocity is independent of the velocity of their
source. The ripples spreading across a pond from the point where a rock
hit the water travel at the same speed regardless of whether you skipped
the rock on the water or lobbed it into the pond at that point. Likewise,
the velocity of the sedan will not be a component of the velocity of the
light waves leaving its headlights. The velocity of the medium, however,
will be a component of the velocity of the waves. A wave is a disturbance
of the medium and is carried along with it. Assuming the medium for
the light waves to be at rest with respect to the highway, the driver going
50 mph will say that the medium is approaching him with 50 mph and
the light with $c + 50$ mph, while the other one will say that the medium
is approaching him with 70 mph and the light with $c + 70$ mph.

If Einstein's postulates are correct, both sets of answers are wrong.
The relativity postulate says that the laws of physics in the frame of

reference of any one of the cars shown in Figure A.1 are the same as in the frame of reference of the highway. The light postulate is an example of such a law. The light postulate thus holds in the frame of reference of both SUVs. The velocity of the sedan in the frame of the SUV going 50 mph is different from its velocity in the frame of the SUV going 70 mph, but according to the light postulate the velocity of the sedan does not affect the velocity of the light coming from its headlights. According to the postulates of special relativity, both SUV drivers will find that the light from the sedan's headlights, or any other light for that matter, approaches them at the same speed c!

This example offers some insight into why the combination of the two postulates seems to lead to contradictions. The light postulate suggests that light is a wave in some medium. This is how light is pictured in the electrodynamics of Maxwell and Lorentz. They saw light as an electromagnetic wave in the ether, a substance thought to fill all of space with no internal motion like a perfectly calm sea filling every nook and cranny of the universe. This is why adherents of this theory expected the relativity principle to break down. An observer in uniform motion through the ether is not equivalent to an observer at rest in the ether. A wave in the ether only moves with the same constant velocity in all directions with respect to the latter. Yet experiment suggested that the relativity principle holds for light as well. The relativity principle is compatible with light consisting of particles moving through empty space. But then the velocity of light depends on the velocity of the source from which the light particles are emitted, which contradicts the light postulate. If we insist on having both the relativity and the light postulate, light, it seems, can neither be a particle nor a wave.

Figure A.2 illustrates the problem of combining Einstein's two postulates in a slightly different way. We have two observers, Al and Bob, moving with respect to one another (like the drivers of the SUVs in Figure A.1). They examine a flash of light emitted by a lightbulb, a cork shooting out of a champagne bottle, and a bullet fired from a gun. Common sense tells us that the light, cork, and bullet will have different velocities for Al and Bob. Special relativity confirms this in the cases of the cork and the bullet, even though, as we shall see, it calls for corrections of the commonsense values $v + v_{cork}$ and $v + v_{bullet}$ shown in Figure A.2. In the case of light, however, it is a direct consequence of Einstein's two postulates that Al and Bob, despite being in motion with respect to one another, register the exact same velocity! How can this possibly be? It is clear that something will have to give.

What Einstein showed in his 1905 paper is that once we accept his two postulates – as we should given all the empirical evidence backing them up – we have to give up some of our commonsense ideas about

Figure A.2. *The postulates and adding velocities.* Drawn by Laurent Taudin.

space and time. Now it is one thing to concede when confronted with the relentless logic of Einstein's 1905 paper that our old ideas were nothing but prejudices; getting comfortable with the new ideas that Einstein put in their place is a different matter. Here the 1905 paper is of little help. It does not tell us how to visualize the new relativistic ideas about space and time. Such visualization was provided a few years later by Minkowski (1909). Minkowski's geometrical formulation of special relativity is still standard today. It took Einstein several years to appreciate this contribution (Pais 1982, 152). In Section 2 we shall turn to Minkowski's geometry of relativistic space-time. In the remainder of this section we stay true to the spirit of Einstein's more abstract approach.

In a nutshell, the argument in Sections 1.2–1.5 is as follows. A direct consequence of the two postulates[3] is that two inertial observers moving with respect to one another will disagree on whether events at different locations happen at the same time or not. This phenomenon is called the *relativity of simultaneity* (Section 1.2). Judgments about the simultaneity of events at different locations are involved in measuring the rate of moving clocks and the length of moving rods. As a consequence, we find that moving clocks must tick at a lower rate than those same clocks at rest (Section 1.3) and that the length of moving rods must be less than the length of those same rods at rest (Section 1.4). Otherwise we end up with violations of the relativity principle. What makes these phenomena – called *time dilation* and *length contraction*, respectively – especially baffling is that which clocks and which rods are moving and which ones are at rest depends on whose point of view we adopt. Two observers in relative motion will both claim that the other observer's rods are contracted and that the other observer's clocks run slow. As we shall see in Section 2, Minkowski's geometrical interpretation of length contraction and time dilation makes this much easier to understand. We conclude Section 1 by examining the consequences of these new and unexpected phenomena for the addition of velocities (Section 1.5). This will resolve the apparent contradictions illustrated in Figure A.2.

Qualitatively, this resolution goes as follows. Once again consider Figure A.2. According to Al, the cork, bullet, and light flash have velocities v_{cork}, v_{bullet}, and c, respectively. According to Bob, however, Al has determined these velocities using rods that are contracted and clocks that are not properly synchronized and are running slow to boot. If Bob wants to know how fast the objects are moving with respect to him given how fast they are moving with respect to Al, he cannot simply add the velocities of the objects reported by Al to Al's own velocity v. He first needs to correct Al's results. The general rule for adding velocities measured by two observers in relative motion that takes into account such corrections is called the *relativistic addition theorem for velocities*. In the case of light, it turns out, the corrected value for the velocity reported by Al is $c - v$ (see note 15). Adding Al's own velocity to this corrected value, Bob finds that the light is moving with velocity c with respect to him as well. In other words, even though they are in motion with respect to one another, both observers find that one and the same light flash is moving with velocity c with respect to them. This takes care of the problem brought out with the help of Figure A.2.

The analysis of Figure A.1 suggested that light could neither be a wave nor a particle if we accept Einstein's two postulates. It turns out, however, that special relativity is compatible with both views. Picture light as a wave in the ether of nineteenth-century physics. For an observer at rest in the ether, the light will have velocity c. Because of the way velocities are added in special relativity, however, it will have the exact same velocity for any observer in uniform motion through the ether. This means that it is impossible to tell with respect to which observer the ether is truly at rest. For this and other reasons, Einstein preferred to do away with the ether altogether, calling it "superfluous" in the introduction of his 1905 paper. Like Einstein, we now think of electromagnetic waves as propagating through empty space rather than through an ether. If we picture light as a particle emitted with velocity c by the headlights of the sedan, we need to add the velocity of the sedan to the velocity of the light. But because of the way velocities are added in special relativity, the result will once again be that these particles move with velocity c with respect to the two SUVs as well. Before we factored in the relativity of simultaneity, length contraction, and time dilation, the wave picture of light and the particle picture of light gave different answers for the velocities with which light from the sedan's headlights is approaching the two SUVs. Once these three phenomena are taken into the account both pictures give the same result, which, moreover, is precisely the result we found on the basis of a direct application of the postulates. Both the wave and the particle picture are thus compatible with the postulates. Special relativity is agnostic about the nature of light.

1.2. The Relativity of Simultaneity

Whether two events occurring at different locations happen simultaneously or one after the other depends on the state of motion of the person making the call. This is the key insight of special relativity. It dawned on Einstein about six weeks before he published his theory, and it made everything fall into place.[4]

Consider Figure A.3. Bob is standing exactly in the middle of a railroad car moving to the right at constant velocity v. Al is standing by the tracks. To both ends of the railroad car a lightbulb is attached, labeled L_1 and L_2, respectively. Both lightbulbs flash once. At exactly the moment that Al and Bob come face to face with one another (see the solid line connecting them), the two flashes reach Bob. Al and Bob agree that these two flashes hit Bob at the same time. They will always agree on what happens at one and the same location at one and the same instant. They will not agree, however, on whether the lightbulbs flashed at the same

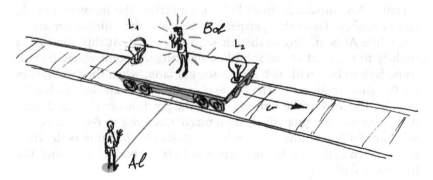

Figure A.3. *Two light flashes reach Bob at the same time.* Drawn by Laurent Taudin.

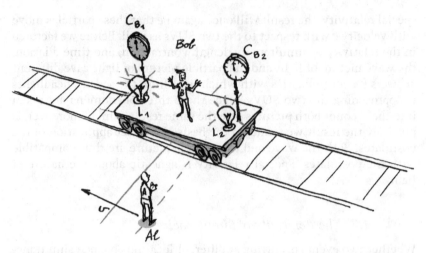

Figure A.4. *According to Bob, lightbulbs* L_1 *and* L_2 *flashed at the same time.* Drawn by Laurent Taudin.

time. The flashing of L_1 and the flashing of L_2 are events happening at different locations. Appealing to Einstein's two postulates, Bob will say that these two events happened simultaneously. Appealing to the same postulates, Al will say that L_1 flashed before L_2.

Bob offers the following impeccable argument:

I am an inertial observer, so the laws of nature are the same for me as they are for any other inertial observer. The light postulate is one such law. Hence, the light flashes from L_1 and L_2 have velocity c with respect to me. I am standing halfway between L_1 and L_2. Hence, the light flashes from L_1 and L_2 had the same distance to cover to get to me. They hit me at the same time. *Ergo*, they must have left L_1 and L_2 a little earlier at the same time.

Figure A.4 illustrates, from Bob's perspective, the moment that L_1 and L_2 flashed. From this perspective, Bob and the railroad car are at rest while Al is moving to the left with velocity v. At this point Al is slightly to the right of the solid line marking the point where he will come face to face with Bob a split second later, when the two flashes hit Bob (see Figure A.3).[5] The figure also shows two clocks, labeled C_{B_1} and C_{B_2}, fastened to the railroad car at the positions of the lightbulbs. If these clocks are properly synchronized according to Bob (hence the subscript "B"), the time on C_{B_1} when L_1 flashes is the same as the time on C_{B_2} when L_2 flashes. In the figure it is 12:00 on both clocks when the lightbulbs flash.

Starting from the same information (the two light flashes hit Bob simultaneously), appealing to the same postulates, and with a logic as impeccable as Bob's, Al reaches a very different conclusion. Al argues as follows:

According to the light postulate, the velocity of light is not affected by the velocity of its source. Hence, the light flashes of the moving lightbulbs L_1 and L_2 both have velocity c with respect to me. Bob, standing in the middle of the railroad car, was rushing away from the flash coming from L_1 and rushing toward the flash coming from L_2. The flash from L_1 thus had a larger distance to cover to get to Bob than the flash from L_2. Yet the two flashes hit Bob at the same time. That means that the flash from L_1 must have started to make its way over to Bob before the flash from L_2 did. *Ergo*, L_1 flashed before L_2.

Figure A.5 illustrates the flashing of the two lightbulbs from Al's perspective. The top half shows the flashing of L_1 at the rear of the railroad car; the bottom half the flashing a split second later of L_2 at the front. It also shows the readings on Bob's clocks C_{B_1} and C_{B_2}. Note that Al agrees with Bob that C_{B_1} read 12:00 when L_1 flashed and, likewise, that C_{B_2} read 12:00 when L_2 flashed. The flashing of a lightbulb and the reading on a clock right where that lightbulb is are part of one event happening at an instant at a particular location. The disagreement between Al and Bob is never about such individual events, but always about what other events are happening elsewhere at the same time. According to Al, the event "L_1 flashes and C_{B_1} reads 12:00" at the rear of the railroad car happened before the event "L_2 flashes and C_{B_2} reads 12:00" at the front. When L_1 flashed, L_2 had not flashed yet and C_{B_2} did not read 12:00 yet but, say, 11:55.[6] Likewise, when L_2 flashed, L_1 had already flashed and C_{B_1} no longer read 12:00 but 12:05. In other words, when Bob has synchronized C_{B_1} and C_{B_2} properly, Al will say that C_{B_1} is five minutes fast compared to C_{B_2}.[7] This is true in general. If one observer synchronizes two clocks at different locations, both at rest with respect to her, another observer with respect to whom she is moving at some constant velocity will find that her rear clock is fast compared to her front clock: the greater the velocity and the farther the two clocks are apart, the greater the discrepancy between the two clocks (the exact formula will be derived in Section 1.5; see Figure A.13).

1.3. Time Dilation: The Rate of Moving Clocks and Other Processes in Systems in Motion

A direct consequence of the relativity of simultaneity is that we have to give up the commonsense notion that clocks in uniform motion run at the same rate as clocks at rest, or, more generally, that processes in systems in uniform motion happen at the same rate as those same processes in systems at rest. Special relativity tells us that processes in

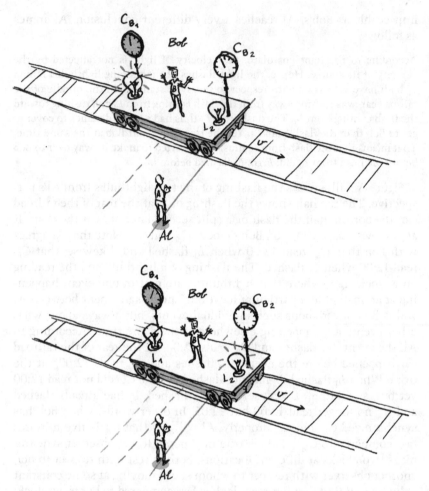

Figure A.5. *According to Al, lightbulb L₁ flashed before lightbulb L₂.* Drawn by Laurent Taudin.

moving systems take longer than in systems at rest. This phenomenon is called *time dilation*. Although it is completely negligible at everyday velocities, time dilation affects all physical processes: mechanical, electrodynamical, and biological ones such as the metabolism of organisms or the aging of human beings (see Section 2.7).

Figure A.6 brings out the problem with the commonsense assumption about the rate of moving clocks. Suppose Bob wants to measure the rate of one of Al's clocks moving to the left with velocity *v* (see the first column of the table in Figure A.6). For this measurement he needs

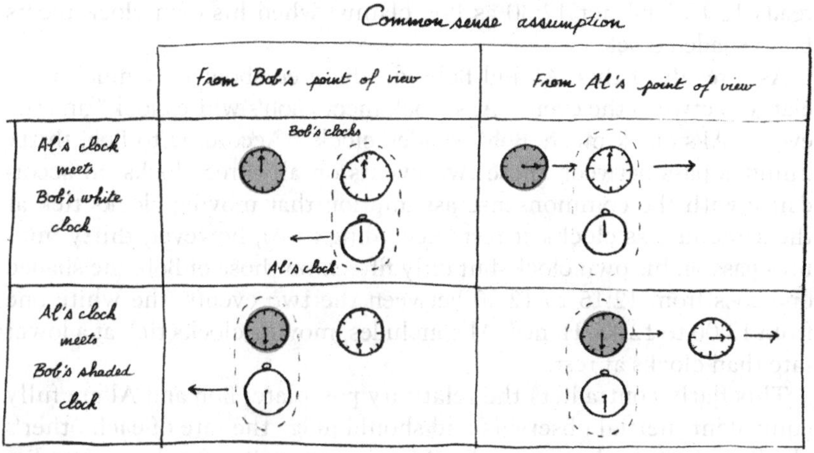

Figure A.6. *Common sense: a moving clock runs at the same rate as an identical clock at rest.* Drawn by Laurent Taudin.

at least two stationary clocks at different locations. These clocks are shown in the figure, one with a white, one with a shaded front. Bob makes sure that they are properly synchronized. When Al's clock passes Bob's white clock at 12:00, Al's clock also happens to read 12:00. Half an hour later, at 12:30 on Bob's clocks, Al's clock passes Bob's shaded clock. Our commonsense assumption tells us that at that point Al's clock will read 12:30 as well.

Now look at the situation from Al's perspective (see the second column of the table in Figure A.6). Al and Bob agree that Al's clock and Bob's white clock both read 12:00 when the two clocks meet. They also agree that Al's clock and Bob's shaded clock both read 12:30 when the two clocks meet. These are events taking place at one instant at one place. They are circled with dashed lines in Figure A.6. What Al and Bob disagree about is what Bob's shaded clock reads when Al's clock meets Bob's white clock and what Bob's white clock reads when Al's clock meets Bob's shaded clock. This, after all, involves judgments about the simultaneity of events at different locations. As we saw in Section 1.2, such judgments depend on one's state of motion. More specifically, Al finds that Bob's clocks, moving to the right with velocity v, are not properly synchronized: the shaded clock at the rear is fast compared to the white clock at the front (see Figure A.5), say, given their velocity and the distance between them, by fifteen minutes. According to Al, Bob's shaded clock therefore reads 12:15 (and not 12:00 as Bob claims) when his own clock meets Bob's white clock and Bob's white clock reads 12:15 (and not 12:30 as Bob claims) when his own clock meets Bob's shaded clock.

As a result of this, Al and Bob will disagree about how much time elapses between the event "Al's clock meets Bob's white clock" and the event "Al's clock meets Bob's shaded clock." According to Bob, thirty minutes pass between these two events on all three clocks, in accordance with the commonsense assumption that moving clocks tick at the same rate as clocks at rest. According to Al, however, thirty minutes pass on his own clock, but only fifteen on those of Bob: the shaded one goes from 12:15 to 12:30 between the two events, the white one from 12:00 to 12:15. Hence, Al concludes, moving clocks tick at a lower rate than clocks at rest.

This flatly contradicts the relativity postulate. Bob and Al are fully equivalent inertial observers and should judge the rate of each other's clocks in exactly the same way. What is responsible for this contradiction between our commonsense assumption about the rate of moving clocks and the relativity postulate is the relativity of simultaneity. Time dilation provides the escape from the contradiction. Figure A.7 shows how this phenomenon restores the symmetry between Al and Bob.

Figure A.7 retains as much as possible from Figure A.6. We assume (a) that, for Al, Bob's shaded clock is fifteen minutes fast compared to Bob's white clock; (b) that Al's clock and Bob's white clock both read 12:00 as they pass each other; and (c) that it takes thirty minutes on Bob's clocks for Al's to get from one to the other. As we just saw, Al's clock cannot read 12:30 when it passes Bob's shaded clock. That is incompatible with

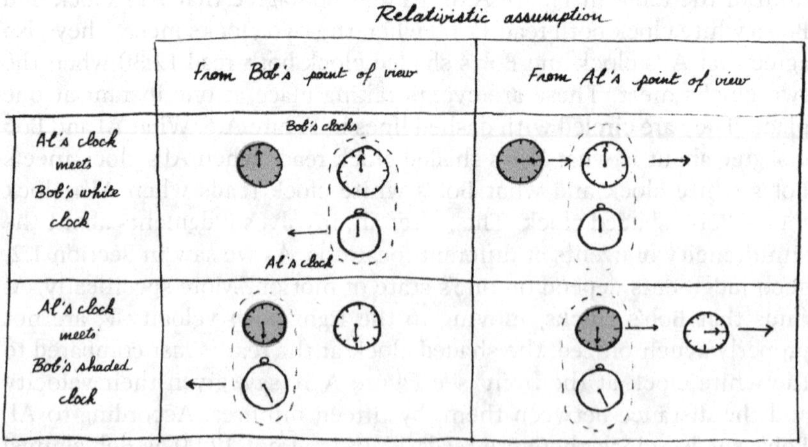

Figure A.7. *Special relativity: moving clocks run slow.* Drawn by Laurent Taudin.

the relativity postulate. So what should Al's clock read at that point? Let x be the as yet unknown amount of time that elapses on Al's clock between the events "Al's clock meets Bob's white clock" and "Al's clock meets Bob's shaded clock." For Bob, the rate of Al's clock is to the rate of his own as x:30. For Al, the rate of Bob's clocks is to the rate of his own as 15:x. Because of the relativity postulate, these two ratios must be the same: x:30 = 15:x. It follows that x^2 = 30·15 = 450, so $x = \sqrt{450} \approx 21$. This is the number used in the construction of Figure A.7.

First consider the situation from Bob's point of view (see the first column of the table in Figure A.7). For Bob, thirty minutes pass on his own clocks between the two circled events, whereas only twenty-one minutes pass on Al's. Bob concludes that the rate of Al's clock is 21/30 or about 70 percent the rate of his own.

Now look at the situation from Al's point of view (see the second column of the table in Figure A.7). For Al, twenty-one minutes pass on his own clock between the two circled events, whereas only fifteen minutes pass on Bob's. Al concludes that the rate of Bob's clocks is 15/21 or about 70 percent the rate of his own.

Time dilation thus restores the symmetry between Al and Bob. The combination of relativity of simultaneity and time dilation ensures that both observers will claim that the other person's clocks tick at a lower rate than their own.

The rate of any process in a system moving with velocity v will be $\sqrt{1 - v^2 / c^2}$ times the rate of that same process in the system at rest. This factor equals 1 for $v = 0$, steadily decreases as v increases, and goes to 0 as v approaches c, the speed of light. A simple way to derive this factor is by examining a so-called light clock (Figure A.8). This clock works by having a light signal go back and forth between two mirrors a distance L apart. One tells time by counting how many round-trips of the signal fit into the time interval of interest. Consider this clock from the point of view of two observers. For the first observer, the clock is at rest (see the drawing on the left in Figure A.8); for the second, it is moving with a velocity v perpendicular to L (see the drawing on the right in Figure A.8).[8] For the first observer, the signal simply goes up and down and one round-trip takes $2L/c$. For the second observer, the motion of the signal has a horizontal as well as a vertical component. According to the light postulate the net velocity of the signal is c for both observers. This means that, for the second observer, the signal's velocity in the vertical direction is only $c\sqrt{1 - v^2 / c^2}$.[9] For this observer, one round-trip of the signal thus takes longer than $2L/c$, namely $(2L / c) / \sqrt{1 - v^2 / c^2}$. This means that the rate of the light clock in motion is a fraction $\sqrt{1 - v^2 / c^2}$ of the rate of the light clock at rest. What is true for light clocks will be true for any clock or any process subject to laws compatible with the postulates of special relativity.

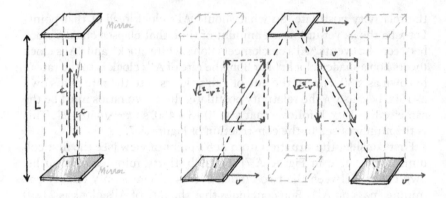

Figure A.8. *Moving clocks run slow by a factor* $\sqrt{1-v^2/c^2}$. Drawn by Laurent Taudin.

1.4. Length Contraction: The Length of Moving Rods and Other Objects in Motion

Another direct consequence of the relativity of simultaneity is that we have to give up the commonsense notion that uniform motion does not affect the length of measuring rods and other objects. According to special relativity, moving objects must be shorter in the direction of motion than those same objects at rest. This effect is called *length contraction*. Like time dilation, it is negligible at everyday velocities, but gets large at velocities approaching the speed of light.

The argument for length contraction in this section will have the same structure as the argument for time dilation in Section 1.3. Because of the relativity of simultaneity, the commonsense assumption about the length of moving objects leads to a contradiction with the relativity postulate. Length contraction provides the escape from this contradiction.

Figure A.9 brings out the problem with the commonsense assumption. Suppose Al and Bob are given identical measuring rods. One half of each rod is white, the other half is shaded. Bob attaches clocks to both ends of his rod and carefully synchronizes them. For Bob, Al's rod is moving with velocity v to the left. If the length of rods does not depend on their velocity, there will be one moment in time such that the two rods line up perfectly. This is illustrated in the picture on the left in Figure A.9. At 12:00 on Bob's clocks both the white ends and the shaded ends of the rods meet.

Now look at this situation from Al's perspective (see the drawings on the right in Figure A.9). Bob's rod and the two clocks are moving with velocity v to the right. Al and Bob agree that the shaded ends meet when

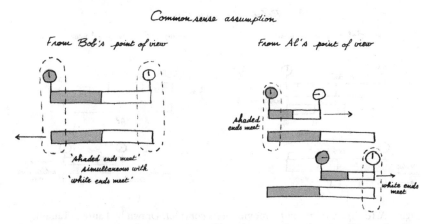

Figure A.9. *Common sense: a moving rod has the same length as an identical rod at rest.* Drawn by Laurent Taudin.

Bob's shaded clock reads 12:00 and that the white ends meet when Bob's white clock reads 12:00. But Al does not agree with Bob that the events "shaded ends meet" and "white ends meet" happen simultaneously. According to Al, Bob's shaded clock is fast compared to his white clock by, say, fifteen minutes. When the shaded ends meet at 12:00 on the shaded clock, it is only 11:45 on the white clock. The white ends only meet when the white clock reads 12:00, at which point the shaded one reads 12:15. The way Al sees it, the white ends have yet to meet when the shaded ends meet and the shaded ends have already met when the white ends meet.

Al and Bob will therefore draw very different conclusions about the length of moving rods. According to Bob, moving rods are equally long as rods at rest. According to Al, however, moving rods are shorter than rods at rest. This flatly contradicts the relativity postulate. Bob and Al are fully equivalent inertial observers and they should judge the length of moving rods in exactly the same way. Length contraction restores the symmetry between Al and Bob.

This is shown in Figure A.10, which not only illustrates the effect of length contraction in this situation but also the effects of relativity of simultaneity and time dilation. Assume that the relative velocity v of Al and Bob and the length of their identical rods are such that both of them will say that (a) the length of the other person's rods is two-thirds the length of their own, (b) the rate of the other person's clocks is two-thirds the rate of their own (as we shall see shortly, these ratios must be the same), and (c) a clock attached to the rear of the other person's rod is twenty-five minutes fast compared to the clock attached to

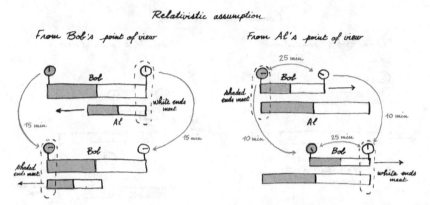

Figure A.10. *Special relativity: moving rods contract.* Drawn by Laurent Taudin.

the front (how we got that number will be shown). These are the numbers used in the construction of Figure A.10.

First, consider the situation from Bob's point of view (see the drawings on the left in Figure A.10). For Bob, Al's rod is moving to the left at velocity v and its length is only two-thirds the length of his own. The white ends meet at 12:00; fifteen minutes later, at 12:15, the shaded ends meet. Both events are circled with dashed lines.

Now consider the situation from Al's point of view (see the drawings on the right in Figure A.10). Al agrees with Bob that the white ends meet when Bob's white clock reads 12:00, and that the shaded ends meet when Bob's shaded clock reads 12:15. According to Al, however, the shaded clock is twenty-five minutes fast compared to the white clock. So, according to Al, when the white clock reads 12:00, it is already 12:25 on the shaded clock, and when the shaded clock reads 12:15, it is only 11:50 on the white clock. That means that for Al the order of the events "white ends meet" and "shaded ends meet" is just the reverse of their order for Bob. For Al, the shaded ends meet first (at 12:15 on the shaded and 11:50 on the white clock) and then the white ends meet (at 12:25 on the shaded and 12:00 on the white clock). The ten minutes that pass on Bob's clocks between these two events (from 12:15 to 12:25 on the shaded clock and from 11:50 to 12:00 on the white one) correspond to fifteen minutes on Al's own clocks (not shown in the figure).[10] For Al, after all, Bob's clocks run slow by a factor of two-thirds. The situation of Al and Bob is thus fully symmetric. For both of them the front end of the other person's rod meets the corresponding end of their own rod fifteen minutes after the rear end meets the corresponding end of their own rod. If Bob found that the length of Al's rod is two-thirds the length of his own, Al will likewise find that the length of Bob's is two-thirds the length of his own.

Once we assume that between the events "white ends meet" and "shaded ends meet" fifteen minutes pass on clocks attached to the rod at rest and ten minutes pass on clocks attached to the moving rod (where rest and motion are relative), we know that of the two moving clocks the rear one must be fast compared to the front one by twenty-five minutes. This can be seen directly in Figure A.10. According to Bob, his shaded clock reads 12:00 when his white clock reads 12:00. According to Al, however, Bob's shaded clock reads 15 + 10 = 25 minutes *after* 12:00 when Bob's white clock reads 12:00 (follow the arrows connecting the three times shown on the shaded clock in Figure A.10).

The upshot of our analysis of Figure A.10 is that length contraction restores the symmetry between Al and Bob that was broken in Figure A.9. The combination of relativity of simultaneity, time dilation, and length contraction ensures that both observers will claim that the other person's rods are shorter than their own.

The length of any object moving with velocity v will be shortened by a factor $\sqrt{1 - v^2/c^2}$ in the direction of motion compared to that same object at rest. This factor can be derived by considering the light clock introduced in Figure A.8 again. Figure A.11 shows this clock from the point of view of an inertial observer with respect to whom it moves at velocity v in the direction parallel to the line connecting the two mirrors. The light clock, like all other objects, will be shortened in the direction of motion and the distance between the two mirrors will only be xL, where x is some yet-to-be determined factor between 0 and 1. When the velocity v is parallel to L, the outbound leg of the light signal's round-trip between the two mirrors will take longer than the inbound leg. During the outbound leg, the light signal is heading for a mirror moving away from it; during the inbound leg, it is heading for a mirror moving toward it.

The round-trip will thus take a total time of

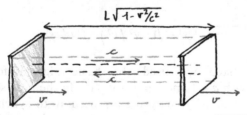

Figure A.11. *Moving rods contract by a factor* $\sqrt{1 - v^2/c^2}$. Drawn by Laurent Taudin.

$$\frac{xL}{c-v}+\frac{xL}{c+v}=\frac{xL(c+v)+xL(c-v)}{c^2-v^2}=\frac{2xLc}{c^2\left(1-v^2/c^2\right)}=\left(\frac{2L}{c}\right)\frac{x}{1-v^2/c^2}.$$

This result allows us to determine the factor by which moving objects must contract. The rate of a clock in motion should only depend on the magnitude of its velocity, not on its direction. In the case of the light clock, it should therefore not make any difference whether it is perpendicular or parallel to L. As we saw in Section 1.3, the rate of a light clock moving perpendicular to L is a fraction $\sqrt{1-v^2/c^2}$ of its rate at rest. The equation above shows that the rate of a light clock moving parallel to L is a fraction $(1-v^2/c^2)/x$ of its rate at rest. Since these two fractions must be the same, the factor x by which moving objects contract in their direction of motion must be equal to $\sqrt{1-v^2/c^2}$.[11]

We can now also calculate by how much two clocks a certain distance apart are out of sync for an observer with respect to whom these clocks are moving. As we saw in Figure A.10, according to Al, Bob's rear clock is fast compared to his front clock by 15 + 10 minutes. In general, it will be fast by the time x between the events "white ends meet" and "shaded ends meet" as measured on a clock at rest *plus* the time $x\sqrt{1-v^2/c^2}$ between these same two events (albeit in the opposite order) as measured on a moving clock running slow:

$$x\left(1+\sqrt{1-v^2/c^2}\right).$$

The time x is the difference in length between the rod at rest (length L) and the contracted moving rod (length $L\sqrt{1-v^2/c^2}$) divided by the velocity v of the moving rod:

$$x=\frac{L}{v}\left(1-\sqrt{1-v^2/c^2}\right).$$

Inserting this last expression in the one before it, we find:

$$\frac{L}{v}\left(1-\sqrt{1-v^2/c^2}\right)\left(1+\sqrt{1-v^2/c^2}\right)=L(v/c^2),$$

where in the last step we used the basic algebra formula $(a-b)(a+b)=a^2-b^2$ with $a=1$ and $b=\sqrt{1-v^2/c^2}$. In the next section, we will rederive this result by analyzing a procedure to synchronize clocks both from the perspective of someone at rest and from the perspective of someone moving with respect to these clocks.

1.5. *The Addition of Velocities*

With the three consequences of the postulates of relativity derived in Sections 1.2–1.4 (relativity of simultaneity, time dilation, and length contraction), we have all the necessary ingredients to explain how two inertial observers moving with respect to one another can both find that light has velocity c with respect to them.

Consider the situation in Figure A.12. As in Figures A.3–A.5, Bob is standing on a railroad car moving to the right with respect to Al at a constant velocity v. On the railroad car is a remote-control car also moving to the right with some constant velocity. According to Bob, the car is moving with velocity u with respect to the railroad car. What is the velocity w of the car with respect to Al? The commonsense answer would be $u + v$. This answer is wrong. Al cannot accept the velocity u reported by Bob. As far as Al is concerned, Bob has determined this velocity with rods that are contracted and clocks that run slow and are out of sync. Al needs to correct u for all three of these effects before he can add it to v.

In Sections 1.3 and 1.4 we saw that moving clocks run slow by a factor $\sqrt{1 - v^2 / c^2}$ and that moving rods contract by that same factor. To calculate Al's corrections to velocities reported by Bob we also need to know by how much two clocks are out of sync when they are both moving at some velocity v a certain distance apart.

Figure A.13 shows how Bob checks whether his clocks at P and Q are properly synchronized. At $t = 0$ on the clock at Q he sends out a light signal from Q to P. If the distance between P and Q is D (according to Bob), the travel time of the signal is D/c (according to Bob). Hence, Bob concludes that his clocks at P and Q are properly synchronized if the one at P reads $t = D/c$ the moment the signal arrives at P.

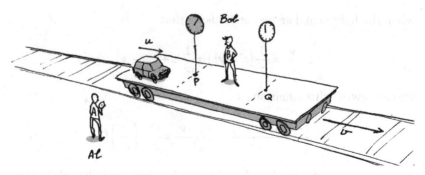

Figure A.12. *The velocity of the remote-control car according to Al and Bob.* Drawn by Laurent Taudin.

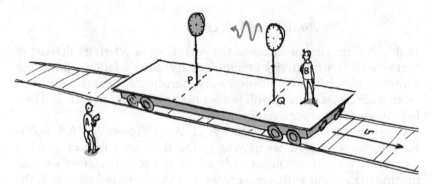

Figure A.13. *Bob checks whether his clocks are properly synchronized.* Drawn by Laurent Taudin.

Consider Bob's synchronization check from Al's perspective. According to Al, the travel time of the signal is less than D/c. Bob measures the distance between P and Q with contracted rods, so the real distance is only $D\sqrt{1 - v^2 / c^2}$. Moreover, the clock at P is rushing toward the light signal with velocity v. According to Al, the travel time is therefore only:

$$\frac{D\sqrt{1 - v^2 / c^2}}{c + v}.$$

Since Bob's clocks, according to Al, run slow, they only register a fraction $\sqrt{1 - v^2 / c^2}$ of this travel time. Al concludes that the clocks at P and Q are properly synchronized if the one at P reads

$$t = \frac{D\left(1 - v^2 / c^2\right)}{c + v}$$

when the light signal arrives at P. Using that

$$1 - \frac{v^2}{c^2} = \frac{1}{c^2}\left(c^2 - v^2\right) = \frac{1}{c^2}(c + v)(c - v),$$

we can rewrite this equation as

$$t = \frac{D}{c^2}\frac{(c + v)(c - v)}{c + v} = \frac{D(c - v)}{c^2} = \frac{D}{c} - \left(\frac{v}{c^2}\right)D.$$

If Bob has properly synchronized his clocks, the one at P will, in fact, read $t = D/c$ when the light signal arrives. Al concludes that, since it *should* read $(v/c^2)D$ *before* D/c when the light signal arrives, the clock at P is *fast* compared to the one at Q by $(v/c^2)D$.

Now return to Figure A.12. To determine the velocity of the remote-control car with respect to the railroad car, Bob measures the time T it takes the car to get from P to Q and divides the result into the distance D between these two points. To find T, Bob subtracts the reading on his clock at P when the car passes P from the reading on his clock at Q when the car passes Q. According to Al, the clock at P is $(v/c^2)D$ fast compared to one at Q. Al therefore has to *add* the difference to the time reported by Bob.[12] He then needs to divide the result by $\sqrt{1 - v^2/c^2}$ since Bob is measuring times with clocks that run slow. According to Al, it thus takes the car longer than T to get from P to Q, namely

$$\frac{T + (v/c^2)D}{\sqrt{1 - v^2/c^2}}.$$

Finally, Al has to multiply the distance D reported by Bob by $\sqrt{1 - v^2/c^2}$ since Bob is measuring distances with rods that are contracted. Dividing this shorter distance by the longer time given by the preceding formula, Al finds that the velocity of the remote-control car with respect to the railroad car is less than the velocity $u = D/T$ reported by Bob, namely

$$u_{\text{corrected}} = \frac{D(1 - v^2/c^2)}{T + (v/c^2)D}.$$

Dividing numerator and denominator by T and using that $D/T = u$, we find[13]

$$u_{\text{corrected}} = \frac{u(1 - v^2/c^2)}{1 + uv/c^2}.$$

Adding the velocity v of the railroad car, Al finds that the velocity w of the remote-control car with respect to him is not $u + v$, as suggested by common sense, but:

$$w = u_{\text{corrected}} + v = \frac{u(1 - v^2/c^2) + v(1 + uv/c^2)}{1 + uv/c^2} = \frac{u + v}{1 + uv/c^2}.$$

This result is known as the *relativistic addition theorem* of velocities. It gives the rule for the composition of velocities such as u and v measured by different inertial observers. Adding such velocities directly would be like adding apples and oranges. The addition theorem is such that as long as the velocities u and v are both subluminal (i.e., less than the velocity of light) their composite $u_{\text{corrected}} + v$ is also subluminal. This means that an object can never be accelerated from subluminal to superluminal velocities and that its inertial mass, a measure of its resistance to acceleration, must increase without limit as its velocity approaches the speed of light.[14]

Suppose that Bob determines the velocity of the light from the headlights of the remote-control car. In that case, $u = c$. Inserting this into the preceding formula, we find that this light also has velocity c with respect to Al:[15]

$$u_{\text{corrected}} + v = \frac{c + v}{1 + v/c} = \frac{c(1 + v/c)}{1 + v/c} = c.$$

This is just as it should be according to the postulates of special relativity.

2. SPECIAL RELATIVITY AND MINKOWSKI SPACE-TIME

2.1. Minkowski or Space-Time Diagrams

Figure A.14 shows a series of snapshots illustrating, from Al's point of view, an experiment like the one analyzed in Section 1.2 with Bob standing on a railroad car moving to the right at velocity v. This sequence of snapshots should be read from bottom to top.

The first snapshot shows the flashing of the lightbulb at the rear of the railroad car. The next four show the light flash from this lightbulb catching up with Bob, who is rushing away from it. The third shows the flashing of the lightbulb at the front. The next two show the light flash from this lightbulb making its way over to Bob, who is rushing toward it. The two light flashes both hit Bob in the final snapshot. Bob is still somewhat to the left of Al when this happens.

In Figures A.15 and A.16, the snapshots of Figure A.14 are used to construct a *Minkowski diagram* or *space-time diagram* for this situation. These diagrams picture the way space and time must be in a world in accordance with the postulates of special relativity. This relativistic space-time is called *Minkowski space-time* or, to distinguish it from the curved space-times of general relativity, *flat space-time*. It is standard practice to suppress two spatial dimensions in space-time diagrams. That leaves only one spatial dimension in addition to the time dimension, so motion can only be to the left or to the right. This suffices to illustrate all salient features of Minkowski space-time.

We begin our construction of a space-time diagram for the situation shown in Figure A.14 by picking space and time axes for a frame of

Figure A.14. *An experiment like the one in Figures A.3–A.5 from Al's perspective.*
Drawn by Laurent Taudin.

reference in which Al is at rest. Let the event "lightbulb L_1 flashes"
be the origin of this space-time coordinate system. Al's space and time
axes are the horizontal and the vertical lines through the origin. We then
trace the trajectories of the various elements pictured in Figure A.14
through the space-time spanned by these two axes: the observers Al
and Bob, the lightbulbs L_1 and L_2, and the light flashes emitted by them.
Such trajectories are called *world lines*. In general, world lines will not
be straight. Only those of objects at rest or in uniform motion will be.
Vertical lines are the world lines of objects at rest in Al's frame. The
world line of Al provides an example of this. The other world lines in
Figure A.15 are tilted, that is, they are at an angle to the time axis of Al's
frame. These are the world lines of objects moving at constant velocity
in Al's frame, to the right if the world line is tilted to the right, to the
left if it is tilted to the left. The greater the tilt, that is, the greater the
angle between the world line and the time axis, the greater the velocity.
The world lines of Bob and the lightbulbs L_1 and L_2 are tilted to the right
at the same angle. They are parallel to one another. This reflects that
these three objects move at the same velocity. The world lines of the
light flashes from L_1 and L_2 are tilted at a larger angle, to the left and the
right, respectively. We shall use units for measuring times and distances

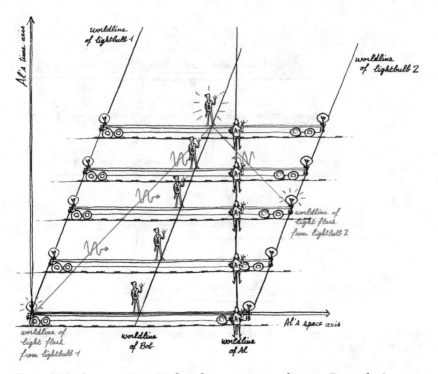

Figure A.15. *Constructing a Minkowski or space-time diagram.* Drawn by Laurent Taudin.

such that the world lines of light will always be tilted at 45°. Where two world lines intersect, the corresponding objects meet. For instance, the world lines of the light flashes from L_1 and L_2 intersect Bob's world line at the event "the two light flashes hit Bob."

The icons depicting Al, Bob, the lightbulbs, and the light signals in Figure A.15 are nothing but window dressing in the end. Stripping Figure A.15 of such unnecessary detail, we are left with the space-time diagram shown in Figure A.16. Every horizontal slice of Figure A.16 corresponds to a snapshot like the ones shown in Figure A.14.

In the internationally accepted system of units, distance is measured in meters and time in seconds. In these units c, the velocity of light, is about $3 \cdot 10^8$ m/s. Following standard practice in special relativity, we use a different time unit, which can be called – although the name is not commonly used – the "light meter" (cf., e.g., Mermin 1968, 180). One light meter is defined as the time it takes light to travel one meter. It follows that one second is about $3 \cdot 10^8$ light meters. The light meter is thus a much smaller unit of time than the second. To convert seconds

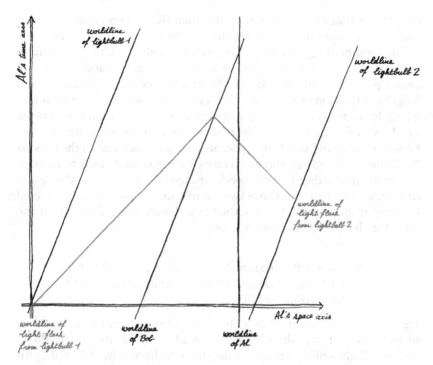

Figure A.16. *Minkowski or space-time diagram.* Drawn by Laurent Taudin.

to light meters we need to multiply by $c \approx 3 \cdot 10^8$ m/s: t seconds correspond to ct light meters. The result of multiplying seconds and meters
per second is meters. If distances are measured in meters and times
in light meters, then both quantities have the same dimension and
can meaningfully be added to one another. As we shall see in Sections
2.5–2.6, this is an important advantage of using light meters. Another
advantage is that the velocity of light in these units is equal to 1. From
the definition of a light meter it follows that light travels one meter per
light meter. This is why the world lines of light are always tilted at 45°
in our space-time diagrams.[16]

   In the remainder of this section we develop the geometry of Minkowski
space-time with the help of space-time diagrams. In Section 2.2, the
key insight of special relativity, the relativity of simultaneity, is given
a geometrical representation: different inertial observers have different
ways of carving up space-time into slices of simultaneous events. In
a space-time diagram one can see at a glance how it can be that light
has the same velocity for all inertial observers. They also make it

easy to see that objects moving faster than light in one frame of reference go backward in time in another (Section 2.3). This amounts to a
strong argument against the existence of so-called tachyons, particles
moving faster than light, for they would open up a Pandora's box of
causal paradoxes. In Section 2.4, the arguments for time dilation and
length contraction of Sections 1.3–1.4 are rephrased in geometrical language. In Sections 2.5–2.6, the spatiotemporal distance between events
in Minkowski space-time is introduced in analogy with the distance
between points in Euclidean space. Section 2.7 is devoted to the famous
"twin paradox," which shows vividly that the time it takes to go from
one event to another in Minkowski space-time depends on the space-
time trajectory between those two events, just as the distance covered
in going from one point to another in ordinary space depends on the
trajectory between those two points.

2.2. Relativity of Simultaneity: Different Observers Carving Minkowski Space-Time into Different Simultaneity Slices

Figure A.17 once again shows the space-time diagram corresponding
to the series of snapshots in Figure A.14. The events "lightbulb L_1
flashes,""lightbulb L_2 flashes," and "the light flashes from L_1 and L_2 hit
Bob" are labeled O, P, and Q, respectively. The event O is the origin
of Al's space-time coordinate system. The point R on Al's space axis
and the point S on his time axis mark the space and time coordinates
that Al assigns to the event Q. According to Al, Q happens a distance
OR to the right of O and a time OS after O. According to Al, the light
signal from O to Q covers the distance OR in a time OS. In the figure
OR (representing meters) is equal to OS (representing light meters). The
light thus has the velocity OR/OS = 1 meter/light meter, the velocity of
light, with respect to Al.

Figure A.17 also shows the space and time axes for Bob and the points
U and V marking the space and time coordinates of the event Q for Bob.
The origin of Bob's space-time coordinate system coincides with the
origin of Al's. How do we find the space and time axes for Bob?

The time axis is a line through O that, according to Bob, is purely in
the time direction. In other words, it is a line connecting events that,
according to Bob, all happen at the same location. This means that Bob's
time axis is a line through O parallel to Bob's world line. Bob's time axis
is tilted to the right of Al's time axis at an angle $\angle SOV$.

Bob's space axis is a line through O that, according to Bob, is purely in
the spatial direction. In other words, it is a line connecting events that,
according to Bob, all happen at the same time. According to Bob, the

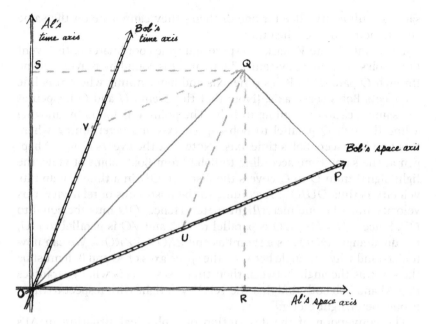

Figure A.17. *Space and time axes for Al and Bob.* Drawn by Laurent Taudin.

events O ("L_1 flashes") and P ("L_2 flashes") happen at the same time (see Figures A.3–A.4). Bob's space axis is therefore the line through O and P. This line is tilted with respect to Al's space axis at an angle $\angle ROP$. As we shall see shortly, Bob's space axis must be tilted with respect to Al's space axis at the same angle as Bob's time axis with respect to Al's time axis ($\angle ROP = \angle SOV$).

For Bob, as for Al, the time direction is orthogonal (i.e., perpendicular) to the space direction: $\angle VOU = \angle SOR = 90°$. In Figure A.17, however, the angle between Bob's space and time axes is acute. This is because the geometry of Minkowski space-time is not the same as ordinary Euclidean geometry. Using a Euclidean sheet of paper to represent a two-dimensional Minkowski space-time, we cannot capture all features of the latter. An example of this complication is that the angle $\angle VOU$ between Bob's space and time axes, which is a right angle in space-time, is represented by an acute angle in Figure A.17.

The different directions of the space axes of Al and Bob reflect their disagreements about simultaneity. Al and Bob each carve up Minkowski space-time into slices of simultaneous events by drawing lines parallel to their own space axis. If two events at different locations lie on the same simultaneity slice for one of them, they cannot lie on the same simultaneity slice for the other.

The points U and V mark the space and time coordinates of the event Q in Bob's coordinate system. The point U is found by drawing a line through Q parallel to Bob's time axis and determining where this line intersects Bob's space axis. (Note that the events U and Q happen at the same location according to Bob.) The point V is found by drawing a line through Q parallel to Bob's space axis and determining where this line intersects Bob's time axis. (Note that the events V and Q happen at the same time according to Bob.) From Bob's point of view, the light signal from O to Q covers the distance OU in a time OV and its velocity is thus OU/OV. According to the postulates of relativity, this velocity must be one meter/light meter. Hence, OU must be equal to OV. Since $OU = OV$, UQ is parallel to OV, and VQ is parallel to OU, the quadrangle $OUQV$ is a rhombus and $\angle SOV = \angle ROU$. We can now understand why the angle between the space axes of Al and Bob must be the same as the angle between their time axes: this is what guarantees that Al and Bob both find that the velocity of light with respect to them is one meter/light meter.[17]

The conversion of the description of a physical situation in Al's space-time coordinate system to a description of that same situation in Bob's space-time coordinate system is an example of a (passive) *Lorentz transformation*. The same transformation equations can be used to turn the description of a system at rest into the description of a system in motion in the same space-time coordinate system (*active Lorentz transformation* or *Lorentz boost*). Lorentz had introduced the transformation equations in this latter sense before 1905 but the former sense had eluded him.[18] The basic transformation equations for the space-time coordinates incorporate the relativity of simultaneity, time dilation, and length contraction. The relativity principle requires that all physical laws can be expressed in the same way in any of the space-time coordinate systems related by Lorentz transformations. Special relativity requires all physical laws to be *Lorentz invariant*. This, in turn, guarantees that all physical systems obeying these laws will exhibit the effects deduced from the postulates in Section 1.

There is nothing sacred about representing Al's space and time axes by horizontal and vertical lines. We might just as well represent Bob's space and time axes by horizontal and vertical lines, in which case we arrive at the space-time diagram in Figure A.18. As in Figure A.17, the events O ("L_1 flashes"), P ("L_2 flashes"), and Q ("the light flashes hit Bob") are labeled, and so are the points R, S, U, and V marking the coordinates of the event Q on the space and time axes of Al and Bob.

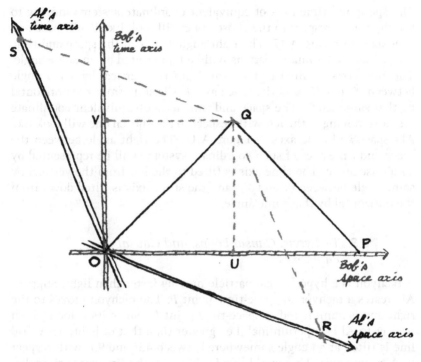

Figure A.18. *Space-time diagram with the space and time axes of Bob represented by horizontal and vertical lines.* Drawn by Laurent Taudin.

Al is moving to the left with respect to Bob at velocity v, so Al's time axis is tilted to the left at the same angle that Bob's time axis is tilted to the right in Figure A.17. Al's space axis must be tilted at that same angle with respect to Bob's space axis ($\angle SOV = \angle ROU$) so that the quadrangle $ORQS$ is a rhombus and OR/OS, the velocity of the light signal from O to Q according to Al, equals one meter/light meter. Note that O and P lie on the same horizontal line, which in this space-diagram is what tells us that these events ("L_1 flashes" and "L_2 flashes") are simultaneous for Bob. For Al, P comes after O: the simultaneity slice containing P (a line through P parallel to Al's space axis) lies above the simultaneity slice containing O (Al's space axis).

With the help of Figures A.17 and A.18 a complete inventory can be made of two-dimensional space-time coordinate systems with origin O and orthogonal axes moving with respect to one another at arbitrary but constant subluminal velocities. Choose one such coordinate system and represent its space and time axes by horizontal and vertical lines.

The space and time axes of equivalent coordinate systems moving to the right with respect to the chosen one will look like Bob's space and time axes in Figure A.17. The right angle between the space and time axes of such coordinate systems will be represented by an acute angle. The time axis is tilted to the right from the vertical by some angle between 0° and 45° and the space axis is tilted up from the horizontal by that same angle. The space and time axes of equivalent coordinate systems moving to the left with respect to the chosen one will look like Al's space and time axes in Figure A.18. The right angle between the space and time axes of such coordinate systems will be represented by an obtuse angle. The time axis is tilted to the left from the vertical by some angle between 0° and 45° and the space axis is tilted down from the horizontal by that same angle.

2.3. Tachyons, Causal Loops, and Causal Paradoxes

A tachyon is a hypothetical particle moving faster than light. Suppose Al creates a tachyon at space-time point P. The tachyon moves to the right and is annihilated at space-time point Q. Since its velocity with respect to Al is superluminal (i.e., greater than that of light), its world line is tilted at an angle somewhere between 45° and 90° with respect to Al's time axis. This world line is shown in the first space-time diagram in Figure A.19. The space and time axes of Al are represented by horizontal and vertical lines, those of Bob by tilted lines. The event P ("tachyon is created") is chosen as the origin of both coordinate systems. Bob is moving to the right with respect to Al. The velocities of Bob and the tachyon with respect to Al are such that the world line of the tachyon lies between the space axes of Al and Bob. For Bob, the tachyon is therefore annihilated (at Q) before it is created (at P)! For Bob, the tachyon is not only moving faster than light, it is also moving backward in time!

The second space-time diagram in Figure A.19 shows the same situation but now Bob's space and time axes are represented by horizontal and vertical lines. The tachyon is created at P, travels backward in time (for Bob), and is annihilated at Q. If it is possible to create a tachyon at P, it should also be possible to create one at Q. There are no preferred points in space (i.e., space is homogeneous). Likewise, if it is possible to create a tachyon moving to the right, it should also be possible to create one moving to the left. There are no preferred directions in space (i.e., space is isotropic). Hence, if it is possible to have a tachyon going from P to Q, it should also be possible to have one just like it going from Q

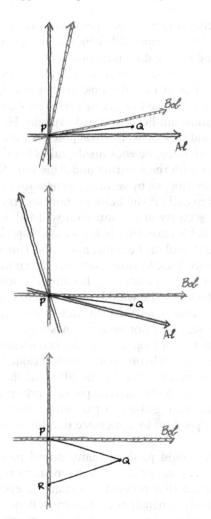

Figure A.19. *Tachyons and causal paradoxes.* Drawn by Laurent Taudin.

to R, where R is an event that, for Bob, happened at the same location as P two lifetimes of these tachyons before P. This is shown in the third space-time diagram in Figure A.19. If Bob's world line goes through R and P, he can use these tachyons to send information acquired at P to his former self at R. The figure $PQRP$ forms what is called a *causal loop*. The presence of causal loops puts tight constraints on what can happen in the space-time points that are part of them. Given a causal loop it is easy to construct scenarios that lead to *causal paradoxes*, scenarios in which we are driven to the logically impossible conclusion that one and the same state of affairs both does and does not obtain.

An example may help to appreciate what it would take to rule out causal paradoxes in the presence of causal loops. With the help of a tachyon phone, a gadget for sending and receiving messages carried by tachyons, Bob is planning a museum heist in broad daylight. He parks his car outside the museum and checks his tachyon phone for messages from his future self. If there are none, he goes inside, takes his favorite painting, and walks out again with the painting under his arm. Should the alarm go off or should he be stopped by security, he uses his tachyon phone to tell his former self to call off the heist for now and try again later. One day, Bob reasons, security at the museum will fail and the heist will succeed. Clearly, Bob's reasoning is fallacious. Suppose the alarm does go off, as it probably will during most attempts. This would prompt Bob to send his message back in time. But then he would have received it before he entered the museum and, heeding the warning, would have driven off; in which case he would not have sent his message to begin with. This scenario is not just physically but logically impossible. Bob cannot both send and not send his message. This particular chain of events on the causal loop must be broken somewhere. For some reason, for instance, his tachyon phone malfunctioned. Or it did work and he did get the message but Bob foolhardily went ahead and tried to steal the painting anyway. If the tachyon phone works properly and Bob sticks to his plan, something else must go wrong. Bob cannot possibly have left the scene tipped off by a message that he is never in a position to send.

If there were tachyons, we could produce many causal paradoxes such as Bob's bungled heist. The existence of tachyons thus requires elaborate conspiracies in nature that prevent the chain of events in such loops to turn into logically contradictory scenarios. Believing in such conspiracies is the price we have to pay for believing in tachyons. For most physicists this price is too high. There is broad consensus that tachyons do not exist and that no signal capable of transmitting information from one place to another can travel faster than the speed of light.

2.4. Time Dilation and Length Contraction: Different Observers Using Different Line Segments to Determine the Rate of Clocks and the Length of Rods

The space-time diagrams in Figures A.20–A.23 in this section are the counterparts of Figures A.6–A.7 and A.9–A.10 that formed the basis for

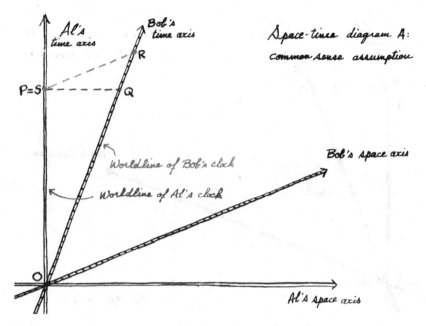

Figure A.20. *Common sense: a moving clock ticks at the same rate as an identical clock at rest.* Drawn by Laurent Taudin.

the arguments for time dilation and length contraction in Sections 1.3–1.4. This section offers geometrical versions of those arguments.

Figure A.20 brings out the problem with the commonsense assumption that moving clocks tick at the same rate as clocks at rest. Al and Bob have identical clocks with world lines that coincide with their respective time axes. Al and Bob both pick the event *O* as their zero point of time. Let *P* be the point on Al's time axis where Al's clock reads one light meter and let *R* be the point on Bob's time axis that, according to Bob, is simultaneous with *P*. Similarly, let *Q* be the point on Bob's time axis where Bob's clock reads one light meter and let *S* be the point on Al's time axis that, according to Al, is simultaneous with *Q*.

On the basis of the commonsense assumption about moving clocks, Al would insist that *S* (simultaneous to *Q* according to Al) coincides with *P*, as shown in Figure A.20. In that case, after all, the two clocks would take the same time to go from reading 0 to reading 1 for Al: *OS* = *OP*. For Bob, however, Al's clock would then take longer to go from 0 to 1 than his own clock: *OR* > *OQ*. Hence, according to Bob, Al's clock would run slow.

On the basis of the same commonsense assumption about moving clocks, Bob would insist that *R* (simultaneous to *P* according to Bob)

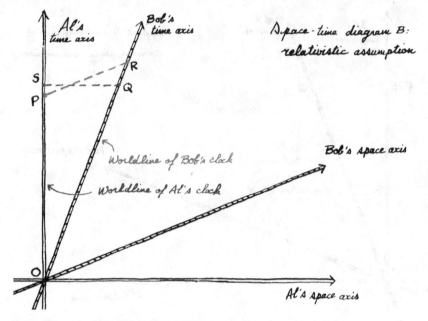

Figure A.21. *Special relativity: moving clocks run slow.* Drawn by Laurent Taudin.

coincides with Q. In that case, the two clocks would take the same time to go from 0 to 1 for Bob: OR = OQ. For Al, however, Bob's clock would then take longer to go from 0 to 1 than his own clock: S would be above P which means that OS > OP. Hence, according to Al, Bob's clock would run slow.

Either way we run into a contradiction with the relativity principle. Al and Bob are equivalent inertial observers. They should judge the rate of moving clocks the same way.

The solution to the problem is to choose P and Q in such a way that the line segments PR and QS intersect somewhere in between the time axes of Al and Bob, as shown in Figure A.21. In that case, Al and Bob both conclude that the other person's clock runs slow, Al because OS > OP, and Bob because OR > OQ. Time dilation thus restores the symmetry between Al and Bob.

Figure A.21 also provides the solution to the puzzle of how Al and Bob can both claim that the other person's clock runs slow. They agree that OS > OP and that OR > OQ, but they disagree about which line segments represent the time elapsed between the readings 0 and 1 on their two clocks, OP and OS on Al's time axis or OQ and OR on Bob's.

Although the space-time diagram in Figure A.21 solves this puzzle, it seems to leave us with another. Since the clocks of Al and Bob are

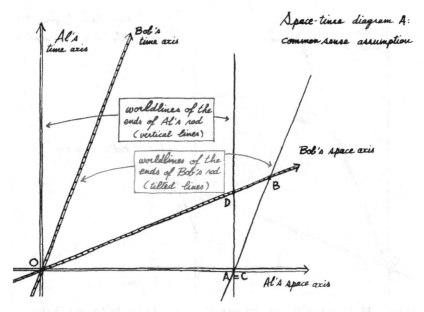

Figure A.22. *Common sense: a moving rod has the same length as an identical rod at rest.* Drawn by Laurent Taudin.

identical, the line segments on their own time axes representing the time elapsed between the readings 0 and 1 on their own clocks must be equally long, that is, $OP = OQ$. In Figure A.21, however, OQ is longer than OP. This is another feature of Minkowski space-time that space-time diagrams on Euclidean sheets of paper fail to capture. One light meter on a tilted time axis is represented by a larger line segment than one light meter on a vertical axis. The more tilted the axis, the larger the segment corresponding to one unit of time.

A geometrical argument fully analogous to the one for time dilation can be given for length contraction. Consider Figure A.22. Al and Bob have identical rods – each half white, half shaded (cf. Figures A.9–A.10). The shaded ends meet at the common origin O of their space-time coordinate systems. The world lines of the shaded ends of the rods thus coincide with the time axes of Al and Bob. The world lines of the white ends are parallel to these time axes. Call the points where the world line of the white end of Al's rod intersects the space axes of Al and Bob, A and D, respectively, and the points where the world line of the white end of Bob's rod intersects the space axes of Al and Bob, C and B, respectively. Consider the segments AD and BC of the world lines of the white ends of the two rods.

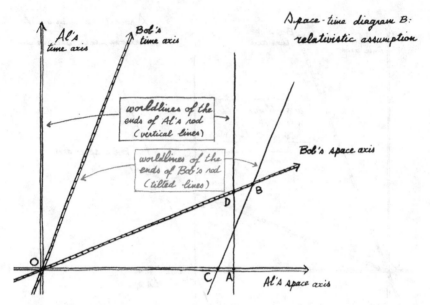

Figure A.23. *Special relativity: moving rods contract.* Drawn by Laurent Taudin.

On the basis of the commonsense assumption that moving rods have the same length as rods in motion, Al would insist that AD and BC intersect at $A = C$, as shown in Figure A.22. In that case, after all, the two rods would be equally long for Al: $OA = OC$. But then Bob would say that his rod is longer than Al's: $OB > OD$.

On the basis of the same commonsense assumption about moving rods, Bob would insist that AD and BC intersect at $B = D$. In that case, the two rods would be equally long for Bob: $OB = OD$. But then Al would say that his rod is longer than Bob's: C would be to the left of A which means that $OA > OC$.

Either way we run into a contradiction with the relativity principle. Al and Bob are equivalent inertial observers. They should judge the length of moving rods the same way.

The solution to the problem is to choose A and B in such a way that AD and BC intersect somewhere in between the space axes of Al and Bob, as shown in Figure A.23. In this case, Al and Bob both conclude that their own rod is longer than the other person's rod, Al because $OA > OC$, and Bob because $OB > OD$. Length contraction thus restores the symmetry between Al and Bob.

Figure A.23 makes it clear how Al and Bob can both claim that the other person's rod is contracted. They agree that $OA > OC$ and that $OB > OD$, but they disagree about which line segments represent the lengths of the two rods, *OA* and *OC* on Al's space axis or *OB* and *OD* on Bob's.

Since the rods of Al and Bob are identical, the line segments representing the length of their own rods on their own space axes must be equally long, that is, *OA* = *OB*. In Figure A.23, however, *OB* is longer than *OA*. This is the same complication that we encountered with time units in Figure A.21. One meter on a tilted space axis is represented by a larger line segment than one meter on a horizontal axis. The more tilted the axis, the larger the segment corresponding to one unit of distance.

As a result, we have to be careful comparing the length of segments of lines that are not either parallel or orthogonal to one another. The arguments in this section all turned on the comparison of the length of segments of the same line: (*OP,OS*) and (*OQ,OR*) in Figures A.20–A.21; (*OA,OC*) and (*OB,OD*) in Figures A.22–A.23. In Section 2.2, we already saw that the light postulate requires segments representing meters on a space axis and segments representing light meters on the time axis orthogonal to that space axis to be equally long.

2.5. Euclidean Geometry and the Geometry of Minkowski Space-Time

Figure A.24 shows, from the point of view of Al, a measuring rod of Bob passing by an identical measuring rod of Al at some constant velocity (cf. the drawing on the right in Figure A.10 in Section 1.4). The events "shaded ends meet" and "white ends meet" are labeled *P* and *Q*, respectively. On the right, these events are shown in a space-time diagram. (In Figure A.23, *P* would be the origin *O* and *Q* would be the point where

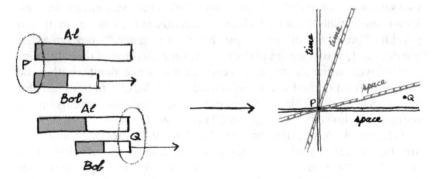

Figure A.24. *Different perspectives on two events in Minkowski space-time.* Drawn by Laurent Taudin.

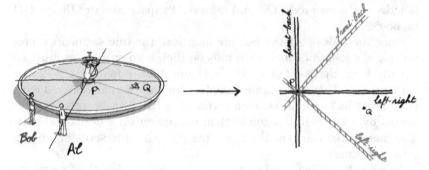

Figure A.25. *Different perspectives on two points in Euclidean space.* Drawn by Laurent Taudin.

AD and *BC* intersect.) Al's space and time axes are represented by horizontal and vertical lines, Bob's by tilted lines.

As we saw in Section 1.4, Al and Bob disagree about the order in which these two events take place. For Al, the shaded ends meet before the white ends meet; for Bob, it is just the other way around (cf. Figure A.10). This is brought out clearly in the space-time diagram on the right in Figure A.24. In any space-time coordinate system, the space axis consists of all events simultaneous to the origin in that coordinate system. The event *Q* lies above Al's space axis but below Bob's. Hence, *Q* happens after *P* for Al, but before *P* for Bob.

This situation in space-time is closely analogous to situations in ordinary space such as the one illustrated in Figure A.25. Al and Bob are standing at the perimeter of a basin with a fountain at its center *P* and a duck at *Q*. Because they are looking at the basin from different angles, Al and Bob have a different sense of "back-front" and "left-right." On the right, two sets of orthogonal "front-back" and "left-right" axes are shown, one for Al, one for Bob. They both choose the center of the basin as their zero point for "front-back distance" and "left-right distance." They agree that the duck at *Q* is to the right of the fountain at *P*. Because of their different perspectives on the situation, however, they disagree about whether the duck is in front of the fountain or behind it. For Al, *Q* is in front of *P*; for Bob, *Q* is behind *P*. This is brought out in the coordinate systems on the right: *Q* is below Al's "left-right axis" but above Bob's.

In Figure A.26, the diagrams on the right of Figures A.24 and A.25 are shown next to each other. Consider the one on the left first. With the help of the various line segments with endpoint *Q* perpendicular to the two sets of orthogonal axes, we can read off the "front-back distance" and "left-right distance" between *P* and *Q* for both Al and Bob.

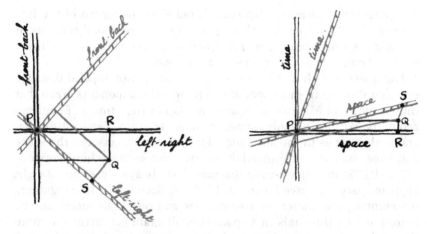

Figure A.26. *Distance in Euclidean space and in Minkowski space-time.* Drawn by Laurent Taudin.

According to Al, Q is QR to the front and PR to the right of P; according to Bob, Q is QS to the back and PS to the right of P.

The total distance between P and Q is a combination of "left-right distance" and "front-back distance." It can be computed from these two components with the help of the Pythagorean theorem. The rule for computing distances from these orthogonal components is:

$$(\text{total distance})^2 = (\text{front-back distance})^2 + (\text{left-right distance})^2.$$

Al applies this rule to the right-angled triangle PRQ:

$$PQ^2 = RQ^2 + PR^2;$$

Bob applies it to the right-angled triangle PSQ:

$$PQ^2 = SQ^2 + PS^2.$$

These are two equivalent ways of computing the same distance PQ. Al and Bob agree on the total distance between P and Q; they only disagree about how the total distance breaks down into a "front-back" and a "left-right" component.

A similar result holds in Minkowski space-time, as is illustrated on the right in Figure A.26. As on the left, four line segments with endpoint Q perpendicular to the two sets of orthogonal axes are drawn. In this case these are Al and Bob's space and time axes rather than their "front-back" and "left-right" axes. Al and Bob agree that the total "spatiotemporal distance" between P and Q is the length of the line segment PQ. However, according to Al, PQ has a temporal component QR and a spatial component PR, whereas, according to Bob, it has a temporal component QS and a spatial component PS.

The question is how to compute the total spatiotemporal distance between P and Q from its spatial and temporal components given that the geometry of Minkowski space-time is not the same as Euclidean geometry. We have already encountered two features of space-time diagrams alerting us to the difference. First, the angle between the space and time axes of some inertial observer can be anywhere between 45° and 135° in a space-time diagram, but always represents a right angle in space-time (see Figures A.17–A.18). Second, the line segments representing one meter on a space axis and one light meter on the corresponding time axis in a space-time diagram get larger the more the angle between the space and time axes in the diagram deviates from 90° (see Figures A.21 and A.23). Given these differences between Minkowski space-time and ordinary Euclidean space, it need not surprise us that we cannot use the Pythagorean theorem to compute distances in space-time. As we shall see in the next section, however, a theorem very similar to the Pythagorean theorem holds in Minkowski space-time. As a consequence, the rule for computing distances in space and time is also very similar to the rule for computing distances in ordinary space:

(distance in space-time)² = (distance in time)² – (distance in space)².

Note the minus sign. Also note that the subtraction on the right-hand side only makes sense if the quantities "distance in time" and "distance in space" have the same dimension, as they do when the former is measured in light meters and the latter in meters. Al applies this rule to the right-angled triangle PRQ (in which $\angle PRQ$ is the right angle):

$$PQ^2 = RQ^2 - PR^2;$$

Bob applies it to the right-angled triangle PSQ (in which $\angle PSQ$ is the right angle):

$$PQ^2 = SQ^2 - PS^2.$$

The minus sign in the rule for computing spatiotemporal distances is responsible for all the differences between the geometry of Minkowski space-time and ordinary Euclidean geometry. Given the close similarity between the two geometries, the geometry of flat Minkowski space-time is sometimes called pseudo-Euclidean, and should be distinguished

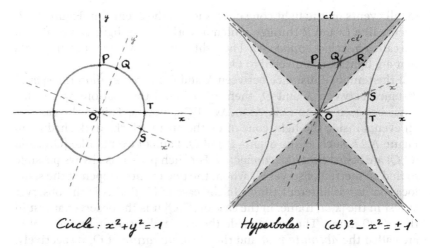

Circle: $x^2 + y^2 = 1$ Hyperbolas: $(ct)^2 - x^2 = \pm 1$

Figure A.27. *The analogy between circles in Euclidean space and hyperbolae in Minkowski space-time.* Drawn by Laurent Taudin.

from the non-(pseudo-)Euclidean geometries of the curved space-times of general relativity.

Consider an arbitrary event O in space-time. Starting from this event, we can group all other events X in space-time in three different classes, characterized by the sign of the quantity Δs^2, the square of the spatiotemporal distance between O and X. The quantity Δs is called the *space-time interval* between O and X. Its square can be positive, negative, or zero:

$$\Delta s^2 > 0, \quad \Delta s^2 < 0, \quad \Delta s^2 = 0.$$

The corresponding regions of space-time are shown in the space-time diagram on the right in Figure A.27. Ignore the shading and the four hyperbolae for the time being, and focus on (a) the events labeled O, P, Q, R, S, and T; (b) the two sets of space and time axes, with coordinates (x,ct) and (x',ct') [where t and t' are measured in seconds and converted to light meters through multiplication by c]; and (c) the dashed lines bisecting the angle between the space axis and the time axis in both coordinate systems. These dashed lines are the world lines of light traveling in opposite directions. They form what is called the *light cone* of O. With two spatial dimensions rather than one, these world lines do form a double cone. The part before O is called the *past light cone*, the part after O is called the *future light cone*. Light cones play a central role in special relativity.

If the temporal distance between an arbitrary point X and O is *equal to* the spatial distance between X and O, then $\Delta s^2 = 0$. This is the case for all events on the light cone of O, such as the event R in Figure A.27. Events like O and R through which a world line of light can be drawn are called *lightlike connected*. The light cone of O is the set of all points that are lightlike connected to O.

If the temporal distance between X and O is *greater than* the spatial distance between X and O, then $\Delta s^2 > 0$ and the spatiotemporal distance between X and O is $\Delta s = \sqrt{\Delta s^2}$ light meters. This is the case for all events inside the light cone of O, the area with the dark shading in Figure A.27, such as the events P and Q. Pairs of events like (O,P) and (O,Q) are called *timelike connected*. For such pairs it is always possible to find an inertial observer for whom the two events happen at the same location but at different times. In the case of (O,P) this is the observer at rest in the (x,ct) frame. In the case of (O,Q) it is the observer at rest in the (x',ct') frame. The areas inside the past and future light cones of O are called the *absolute past* and the *absolute future* of O, respectively. Since nothing travels faster than light (see Section 2.3), no event outside the past light cone of O can have any influence on it and O can have no influence on any event outside of its future light cone. The collection of light cones of all space-time points thus determines what is called the *causal structure* of space-time.

If the temporal distance between X and O is *less than* the spatial distance between X and O, then $\Delta s^2 < 0$ and the spatiotemporal distance between X and O is $\Delta s = \sqrt{-\Delta s^2}$ meters. This is the case for all points outside the light cone of O, the area with the light shading in Figure A.27, such as the events S and T. Pairs of points like (O,S) and (O,T) are called *spacelike connected*. For such pairs it is always possible to find an inertial observer for whom the two events happen at the same time but at different locations. In the case of (O,T) this is the observer at rest in the (x,ct) frame. In the case of (O,S) it is the observer at rest in the (x',ct') frame. The area outside the light cone of O is sometimes called the *absolute elsewhere* of O.

On the left in Figure A.27, a unit circle around the point O in a Euclidean plane is drawn. The center O is the origin of both the (x,y) and (x',y') coordinate systems shown in the figure. The unit circle with O as its center is the set of all points at unit distance from O. The coordinates of these points – such as P, Q, S, and T in the figure – satisfy the equation $x^2 + y^2 = x'^2 + y'^2 = 1$. The line segments OP, OQ, OS, and OT all have length 1, the radius of the unit circle.

The four hyperbolae in the space-time diagram on the right in Figure A.27 are the analogues of the unit circle on the left.

The space-time coordinates of all points at one light meter from O satisfy the equation $(ct)^2 - x^2 = (ct')^2 - x'^2 = 1$. These are the points on the two hyperbolae inside the light cone of O (such as P and Q).

The space-time coordinates of all points at one meter from O satisfy the equation $(ct)^2 - x^2 = (ct')^2 - x'^2 = -1$. These are the points on the two hyperbolae outside the light cone of O (such as S and T).

The line segments OP, OQ, OS, and OT are all of length one in space-time, but in the space-time diagram in Figure A.27 OS and OQ (representing one meter and one light meter on the (x', ct') axes, respectively) are longer than OT and OP (representing one meter and one light meter on the (x,ct) axes, respectively). The formula for computing spatiotemporal distances introduced in this section thus confirms what we found in Section 2.4: units on tilted axes look bigger (see Figures A.21 and A.23).[19]

2.6. The Analogue of the Pythagorean Theorem in Minkowski Space-Time

In this section, we adapt a simple geometrical construction used to prove the Pythagorean theorem in Euclidean space in order to prove the analogue of the theorem (with the plus sign replaced by a minus sign) in Minkowski space-time.[20]

Figure A.28 shows two right-angled triangles ABC, one in Euclidean space and one in Minkowski space-time. For the moment, focus on the Euclidean case on the left. The Pythagorean theorem says that the square of the hypotenuse equals the sum of the squares of the two sides: $AC^2 = AB^2 + BC^2$. To prove the theorem for this triangle, we proceed as follows. We first draw a line segment BD perpendicular to AC. This gives us three similar triangles: ABC, ADB, and BDC. The similarity of these three triangles can be seen upon inspection.

A formal proof of their similarity can be given with the help of the theorem that two triangles that have two angles in common are similar. We shall take this theorem to be intuitively obvious rather than derive it from the postulates of Euclidean geometry. By construction, the three triangles have a right angle in common. They also share the angle $\angle ABD = \angle ACB = \angle DCB$. The proof that the first angle is equal to the second and the third rests on another intuitively obvious theorem of Euclidean geometry, namely that two angles are the same if the sides of these angles are mutually perpendicular. This is the case here. By construction, we have that $AB \perp BC$ and $DB \perp AC$ (where the symbol "$\perp$" stands for "is perpendicular to").

From the similarity of the triangles ABC and ADB, it follows that

$$\frac{AB}{AC} = \frac{AD}{AB} \quad \Rightarrow \quad AB^2 = AC \cdot AD.$$

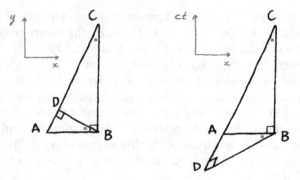

Figure A.28. *The Pythagorean theorem in Euclidean space and its analogue in Minkowski space-time.* Drawn by Laurent Taudin.

From the similarity of the triangles *ABC* and *BDC*, it likewise follows that

$$\frac{BC}{AC} = \frac{DC}{BC} \quad \Rightarrow \quad BC^2 = AC \cdot DC.$$

Adding these two relations, we find:

$$AB^2 + BC^2 = AC \cdot (AD + DC).$$

As can be seen from the figure, $AD + DC = AC$. Substituting this in the preceding equation, we find:

$$AB^2 + BC^2 = AC^2.$$

This proof of the Pythagorean theorem can easily be adapted to show that an analogous theorem holds for the triangle *ABC* in Minkowski space-time on the right of Figure A.28.

   As in the Euclidean triangle on the right in Figure A.28, we construct a line segment *BD* perpendicular to *AC*. In this case, these two line segments certainly do not *look* perpendicular to one another, but that is because the geometry of the paper used to draw the diagram is different from the geometry of Minkowski space-time. In space-time, *BD* is in the direction of the space axis that goes with a time axis in the direction of *AC*. *BD* is thus tilted up from the horizontal by the same angle that *AC* is tilted to the right from the vertical. Hence, *BD* and *AC* are perpendicular to one another. As before, this gives us three similar triangles: *ABC*, *ADB*, and *BDC*. The similarity can no longer be seen upon inspection, but it can be demonstrated in the exact same way that we proved the similarity of the corresponding three triangles in the Euclidean case. The proof rests on theorems that are valid both in Euclidean geometry and in the geometry of Minkowski space-time, namely that two triangles are similar if they have two angles in common and that two angles are equal if their sides are mutually perpendicular. On the basis of the similarity of the triangles ABC, ADB, and BDC, we can write down the exact same relations that we found in the Euclidean case:

$$AB^2 = AC \cdot AD, \quad BC^2 = AC \cdot DC.$$

For the Euclidean triangle ABC on the left in Figure A.28, we arrived at the Pythagorean theorem by *adding* these two relations and using that $AD + DC = AC$. For the triangle ABC on the right in Figure A.28, we arrive at the analogue of the Pythagorean theorem in Minkowski space-time by *subtracting* the first relation from the second and using that $DC - AD = AC$:

$$BC^2 - AB^2 = AC \cdot (DC - AD) = AC^2.$$

This is the basis for the rule for computing spatiotemporal distances in given in Section 2.5:[21] (distance in space-time)^2 = (distance in time)^2 − (distance in space)^2.

2.7. The Twin Paradox

Consider Figures A.29 and A.30. For the time being, ignore the dashed lines and the points M and N in Figure A.29, and focus on the line segments SUR and SAR, the world lines of the identical twin sisters Suzy and Sara. Up to point S (for separation), the world lines of Suzy and Sara coincide. At S, the twins get separated. Suzy remains in the same state of inertial motion she has been in all along. Sara boards a spaceship that takes her away from her twin sister at about 75 percent the speed of light. At A (for acceleration), the spaceship turns around and starts moving back to Suzy, again at 75 percent the speed of light. At R (for reunion) the sisters are reunited. For Suzy, nine years have passed (see the marks on Suzy's world line in Figure A.29 and the marks on her face in Figure A.30). For Sara, only six years have passed (see the marks on Sara's world line in Figure A.29 and her youthful appearance in Figure A.30).[22] Note that the line segments representing one year on the world line of Sara in Figure A.29 are larger than the line segments representing one year on the world line of Suzy. This is in accordance with what we found in Sections 2.4–2.5.

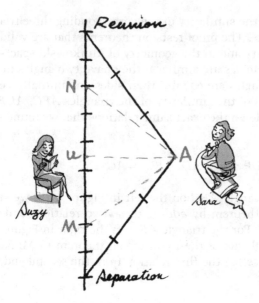

Figure A.29. *Space-time diagram for the twin paradox.* Drawn by Laurent Taudin.

The scenario illustrated in Figures A.29–A.30 is known as the "twin paradox." As the phrase suggests, many people have been baffled by this prediction of special relativity (it has successfully been tested with identical atomic clocks instead of twins). Here is how people tend to get confused. They argue as follows:

Suzy and Sara are two equivalent observers who will both find that the clocks and metabolism of the other run slow, so that the other should age more slowly than she herself. So, according to Suzy, Sara should be younger when they are reunited, whereas, according to Sara, Suzy should be. Clearly, they cannot both be younger than the other. Hence, there must be some other factor that ensures that they are equally old when reunited.

As we just saw, special relativity tells us that Sara will be younger than Suzy. So, what is wrong with the preceding argument? It is the very first step that is wrong. Suzy and Sara are not two equivalent observers. Suzy is an inertial observer, Sara is not. At *A* Sara switches from one inertial state of motion to another. She accelerates.

But, one might object, just as Sara accelerates with respect to Suzy, Suzy accelerates with respect to Sara. Why can they not both claim that the other is accelerating? The answer is that acceleration, unlike velocity, is an absolute, not a relative affair. Perhaps the easiest way to see this is to compare how Suzy and Sara experience the moment that they start to approach each other again. For Suzy there is nothing

Figure A.30. *Suzy and Sara at separation and reunion.* Drawn by Laurent Taudin.

special about this moment. Sara, however, will feel it in her stomach. And if the spaceship were to turn around as abruptly as suggested by Figure A.29, the G forces would kill her. So, Suzy and Sara are not equivalent observers. Sara accelerates and Suzy does not. Therefore, it need not surprise us that Suzy and Sara age a different number of years.

There has been a great deal of confusion about the role of the acceleration at *A*. Some have argued that Sara ages six years during the acceleration so that she ends up being as old as Suzy when the twins are reunited. Others have argued that the acceleration causes the difference in aging of the twins. Both claims are wrong. Here is a simple argument showing that the acceleration cannot be the cause of the difference in aging. Let Sara make a second journey. Again she takes off at 75 percent the speed of light, but now she keeps going for six rather than three years before reversing course. If it were the acceleration that caused the difference in aging, Sara should gain the same number of years on Suzy on both trips. After all, the acceleration part in both trips is exactly the same. According to special relativity, however, Sara gains three years during the first trip and six years during the second. Hence, the acceleration does not cause the difference in aging. But then again, what does?

The answer is that it is simply a matter of the space-time trajectory *SAR* between *S* and *R* being shorter than the trajectory *SUR*. The situation is analogous to the following situation in ordinary space illustrated in Figure A.31. Suppose Suzy and Sara drive from Pittsburgh to Minneapolis. The figure illustrates the two different routes they

Figure A.31. *A spatial analogue of the twin paradox.* Drawn by Laurent Taudin.

take. The first is an essentially straight highway from Pittsburgh to Minneapolis going through Chicago. This route is the analogue of the route SUR in space-time. The other is a terrible detour consisting of an essentially straight highway from Pittsburgh to Memphis and another essentially straight highway from Memphis to Minneapolis. This second route corresponds to the route SAR in space-time. The twin taking this route will not drive straight the whole way. She needs to make a sharp right turn in Memphis. This right turn is the analogue of the acceleration at A in Figure A.29. The second route is clearly longer than the first. This is a minor disanalogy with the situation in space-time: in ordinary space a straight line is the *shortest* route between two points, in space-time a straight line is the *longest* route between two points.

Suppose we do not know at first whether Suzy or Sara took the detour. Then we are told that Sara made a sharp right turn at one point whereas Suzy did not. This tells us immediately that Sara went from Pittsburgh to Minneapolis the long way. Similarly, knowing that Sara accelerated at some point in her space-time journey, whereas Suzy did not, tells us that Sara went from separation to reunion the short way. So the acceleration serves to pick out which one of the twins took the shortest trip in space-time, just as "making a sharp right turn" serves to pick out which one of the twins took the longest trip in space. Nobody in his right mind, however, would claim that the right turn in Memphis causes the difference between Suzy and Sara's odometer readings upon arrival in Minneapolis. What causes the difference is that the route from Pittsburgh to Minneapolis through Memphis is longer than the one through Chicago. The acceleration at A likewise does not cause the age difference between Suzy and Sara. It is just that the space-time journey SAR is shorter than SUR.

To show that the acceleration really does not play a crucial role, we change the logistics of the experiment in such a way that we can establish the age difference between Suzy and Sara at R avoiding the acceleration at A altogether. To this end we introduce the inertial observers Al and Bob. Al's world line is a straight line through S and A. It coincides with Sara's on the first half of her journey (SA). Bob's world line is a straight line through A and R and coincides with Sara's on the second half of her journey (AR). Al times the first half of the journey, Bob the second. Al and Bob both find that they accompanied Sara for three years. So, the space-time journey takes six years. Inertial observer Suzy already told us that the space-time journey takes nine years. When the twins are reunited, Sara is three years younger than Suzy. Notice that we reached this conclusion using inertial observers only.

This maneuver of, in effect, replacing Sara with the inertial observers Al and Bob seems to leave us with another puzzle. Look at the first half of the journey. According to Suzy, Al ages only three years between S and A, which for her is four and a half years. Al and Suzy are two fully equivalent inertial observers. If Suzy finds that Al ages at two-thirds her own rate, then Al finds that Suzy ages at two-thirds his rate. So, according to Al, Suzy only ages two years between S and A, which for him is three years. A similar result is found when we compare Suzy and Bob on the second half of the journey. According to Bob, Suzy ages only two years between A and R, which for Bob is three years. Combining Al and Bob's conclusions, it looks as if Suzy only ages four years between S and R. However, we already know that Suzy ages nine years. What is going on?

The dashed lines and the points marked M and N in Figure A.29 help us answer this question. Al and Bob fail to take into account five years of Suzy's life between S and R because of their disagreement about simultaneity. According to Al, A is simultaneous with M, whereas, according to Bob, A is simultaneous with N. This observation restores the consistency of the whole story. For Al, Suzy aged two years (between S and M) during the three years that he accompanied Sara from S to A. Likewise, for Bob, Suzy aged two years (between N and R) during the three years that he accompanied Sara from A to R. The five years that Suzy aged between M and N should be added to the four years counted by Al and Bob, giving a total of nine years for Suzy's aging between S and R.

The upshot then is that there is nothing mysterious about the twin paradox. All there is to it is that the time that elapses for an observer between two points in space-time depends on the space-time trajectory she takes from one to the other, just as the distance she covers between two points in space depends on the route she takes to go from one to the other.

ACKNOWLEDGMENTS

This appendix grew out of a text I wrote for a class introducing Einstein to nonphysics majors. I am grateful to students in various editions of this class for their feedback; especially to Jackson Graves and others in the 2012 fall semester, who made the observation about Figure A.10 that prompted the derivation at the end of Section 1.4. I also want to thank Suzy and Sara Durkacs for agreeing to be featured in Section 2.7.

NOTES

1 Given that the Earth is rotating on its own axis, a frame at rest with respect to the Earth is only approximately an inertial frame. A frame at rest with respect to the Sun would be an even closer approximation to a true inertial frame.

2 See Chapter 6 for a discussion of Einstein's unsuccessful attempts to generalize the relativity principle from uniform to arbitrary motion.

3 In fact, Einstein made two more assumptions to the effect that space and time are the same everywhere. There are no special times and locations (space and time are *homogeneous*) and no special directions (space and time are *isotropic*). Both assumptions are routinely granted.

4 For a reconstruction of Einstein's path toward special relativity, see Chapter 2.

5 Note that Al is moving at about half the speed of light. To bring out the effects more clearly, the speeds of the observers here and in subsequent figures have been grossly exaggerated.

6 To bring out the effects more clearly, time differences here and in subsequent figures have been grossly exaggerated.

7 "Fast" is meant in the sense of "out of sync" here, not in the sense of "ticking at a higher rate." This distinction becomes important in Section 1.3: relativity of simultaneity and time dilation are two different phenomena.

8 As we shall see in Section 1.4, the analysis gets more complicated when the velocity is not perpendicular to L.

9 According to the Pythagorean theorem (see Section 2.6 for a simple proof), the sum of the squares of the sides of a right-angled triangle is equal to the square of its hypotenuse. Consider the velocity diagram in Figure A.8. Since the net velocity is c and the horizontal velocity is v, the vertical velocity must be $c\sqrt{1 - v^2 / c^2}$ to satisfy the Pythagorean theorem: $v^2 + c^2 (1 - v^2/c^2) = c^2$.

10 To bring out the symmetry between Al and Bob more fully, we can attach clocks to the ends of Al's rod as well, one with a shaded front to the shaded end of his rod, one with a white front to the white end. Al makes sure these clocks are properly synchronized. Once the reading on one of Al's clocks in

one of the four situations shown in Figure A.10 is fixed, all others can be inferred on the basis of the assumptions (a)–(c) made in the constuction of Figure A.10. If Al's white clock, like Bob's white clock, reads 12:00 when the white ends of the rods meet, the readings on Al's two clocks in the four drawings in Figure A.10 (upper-left, lower-left, upper-right, lower-right) will be as follows:

	Left:		Right:	
	Shaded:	White:	Shaded:	White:
Upper:	11:35	12:00	11:45	11:45
Lower:	11:45	12:10	12:00	12:00

11 In 1887, Michelson and Morley tried to detect the Earth's presumed motion through the ether by checking whether the travel time of a light signal going back and forth between two mirrors depends on whether the line connecting these mirrors is parallel or perpendicular to the direction of motion. In accordance with the preceding considerations, they found that it does not. Independently of one another, FitzGerald (1899) and Lorentz (1892) suggested a few years later that this negative result can be accounted for by assuming that objects moving through the ether contract in the direction of motion by the same factor we just found (Brown 2005, ch. 4). The Michelson-Morley experiment has been a staple of popular expositions of special relativity and has great pedagogical value, even though most historians now agree that it played no role in Einstein's development of the theory (Holton 1969, Stachel 1982, Janssen 2002b, and Chapter 2).

12 Think of timing a runner on the 100 meter by subtracting the time read on a clock at the start from the time read on a clock at the finish. If the clock at the start is fast compared to the clock at the finish, the difference needs to be added to the time found in this manner, otherwise average runners would seem to break the world record.

13 Note that $u_{\text{corrected}} = -v$ for $u = -v$: Al and Bob agree on the velocity with which they are moving with respect to one another.

14 It does not follow from the addition theorem that nothing can travel faster than the speed of light, although there are good reasons for believing this to be impossible (see Section 2.3).

15 Inserting $u = c$ into the formula for $u_{\text{corrected}}$, we find:

$$u_{\text{corrected}} = \frac{c(1 - v^2 / c^2)}{1 + v / c} = \frac{c(1 + v / c)(1 - v / c)}{1 + v / c} = c - v.$$

16 The light *meter* as a unit of *time* is fully analogous to the more familiar light *year*, the distance traveled by light in a year, as a unit of *distance*. (Joan Baez thus made a category mistake in the line "a couple of light years ago" of her song *Diamonds and Rust*.) We could likewise introduce the "light second," the distance traveled by light in one second, as our unit of distance. One light second is about $3 \cdot 10^8$ meters. The light second is thus a much larger unit of distance than the meter. To convert meters to light seconds, we need to divide by $c \approx 3 \cdot 10^8$ m/s. The result of dividing meters

by meters per seconds is seconds. If distances are measured in light seconds and time in seconds, then these two quantities once again have the same dimension and can be added to one another. Moreover, the velocity of light in these units is once again equal to one. From the definition of a light second, it follows that light travels one light second per second.

17 The world line OQ of the light signal from O to Q bisects the right angle between Bob's time and the space axes. For Bob, as for Al, the world lines of light are 45° lines.

18 For further discussion, see Janssen (2002b) and Chapter 2.

19 The similarities between Euclidean space and Minkowski space-time led Minkowski to an important conclusion: that the laws of physics stay the same when we switch to a different set of orthogonal axes in space-time corresponding to a different state of inertial motion is no more surprising than that the laws of physics stay the same when we switch to a different set of orthogonal axes in space pointing in different directions. As Minkowski pointed out, in Newtonian mechanics the statement that uniform motion does not make a difference for the laws of physics is something completely different from the statement that orientation in space does not make a difference for the laws of physics. In special relativity, these two statements turn out to be intimately connected. Switching to a different inertial frame and switching to a rotated set of space axes are both nothing but a change of perspective in space-time. Neither affects the laws of physics.

20 The argument in this section is due to Jon Dorling.

21 The triangle on the right in Figure A.28 is made up of the timelike line segments AC and BC and the spacelike line segment AB. A completely analogous argument can be given for the case of a right-angled triangle made up of one timelike line segment and two spacelike ones.

22 Since Sara's velocity is $v = .75c$ she ages at $\sqrt{1 - v^2 / c^2} \approx 2/3$ the rate of Suzy.

Bibliography

Note: Material from the Albert Einstein Archive at the Jewish National and University Library, Jerusalem, is quoted as EA xx–yyy, where xx is the reel and yyy the document number.

Abiko, Seiya. 2000. "Einstein's Kyoto Address: 'How I Created the Theory of Relativity.'" *Historical Studies in the Physical and Biological Sciences* 31: 1–35.

Abraham, Carolyn. 2001. *Possessing Genius: The Bizarre Odyssey of Einstein's Brain.* New York: St. Martin's Press.

Abraham, Max. 1912. "Das Elementargesetz der Gravitation." *Physikalische Zeitschrift* 13: 4–5. Reprinted in English trans. in Renn (2007a, Vol. 3, 331–9).

Adler, Friedrich (1909). "Die Einheit des physikalischen Weltbildes." *Naturwissenschaftliche Wochenschrift* 8: 817–22.

Alexander, H. G., ed. 1956. *The Leibniz-Clarke Correspondence.* Manchester: Manchester University Press.

Allen, Herbert S. 1925. *Photo-Electricity: The Liberation of Electrons by Light.* 2nd ed. London: Longmans, Green and Co.

Allison, Henry E. 2004. *Kant's Transcendental Idealism.* Rev. ed. New Haven, CT: Yale University Press.

Anderson, James L. 1967. *Principles of Relativity Physics.* New York: Academic Press.

Ardelt, Rudolf G. 1984. *Friedrich Adler: Probleme einer Persönlichkeitsentwicklung um die Jahrhundertwende.* Vienna: Österreichischer Bundesverlag.

Ashtekar, Abhay, Cohen, Robert S., Howard, Don, Renn, Jürgen, Sarkar, Sahotra, and Shimony, Abner, eds. 2003. *Revisiting the Foundations of Relativistic Physics.* Dordrecht: Kluwer.

Auletta, Gennaro. 2001. *Foundations and Interpretation of Quantum Mechanics.* Singapore: World Scientific.

Baade, Walter. 1952. "Extragalactic Nebulae: Report to IAU Commission 28." *Transactions of the International Astronomical Union of Astronomy* 8: 397–9.

Bacciagaluppi, Guido, and Valentini, Antony. 2009. *Quantum Theory at the Crossroads: Reconsidering the 1927 Solvay Conference.* Cambridge: Cambridge University Press.

Bach, Hans I. 1984. *The German Jew: A Synthesis of Judaism and Western Civilization, 1730–1930.* Oxford: Oxford University Press.

Badash, Lawrence. 1967. "Nagaoka to Rutherford, 22 February 1911." *Physics Today* 20: 55–60.

Balashov, Yuri. 2002. "Laws of Physics and the Universe." In Balashov and Vizgin (2002, 107–48).

Balashov, Yuri, and Janssen, Michel. 2003. "Presentism and Relativity." *British Journal for the Philosophy of Science* 54: 327–46.

Balashov, Yuri, and Vizgin, Vladimir, eds. 2002. *Einstein Studies in Russia*. Einstein Studies 10. Boston: Birkhäuser.

Banks, Erik C. 2003. *Ernst Mach's World Elements: A Study in Natural Philosophy*. Dordrecht: Kluwer.

Barbour, Julian B. 1992. "Einstein and Mach's Principle." In Eisenstaedt and Kox (1992, 125–53).

2007. "Einstein and Mach's Principle." In Renn (2007a, Vol. 3, 569–604).

Barbour, Julian B., and Pfister, Herbert, eds. 1995. *Mach's Principle: From Newton's Bucket to Quantum Gravity*. Einstein Studies 6. Boston: Birkhäuser.

Barkan, Diana. 1993. "The Witches' Sabbath: The First International Solvay Congress in Physics." *Science in Context* 6: 59–82.

Barkla, Charles G., and White, Margaret P. 1917. "Notes on the Absorption and Scattering of X-rays and the Characteristic Radiation of J-series." *Philosophical Magazine* 34: 270–85.

Bell, John S. 1964. "On the Einstein Podolsky Rosen Paradox." *Physics* 1: 195–200.

Beller, Mara. 1999. *Quantum Dialogue: The Making of a Revolution*. Chicago: University of Chicago Press.

2000. "Kant's Impact on Einstein's Thought." In Howard and Stachel (2000, 83–106).

Beller, Mara, and Fine, Arthur. 1993. "Bohr's Response to EPR." In Faye and Folse (1994, 1–31).

Belousek, Darrin S. 1996. "Einstein's 1927 Unpublished Hidden-Variable Theory: Its Background, Context and Significance." *Studies in History and Philosophy of Modern Physics* 27: 437–61.

Bergia, Silvio. 1993. "Attempts at Unified Field Theories (1919–1955): Alleged Failure and Intrinsic Validation/Refutation Criteria." In Earman et al. (1993, 274–307).

Bergia, Silvio, and Navarro, Luis. 1988. "Recurrences and Continuity in Einstein's Research on Radiation between 1905 and 1916." *Archive for History of Exact Sciences* 38: 79–99.

Bergman, Hugo. 1974. "Personal Remembrance of Albert Einstein." In *Logical and Epistemological Studies in Contemporary Physics*, ed. Robert S. Cohen and Marx W. Wartofsky. Dordrecht: Reidel, 388–94.

Bergmann, Peter. 1949. "Non-linear Field Theories." *Physical Review* 75: 680–6.

1979. "The Quest for Unity: General Relativity and Unitary Field Theories." In Holton and Elkana (1982, 27–38).

Bernstein, Aaron. 1853–7. *Aus dem Reiche der Naturwissenschaft. Für Jedermann aus dem Volke*. 12 vols. Berlin: Besser. Reprinted as *Naturwissenschaftliche Volksbücher*. 20 vols. (1867–9). Berlin: Duncker.

Bernstein, Jeremy. 2006. *Secrets of the Old One: Einstein, 1905*. New York: Copernicus Books.

Bertlmann, Reinhold A., and Zeilinger, Anton, eds. 2002. *Quantum (Un)Speakables: From Bell to Quantum Information*. Berlin and Heidelberg: Springer.

Bird, Graham. 2006. *The Revolutionary Kant: A Commentary on the Critique of Pure Reason*. Chicago: Open Court.

Bird, Kai, and Sherwin, Martin. 2005. *American Prometheus: The Triumph and Tragedy of J. Robert Oppenheimer*. New York: Alfred A. Knopf.

Birkhoff, George D., and Langer, Rudolph E. 1923. *Relativity and Modern Physics*. Cambridge, MA: Harvard University Press.

Blum, Alexander S., Renn, Jürgen, Salisbury, Donald C., Schemmel, Matthias, and Sundermeyer, Kurt. 2012. "1912: A Turning Point on Einstein's Way to General Relativity." *Annalen der Physik* 524.1: A11–A13.

Bohm, David. 1951. *Quantum Theory*. Englewood Cliffs, NJ: Prentice-Hall.

1952. "A Suggested Interpretation of the Quantum Theory in Terms of 'Hidden' Variables." *Physical Review* 85: 166–93.

Bohr, Niels. 1922. "'The Structure of the Atom.' Nobel Lecture, December 11, 1922." In Nobel Foundation (1965, 7–43).

1935a. "Quantum Mechanics and Physical Reality." *Nature* 136: 65.

1935b. "Can Quantum Mechanical Description of Physical Reality Be Considered Complete?" *Physical Review* 48: 696–702.

1939. "The Causality Problem in Atomic Physics." In *New Theories in Physics*. Warsaw: Institut international de coopération intellectuelle, Paris, 11–38.

1949. "Discussion with Einstein on Epistemological Problems in Atomic Physics." In Schilpp (1949, 199–241).

1981. *Collected Works*. Vol. 2. *Works on Atomic Physics (1912–1917)*, ed. Ulrich Hoyer. Amsterdam: North-Holland.

1984. *Collected Works*. Vol. 5. *The Emergence of Quantum Mechanics (Mainly 1924–1926)*, ed. Klaus Stolzenburg. Amsterdam: North Holland.

1985. *Collected Works*. Vol. 6. *Foundations of Quantum Physics I (1926–1932)*, ed. Jørgen Kalckar. Amsterdam: North-Holland.

Bohr, Niels, and Rosenfeld, Léon. 1933. "Zur Frage der Meßbarkeit der elektromagnetischen Feldgrößen." *Det Kongelige Danske Videnskabernes Selskab. Matematisk-fysiske Meddelser* 12: 1–65.

Bohr, Niels, Kramers, Hendrik A., and Slater, John C. 1924. "The Quantum Theory of Radiation." *Philosophical Magazine* 47: 785–822.

Boltzmann, Ludwig. 1897. "Über die Unentbehrlichkeit der Atomistik in der Naturwissenschaft." *Annalen der Physik* 60: 231–47.

1898. *Vorlesungen über Gastheorie*. Leipzig: Barth.

Bondi, Hermann. 1979. "Is 'General Relativity' Necessary for Einstein's Theory of Gravitation?" In *Relativity, Quanta, and Cosmology in the Development of the Scientific Thought of Albert Einstein*. Vol. 1, ed. Mario Pantaleo and Francesco de Finis. New York: Johnson Reprint Corporation, 179–86.

Bondi, Hermann, and Gold, Thomas. 1948. "The Steady-State Theory of the Expanding Universe." *Royal Astronomical Society. Monthly Notices* 108: 252–72.

Bondi, Hermann, Pirani, Felix A. E., and Robinson, Ivor. 1959. "Gravitational Waves in General Relativity. III. Exact Plane Waves." *Proceedings of the Royal Society of London* (Series A) 251: 519–33.

Born, Max. 1926. "Zur Quantenmechanik der Stoßvorgänge." *Zeitschrift für Physik* 37: 863–7.

1962. *Einstein's Theory of Relativity*. New York: Dover.

1971. *The Born-Einstein Letters: Correspondence between Albert Einstein and Max and Hedwig Born from 1916 to 1955*. London: MacMillan. English translation of Einstein et al. (1969).

Born, Max, Heisenberg, Werner, and Jordan, Pascual. 1926. "Zur Quantenmechanik II." *Zeitschrift für Physik* 35: 557–615. English translation in van der Waerden (1968, 321–85).

Bothe, Walther, and Geiger, Hans. 1924. "Ein Weg zur experimentellen Nachprüfung der Theorie von Bohr, Kramers und Slater." *Zeitschrift für Physik* 26: 44.

1925a. "Experimentelles zur Theorie von Bohr, Kramers und Slater." *Die Naturwissenschaften* 13: 440–1.

1925b. "Über das Wesen des Comptoneffekts: ein experimenteller Beitrag zur Theorie der Strahlung." *Zeitschrift für Physik* 32: 639–63.

Brading, Katherine A. 2002. "Which Symmetry? Noether, Weyl, and the Conservation of Electric Charge." *Studies in History and Philosophy of Modern Physics* 33: 3–22.

Brading, Katherine A., and Ryckman, Thomas A. 2008. "Hilbert's 'Foundations of Physics': Gravitation and Electromagnetism within the Axiomatic Method." *Studies in History and Philosophy of Modern Physics* 39: 102–53.

Brenner, Michael. 1996. *The Renaissance of Jewish Culture in Weimar Germany.* New Haven, CT: Yale University Press.

Brian, Denis. 1996. *Einstein: A Life.* New York: Wiley.

Brighouse, Carolyn. 1994. "Spacetime and Holes." In *Proceedings of the 1994 Biennial Meeting of the Philosophy of Science Association*, Vol. 1, ed. David L. Hull, Micky Forbes, and Richard M. Burian. East Lansing, MI: Philosophy of Science Association, 117–25.

Broda, Engelbert. 1983. *Ludwig Boltzmann: Man, Physicist, Philosopher.* Woodbridge, CT: Ox Box Press.

Broglie, Louis de. 1922. "Rayonnement noir et de lumière." *Journal de physique et le Radium* 3: 422–8.

1979. "Correspondance entre Albert Einstein et Louis de Broglie."*Annales de la Fondation Louis de Broglie* 4: 56.

Brown, Harvey R. 2005. *Physical Relativity: Space-Time Structure from a Dynamical Perspective.* Oxford: Oxford University Press.

Brown, Harvey R., and Brading, Katherine A. 2002. "General Covariance from the Perspective of Noether's Theorems." *Diálogos* 79: 59–86.

Brown, Harvey R., and Pooley, Oliver. 2006. "Minkowski Space-Time: A Glorious Non-entity." In *The Ontology of Spacetime*, ed. Dennis Dieks. Amsterdam: Elsevier, 67–92.

Brush, Stephen G. 1976. *The Kind of Motion We Call Heat: A History of the Kinetic Theory of Gases in the 19th Century.* Amsterdam: North-Holland.

1996a. *Fruitful Encounters: The Origin of the Solar System and of the Moon from Chamberlin to Apollo.* Cambridge: Cambridge University Press.

1996b. *Nebulous Earth: The Origin of the Solar System and the Core of the Earth from Laplace to Jeffreys.* Cambridge: Cambridge University Press.

Buchdahl, Gerd. 1969. *Metaphysics and the Philosophy of Science.* Oxford: Blackwell.

Buchwald, Diana K., Renn, Jürgen, and Schlögl, Robert. 2013. "A Note on Einstein's Scratch Notebook of 1910–1913." In *Physics as a Calling, Science for Society*, eds. Ad Maas and Henriette Schatz. Leiden: Leiden University Press, 81–8.

Buchwald, Jed Z. 1994. *The Creation of Scientific Effect: Heinrich Hertz and Electric Waves.* Chicago: University of Chicago Press.

Büttner, Jochen, Renn, Jürgen, and Schemmel, Matthias. 2003. "Exploring the Limits of Classical Physics: Planck, Einstein, and the Structure of a Scientific Revolution." *Studies in History and Philosophy of Modern Physics* 34: 35–59.

Cahnmann, Werner. 1989. *German Jewry: Its History and Sociology.* New Brunswick, NJ, and Oxford: Transaction Publications.

Calaprice, Alice, ed. 1996. *The Quotable Einstein.* Princeton, NJ: Princeton University Press. Revised editions: *The Expanded Quotable Einstein* (2000); *The New Quotable Einstein* (2005); *The Ultimate Quotable Einstein* (2011).

Camilleri, Kristian. 2009. *Heisenberg and the Interpretation of Quantum Mechanics: The Physicist as Philosopher.* Cambridge: Cambridge University Press.

Carnap, Rudolf. 1922. *Der Raum: Ein Beitrag zur Wissenschaftslehre.* Berlin: Reuther & Richard.

1966. *Philosophical Foundations of Physics: An Introduction to the Philosophy of Science,* ed. Martin Gardner. New York: Basic Books. Reprinted as *An Introduction to the Philosophy of Science,* 1974.

Cassidy, David C. 2005. "Einstein and the Quantum Hypothesis." *Annalen der Physik* 14 (Supplement): 15–22. Supplement published in book form as Renn (2005a).

Cassirer, Ernst. 1918. *Kants Leben und Lehre.* Berlin: Bruno Cassirer.

1921. *Zur Einsteinschen Relativitätstheorie.* Berlin: Bruno Cassirer.

Cattani, Carlo, and De Maria, Michelangelo. 1993. "Conservation Laws and Gravitational Waves in General Relativity (1915–1918)." In Earman et al. (1993, 63–87).

Charlier, Carl V. L. 1908. "Wie eine unendliche Welt aufgebaut sein kann." *Arkiv för Matematik, Astronomi och Fysik* 4: 1–15.

Cheng, Ta-Pei. 2013. *Einstein's Physics: Atoms, Quanta, and Relativity Derived, Explained, and Appraised.* Oxford: Oxford University Press.

Clark, George L., and Duane, William. 1923. "The Wave-Lengths of Secondary X-Rays." *National Academy of Science Proceedings* 9: 413–24.

Clark, Ronald W. 1971. *Einstein: The Life and Times.* New York: World Publishing Co.

Clerke, Agnes. 1890. *The System of the Stars.* London: Longmans, Green and Co.

Cohen, I. Bernard, and Whitman, Anne. 1999. *Isaac Newton: The Principia. Mathematical Principles of Natural Philosophy. A New Translation.* Berkeley, Los Angeles, and London: University of California Press.

Cohen, Robert S., and Yehuda, Elkana. 1977. *Hermann von Helmholtz: Epistemological Writings.* Dordrecht: Reidel.

Compton, Arthur H. 1922. "Secondary Radiations Produced by X-Rays, and Some of Their Applications to Physical Problems." *Bulletin of the National Research Council* 4: 1–56. Reprinted in Compton (1973, 321–77).

1923a. "A Quantum Theory of the Scattering of X-Rays by Light Elements." *Physical Review* 21: 483–502. Reprinted in Compton (1973, 382–401).

1923b. "The Total Reflexion of X-Rays." *Philosophical Magazine* 45: 1121–31. Reprinted in Compton (1973, 402–12).

1973. *Scientific Papers of Arthur Holly Compton: X-Ray and Other Studies,* ed. Robert S. Shankland. Chicago: University of Chicago Press.

Compton, Arthur H., and Simon, Alfred W. 1925. "Directed Quanta of Scattered X-Rays." *Physical Review* 26: 289–99. Reprinted in Compton (1973, 508–18).

Compton, Karl T., and Richardson, Owen W. 1913. "The Photoelectric Effect–II." *Philosophical Magazine* 26: 530.

Corry, Leo. 1997. "Hermann Minkowski and the Postulate of Relativity." *Archive for History of Exact Sciences* 51: 273–314.

2004. *David Hilbert and the Axiomatization of Physics, 1898–1918: From Grundlagen der Geometrie to Grundlagen der Physik.* Dordrecht: Kluwer.

Corry, Leo, Renn, Jürgen, and Stachel, John. 1997. "Belated Decision in the Hilbert-Einstein Priority Dispute." *Science* 278: 1270–3.

CPAE 1. *The Collected Papers of Albert Einstein.* Vol. 1, *The Early Years 1879–1902,* eds. John Stachel, Robert Schulmann, David Cassidy, and Jürgen Renn. Princeton, NJ: Princeton University Press, 1987.

CPAE 1E. *The Collected Papers of Albert Einstein.* Vol. 1, *The Early Years, 1879–1902.* English eds. translated by A. Beck, consultant Peter Havas. Princeton, NJ: Princeton University Press, 1987.

CPAE 2. *The Collected Papers of Albert Einstein.* Vol. 2, *The Swiss Years: Writings, 1900–1909,* eds. John Stachel, Robert Schulmann, David C. Cassidy, and Jürgen Renn. Princeton, NJ: Princeton University Press, 1989.

CPAE 2E. *The Collected Papers of Albert Einstein.* Vol. 2, *The Swiss Years: Writings, 1900–1909.* English trans. by Anna Beck, consultant Peter Havas. Princeton, NJ: Princeton University Press, 1989.

CPAE 3. *The Collected Papers of Albert Einstein.* Vol. 3, *The Swiss Years: Writings, 1909–1911,* eds. Martin Klein, A. J. Kox, Jürgen Renn, and Robert Schulmann. Princeton, NJ: Princeton University Press, 1993.

CPAE 3E. *The Collected Papers of Albert Einstein.* Vol. 3, *The Swiss Years: Writings, 1909–1911.* English trans. by Anna Beck, consultant Don Howard. Princeton, NJ: Princeton University Press, 1993.

CPAE 4. *The Collected Papers of Albert Einstein.* Vol. 4, *The Swiss Years: Writings, 1912–1914,* eds. Martin Klein, A. J. Kox, Jürgen Renn, and Robert Schulmann. Princeton, NJ: Princeton University Press, 1995.

CPAE 4E. *The Collected Papers of Albert Einstein.* Vol. 4, *The Swiss Years: Writings, 1912–1914.* English trans. by Anna Beck, consultant Don Howard. Princeton, NJ: Princeton University Press, 1996.

CPAE 5. *The Collected Papers of Albert Einstein.* Vol. 5, *The Swiss Years: Correspondence, 1902–1914,* eds. Martin Klein, A. J. Kox, and Robert Schulmann. Princeton, NJ: Princeton University Press, 1993.

CPAE 5E. *The Collected Papers of Albert Einstein.* Vol. 5, *The Swiss Years: Correspondence, 1902–1914.* English trans. by Anna Beck, consultant Don Howard. Princeton, NJ: Princeton University Press, 1995.

CPAE 6. *The Collected Papers of Albert Einstein.* Vol. 6, *The Berlin Years: Writings, 1914–1917,* eds. A. J. Kox, Martin Klein, and Robert Schulmann. Princeton, NJ: Princeton University Press, 1996.

CPAE 6E. *The Collected Papers of Albert Einstein.* Vol. 6, *The Berlin Years: Writings, 1914–1917.* English trans. by Alfred Engel, consultant Engelbert Schücking. Princeton, NJ: Princeton University Press, 1997.

CPAE 7. *The Collected Papers of Albert Einstein.* Vol. 7, *The Berlin Years: Writings, 1918–1921,* eds. Michel Janssen, Robert Schulmann, József Illy, Christoph Lehner, and Diana Kormos Buchwald. Princeton, NJ: Princeton University Press, 2002.

CPAE 7E. *The Collected Papers of Albert Einstein.* Vol. 7, *The Berlin Years: Writings, 1918–1921.* English trans. by Alfred Engel, consultant Engelbert Schücking. Princeton, NJ: Princeton University Press, 2002.

CPAE 8. *The Collected Papers of Albert Einstein*. Vol. 8, *The Berlin Years: Correspondence, 1914–1918 (Parts A and B)*, eds. Robert Schulmann, A. J. Kox, Michel Janssen, and József Illy. Princeton, NJ: Princeton University Press, 1998.

CPAE 8E. *The Collected Papers of Albert Einstein*. Vol. 8, *The Berlin Years: Correspondence, 1914–1918 (Parts A and B)*. English trans. by Ann M. Hentschel, consultant Klaus Hentschel. Princeton, NJ: Princeton University Press, 1998.

CPAE 9. *The Collected Papers of Albert Einstein*. Vol. 9, *The Berlin Years: Correspondence, January 1919–April 1920*, eds. Diana Kormos Buchwald, Robert Schulmann, József Illy, Daniel J. Kennefick, and Tilman Sauer. Princeton, NJ: Princeton University Press, 2004.

CPAE 9E. *The Collected Papers of Albert Einstein*. Vol. 9, *The Berlin Years: Correspondence, January 1919–April 1920*. English trans. by Ann M. Hentschel, consultant Klaus Hentschel. Princeton, NJ: Princeton University Press, 2004.

CPAE 10. *The Collected Papers of Albert Einstein*. Vol. 10, *The Berlin Years: Correspondence, May–December 1920 and Supplementary Correspondence, 1909–1920*, eds. Diana Kormos Buchwald, Tilman Sauer, Ze'ev Rosenkranz, József Illy, and Iris Virginia Holmes. Princeton, NJ: Princeton University Press, 2006.

CPAE 10E. *The Collected Papers of Albert Einstein*. Vol. 10, *The Berlin Years: Correspondence, May–December 1920 and Supplementary Correspondence, 1909–1920*. English trans. by Ann M. Hentschel, consultant Klaus Hentschel. Princeton, NJ: Princeton University Press, 2006.

CPAE 11. *The Collected Papers of Albert Einstein*. Vol. 11, *Cumulative Index, Bibliography, List of Correspondence, Chronology and Errata to Volumes 1–10*. Compiled by A. J. Kox, Tilman Sauer, Diana Kormos Buchwald, Rudy Hirschmann, Osik Moses, Benjamin Aronin, and Jennifer Stolper. Princeton, NJ: Princeton University Press, 2009.

CPAE 12. *The Collected Papers of Albert Einstein*. Vol. 12, *The Berlin Years: Correspondence, January–December 1921*, eds. Diana Kormos Buchwald, Ze'ev Rosenkranz, Tilman Sauer, József Illy, and Virginia Iris Holmes. Princeton, NJ: Princeton University Press, 2009.

CPAE 12E. *The Collected Papers of Albert Einstein*. Vol. 12, *The Berlin Years: Correspondence, January–December 1921*. English trans. by Ann M. Hentschel, consultant Klaus Hentschel. Princeton, NJ: Princeton University Press, 2009.

CPAE 13. *The Collected Papers of Albert Einstein*. Vol. 13, *The Berlin Years: Writings & Correspondence, January 1922–March 1923*, eds. Diana Kormos Buchwald, József Illy, Ze'ev Rosenkranz, and Tilman Sauer. Princeton, NJ: Princeton University Press, 2012.

CPAE 13E. *The Collected Papers of Albert Einstein*. Vol. 13, *The Berlin Years: Writings & Correspondence, January 1922–March 1923*. English trans. by Ann M. Hentschel and Osik Moses, consultant Klaus Hentschel. Princeton, NJ: Princeton University Press, 2012.

Crelinsten, Jeffrey. 2006. *Einstein's Jury: The Race to Test Relativity*. Princeton, NJ: Princeton University Press.

Curd, Martin, and Cover, J. A., eds. 1998. *Philosophy of Science: The Central Issues*. New York: Norton.

Curtis, Heber. 1921. "The Scale of the Universe." *Bulletin of the National Research Council* 2: 194–217.

Cushing, James T. 1994. *Quantum Mechanics: Historical Contingency and the Copenhagen Hegemony*. Chicago: University of Chicago Press.

Damerow, Peter, Freudenthal, Gideon, McLaughlin, Peter, and Renn, Jürgen. 2004. *Exploring the Limits of Preclassical Mechanics*. 2nd ed. New York: Springer.

Darrigol, Olivier. 1986. "The Origin of Quantized Matter Waves." *Historical Studies in the Physical and Biological Sciences* 16: 197–253.

——— 1988. "Statistics and Combinatorics in Early Quantum Theory I." *Historical Studies in the Physical Sciences* 19: 17–80.

——— 1991. "Statistics and Combinatorics in Early Quantum Theory, II: Early Symptoms of Indistinguishability and Holism." *Historical Studies in the Physical Sciences* 21: 237–98.

——— 1992. *From c-Numbers to q-Numbers: The Classical Analogy in the History of Quantum Theory*. Berkeley: University of California Press.

——— 1993. "Strangeness and Soundness in Louis de Broglie's Early Works." *Physis* 30: 303–72.

——— 1995. "Henri Poincaré's Criticism of *fin de siècle* Electrodynamics." *Studies in History and Philosophy of Modern Physics* 26: 1–44.

——— 2000a. "Continuities and Discontinuities in Planck's *Akt der Verzweiflung*." *Annalen der Physik* 9: 951–60.

——— 2000b. *Electrodynamics from Ampère to Einstein*. Oxford: Oxford University Press.

——— 2001. "The Historians' Disagreements over the Meaning of Planck's Quantum." *Centaurus* 43: 219–39.

——— 2004. "The Mystery of the Einstein-Poincaré Connection." *Isis* 95: 614–26.

Debye, Peter. 1923. "Zerstreuung von Röntgenstrahlen und Quantentheorie." *Physikalische Zeitschrift* 24: 161–6. English trans. in Debye (1954, 80–8).

——— 1954. *The Collected Papers of Peter J. W. Debye*. New York: Interscience.

Debye, Peter, and Sommerfeld, Arnold. 1913. "Theorie des lichtelektrischen Effektes vom Standpunkt des Wirkungsquantums." *Annalen der Physik* 41: 873–930. Reprinted in Sommerfeld (1968, Vol. 4, 78–135).

De Sitter, Willem. 1916a. "On Einstein's Theory of Gravitation, and Its Astronomical Consequences: First Paper." *Royal Astronomical Society. Monthly Notices* 76: 699–728.

——— 1916b. "De relativiteit der rotatie in de theorie van Einstein." *Koninklijke Akademie van Wetenschappen te Amsterdam. Wis- en Natuurkundige Afdeeling. Verslagen van de Gewone Vergaderingen* 25: 499–504. English trans. "On the Relativity of Rotation in Einstein's Theory." *Koninklijke Akademie van Wetenschappen te Amsterdam. Section of Sciences. Proceedings* 19 (1916): 527–32.

——— 1916c. "On Einstein's Theory of Gravitation, and Its Astronomical Consequences: Second Paper." *Royal Astronomical Society. Monthly Notices* 77: 155–84.

——— 1917a. "Over de relativiteit der traagheid: Beschouwingen naar aanleiding van Einsteins laatste hypothese." *Koninklijke Akademie van Wetenschappen te Amsterdam. Wis- en Natuurkundige Afdeeling. Verslagen van de Gewone Vergaderingen* 25: 1268–76. English trans. "On the Relativity of Inertia: Remarks Concerning Einstein's Latest Hypothesis." *Koninklijke Akademie van Wetenschappen te Amsterdam. Section of Sciences. Proceedings* 19 (1917): 1217–25.

——— 1917b. "Over de kromming der ruimte." *Koninklijke Akademie van Wetenschappen te Amsterdam. Wis- en Natuurkundige Afdeeling. Verslagen van de Gewone Vergaderingen* 26: 222–36. English trans. "On the Curvature of Space." *Koninklijke Akademie van Wetenschappen te Amsterdam. Section of Sciences. Proceedings* 20 (1917): 229–43.

——— 1917c. "On Einstein's Theory of Gravitation, and Its Astronomical Consequences: Third Paper." *Royal Astronomical Society. Monthly Notices* 78: 3–28.

1930. "On the Distances and Radial Velocities of Extra-Galactic Nebulae, and the Explanation of the Latter by the Relativity Theory of Inertia." *National Academy of Science. Proceedings* 16: 474–88.

1931. "The Expanding Universe." *Scientia* 49: 1–10.

Dieks, Dennis. 2006. "Another Look at General Covariance and the Equivalence of Reference Frames." *Studies in History and Philosophy of Modern Physics* 37: 174–91.

Dirac, Paul Adrien Maurice. 1930. *The Principles of Quantum Mechanics*. Oxford: Clarendon Press.

DiSalle, Robert. 1991. "Conventionalism and the Origins of the Inertial Frame Concept." *PSA 1990: Proceedings of the 1990 Biennial Meeting of the Philosophy of Science Association*, ed. Linda Wessels, Arthur Fine, and Micky Forbes. East Lansing, MI: Philosophy of Science Association, 139–47.

1995. "Spacetime Theory as Physical Geometry." *Erkenntnis* 42: 317–37.

2006. *Understanding Space-Time: The Philosophical Development of Physics from Newton to Einstein*. Cambridge: Cambridge University Press.

Domski, Mary, and Dickson, Michael, eds. 2010. *Discourse on a New Method: Reinvigorating the Marriage of History and Philosophy of Science*. La Salle, IL: Open Court.

Dorling, Jon. 1978. "Did Einstein Need General Relativity to Solve the Problem of Absolute Space? Or Had the Problem Already Been Solved by Special Relativity?" *British Journal for the Philosophy of Science* 29: 311–23.

Dresden, Max. 1987. *H. A. Kramers: Between Tradition and Revolution*. Berlin: Springer.

Duane, William, and Hunt, Franklin L. 1915. "On X-Ray Wave Length." *Physical Review* 6: 166–77.

Duhem, Pierre. 1906. *La Théorie physique, son objet et sa structure*. Paris: Chevalier & Rivière.

1908. *Ziel und Struktur der physikalischen Theorien*. Trans. of Duhem (1906) by Friedrich Adler, foreword by Ernst Mach. Leipzig: Barth.

1954. *The Aim and Structure of Physical Theory*. English trans. of 2nd ed. of Duhem (1906) by Philip P. Wiener, foreword by Louis de Broglie. Princeton, NJ: Princeton University Press.

Dühring, Eugen. 1873. *Kritische Geschichte der allgemeinen Principien der Mechanik*. Berlin: Theobald Grieben.

Dukas, Helen, and Hoffmann, Banesh. 1979. *Albert Einstein: The Human Side. New Glimpses from His Archives*. Princeton, NJ: Princeton University Press.

Duncan, Anthony, and Janssen, Michel. 2007. "On the Verge of *Umdeutung* in Minnesota: Van Vleck and the Correspondence Principle." 2 Parts. *Archive for History of Exact Sciences* 61: 553–671.

2008. "Pascual Jordan's Resolution of the Conundrum of the Wave-Particle Duality of Light." *Studies in History and Philosophy of Modern Physics* 39: 634–66.

2009. "From Canonical Transformations to Transformation Theory, 1926–1927: The Road to Jordan's *Neue Begründung*." *Studies in History and Philosophy of Modern Physics* 40: 352–62.

Earman, John. 1989. *World Enough and Space-Time: Absolute versus Relational Theories of Space and Time*. Cambridge, MA: MIT Press.

1995. *Bangs, Crunches, Whimpers, and Shrieks: Singularities and Acausalities in Relativistic Spacetimes*. Oxford: Oxford University Press.

1999. "The Penrose-Hawking Singularity Theorems: History and Implications." In Goenner et al. (1999, 235–70).

2001. "Lambda: The Constant That Refuses to Die." *Archive for the History of Exact Sciences* 55: 189–220.

2003. "The Cosmological Constant, the Fate of the Universe, Unimodular Gravity, and All That." *Studies in History and Philosophy of Modern Physics* 55: 559–77.

Earman, John, and Eisenstaedt, Jean. 1999. "Einstein and Singularities." *Studies in History and Philosophy of Modern Physics* 30: 185–235.

Earman, John, and Glymour, Clark. 1978. "Lost in the Tensors: Einstein's Struggle with Covariance Principles 1912–1916." *Studies in History and Philosophy of Science* 9: 251–78.

1980a. "Relativity and Eclipses: The British Eclipse Expeditions of 1919 and Their Predecessors." *Studies in History and Philosophy of Science* 11: 49–85.

1980b. "The Gravitational Red Shift as a Test of General Relativity: History and Analysis." *Studies in History and Philosophy of Science* 11: 175–214.

Earman, John, and Janssen, Michel. 1993. "Einstein's Explanation of the Motion of Mercury's Perihelion." In Earman et al. (1993, 129–72).

Earman, John, and Norton, John D. 1987. "What Price Space-Time Substantivalism? The Hole Story." *British Journal for the Philosophy of Science* 38: 515–25.

Earman, John, Glymour, Clark, and Rynasiewicz, Robert. 1983. "On Writing the History of Special Relativity." In *PSA 1982: Proceedings of the 1982 Biennial Meeting of the Philosophy of Science Association*, ed. Peter D. Asquith and Thomas Nickles. East Lansing, MI: Philosophy of Science Association, 403–16.

Earman, John, Janssen, Michel, and Norton, John D., eds. 1993. *The Attraction of Gravitation*. Einstein Studies 5. Boston: Birkhäuser.

Eddington, Arthur S. 1914. *Stellar Movements and the Structure of the Universe*. London: MacMillan.

1921. "A Generalization of Weyl's Theory of the Electromagnetic and Gravitational Fields." *Proceedings of the Royal Society of London* (Series A) 99: 104–22.

1922. "The Propagation of Gravitational Waves." *Proceedings of the Royal Society of London* (Series A) 102: 268–82.

1923. *The Mathematical Theory of Relativity*. Cambridge: Cambridge University Press.

1930. "On the Instability of Einstein's Spherical World." *Royal Astronomical Society. Monthly Notices* 90: 668–78.

1933. *The Expanding Universe*. New York: MacMillan.

1936. *The Relativity Theory of Protons and Electrons*. Cambridge: Cambridge University Press.

Ehrenfest, Paul. 1905. "Über die physikalischen Voraussetzungen der Planck'schen Theorie der irreversiblen Strahlungsvorgänge." *Sitzungsberichte der Kaiserlichen Akademie der Wissenschaften* (Wien) *Mathematisch-Naturwissenschaftliche Klasse* 114: 1301–14.

1906. "Zur Planckschen Strahlungstheorie." *Physikalische Zeitschrift* 7: 528–32.

For Einstein's works published in the period 1901–21 the short titles assigned to them in *The Collected Papers of Albert Einstein* are used. Unless noted otherwise, page numbers refer to the original publication.

Einstein, Albert. 1901. "Folgerungen aus den Capillaritätserscheinungen." *Annalen der Physik* 4: 513–23. Reprinted in CPAE 2, Doc. 1.

1902a. "Über die thermodynamische Theorie der Potentialdifferenz zwischen Metallen und vollständig dissociirten Lösungen ihrer Salze, und über eine elektrische Methode zur Erforschung der Molekularkräfte." *Annalen der Physik* 8: 798–814. Reprinted in CPAE 2, Doc. 2.

1902b. "Kinetische Theorie des Wärmegleichgewichtes und des zweiten Hauptsatzes der Thermodynamik." *Annalen der Physik* 9: 417–33. Reprinted in CPAE 2, Doc. 3.

1903. "Eine Theorie der Grundlagen der Thermodynamik." *Annalen der Physik* 11: 170–87. Reprinted in CPAE 2, Doc. 4.

1904. "Zur allgemeinen molekularen Theorie der Wärme." *Annalen der Physik* 14: 354–62. Reprinted in CPAE 2, Doc. 5.

1905i. "Über einen die Erzeugung und die Verwandlung des Lichtes betreffenden heuristischen Gesichtspunkt." *Annalen der Physik* 17: 132–48. Reprinted in CPAE 2, Doc. 14.

1905j. *Eine neue Bestimmung der Moleküldimensionen: Inaugural-Dissertation zur Erlangung der philosophischen Doktorwürde der hohen Philosophischen Fakultät (Mathematisch-Naturwissenschaftliche Sektion) der Universität Zürich.* Bern: Buchdruckerei K. J. Wyss. Reprinted in CPAE 2, Doc. 15.

1905k. "Über die von der molekularkinetischen Theorie der Wärme geforderte Bewegung von in ruhenden Flüssigkeiten suspendierten Teilchen." *Annalen der Physik* 17: 549–60. Reprinted in CPAE 2, Doc. 16.

1905r. "Zur Elektrodynamik bewegter Körper." *Annalen der Physik* 17: 891–921. Reprinted in CPAE 2, Doc. 23. English trans. in Einstein et al. (1952, 37–65).

1905s. "Ist die Trägheit eines Körpers von seinem Energieinhalt abhängig?" *Annalen der Physik* 18: 639–41. Reprinted in CPAE 2, Doc. 24.

1906a. "Eine neue Bestimmung der Moleküldimensionen." *Annalen der Physik* 19: 289–305. Reprinted in CPAE 2, Doc. 15.

1906b. "Zur Theorie der Brownschen Bewegung." *Annalen der Physik* 19: 371–81. Reprinted in CPAE 2, Doc. 32.

1906d. "Zur Theorie der Lichterzeugung und Lichtabsorption." *Annalen der Physik* 20: 199–206. Reprinted in CPAE 2, Doc. 34.

1907a. "Die Plancksche Theorie der Strahlung und die Theorie der spezifischen Wärme." *Annalen der Physik* 22: 180–90. Reprinted in CPAE 2, Doc. 38.

1907b. "Über die Gültigkeitsgrenze des Satzes vom thermodynamischen Gleichgewicht und über die Möglichkeit einer neuen Bestimmung der Elementarquanta." *Annalen der Physik* 22: 569–72. Reprinted in CPAE 2, Doc. 39.

1907c. "Theoretische Bemerkungen über die Brownsche Bewegung." *Zeitschrift für Elektrochemie und angewandte physikalische Chemie* 13: 41–2. Reprinted in CPAE 2, Doc. 40.

1907j. "Über das Relativitätsprinzip und die aus demselben gezogenen Folgerungen." *Jahrbuch für Radioaktivität und Elektronik* 4: 411–62. Reprinted in CPAE 2, Doc. 47.

1908b. "Berichtigungen zu der Arbeit: 'Über das Relativitätsprinzip und die aus demselben gezogenen Folgerungen.'" *Jahrbuch der Radioaktivität und Elektronik* 5: 98–9. Reprinted in CPAE 2, Doc. 49.

1908c. "Elementare Theorie der Brownschen Bewegung." *Zeitschrift für Elektrochemie und angewandte physikalische Chemie* 14: 235–9. Reprinted in CPAE 2, Doc. 50.

1909b. "Zum gegenwärtigen Stand des Strahlungsproblems." *Physikalische Zeitschrift* 10: 185–93. Reprinted in CPAE 2, Doc. 56.

1909c. "Über die Entwicklung unserer Anschauungen über das Wesen und die Konstitution der Strahlung." *Physikalische Zeitschrift* 10: 817–25. Reprinted in CPAE 2, Doc. 60.

1910d. "Theorie der Opaleszenz von homogenen Flüssigkeiten und Flüssigkeitsgemischen in der Nähe des kritischen Zustandes." *Annalen der Physik* 33: 1275–98. Reprinted in CPAE 3, Doc. 9.

1911c. "Bemerkungen zu den P. Hertzschen Arbeiten: 'Über die mechanischen Grundlagen der Thermodynamik.'" *Annalen der Physik* 34: 175–6. Reprinted in CPAE 3, Doc. 10.

1911e. "Berichtigung zu meiner Arbeit: 'Eine neue Bestimmung der Moleküldimensionen.'" *Annalen der Physik* 34: 591–2. Reprinted in CPAE 3, Doc. 15.

1911h. "Über den Einfluß der Schwerkraft auf die Ausbreitung des Lichtes." *Annalen der Physik* 35: 898–908. Reprinted in CPAE 3, Doc. 23. English trans. in Einstein et al. (1952, 99–108).

1912a. "L'état actuel du problème des chaleurs spécifiques." In Langevin and de Broglie (1912, 407–35). German trans.: Einstein (1914a).

1912b. "Thermodynamische Begründung des photochemischen Äquivalenzgesetzes." *Annalen der Physik* 35: 820–38. Reprinted in CPAE 4, Doc. 2.

1912c. "Lichtgeschwindigkeit und Statik des Gravitationsfeldes." *Annalen der Physik* 38: 355–69. Reprinted in CPAE 4, Doc. 3.

1912d. "Zur Theorie des statischen Gravitationsfeldes." *Annalen der Physik* 38: 443–58. Reprinted in CPAE 4, Doc. 4.

1912e. "Gibt es eine Gravitationswirkung, die der elektrodynamischen Induktionswirkung analog ist?" *Vierteljahrsschrift für gerichtliche Medizin und öffentliches Sanitätswesen* 44: 37–40. Reprinted in CPAE 4, Doc. 7.

1912f. "Nachtrag zu meiner Arbeit: 'Thermodynamische Begründung des Photochemischen Äquivalenzgesetzes.'" *Annalen der Physik* 38: 881–4. Reprinted in CPAE 4, Doc. 5.

1912h. "Relativität und Gravitation. Erwiderung auf eine Bemerkung von M. Abraham." *Annalen der Physik* 38: 1059–64. Reprinted in CPAE 4, Doc. 8.

1913c. "Zum gegenwärtigen Stande des Gravitationsproblems." *Physikalische Zeitschrift* 14: 1249–62. Reprinted in CPAE 4, Doc. 17.

1914a. "Zum gegenwärtigen Stande des Problems der spezifischen Wärme." In Eucken (1914, 330–52). Reprinted in CPAE 3, Doc. 26.

1914d. "Bemerkungen." *Zeitschrift für Mathematik und Physik* 62: 260–1. Reprinted in CPAE 4, Doc. 26.

1914e. "Prinzipielles zur verallgemeinerten Relativitätstheorie." *Physikalische Zeitschrift* 15: 176–80. Reprinted in CPAE 4, Doc. 25.

1914k. "Antrittsrede." *Königlich Preußische Akademie der Wissenschaften* (Berlin). *Sitzungsberichte*: 739–42. Reprinted in CPAE 6, Doc. 3.

1914n. "Beiträge zur Quantentheorie." *Deutsche Physikalische Gesellschaft. Verhandlungen* 16: 820–8. Reprinted in CPAE 6, Doc. 5.

1914o. "Die formale Grundlage der allgemeinen Relativitätstheorie." *Königlich Preußische Akademie der Wissenschaften* (Berlin). *Sitzungsberichte*: 1030–85. Reprinted in CPAE 6, Doc. 9.

1915a. "Theoretische Atomistik." In *Die Kultur der Gegenwart: Ihre Entwicklung und ihre Ziele*. Vol. 1, Part 3, Sec. 3, *Physik*, ed. Emil Warburg. Leipzig: Teubner, 251–63. Reprinted in CPAE 4, Doc. 20.

1915c. "Experimenteller Nachweis der Ampèreschen Molekularströme." *Die Naturwissenschaften* 3: 237–8. Reprinted in CPAE 6, Doc. 15.

1915f. "Zur allgemeinen Relativitätstheorie." *Königlich Preußische Akademie der Wissenschaften* (Berlin). *Sitzungsberichte*: 778–6. Reprinted in CPAE 6, Doc. 21.

1915g. "Zur allgemeinen Relativitätstheorie (Nachtrag)." *Königlich Preußische Akademie der Wissenschaften* (Berlin). *Sitzungsberichte*: 799–801. Reprinted in CPAE 6, Doc. 22.

1915h. "Erklärung der Perihelbewegung des Merkur aus der allgemeinen Relativitätstheorie." *Königlich Preußische Akademie der Wissenschaften* (Berlin). *Sitzungsberichte*: 831–9. Reprinted in CPAE 6, Doc. 24.

1915i. "Die Feldgleichungen der Gravitation." *Königlich Preußische Akademie der Wissenschaften* (Berlin). *Sitzungsberichte*: 844–7. Reprinted in CPAE 6, Doc. 25.

1916c. "Ernst Mach." *Physikalische Zeitschrift* 17: 101–4. Reprinted in CPAE 6, Doc. 29. English trans. in *Ernst Mach – A Deeper Look: Documents and New Perspectives*, ed. John Blackmore. Dordrecht: Kluwer, 1992, 154–9.

1916e. "Die Grundlage der allgemeinen Relativitätstheorie." *Annalen der Physik* 49: 769–822. Reprinted in CPAE 6, Doc. 30. English trans. in Einstein et al. (1952, 111–64).

1916g. "Näherungsweise Integration der Feldgleichungen der Gravitation." *Königlich Preußische Akademie der Wissenschaften* (Berlin). *Sitzungsberichte*: 668–96. Reprinted in CPAE 6, Doc. 32.

1916j. "Strahlungs-Emission und -Absorption nach der Quantentheorie." *Deutsche Physikalische Gesellschaft. Verhandlungen* 18: 318–23. Reprinted in CPAE 6, Doc. 34.

1916n. "Zur Quantentheorie der Strahlung." *Physikalische Gesellschaft Zürich. Mitteilungen* 18: 47–62. Reprinted as Einstein (1917c). Reprinted in CPAE 6, Doc. 38.

1916o. "Hamiltonsches Prinzip und allgemeine Relativitätstheorie." *Königlich Preußische Akademie der Wissenschaften* (Berlin). *Sitzungsberichte*: 1111–16. Reprinted in CPAE 6, Doc. 41.

1917a. *Über die spezielle und die allgemeine Relativitätstheorie (Gemeinverständlich)*. Braunschweig: Vieweg. Reprinted in CPAE 6, Doc. 42. English trans. in Einstein (1959).

1917b. "Kosmologische Betrachtungen zur allgemeinen Relativitätstheorie." *Königlich Preußische Akademie der Wissenschaften* (Berlin). *Sitzungsberichte*: 142–52. Reprinted in CPAE 6, Doc. 43. English trans. in Einstein et al. (1952, 177–88).

1917c. "Zur Quantentheorie der Strahlung." *Physikalische Zeitschrift* 18: 121–8. Reprint of Einstein (1916n). English trans. in van der Waerden (1967, 63–77).

1917h. "Der Angst-Traum." *Berliner Tageblatt*. December 25, 1917, morning edition, 3rd supplement, 1. Reprinted in CPAE 6, Doc. 49.

1918a. "Über Gravitationswellen." *Königlich Preußische Akademie der Wissenschaften* (Berlin). *Sitzungsberichte*: 154–67. Reprinted in CPAE 7, Doc. 1.

1918c. "Kritisches zu einer von Hrn. De Sitter gegebenen Lösung der Gravitationsgleichungen." *Königlich Preußische Akademie der Wissenschaften* (Berlin). *Sitzungsberichte*: 270–2. Reprinted in CPAE 7, Doc. 5.

1918d. "Bemerkung zu Herrn Schrödinger's Notiz 'Über ein Lösungssystem der allgemein kovarianten Gravitationsgleichungen.'" *Physikalische Zeitschrift* 19: 165–6. Reprinted in CPAE 7, Doc. 3.

1918e. "Prinzipielles zur allgemeinen Relativitätstheorie." *Annalen der Physik* 55: 241–4. Reprinted in CPAE 7, Doc. 4.

1918f. "Der Energiesatz in der allgemeinen Relativitätstheorie." *Königlich Preußische Akademie der Wissenschaften* (Berlin). *Sitzungsberichte*: 448–59. Reprinted in CPAE 7, Doc. 9.

1918g. "Nachtrag." *Königlich Preußische Akademie der Wissenschaften* (Berlin). *Sitzungsberichte*: 478. Reprinted in CPAE 7, Doc. 8.

1918h. "Hermann Weyl, Raum–Zeit–Materie. Vorlesungen über allgemeine Relativitätstheorie. Berlin, Julius Springer, 1918." *Die Naturwissenschaften* 6: 373. Reprinted in CPAE 7, Doc. 10.

1918j. "Motive des Forschens." In *Zu Max Plancks sechzigstem Geburtstag*, eds. Emil Warburg, Max von Laue, Arnold Sommerfeld, and Albert Einstein. Karlsruhe I. B.: C. F. Müller, 29–32. Reprinted in CPAE 7, Doc. 7. English trans. in Einstein (1954, 224–7).

1918k. "Dialog über Einwände gegen die Relativitätstheorie." *Die Naturwissenschaften* 6: 697–702. Reprinted in CPAE 7, Doc. 13.

1919a. "Spielen Gravitationsfelder im Aufbau der materiellen Elementarteilchen eine wesentliche Rolle?" *Preussische Akademie der Wissenschaften* (Berlin). *Sitzungsberichte*: 349–56. Reprinted in CPAE 7, Doc. 17. English trans. in Einstein et al. (1952, 191–8).

1919b. "Bemerkungen über periodische Schwankungen der Mondlänge, welche bisher nach der Newtonschen Mechanik nicht erklärbar schienen." *Preußische Akademie der Wissenschaften* (Berlin). *Physikalisch-Mathematische Klasse*. *Sitzungsberichte*: 433–6. Reprinted in CPAE 7, Doc. 18.

1919f. "Time, Space, and Gravitation." *The Times* (London). November 28, 1919: 13–14. Reprinted in CPAE 7, Doc. 26.

1919g. "Induktion und Deduktion in der Physik." *Berliner Tageblatt*. December 25, 1919. Reprinted in CPAE 7, Doc. 28.

1919h. "Die Zuwanderung aus dem Osten." *Berliner Tageblatt*. December 30, 1919, morning edition, 2. Reprinted in CPAE 7, Doc. 29.

1920f. "Meine Antwort. Über die anti-relativitätstheoretische G.m.b.H." *Berliner Tageblatt*. August 27, 1920, morning edition. Reprinted in CPAE 7, Doc. 45.

1920h. "A Confession." *Israelitisches Wochenblatt für die Schweiz*, September 24, 1920, 10. Reprinted in CPAE 7, Doc. 37.

1920j. *Äther und Relativitätstheorie*. Berlin: Springer. Reprinted in CPAE 7, Doc. 38. English trans. in Einstein (1983, 1–24).

1920k. "Antwort auf die vorstehende Betrachtung [Reichenbächer 1920]." *Die Naturwissenschaften* 8: 1010–11. Reprinted in CPAE 7, Doc. 49.

1921c. *Geometrie und Erfahrung*. Berlin: Springer. Reprinted in CPAE 7, Doc. 52. English trans. in Einstein (1954, 12–28).

1921e. "Über eine naheliegende Ergänzung des Fundamentes der allgemeinen Relativitätstheorie." *Preußische Akademie der Wissenschaften* (Berlin). *Sitzungsberichte*: 261–4. Reprinted in CPAE 7, Doc. 54.

1921h. "Wie ich Zionist wurde." *Jüdische Rundschau*. June 21, 1921, no. 49, 351–2. Reprinted in CPAE 7, Doc. 57.

1922a. "Über ein den Elementarprozeß der Lichtemission betreffendes Experiment." *Preußische Akademie der Wissenschaften* (Berlin). *Sitzungsberichte*: 882–3. Reprinted in CPAE 7, Doc. 68.

1922c. *Vier Vorlesungen über Relativitätstheorie*. Braunschweig: Vieweg. Reprinted in CPAE 7, Doc. 71. English trans. of later edition: Einstein (1956a).

1922d. "La Théorie de la Relativité." *Interventions sur la théorie de la Relativité par X. Leon, P. Langevin, J. Hadamard, A. Einstein, E. Cartan, P. Painlevé, P. Levy, J. Perrin, J. Becquerel, L. Brunschvicg, E. Le Roy, H. Bergson, E. Meyerson, et H. Pieron. Bulletin de la Société Française de Philosophie, séance du 6 avril 1922* XVII: 91–113.

1922e. "Bemerkung zu der Arbeit von A. Friedmann: 'Über die Krümmung des Raumes.'" *Zeitschrift für Physik* 11: 326.

1922i. "In Memoriam Walther Rathenau." *Neue Rundschau* 33: 815–6. Reprinted in CPAE 13, Doc. 317.

1923a. "Zur allgemeinen Relativitätstheorie." *Preußische Akademie der Wissenschaften* (Berlin). *Sitzungsberichte*: 32–8.

1923b. "Bemerkung zu meiner Arbeit 'Zur allgemeinen Relativitätstheorie.'" *Preußische Akademie der Wissenschaften* (Berlin). *Sitzungsberichte*: 76–7.

1923c. "Zur affinen Feldtheorie." *Preußische Akademie der Wissenschaften* (Berlin). *Sitzungsberichte*: 137–40.

1923d. "Bietet die Feldtheorie Möglichkeiten für die Lösung des Quantenproblems?" *Preußische Akademie der Wissenschaften* (Berlin). *Sitzungsberichte*: 359–64.

1923e. "The Theory of the Affine Field." *Nature* 112: 448–9.

1923f. "Grundgedanken und Probleme der Relativitätstheorie." In *Le Prix Nobel en 1921–1922*. Stockholm: Imprimerie Royale. English trans. "Fundamental Ideas and Problems of the Theory of Relativity" in Nobel Foundation (1967, 482–90).

1923g. "Notiz zu der Bemerkung zu der Arbeit von A. Friedmann 'Über die Krümmung des Raumes.'" *Zeitschrift für Physik* 16: 228.

1924a. "Quantentheorie des einatomigen idealen Gases." *Preußische Akademie der Wissenschaften* (Berlin). *Sitzungsberichte*: 261–7.

1924b. "Elsbachs Buch: Kant und Einstein." *Deutsche Literaturzeitung* 45: 1685–92.

1924c. "Das Komptonsche Experiment: Ist die Wissenschaft um Ihrer Selbst Willen Da?" *Berliner Tageblatt*. April 20, 1924, 1. Beiblatt.

1924d. "Review of J. Winternitz's 'Relativitätstheorie und Erkenntnislehre.'" *Deutsche Literaturzeitung* 149: 20–2.

1925a. "Quantentheorie des einatomigen idealen Gases. Zweite Abhandlung." *Preußische Akademie der Wissenschaften* (Berlin). *Sitzungsberichte*: 3–14.

1925b. "Zur Quantentheorie des idealen Gases." *Preußische Akademie der Wissenschaften* (Berlin). *Sitzungsberichte*: 18–25.

1925c. "Nichteuklidische Geometrie und Physik." *Die Neue Rundschau* 1: 16–20.

1925d. "Einheitliche Feldtheorie von Gravitation und Elektrizität." *Preußische Akademie der Wissenschaften* (Berlin). *Sitzungsberichte*: 414–19.

1927a. "Über die formale Beziehung des Riemannschen Krümmungstensors zu den Feldgleichungen der Gravitation." *Mathematische Annalen* 97: 99–103.

1927b. "Zu Kaluzas Theorie des Zusammenhangs von Gravitation und Elektrizität. Erste Mitteilung." *Preußische Akademie der Wissenschaften* (Berlin). *Sitzungsberichte*: 23–5.

1927c. "Zu Kaluzas Theorie des Zusammenhangs von Gravitation und Elektrizität. Zweite Mitteilung." *Preußische Akademie der Wissenschaften* (Berlin). *Sitzungsberichte*: 26–30.

1928a. "Riemanngeometrie mit Aufrechterhaltung des Begriffes des Fern-Parallelismus." *Preußische Akademie der Wissenschaften* (Berlin). *Sitzungsberichte*: 217–21.

1928b. "Neue Möglichkeit für eine einheitliche Feldtheorie von Gravitation und Elektrizität." *Preußische Akademie der Wissenschaften* (Berlin). *Sitzungsberichte*: 224–7.

1929. "The New Field Theory, II: The Structure of Space-Time." *The Times* (London). February 5, 1929. Reprinted in *The Observatory*, no. 659 (April), 114–18.

1931a. *About Zionism, Speeches and Letters*, ed. and trans. Leon Simon. New York: Macmillan.

1931b. "Zum kosmologischen Problem der allgemeinen Relativitätstheorie." *Preußische Akademie der Wissenschaften* (Berlin). *Sitzungsberichte*: 235–7.

1933a. *The Fight against War*, ed. Alfred Lief. New York: John Day.

1933b. *On the Method of Theoretical Physics*. Oxford: Oxford University Press. Reprinted in slightly different version in Einstein (1954, 270–6). A German manuscript is available (EA 1–114) under the title: "Zur Methode der theoretischen Physik."

1933c. *Origins of the General Theory of Relativity*. Glasgow: Jackson. Reprinted in slightly different version, under the title "Notes on the Origin of the General Theory of Relativity," in Einstein (1954, 285–90). A German manuscript is available (EA 78–668) under the title "Einiges über die Entstehung der allgemeinen Relativitätstheorie."

1934. *Mein Weltbild*, ed. Carl Seelig. Amsterdam: Querido. English trans. in Einstein (1954).

1936. "Physik und Realität." *Journal of the Franklin Institute* 221: 313–37. English trans. "Physics and Reality" in Einstein (1954, 290–323).

1941. "Demonstration of the Non-Existence of Gravitational Fields with a Non-Vanishing Total Mass Free of Singularities." *Tucuman Universidad Nacional. Revista* A2: 11–15.

1943. "Bivector Fields, II." *Annals of Mathematics* 45: 15–23.

1944. "Bemerkungen zu Bertrand Russell's Erkenntnis-Theorie." In *The Philosophy of Bertrand Russell*, ed. Paul A. Schilpp. Evanston, IL: Northwestern University Press, 278–91. Trans. as "Remarks on Bertrand Russell's Theory of Knowledge," in Einstein (1954, 18–24).

1945. "Generalization of the Relativistic Theory of Gravitation." *Annals of Mathematics* 46: 578–84.

1946a. *Testimony before the Anglo-American Committee of Inquiry on Jewish Problems in Palestine and Europe. Washington, D.C. State Department Building, 11 January 1946*. Washington, DC: Ward & Paul (Electreporter, Inc.).

1946b. "An Elementary Derivation of the Equivalence of Mass and Energy." In Einstein (1950c).

1948a. "Generalized Theory of Gravitation." *Reviews of Modern Physics* 20: 35–9.

1948b. "Quanten-Mechanik und Wirklichkeit." *Dialectica* 2: 320–4. Trans. as "Quantum Mechanics and Reality," in Born (1971, 168–72).

1949a. "Autobiographical Notes." In Schilpp (1949, 2–94). Reprinted in Einstein (1979).

1949b. "Reply to Criticism: Remarks Concerning the Essays Brought Together in This Co-operative Volume." In Schilpp. (1949, 665–88).

1950a. "On the Generalized Theory of Gravitation." *Scientific American* 182: 13–17.

1950b. "The Bianchi Identities in the Generalized Theory of Gravitation." *Canadian Journal of Mathematics* 2: 120–8.

1950c. *Out of My Later Years.* New York: Philosophical Library. Reprint: New York: Wings Books, 1993.

1951. "Letter from Dr. Albert Einstein." In *Essay in Physics*, ed. Herbert L. Samuel, 135–45. Oxford: Blackwell.

1953. "Elementare Überlegungen zur Interpretation der Grundlagen der Quanten-Mechanik." In *Scientific Papers Presented to Max Born*. London: Oliver and Boyd, 33–40.

1954. *Ideas and Opinions.* New York: Crown. Reprint: New York: Modern Library, 1994.

1955. "Erinnerungen-Souvenirs." *Schweizerische Hochschulzeitung* 28 (Sonderheft): 145–53. Reprinted as "Autobiographische Skizze" in Seelig (1956, 9–17).

1956a. *The Meaning of Relativity.* Princeton, NJ: Princeton University Press. Trans. of later edition of Einstein (1922c).

1956b. *Albert Einstein: Lettres à Maurice Solovine.* Paris: Gauthier-Villars. English translation: *Albert Einstein: Letters to Solovine.* Translated by Wade Baskin. New York: Philosophical Library, 1987.

1959. *Relativity. The Special and the General Theory: A Clear Explanation That Anyone Can Understand.* New York: Crown Publishers. Trans. of later edition of Einstein (1917a).

1979. *Autobiographical Notes: A Centennial Edition.* LaSalle, IL: Open Court. Reprint of Einstein (1949a).

1983. *Sidelights on Relativity.* New York: Dover.

1989. *Oeuvres choisies d'Albert Einstein*, eds. Françoise Balibar, Olivier Darrigol, and Bruno Jech. Paris: Seuil, CNRS.

Einstein, Albert, and Bargmann, Valentin. 1943. "Bivector Fields, I." *Annals of Mathematics* 45: 1–14.

Einstein, Albert, and Bergmann, Peter G. 1938. "Generalization of Kaluza's Theory of Electricity." *Annals of Mathematics* 39: 683–701.

Einstein, Albert, and Besso, Michele. 1972. *Albert Einstein – Michele Besso. Correspondance 1903–1955*, ed. Pierre Speziali. Paris: Hermann.

Einstein, Albert, and de Haas, Wander. 1915a. "Experimenteller Nachweis der Ampèreschen Molekularströme." *Deutsche Physikalische Gesellschaft. Verhandlungen* 17: 152–70. Reprinted in CPAE 6, Doc. 13.

Einstein, Albert, and de Sitter, Willem. 1932. "On the Relation between the Expansion and the Mean Density of the Universe." *National Academy of Science. Proceedings*: 213–14.

Einstein, Albert, and Fokker, Adriaan D. 1914. "Die Nordströmsche Gravitationstheorie vom Standpunkt des absoluten Differentialkalküls." *Annalen der Physik* 44: 321–8. Reprinted in CPAE 4, Doc. 28.

Einstein, Albert, and Grommer, Jakob. 1923. "Beweis der Nichtexistenz eines überall regulären zentrisch symmetrischen Feldes nach der Feld-Theorie von Th. Kaluza." *Scripta Universitatis atque Bibliothecae Hierosolymitanarum: Mathematica et Physica* 1: 1–5.

Einstein, Albert, and Grossmann, Marcel. 1913. *Entwurf einer verallgemeinerten Relativitätstheorie und einer Theorie der Gravitation.* Leipzig: Teubner. Reprinted in CPAE 4, Doc. 13.

1914a. "Entwurf einer verallgemeinerten Relativitätstheorie und einer Theorie der Gravitation." *Zeitschrift für Mathematik und Physik* 62: 225–59. Reprint of Einstein and Grossmann (1913).

1914b. "Kovarianzeigenschaften der Feldgleichungen der auf die verallgemeinerte Relativitätstheorie gegründeten Gravitationstheorie." *Zeitschrift für Mathematik und Physik* 63: 215–25. Reprinted in CPAE 6, Doc. 2.

Einstein, Albert, and Hopf, Ludwig. 1910. "Statistische Untersuchung der Bewegung eines Resonators in einem Strahlungsfeld." *Annalen der Physik* 33: 1105–15. Reprinted in CPAE 3, Doc. 8.

Einstein, Albert, and Infeld, Leopold. 1938. *The Evolution of Physics.* New York: Simon and Schuster.

Einstein, Albert, and Kaufmann, Bruria. 1954. "Algebraic Properties of the Field in the Relativistic Theory of the Asymmetric Field." *Annals of Mathematics* 59: 230–44.

1955. "A New Form of the General Relativistic Field Equations." *Annals of Mathematics* 62: 128–38.

Einstein, Albert, and Mayer, Walter. 1931. "Einheitliche Theorie von Gravitation und Elektrizität." *Preußische Akademie der Wissenschaften* (Berlin). *Physikalisch-Mathematische Klasse. Sitzungsberichte*: 541–57.

1932a. "Einheitliche Theorie von Gravitation und Elektrizität: Zweite Abhandlung." *Preußische Akademie der Wissenschaften* (Berlin). *Physikalisch-Mathematische Klasse. Sitzungsberichte*: 130–7.

1932b. "Semi-Vektoren und Spinoren." *Preußische Akademie der Wissenschaften* (Berlin). *Physikalisch-Mathematische Klasse. Sitzungsberichte*: 522–50.

1933a. "Die Diracgleichungen für Semi-Vektoren." *Koninklijke Akademie van Wetenschappen te Amsterdam. Section of Sciences. Proceedings* 36: 497–516.

1933b. "Spaltung der natürlichsten Feldgleichungen für Semi-Vektoren in Spinor-Gleichungen vom Dirac'schen Typus." *Koninklijke Akademie van Wetenschappen te Amsterdam. Section of Sciences. Proceedings* 36: 615–19.

1934. "Darstellung der Semi-Vektoren als gewöhnliche Vektoren von besonderem Differentiations Charakter." *Annals of Mathematics* 35: 104–10.

Einstein, Albert, and Pauli, Wolfgang. 1943. "On the Non-Existence of Regular Stationary Solutions of Relativistic Field Equations." *Annals of Mathematics* 44: 131–7.

Einstein, Albert, and Rosen, Nathan. 1937. "On Gravitational Waves." *Journal of the Franklin Institute* 223: 43–54.

Einstein, Albert, and Sommerfeld, Arnold. 1968. *Briefwechsel*, ed. Armin Hermann. Basel: Schwabe.

Einstein, Albert, and Stern, Otto. 1913. "Einige Argumente für die Annahme einer molekularen Agitation beim absoluten Nullpunkt." *Annalen der Physik* 40: 551–60. Reprinted in CPAE 4, Doc. 11.

Einstein, Albert, and Straus, Ernst G. 1945. "The Influence of the Expansion of Space on the Gravitation Fields Surrounding the Individual Stars." *Reviews of Modern Physics* 17: 120–4.

1946. "Generalization of the Relativistic Theory of Gravitation. II." *Annals of Mathematics* 47: 731–41.

Einstein, Albert, Bargmann, Valentine, and Bergmann, Peter G. 1941. "Five-Dimensional Representation of Gravitation and Electricity." In *Theodore von Karman Anniversary Volume*. Pasadena: California Institute of Technology, 221–5.

Einstein, Albert, Born, Max, and Born, Hedwig. 1969. *Briefwechsel 1916–1955*. München: Nymphenburger Verlagshandlung. English trans. in Born (1971).

Einstein, Albert, Infeld, Leopold, and Hoffmann, Banesh. 1938. "Gravitational Equations and the Problem of Motion." *Annals of Mathematics* 39: 65–100.

Einstein, Albert, Podolsky, Boris, and Rosen, Nathan. 1935. "Can Quantum-Mechanical Description of Reality Be Considered Complete?" *Physical Review* 47: 777–80.

Einstein, Albert, Tolman, Richard C., and Podolsky, Boris. 1931. "Knowledge of Past and Future in Quantum Mechanics." *Physical Review* 37: 780–1.

Einstein, Albert, Lorentz, Hendrik A., Minkowski, Hermann, and Weyl, Hermann. 1952. *The Principle of Relativity*. New York: Dover.

Einstein, Albert et al. 1913. "Diskussion." *Physikalische Zeitschrift* 14: 1262–6. Reprinted in CPAE 4, Doc. 18.

1914. "Diskussion." In Eucken (1914, 353–64). Reprinted in CPAE 3, Doc. 27.

1920. "Diskussion." *Physikalische Zeitschrift* 21: 650–1, 662, 666–8. Reprinted in CPAE 7, Doc. 46

Eisenstaedt, Jean. 1989. "Cosmology: A Space for Thought on General Relativity." In *Foundations of Big Bang Cosmology*, ed. F. Walter Meyerstein. Singapore: World Scientific, 271–95.

1993. "Lemaître and the Schwarzschild Solution." In Earman et al. (1993, 353–89).

2006. *The Curious History of Relativity: How Einstein's Theory of Gravity Was Lost and Found Again*. Princeton, NJ: Princeton University Press.

Eisenstaedt, Jean, and Kox, A. J., eds. 1992. *Studies in the History of General Relativity*. Einstein Studies 3. Boston: Birkhäuser.

Eisinger, Josef. 2011. *Einstein on the Road*. New York: Prometheus Books.

Ellis, Charles D. 1926. "The Light-Quantum Theory." *Nature* 117: 896.

Ellis, George F. R. 1989. "The Expanding Universe: A History of Cosmology from 1917 to 1960." In Howard and Stachel (1989, 367–431).

Ellis, George F. R., and Rothman, T. 1993. "Lost Horizons." *American Journal of Physics* 61: 883–93.

Elsbach, Alfred. 1924. *Kant und Einstein: Untersuchungen über das Verhältnis der modernen Erkenntnistheorie zur Relativitätstheorie*. Berlin and Leipzig: Walter de Gruyter.

Elzinga, Aant. 2006. *Einstein's Nobel Prize: A Glimpse behind Closed Doors: The Archival Evidence*. Sagamore Beach, MA: Science History Publication/USA.

Engler, Fynn Ole, and Renn, Jürgen. 2013. "Hume, Einstein und Schlick über die Objektivität der Wissenschaft." In *Moritz Schlick – Die Rostocker Jahre und ihr Einfluss auf die Wiener Zeit (Schlickiana* Vol. 6), eds. Fynn Ole Engler and Mathias Iven. Leipzig: Universitätsverlag, 70–97.

Enz, Charles P. 2002. *No Time to Be Brief: A Scientific Biography of Wolfgang Pauli*. Oxford: Oxford University Press.

Eötvös, Roland. 1890. "A föld vonzása különböző anyagroka." *Akadémiai Értesítő* 2:108–10, German trans. "Über die Anziehung der Erde auf verschiedene Substanzen." *Mathematische und Naturwissenschaftliche Berichte aus Ungarn* 8 (1891): 65–8, 108–10.

Eucken, Arnold, ed. 1914. *Die Theorie der Strahlung und Quanten: Verhandlungen auf einer von E. Solvay einberufenen Zusammenkunft (30. Oktober bis 3. November 1911). Mit einem Anhange über die Entwicklung der Quantentheorie vom Herbst 1911 bis zum Sommer 1913.* Halle an der Saale: Wilhelm Knapp.

Everett, Hugh. 1973. "The Theory of the Universal Wave Function." In *The Many-Worlds Interpretation of Quantum Mechanics*, eds. Bryce DeWitt and Neill Graham. Princeton, NJ: Princeton University Press, 3–140.

Everitt, C. W. F., et al. 2011. "Gravity Probe B: Final Results of a Space Experiment to Test General Relativity." *Physical Review Letters* 106: 221101.

Faye, Jan, and Folse, Henry J., eds. 1994. *Niels Bohr and Contemporary Philosophy.* Dordrecht: Kluwer.

Feuer, Lewis S. 1982. *Einstein and the Generations of Science.* 2nd ed. New Brunswick, NJ: Transaction Books.

Feynman, Richard P., and Leighton, Ralph. 1985. *Surely You're Joking, Mr. Feynman.* New York: Norton.

Fine, Arthur. 1986. "Einstein's Realism." In Fine (1996, 86–111).

FitzGerald, George F. 1889. "The Ether and the Earth's Atmosphere." *Science* 13: 390.

———. 1996. *The Shaky Game: Einstein, Realism and the Quantum Theory.* 2nd ed. Chicago: University of Chicago Press.

Flückiger, Max. 1974. *Albert Einstein in Bern: Das Ringen um ein neues Weltbild. Eine dokumentarische Darstellung über den Aufstieg eines Genies.* Bern: Haupt.

Fock, Vladimir A. 1959. *Theory of Space, Time and Gravitation.* New York: Pergamon.

Fölsing, Albrecht. 1993. *Albert Einstein: Eine Biographie.* Frankfurt am Main: Suhrkamp. English translation: *Albert Einstein: A Biography.* Trans. Ewald Osers. New York: Viking, 1997.

Föppl, Abraham. 1894. *Einführung in die Maxwellsche Theorie der Elektrizität.* Leipzig: Teubner.

Frank, Philipp. 1907. "Kausalgesetz und Erfahrung." *Annalen der Naturphilosophie* 6: 443–50.

———. 1947. *Einstein: His Life and Times.* New York: Alfred Knopf.

———. 1949a. "Historical Background." In *Modern Science and Its Philosophy*, ed. Philipp Frank. Cambridge, MA: Harvard University Press, 1–52.

———. 1949b. "Einstein's Philosophy of Science." *Reviews of Modern Physics* 21: 349–55.

———. 1949c. "Einstein, Mach, and Logical Positivism." In Schilpp (1949, 271–86).

French, Anthony P., ed. 1979. *Einstein: A Centenary Volume.* Cambridge, MA: Harvard University Press.

Freundlich, Erwin. 1916. *Die Grundlagen der Einsteinschen Gravitationstheorie.* Berlin: Springer.

Friedman, Michael. 1983. *Foundations of Space-Time Theories: Relativistic Physics and Philosophy of Science.* Princeton, NJ: Princeton University Press.

———. 1997. "Helmholtz's *Zeichentheorie* and Schlick's *Allgemeine Erkenntnislehre*: Early Logical Empiricism and Its Nineteenth-Century Background." *Philosophical Topics* 25: 19–50.

———. 1999. *Reconsidering Logical Positivism.* Cambridge: Cambridge University Press.

———. 2000. "Geometry, Construction, and Intuition in Kant and His Successors." In *Between Logic and Intuition: Essays in Honor of Charles Parsons*, ed. Gila Scher and Richard Tieszen. Cambridge: Cambridge University Press.

2002a. "Geometry as a Branch of Physics: Background and Context for Einstein's 'Geometry and Experience.'" In Malament (2002, 193–229).

2002b. "Physics, Philosophy, and the Foundations of Geometry." *Diálogos* 37: 121–42.

2008a. "Space, Time, and Geometry: Einstein and Logical Empiricism." In Galison et al. (2008, 205–16).

2008b. "Einstein, Kant, and the A Priori." In Massimi (2008, 95–112).

Friedman, Robert Marc. 2001. *The Politics of Excellence: Behind the Nobel Prize in Science.* New York: Henry Holt.

Friedmann, Alexander. 1922. "Über die Krümmung des Raumes." *Zeitschrift für Physik* 10: 377–86.

1924. "Über die Möglichkeit einer Welt mit Konstanter negativer Krümmung des Raumes." *Zeitschrift für Physik* 21: 326–32.

Friedrich, Walter, Knipping, Paul, and Laue, Max. 1912. "Interferenz-Erscheinungen bei Röntgenstrahlen." *Bayerische Akademie der Wissenschaften, Sitzungsberichte*: 303–22. Reprinted in von Laue (1961, 183–207).

Gale, George. 2002. "Cosmology: Methodological Debates in the 1930s and 1940s." *The Stanford Encyclopedia of Philosophy* (Spring 2002 ed.), ed. Edward N. Zalta, http://plato.stanford.edu/entries/cosmology-30s/.

Gale, George, and Shanks, Niall. 1996. "Methodology and the Birth of Modern Cosmological Inquiry." *Studies in History and Philosophy of Modern Physics* 27: 279–96.

Gale, George, and Urani, John. 1993. "Philosophical Midwifery and the Birthpangs of Modern Cosmology." *American Journal of Physics* 61: 66–73.

Galison, Peter. 1987. *How Experiments End.* Chicago: University of Chicago Press.

2003. *Einstein's Clocks and Poincaré's Maps: Empires of Time.* New York: Norton.

2008. "The Assassin of Relativity." In Galison et al. (2008, 185–204).

Galison, Peter, Holton, Gerald, and Schweber, Silvan, eds. 2008. *Einstein for the 21st Century: His Legacy in Science, Art, and Modern Culture.* Princeton, NJ: Princeton University Press.

Gamow, George. 1970. *My World Line.* New York: Viking Press.

Gearhart, Clayton A. 1990. "Einstein before 1905: The Early Papers on Statistical Mechanics." *American Journal of Physics* 58: 460–80.

2002. "Planck, the Quantum, and the Historians." *Physics in Perspective* 4: 170–215.

Gentner, Dedre, and Stevens, Albert L. 1983. *Mental Models.* Hillsdale, NJ: Erlbaum.

Gibbs, Josiah W. 1902. *Elementary Principles in Statistical Mechanics.* New York: Scribner.

Giulini, Domenico. 2006. "What Is (Not) Wrong with Scalar Gravity?" *PhilSci archive*, http://philsci-archive.pitt.edu/archive/00003069.

Giulini, Domenico, and Straumann, Norbert. 2006. "Einstein's Impact on the Physics of the Twentieth Century." *Studies in History and Philosophy of Modern Physics* 37: 115–73.

Goenner, Hubert F. M. 2001. "Weyl's Contributions to Cosmology." In Scholz (2001, 105–37).

2004. "On the History of Unified Field Theories." *Living Reviews in Relativity* 7, http://www.livingreviews.org/lrr-2004-2.

2005. *Einstein in Berlin: 1914–1933.* Berlin: Beck.

Goenner, Hubert F. M., and Wünsch, Daniela. 2003. "Kaluza's and Klein's Contributions to Kaluza-Klein-Theory." *Max Planck Institute for the History of Science* (Berlin), preprint 235.

Goenner, Hubert F. M., Renn, Jürgen, Ritter, Jim, and Sauer, Tilman, eds. 1999. *The Expanding Worlds of General Relativity.* Einstein Studies 7. Boston: Birkhäuser.

Goldstein, Catherine, and Ritter, Jim. 2003. "The Varieties of Unity: Sounding Unified Field Theories 1920–1930." In Ashtekar et al. (2003, 93–149).

Gray, Jeremy, ed. 1999. *The Symbolic Universe: Geometry and Physics, 1890–1930.* Oxford: Oxford University Press.

Grundmann, Siegfried. 1998. *Einsteins Akte.* Berlin: Springer.

Gülzow, Erwin. 1969. *Der Bund "Neues Vaterland": Probleme der bürgerlich-pazifistischen Demokratie im ersten Weltkrieg (1914–1918).* PhD diss. Humboldt University (Berlin).

Guth, Alan. 1981. "Inflationary Universe: A Possible Solution for the Horizon and Flatness Problems." *Physical Review* 23: 347–56.

Haller, Rudolf. 1985. "Der erste Wiener Kreis." *Erkenntnis* 22: 341–58.

Hallwachs, Wilhelm. 1916. "Die Lichtelektrizität." In Marx (1916a, 245–563, 595–618).

Hecht, Hartmut, and Hoffmann, Dieter. 1982. "Die Berufung Hans Reichenbachs an die Berliner Universität." *Deutsche Zeitschrift für Philosophie* 30: 651–62.

Heckmann, Otto. 1942. *Theorien der Kosmologie.* Berlin: Springer.

Heckmann, Otto, and Schücking, Engelbert. 1955. "Bemerkungen zur Newtonschen Kosmologie. I." *Zeitschrift für Astrophysik* 38: 95–109.

1956a. "Bemerkungen zur Newtonschen Kosmologie. II." *Zeitschrift für Astrophysik* 40: 81–92.

1956b. "Reply to Layzer." *The Observatory* 76: 74–5.

Heilbron, John L. 1986. *The Dilemmas of an Upright Man: Max Planck as Spokesman for German Science.* Berkeley: University of California Press.

Heisenberg, Werner. 1926. "Quantenmechanik." *Die Naturwissenschaften* 14: 989–94. Reprinted in Heisenberg (1984, 52–7).

1927. "Über den anschaulichen Inhalt der quantentheoretischen Kinematik und Mechanik." *Zeitschrift für Physik* 43: 172–98.

1969. *Der Teil und das Ganze: Gespräche im Umkreis der Atomphysik.* Munich: Piper.

1984. *Collected Works. Series B. Scientific Review Papers, Talks, and Books,* ed. Walter Blum, Hans-Peter Dürr, and Helmut Rechenberg. New York: Springer.

Helmholtz, Hermann von. 1868. "Über die Tatsachen, die der Geometrie zum Grunde liegen." *Nachrichten von der Königlichen Gesellschaft der Wissenschaften und der Georg-August-Universität aus dem Jahre 1868* 9: 193–221. English trans. in Cohen and Elkana (1977, 39–71).

1921. *Schriften zur Erkenntnistheorie,* ed. Paul Hertz and Moritz Schlick. Berlin: Springer.

Hendricks, Vincent F., Jørgensen, Klaus Frovin, Lützen, Jesper, and Pedersen, Stig Andur, eds. 2006. *Interactions: Mathematics, Physics and Philosophy, 1860–1930.* Dordrecht: Springer.

Hentschel, Klaus. 1990. *Interpretationen und Fehlinterpretationen der speziellen und der allgemeinen Relativitätstheorie durch Zeitgenossen Albert Einsteins.* Basel: Birkhäuser.

1992. *Der Einstein-Turm: Erwin F. Freundlich und die Relativitätstheorie; Ansätze zu einer "dichten Beschreibung" von institutionellen, biographischen und theoriegeschichtlichen Aspekten.* Heidelberg, Berlin, and New York: Spektrum Akademie Verlag. English translation: *The Einstein Tower: An Intertexture of Dynamic Construction, Relativity Theory, and Astronomy.* Translated by Ann M. Hentschel. Stanford, CA: Stanford University Press, 1997.

1993. "The Conversion of St. John: A Case Study on the Interplay of Theory and Experiment." *Science in Context* 6: 137–94.

Hiebert, Erwin N. 1978. "Nernst, Hermann Walther." In *Dictionary of Scientific Biography*, Vol. 15 (Supplement I), ed. Charles C. Gillispie. New York: Scribner, 432–53.

Highfield, Roger, and Carter, Paul. 1993. *The Private Lives of Albert Einstein.* New York: St. Martin's Press.

Hilbert, David. 1899. *Grundlagen der Geometrie.* Leipzig: Teubner. English trans. (from 10th ed. 1968) *Foundations of Geometry.* Chicago: Open Court, 1971.

1915. "Die Grundlagen der Physik. (Erste Mitteilung)." *Königliche Gesellschaft der Wissenschaften zu Göttingen. Mathematisch-physikalische Klasse. Nachrichten:* 395–407.

1924. "Die Grundlagen der Physik." *Mathematische Annalen* 92: 1–32.

1992. *Natur und mathematisches Erkennen: Vorlesungen, gehalten 1919–1920 in Göttingen. Nach der Ausarbeitung von Paul Bernays,* ed. David Rowe. Boston: Birkhäuser.

Hoefer, Carl. 1994. "Einstein's Struggle for a Machian Gravitation Theory." *Studies in History and Philosophy of Science* 25: 287–335.

1995. "Einstein's Formulations of Mach's Principles." In Barbour and Pfister (1995, 67–87).

Hoffmann, Banesh. 1972. *Albert Einstein: Creator and Rebel.* New York: Viking Press, 1972.

Hoffmann, Dieter. 2001. "On the Experimental Context of Planck's Quantum Theory." *Centaurus* 43: 240–59.

Holton, Gerald. 1968. "Mach, Einstein, and the Search for Reality." *Daedalus* 97: 636–73. Reprinted in Holton (1988, 237–77).

1969. "Einstein, Michelson, and the 'Crucial' Experiment." *Isis* 60: 133–97. Reprinted in Holton (1988, 279–370).

1988. *Thematic Origins of Scientific Thought: Kepler to Einstein.* Rev. ed. (1st ed.: 1973). Cambridge, MA: Harvard University Press.

Holton, Gerald, and Elkana, Yehuda, eds. 1982. *Albert Einstein: Historical and Cultural Perspectives.* Princeton, NJ: Princeton University Press. Reprint New York: Dover, 1992.

Home, Dipankar, and Whitaker, Andrew. 2007. *Einstein's Struggles with Quantum Theory.* New York: Springer.

Hon, Giora. 2003. "From Propagation to Structure: The Experimental Technique of Bombardment as a Contributing Factor to the Emerging Quantum Physics." *Physics in Perspective* 5: 150–73.

Hoskin, Michael. 1976. "The 'Great Debate': What Really Happened." *Journal for the History of Astronomy* 7: 169–82.

Howard, Don. 1984. "Realism and Conventionalism in Einstein's Philosophy of Science: The Einstein-Schlick Correspondence." *Philosophia Naturalis* 21: 616–29.

1985. "Einstein on Locality and Separability." *Studies in History and Philosophy of Science* 16: 171–201.

1990a. "Einstein and Duhem." *Synthese* 83: 363–84.

1990b. "*Nicht sein kann was nicht sein darf*: Or the Prehistory of EPR, 1909–1935: Einstein's Early Worries about the Quantum Mechanics of Composite Systems." In *Sixty-Two Years of Uncertainty: Historical, Philosophical, Physics Inquiries into the Foundations of Quantum Physics*, ed. Arthur I. Miller. New York: Plenum, 61–111.

1992. "Einstein and *Eindeutigkeit*: A Neglected Theme in the Philosophical Background to General Relativity." In Eisenstaedt and Kox (1992, 154–243).

1993. "Was Einstein Really a Realist?" *Perspectives on Science: Historical, Philosophical, Social* 1: 204–51.

1994. "Einstein, Kant, and the Origins of Logical Empiricism." In *Language, Logic, and the Structure of Scientific Theories*, eds. Wesley Salmon and Gereon Wolters. Pittsburgh: University of Pittsburgh Press, 45–105.

1996. "Relativity, *Eindeutigkeit*, and Monomorphism: Rudolf Carnap and the Development of the Categoricity Concept in Formal Semantics." In *Origins of Logical Empiricism*, eds. Ronal N. Giere and Alan Richardson. Minneapolis: University of Minnesota Press, 115–64.

1997. "A Peek behind the Veil of Maya: Einstein, Schopenhauer, and the Historical Background of the Conception of Space as a Ground for the Individuation of Physical Systems." In *The Cosmos of Science: Essays of Exploration*, eds. John Earman and John D. Norton. Pittsburgh: University of Pittsburgh Press, 87–150.

1998. "Astride the Divided Line: Platonism, Empiricism, and Einstein's Epistemological Opportunism." In *Idealization in Contemporary Physics*, ed. Niall Shanks. Amsterdam and Atlanta: Rodopi, 143–63.

1999. "Point Coincidences and Pointer Coincidences: Einstein on the Invariant Content of Space-Time Theories." In Goenner et al. (1999, 463–500).

2003. "Two Left Turns Make a Right: On the Curious Political Career of North American Philosophy of Science at Mid-century." In *Logical Empiricism in North America*, eds. Alan Richardson and Gary Hardcastle. Minneapolis: University of Minnesota Press, 25–93.

2004a. "Einstein's Philosophy of Science." *The Stanford Encyclopedia of Philosophy* (Spring 2004 ed.), ed. Edward N. Zalta, http://plato.stanford.edu/archives/spr2004/entries/einstein-philscience/.

2004b. "Fisica e filosofia della scienza all'alba del XX secolo." In *Storia della Scienza*. Vol. 8. Rome: Istituto della Enciclopedia Italiana, 3–16.

2004c. "Einstein and Frank." Paper Presented at the Symposium, "Philipp Frank: Wien, Prag, Boston," September 27–28, 2004. Vienna, Austria.

2010. "'Let me briefly indicate why I do not find this standpoint natural.' Einstein, General Relativity, and the Contingent a Priori." In Domski and Dickson (2010, 333–55).

Howard, Don, and Norton, John D. 1993. "Out of the Labyrinth? Einstein, Hertz, and the Göttingen Answer to the Hole Argument." In Earman et al. (1993, 30–62).

Howard, Don, and Stachel, John, eds. 1989. *Einstein and the History of General Relativity*. Einstein Studies 1. Boston: Birkhäuser.

eds. 2000. *Einstein: The Formative Years, 1879–1909*. Einstein Studies 8. Boston: Birkhäuser.

Hoyle, Fred. 1948. "A New Model for the Expanding Universe." *Royal Astronomical Society. Monthly Notices* 108: 372–82.

Hubble, Edwin P. 1929. "A Relation between Distance and Radial Velocity among Extra-Galactic Nebulae." *National Academy of Science. Proceedings* 15: 168–73.

Huggett, Nick, ed. 2000. *Space from Zeno to Einstein: Classic Readings with a Contemporary Commentary*. Cambridge, MA: MIT Press.

Hughes, Arthur L. 1914. *Photo-Electricity*. Cambridge: Cambridge University Press.

Illy, József. 2012. *The Practical Einstein: Experiments, Patents, Inventions*. Baltimore, MD: Johns Hopkins University Press.

Infeld, Leopold. 1941. *Quest: The Evolution of a Physicist*. London: Gollancz.

——— 1978. *Why I Left Canada: Reflections on Science and Politics*. Montreal: McGill University Press.

Isaacson, Walter. 2007. *Einstein: His Life and Universe*. New York: Simon & Schuster.

Jaki, Stanley L. 1990. "The Gravitational Paradox of an Infinite Universe." In *Cosmos in Transition*, ed. Stanley L. Jaki. Tucson: Pachart, 189–212.

Jammer, Max. 1966. *The Conceptual Development of Quantum Mechanics*. New York: McGraw-Hill.

——— 1974. *The Philosophy of Quantum Mechanics: The Interpretations of Quantum Mechanics in Historical Perspective*. New York: Wiley.

——— 1999. *Einstein and Religion: Physics and Theology*. Princeton, NJ: Princeton University Press.

Janssen, Michel. 1992. "H. A. Lorentz's Attempt to Give a Coordinate-Free Formulation of the General Theory of Relativity." In Eisenstaedt and Kox (1992, 344–63).

——— 1995. *A Comparison between Lorentz's Ether Theory and Special Relativity in the Light of the Experiments of Trouton and Noble*. PhD diss. University of Pittsburgh.

——— 1999. "Rotation as the Nemesis of Einstein's *Entwurf* Theory." In Goenner et al. (1999, 127–57).

——— 2002a. "*COI* Stories: Explanation and Evidence in the History of Science." *Perspectives on Science* 10: 457–522.

——— 2002b. "Reconsidering a Scientific Revolution: The Case of Einstein versus Lorentz." *Physics in Perspective* 4: 421–46.

——— 2003a. *A Glimpse behind the Curtain of the Wizard/Un coup d'oeil derrière le rideau du magicien*. Paris: Scriptura and Aristophile.

——— 2003b. "The Trouton Experiment, $E = mc^2$, and a Slice of Minkowski Spacetime." In Ashtekar et al. (2003, 27–54).

——— 2004. "Relativity." In *New Dictionary of the History of Ideas*, ed. Maryanne Cline Horowitz et al. New York: Charles Scribner's Sons, 2039–47.

——— 2005. "Of Pots and Holes: Einstein's Bumpy Road to General Relativity."*Annalen der Physik* 14: 58–85. Supplement published in book form as Renn (2005a).

——— 2006. *Einstein: The Old Sage and the Young Turk*. Slides for a colloquium at the University of Wisconsin–Barron County, October 6, 2006, https://netfiles.umn.edu/users/janss011/home%20page/sage.pdf.

——— 2007. "What Did Einstein Know and When Did He Know It? A Besso Memo Dated August 1913." In Renn (2007a, Vol. 2, 785–837).

——— 2009. "Drawing the Line between Kinematics and Dynamics in Special Relativity." *Studies in History and Philosophy of Modern Physics* 40: 26–52.

2012. "The Twins and the Bucket: How Einstein Made Gravity Rather Than Motion Relative in General Relativity." *Studies in History and Philosophy of Modern Physics* 43: 159–75.

Janssen, Michel, and Mecklenburg, Matthew. 2006. "From Classical to Relativistic Mechanics: Electromagnetic Models of the Electron." In Hendricks et al. (2006, 65–134).

Janssen, Michel, and Renn, Jürgen. 2007. "Untying the Knot: How Einstein Found His Way Back to Field Equations Discarded in the Zurich Notebook." In Renn (2007a, Vol. 2, 839–925).

Jeans, James H. 1910. "On the Motion of a Particle about a Doublet." *Philosophical Magazine* 20: 380–2.

1919. *Problems of Cosmogony and Stellar Dynamics*. Cambridge: Cambridge University Press.

Jenkin, John. 2002. "G. E. M. Jauncey and the Compton Effect." *Physics in Perspective* 4: 320–32.

Jerome, Fred. 2002. *The Einstein File: J. Edgar Hoover's Secret War against the World's Most Famous Scientist*. New York: St. Martin's Press.

Jordan, Pascual. 1934. "Über den positivistischen Begriff der Wirklichkeit." *Die Naturwissenschaften* 22: 485–90.

Kaluza, Theodor. 1921. "Zum Unitätsproblem der Physik." *Preußische Akademie der Wissenschaften* (Berlin). *Sitzungsberichte*: 966–72.

Kangro, Hans. 1970. *Vorgeschichte des Planckschen Strahlungsgesetzes: Messungen und Theorien der spektralen Energieverteilung bis zur Begründung der Quantenhypothese*. Wiesbaden: Steiner.

Kant, Immanuel. 1781/1787. *Kritik der reinen Vernunft*. Riga, Latvia: Hartknoch. English trans. *Critique of Pure Reason*. Cambridge: Cambridge University Press, 1997.

Kapteyn, Jacobus C., and van Rhijn, Pieter J. 1920. "On the Distribution of the Stars in Space Especially in the High Galactic Latitudes." *Astrophysical Journal* 52: 23–38.

Kaufmann, Bruria. 1956. "Mathematical Structure of the Nonsymmetric Field Theory." In *Fünfzig Jahre Relativitätstheorie: Cinquantenaire de la Théorie de la Relativité. Jubilee of Relativity Theory. (Helvetica Physica Acta Supplementum IV)*, eds. A. Mercier and M. Kervaire. Basel: Birkhäuser, 227–38.

Kennedy, Robert E. 2012. *A Student's Guide to Einstein's Major Papers*. Oxford: Oxford University Press.

Kennefick, Daniel J. 2007. *Traveling at the Speed of Thought: Einstein and the Quest for Gravitational Waves*. Princeton, NJ: Princeton University Press.

2009. "Testing Relativity from the 1919 Eclipse: A Question of Bias." *Physics Today* 62.3: 37–42.

2012. "Not Only Because of Theory: Dyson, Eddington, and the Competing Myths of the 1919 Eclipse Expedition." In Lehner et al. (2009, 201–32).

Klein, Felix. 1909. *Elementarmathematik vom höheren Standpunkt aus*. Part 2: *Geometrie*. Leipzig: Teubner. English trans. *Elementary Mathematics from an Advanced Standpoint: Geometry*. New York: Dover, 1939.

1918. "Über die Integralform der Erhaltungssätze und die Theorie der räumlich-geschlossenen Welt." *Königliche Gesellschaft der Wissenschaften zu Göttingen, Mathematisch-physikalische Klasse. Nachrichten*: 393–423.

Klein, Martin J. 1962. "Max Planck and the Beginnings of the Quantum Theory." *Archive for History of Exact Sciences* 1: 459–79.

1963a. "Planck, Entropy, and Quanta, 1901–1906." *The Natural Philosopher* 1: 83–108.

1963b. "Einstein's First Paper on Quanta." *The Natural Philosopher* 2: 59–86.

1964. "Einstein and the Wave-Particle Duality." *The Natural Philosopher* 3: 3–49.

1965. "Einstein, Specific Heats, and the Early Quantum Theory." *Science* 148: 173–80.

1966. "Thermodynamics and Quanta in Planck's Work." *Physics Today* 19: 23–32.

1967. "Thermodynamics in Einstein's Thought." *Science* 157: 509–16.

1970a. "The First Phase of the Bohr-Einstein Dialogue." *Historical Studies in the Physical Sciences* 2: 1–39.

1970b. *Paul Ehrenfest.* Vol 1. *The Making of a Theoretical Physicist.* Amsterdam: North Holland Publishing.

1977. "The Beginnings of the Quantum Theory." In *Storia della fisica del XX secolo,* ed. C. Weiner. New York: Academic Press, 1–39.

1979. "Einstein and the Development of Quantum Physics." In French (1979, 133–51).

1982. "Fluctuations and Statistical Physics in Einstein's Early Work." In Holton and Elkana (1982, 39–58).

Kox, A. J. 1988. "Hendrik Antoon Lorentz, the Ether, and the General Theory of Relativity." *Archive for History of Exact Sciences* 38: 67–78. Reprinted in Howard and Stachel (1989, 201–12).

Kox, A. J., and Eisenstaedt, Jean, eds. 2005. *The Universe of General Relativity.* Einstein Studies 11. Boston: Birkhäuser.

Kraft, Victor. 1950. *Der Wiener Kreis: Der Ursprung des Neopositivismus. Ein Kapitel der jüngsten Philosophiegeschichte.* Vienna: Springer.

Kragh, Helge. 1996. *Cosmology and Controversy: The Historical Development of Two Theories of the Universe.* Princeton, NJ: Princeton University Press.

1999. *Quantum Generations: A History of Physics in the Twentieth Century.* Princeton, NJ: Princeton University Press.

Kretschmann, Erich. 1915. "Über die prinzipielle Bestimmbarkeit der berechtigten Bezugssysteme beliebiger Relativitätstheorien (I)." *Annalen der Physik* 48: 907–42.

1917. "Über den physikalischen Sinn der Relativitätspostulate: A. Einsteins neue und seine ursprüngliche Relativitätstheorie." *Annalen der Physik* 53: 575–614.

Krüger, Lorenz, ed. 1994. *Universalgenie Helmholtz: Rückblick nach 100 Jahren.* Berlin: Akademie Verlag.

Kuhn, Thomas S. 1978. *Black-body Theory and the Quantum Discontinuity, 1894–1912.* New York: Oxford University Press; Oxford: Clarendon Press. Reprint (with a new afterword): Chicago: The University of Chicago Press, 1987.

2012. *The Structure of Scientific Revolutions.* 4th ed. (1st ed.: 1962). Chicago: The University of Chicago Press.

Lacki, Jan. 2004. "The Puzzle of Canonical Transformations in Early Quantum Mechanics." *Studies In History and Philosophy of Modern Physics* 35: 317–44.

Ladenburg, Erich. 1907. "Über Anfangsgeschwindigkeit und Menge der photoelektrischen Elektronen in ihrem Zusammenhange mit der Wellenlänge des auslösenden Lichtes." *Deutsche Physikalische Gesellschaft. Berichte* 9: 504–15. Reprinted in *Physikalische Zeitschrift* 8 (1907): 590–4.

Lanczos, Cornelius. 1922. "Bemerkung zur de Sitterschen Welt." *Physikalische Zeitschrift* 23: 539–43.

1923. "Über die Rotverschiebung in der de Sitterschen Welt." *Zeitschrift für Physik* 17: 168–89.

1974. *The Einstein Decade (1905–1915)*. New York: Academic Press.

1998. *Collected Published Papers with Commentaries*. Raleigh: North Carolina State University Press.

Lange, Friedrich Albert. 1873–5. *Geschichte des Materialismus und Kritik seiner Bedeutung in der Gegenwart*. 2nd ed. Iserlohn: J. Baedeker.

Langevin, Paul, and Broglie, M. de, eds. 1912. *La théorie du rayonnement et les quanta*. Brussels: Gauthier-Villars.

Laue, Max. 1911a. "Zur Dynamik der Relativitätstheorie." *Annalen der Physik* 35: 524–42. Reprinted in von Laue (1961, Vol. 1, 135–53).

1911b. *Das Relativitätsprinzip*. Braunschweig: Vieweg.

1912. "Eine quantitative Prüfung der Theorie für die Interferenz-Erscheinungen bei Röntgenstrahlen." *Bayerische Akademie der Wissenschaften. Sitzungsberichte*: 363–73. Reprinted in von Laue (1961, Vol. 1, 208–18).

Laue, Max von. 1961. *Gesammelte Schriften und Vorträge*. 3 vols. Braunschweig: Vieweg.

Laymon, Ronald. 1978. "Newton's Bucket Experiment." *Journal of the History of Philosophy* 16: 399–413.

Layzer, David. 1954. "On the Significance of Newtonian Cosmology." *Astrophysical Journal* 59: 268–70.

1956. "Newtonian Cosmology." *The Observatory* 76: 73–4.

Lehner, Christoph. 2005. "Einstein and the Principle of General Relativity, 1916–1921." In Kox and Eisenstaedt (2005, 103–8).

Lehner, Christoph, Renn, Jürgen, and Schemmel, Matthias, eds. 2012. *Einstein and the Changing World Views of Physics*. Einstein Studies 12. Boston: Birkhäuser.

Lemaître, Georges. 1925. "Note on de Sitter's Universe." *Journal of Mathematical Physics* 4: 188–92.

1927. "Un universe homogène de masse constante et de rayon croissant, rendant compte de la vitesse radial des nébuleuses extra-galactiques." *Annales de la Société Scientifique de Bruxelles*: 49–56.

1931. "Expansion of the Universe: A Homogeneous Universe of Constant Mass and Increasing Radius Accounting for the Radial Velocity of Extra-Galactic Nebulae." *Royal Astronomical Society. Monthly Notices* 91: 483–90.

1932. "L'Universe en expansion." *Publication du Laboratoire d'Astronomie et de Géodésie de l'Université de Louvain* 9: 171–205.

Lenard, Philipp. 1902. "Über die lichtelektrische Wirkung." *Annalen der Physik* 8: 149–98. Reprinted in Lenard (1944, 251–90).

1944. *Wissenschaftliche Abhandlungen*. Vol. 3. *Kathodenstrahlen, Elektronen, Wirkungen Ultravioletten Lichtes*. Leipzig: S. Hirzel.

Lense, Josef, and Thirring, Hans. 1918. "Über den Einfluß der Eigenrotation der Zentralkörper auf die Bewegung der Planeten und Monde nach der Einsteinschen Gravitationstheorie." *Physikalische Zeitschrift* 19: 156–63.

Lepeltier, Thomas. 2006. "Edward Milne's Influence on Modern Cosmology." *Annals of Science* 63: 471–81.

Levenson, Thomas. 2003. *Einstein in Berlin*. New York: Bantam Books.

Levi-Civita, Tullio. 1916. "Nozione di parallelismo in una varietà qualunque e conseguente specificazione geometrica della curvatura riemanniana." *Circolo Matematico di Palermo. Rendiconti* 42: 173–204. English trans. of excerpts in Renn (2007a, Vol. 4, 1081–8).

Lorentz, Hendrik A. 1892. "De relatieve beweging van de aarde en den aether." *Koninklijke Akademie van Wetenschappen te Amsterdam. Wis- en Natuurkundige Afdeeling. Verslagen van de Gewone Vergaderingen* 1: 74–9. English trans. in Lorentz (1935–9, Vol. 4, 219–23).

1895. *Versuch einer Theorie der electrischen und optischen Erscheinungen in bewegten Körpern*. Leiden: Brill. Reprinted in Lorentz (1935–9, Vol. 5, 1–138).

1904. "Electromagnetische verschijnselen in een stelsel dat zich met willekeurige snelheid, kleiner dan die van het licht, beweegt." *Koninklijke Akademie van Wetenschappen te Amsterdam. Wis- en Natuurkundige Afdeeling. Verslagen van de Gewone Vergaderingen* 12: 986–1009. English trans. "Electromagnetic Phenomena in Systems Moving with Any Velocity Less Than That of Light." *Koninklijke Akademie van Wetenschappen te Amsterdam. Section of Sciences. Proceedings* 6 (1904): 809–31. Reprinted in Lorentz (1935–9, Vol. 5, 172–97) and (without the last section) in Einstein et al. (1952, 11–34).

1909. *The Theory of Electrons and Its Applications to the Phenomena of Light and Radiant Heat*. New York: Columbia University, Ernest Kempton Adams Fund for Physical Research.

1910a. "Die Hypothese der Lichtquanten." *Physikalische Zeitschrift* 11: 349–54. Reprinted in Lorentz (1935–9, Vol. 7, 374–84).

1910b. "Alte und neue Fragen der Physik." *Physikalische Zeitschrift* 11: 1234–57. Reprinted in Lorentz (1935–9, Vol. 7, 205–57).

1915 [1909]. *The Theory of Electrons*. 2nd ed. New York: Dover.

1916. *Les théories statistiques en thermodynamique*. Leipzig: Teubner.

1927. *Problems of Modern Physics*. Boston: Ginn.

1934–9. *Collected Papers*. 9 Vols. Eds. P. Zeeman and A. D. Fokker. The Hague: Nijhoff.

2008. *The Scientific Correspondence of H. A. Lorentz*, ed. A. J. Kox. New York: Springer.

Mach, Ernst. 1883. *Die Mechanik in ihrer Entwickelung: Historisch-kritisch dargestellt*. Leipzig: Brockhaus.

1896. *Die Principien der Wärmelehre: Historisch-kritisch entwickelt*. Leipzig: Barth.

1906. *Erkenntnis und Irrtum: Skizzen zur Psychologie der Forschung*. 2nd ed. Leipzig: Barth.

1910. "Die Leitgedanken meiner naturwissenschaftlichen Erkenntnislehre und ihre Aufnahme durch die Zeitgenossen." *Scientia* 7: 225–40. Reprinted in *Physikalische Zeitschrift* 11 (1910): 599–606.

1960. *The Science of Mechanics: A Critical and Historical Account of Its Development*. La Salle, IL: Open Court. Trans. of later edition of Mach (1883).

Macpherson, Hector. 1929. *Modern Cosmologies: A Historical Sketch of Researches and Theories Concerning the Structure of the Universe*. London: Oxford University Press.

Majer, Ulrich, and Sauer, Tilman. 2005. "Hilbert's 'World Equations' and His Vision of a Unified Science." In Kox and Eisenstaedt (2005, 259–76).

Malament, David B. 1995. "Is Newtonian Cosmology Really Inconsistent?" *Philosophy of Science* 62: 489–510.

ed. 2002. *Reading Natural Philosophy: Essays in the History and Philosophy of Science and Mathematics to Honor Howard Stein on His 70th Birthday*. Chicago: Open Court.

Marianoff, Dimitri, and Wayne, Palma. 1944. *Einstein: An Intimate Study of a Great Man*. New York: Doubleday.

Martínez, Alberto A. 2005. "Handling Evidence in History: The Case of Einstein's Wife." *School Science Review* 86.316: 49–56.

2009. *Kinematics: The Lost Origins of Einstein's Relativity*. Baltimore, MD: Johns Hopkins University Press.

2011. *Science Secrets: The Truth about Darwin's Finches, Einstein's Wife, and Other Myths*. Pittsburgh: University of Pittsburgh Press.

Marx, Erich, ed. 1916a. *Handbuch der Radiologie*. Vol. 3. Leipzig: Akademische Verlagsgesellschaft.

1916b. "Entwickelung der Lichtelektrizität von Januar 1914 bis Oktober 1915." In Marx (1916a, 565–88).

Massimi, Michela, ed. 2008. *Kant and the Philosophy of Science Today*. Cambridge: Cambridge University Press.

Maudlin, Tim. 1990. "Substances and Space-Time: What Aristotle Would Have Said to Einstein." *Studies in History and Philosophy of Science* 21: 531–61.

Maxwell, James C. 1876. *Matter and Motion*. London: Society for Promoting Christian Knowledge.

McCormmach, Russell. 1970. "H. A. Lorentz and the Electromagnetic View of Nature." *Isis* 61: 459–97.

McCrea, William H. 1955. "On the Significance of Newtonian Cosmology." *Astrophysical Journal* 60: 271–4.

McCrea, William H., and Milne, Edward A. 1934. "Newtonian Universes and the Curvature of Space." *Quarterly Journal of Mathematics* 5: 73–80.

Mermin, N. David. 1968. *Space and Time in Special Relativity*. New York: McGraw-Hill. Reprint: Prospect Heights, IL: Waveland Press, 1989.

2011. "Understanding Einstein's 1905 Derivation of $E=mc^2$." *Studies in History and Philosophy of Modern Physics* 42: 1–2.

2012. "Reply to Ohanian's Comment." *Studies in History and Philosophy of Modern Physics* 43: 218–19.

Michelson, Albert A. 1881. "The Relative Motion of the Earth and the Luminiferous Ether." *American Journal of Science* 22: 120–9. Reprinted in facsimile in Swenson (1972, 249–58).

Michelson, Albert A., and Morley, Edward W. 1886. "Influence of Motion of the Medium on the Velocity of Light." *American Journal of Science* 31: 377–86. Reprinted in facsimile in Swenson (1972, 261–70).

1887. "On the Relative Motion of the Earth and the Luminiferous Ether." *American Journal of Science* 34: 333–5. Reprinted in facsimile in Swenson (1972, 273–85).

Mie, Gustav. 1912a. "Grundlagen einer Theorie der Materie. Erste Mitteilung." *Annalen der Physik* 37: 511–34.

1912b. "Grundlagen einer Theorie der Materie. Zweite Mitteilung." *Annalen der Physik* 39: 1–40.

1913. "Grundlagen einer Theorie der Materie. Dritte Mitteilung."*Annalen der Physik* 40: 1–66. English trans. of ch. 5 in Renn (2007a, Vol. 4, 663–96).

1914. "Bemerkungen zu der Einsteinschen Gravitationstheorie." *Physikalische Zeitschrift* 15: 115–22 (I); 169–76 (II). English trans. in Renn (2007a, Vol. 4, 699–728).

1917. "Die Einsteinsche Gravitationstheorie und das Problem der Materie." *Physikalische Zeitschrift* 18: 551–6 (I); 574–80 (II); 596–602 (III).

Miller, Arthur I. 1981. *Albert Einstein's Special Theory of Relativity: Emergence (1905) and Early Interpretation (1905–1911)*. Reading, MA: Addison-Wesley. Reprint: New York: Springer, 1998.

— 2000. *Insights of Genius: Imagery and Creativity in Science and Art*. Cambridge, MA: MIT Press.

— 2001. *Einstein, Picasso: Space, Time, and the Beauty That Causes Havoc*. New York: Basic Books.

Millikan, Robert A. 1913a. "On the Elementary Electrical Charge and the Avogadro Constant." *Physical Review* 2: 109–43.

— 1913b. "Atomic Theories of Radiation." *Science* 37: 119–33.

— 1916. "A Direct Photoelectric Determination of Planck's 'h.'" *Physical Review* 7: 355–88.

— 1917. *The Electron: Its Isolation and Measurement and the Determination of Some of Its Properties*. Chicago: University of Chicago Press.

— 1923. "The Electron and the Light-Quant from the Experimental Point of View." In Nobel Foundation (1965, 54–66).

— 1950. *The Autobiography of Robert A. Millikan*. New York: Prentice-Hall.

Millikan, Robert A., and Gale, Henry Gordon. 1906. *A First Course in Physics*. Boston: Ginn.

Milne, Edward A. 1934. "A Newtonian Expanding Universe." *Quarterly Journal of Mathematics* 5: 64–72.

— 1935. *Relativity, Gravitation, and World–Structure*. Oxford: Clarendon Press.

Minkowski, Hermann. 1909. "Raum und Zeit." *Physikalische Zeitschrift* 10: 104–11. English trans. in Einstein et al. (1952, 75–91).

Mises, Richard von. 1938. *Ernst Mach und die empiristische Weltauffassung*. The Hague: W. P. van Stockum & Zoon.

— 1939. *Kleines Lehrbuch des Positivismus: Einführung in die empiristische Wissenschaftsauffassung*. The Hague: W. P. van Stockum & Zoon.

Misner, Charles W., Thorne, Kip S., and Wheeler, John A. 1973. *Gravitation*. San Francisco: Freeman.

Moore, Walter. 1994. *A Life of Erwin Schrödinger*. Cambridge: Cambridge University Press.

Moszkowski, Alexander. 1921. *Einstein: Einblicke in seine Gedankenwelt; gemeinverständliche Betrachtungen über die Relativitätstheorie und ein neues Weltsystem; entwickelt aus Gesprächen mit Einstein*. Hamburg: Hoffmann und Campe.

Mulder, Henk, and van de Velde-Schlick, Barbara, eds. 1979. *Moritz Schlick: Philosophical Papers*. Vol. 1. Dordrecht: Reidel.

Müller-Kumbhaar, Heiner, and Wagner, Hermann F. 2001. *... und Er würfelt doch: Von der Erforschung des ganz Grossen, des ganz Kleinen und der ganz vielen Dinge*. New York: Wiley.

Murdoch, Dugald. 1987. *Niels Bohr's Philosophy of Physics*. Cambridge: Cambridge University Press.

Nagel, Ernest. 1950. "Einstein's Philosophy of Science." *The Kenyon Review* 12.3: 520–31. Reprinted in *Logic without Metaphysics and Other Essays in the Philosophy of Science*, ed. E. Nagel. Glencoe, IL: Free Press, 1956, 287–301.

Nathan, Otto, and Norden, Heinz. 1960. *Einstein on Peace*. New York: Avenel Books.

Natorp, Paul. 1910. *Die logischen Grundlagen der exakten Wissenschaften*. Leipzig: Teubner.

Needell, Allan A. 1980. *Irreversibility and the Failure of Classical Dynamics: Max Planck's Work on the Quantum Theory, 1900–1915.* PhD diss. Yale University.

Neffe, Jürgen. 2005. *Einstein: Eine Biographie.* Berlin: Rowohlt. English translation: *Einstein: A Biography.* Translated by Shelley Frisch. New York: Farrar, Straus, and Giroux, 2007.

Neiman, Susan. 1994. *The Unity of Reason: Re-Reading Kant.* New York: Oxford University Press.

Neumann, Johann von. 1932. *Mathematische Grundlagen der Quantenmechanik.* Berlin: Springer.

Neurath, Otto, Hahn, Hans, and Carnap, Rudolf. 1929. *Wissenschaftliche Weltauffassung: Der Wiener Kreis.* Vienna: Artur Wolf.

Newton, Sir Isaac. 1959–77. *The Correspondence of Sir Isaac Newton,* ed. Herbert W. Turnbull (Vols. 1–3), J. W. Scott (Vol. 4), A. Rupert Hall, and Laura Tilling (Vols. 5–7). Cambridge: Cambridge University Press.

Nicolai, Friedrich. 1917. *Die Biologie des Krieges: Betrachtungen eines deutschen Naturforschers.* Zürich: Orell Füssli.

Noether, Emmy. 1918. "Invariante Variationsprobleme." *Königliche Gesellschaft der Wissenschaften zu Göttingen. Mathematisch-Physikalische Klasse. Nachrichten:* 235–57.

Nordström, Gunnar. 1912. "Relativitätsprinzip und Gravitation." *Physikalische Zeitschrift* 13: 1126–9. English trans. in Renn (2007a, Vol. 3, 489–97).

1913a. "Träge und schwere Masse in der Relativitätsmechanik." *Annalen der Physik* 40: 856–78. English trans. in Renn (2007a, Vol. 3, 499–521).

1913b. "Zur Theorie der Gravitation vom Standpunkt des Relativitätsprinzips." *Annalen der Physik* 42: 533–54. English trans. in Renn (2007a, Vol. 3, 523–42).

North, John. 1965. *The Measure of the Universe.* Oxford: Oxford University Press.

1995. *The Norton History of Astronomy and Cosmology.* New York: Norton.

Norton, John D. 1984. "How Einstein Found His Field Equations: 1912–1915." *Historical Studies in the Physical Sciences* 14: 253–316. Reprinted in Howard and Stachel (1989, 101–59).

1985. "What Was Einstein's Principle of Equivalence?" *Studies in History and Philosophy of Science* 16: 203–46. Reprinted in Howard and Stachel (1989, 5–47).

1987. "Einstein, the Hole Argument and the Reality of Space." In *Measurement, Realism and Objectivity,* ed. John Forge. Dordrecht: Reidel, 153–88.

1992a. "The Physical Content of General Covariance." In Eisenstaedt and Kox (1992, 281–315).

1992b. "Einstein, Nordström and the Early Demise of Scalar, Lorentz Covariant Theories of Gravitation." *Archive for the History of Exact Sciences* 45: 17–94. Reprinted in Renn (2007a, Vol. 3, 413–87).

1993a. "Einstein and Nordström: Some Lesser-Known Thought Experiments in Gravitation." In Earman et al. (1993, Vol. 5, 3–29).

1993b. "General Covariance and the Foundations of General Relativity: Eight Decades of Dispute." *Reports on Progress in Physics* 56: 791–858.

1995. "The Force of Newtonian Cosmology: Acceleration is Relative." *Philosophy of Science* 62: 511–22.

1995b. "Mach's Principle before Einstein." In Barbour and Pfister (1995, 9–57).

1999a. "Geometries in Collision: Einstein, Klein, and Riemann." In Gray (1999, 128–44).

1999b. "The Cosmological Woes of Newtonian Gravitation Theory." In Goenner et al. (1999, 271–323).

2000. "'Nature is the Realisation of the Simplest Conceivable Mathematical Ideas': Einstein and the Canon of Mathematical Simplicity." *Studies in History and Philosophy of Modern Physics* 31: 135–70.

2003. "The N-Stein Family." In Ashtekar et al. (2003, 55–67).

2004. "Einstein's Investigations of Galilean Covariant Electrodynamics prior to 1905." *Archive for History of Exact Sciences* 59: 45–105.

2006. "Atoms, Entropy, Quanta: Einstein's Miraculous Argument of 1905." *Studies in History and Philosophy of Modern Physics* 37: 71–100.

2007. "What Was Einstein's 'Fatal Prejudice'?" In Renn (2007a, Vol. 2, 715–83).

2008. "The Hole Argument." *The Stanford Encyclopedia of Philosophy* (Fall 2008 ed.), ed. Edward N. Zalta, http://plato.stanford.edu/archives/fall2008/entries/spacetime-holearg/.

2010. "How Hume and Mach Helped Einstein Find Special Relativity." In Domski and Dickson (2010, 359–86).

Nozick, Robert. 2001. *Invariances: the Structure of the Objective World*. Cambridge, MA: Harvard University Press.

Nye, Mary Jo. 1972. *Molecular Reality: A Perspective on the Scientific Work of Jean Perrin*. London: Macdonald.

Ohanian, Hans C. 2009. "Did Einstein Prove $E=mc^2$?" *Studies in History and Philosophy of Modern Physics* 40: 167–73.

2012. "A Comment on Mermin's 'Understanding Einstein's 1905 Derivation of $E = mc^2$.'" *Studies in History and Philosophy of Modern Physics* 43: 215–17.

Oppenheimer, J. Robert. 1979. "On Albert Einstein." In French (1979, 44–9).

O'Raifeartaigh, Lochlainn, and Straumann, Norbert. 2000. "Gauge Theory: Historical Origins and Some Modern Developments." *Reviews of Modern Physics* 72: 1–23.

Overbye, Dennis. 2000. *Einstein in Love: A Scientific Romance*. New York: Viking.

Pais, Abraham. 1979. "Einstein and the Quantum Theory." *Reviews of Modern Physics* 51: 863–914.

1982. '*Subtle is the Lord ...*' *The Science and Life of Albert Einstein*. Oxford: Oxford University Press.

1994. *Einstein Lived Here*. Princeton, NJ: Princeton University Press.

1997. *A Tale of Two Continents: A Physicist's Life in a Turbulent World*. Princeton, NJ: Princeton University Press.

Paterniti, Michael. 2000. *Driving Mr. Albert: A Trip across America with Einstein's Brain*. New York: Dial Press.

Paty, Michel. 1993. *Einstein philosophe: La physique comme pratique philosophique*. Paris: Presses Universitaires de France.

Paul, Erich. 1993. *The Milky Way Galaxy and Statistical Cosmology, 1890–1924*. Cambridge: Cambridge University Press.

Pauli, Wolfgang. 1921. "Relativitätstheorie." In *Encyklopädie der mathematischen Wissenschaften, mit Einschluss ihrer Anwendungen*. Vol. 5, *Physik*, Part 2, ed. Arnold Sommerfeld. Leipzig: Teubner, 539–775. Reprinted in facsimile: *Relativitätstheorie*, ed. Domenico Giulini. Berlin: Springer, 2000. English trans. *Theory of Relativity*. London: Pergamon, 1958. Reprinted: New York: Dover, 1981.

1979. *Scientific Correspondence with Bohr, Einstein, Heisenberg, a.o.* Vol. 1, *1919–1929*, eds. Armin Hermann, Karl von Meyenn, and V. F. Weisskopf. New York: Springer.

1985. *Scientific Correspondence with Bohr, Einstein, Heisenberg, a.o.* Vol. 2, *1930–1939*, eds. Armin Hermann, Karl von Meyenn, and V. F. Weisskopf. New York: Springer.

Peebles, P. James E. 1980. *Large-scale Structure of the Universe.* Princeton, NJ: Princeton University Press.

Perovic, Slobodan. 2006. "Schrödinger's Interpretation of Quantum Mechanics and the Relevance of Bohr's Experimental Critique." *Studies in History and Philosophy of Modern Physics* 37: 275–97.

Petzoldt, Joseph. 1895. "Das Gesetz der Eindeutigkeit." *Vierteljahrsschrift für wissenschaftliche Philosophie und Soziologie* 19: 146–203.

Pfister, Herbert. 2007. "On the History of the So-called Lense-Thirring Effect." *General Relativity and Gravitation* 39: 1735–48.

Planck, Max. 1900a. "Über irreversible Strahlungsvorgänge." *Annalen der Physik* 1: 69–122.

1900b. "Zur Theorie des Gesetzes der Energievertheilung im Normalspektrum." *Deutsche Physikalische Gesellschaft. Verhandlungen* 2: 237–45.

1901a. "Über das Gesetz der Energieverteilung im Normalspectrum." *Annalen der Physik* 43: 553–63.

1901b. "Über die Elementarquanta der Materie und Elektricität." *Annalen der Physik* 4: 564–6.

1909. "Die Einheit des physikalischen Weltbildes." *Physikalische Zeitschrift* 10: 62–75.

1910. "Zur Machschen Theorie der physikalischen Erkenntnis: Eine Erwiderung." *Physikalische Zeitschrift* 11: 1186–90.

1912. "La loi du rayonnement et l'hypothèse des quantités élémentaires d'action." In Langevin and de Broglie (1912, 93–114).

1915. *Eight Lectures on Theoretical Physics.* New York: Columbia University Press.

1920. *Die Entstehung und bisherige Entwicklung der Quantentheorie. Nobelvortrag.* Braunschweig: Vieweg. Reprinted in *Physikalische Abhandlungen und Vorträge* (Braunschweig: Vieweg, 1958), Vol. 3, 121–36. Page references are to this reprint.

Pohl, Robert W., and Pringsheim, Peter. 1913. "On the Long-Wave Limits of the Normal Photoelectric Effect." *Philosophical Magazine* 26: 1017–24.

1914. *Die lichtelektrischen Erscheinungen.* Braunschweig: Vieweg.

Poincaré, Henri. 1898. "La mesure du temps." *Revue de Métaphysique et de Morale* 6: 1–13. English trans. "The Measure of Time" in Poincaré (1913, 223–34).

1902a. *La Science et l'hypothèse.* Paris: E. Flammarion. English trans. "Science and Hypothesis" in Poincaré (1913, 27–197).

1902b. "L'expérience et la géométrie." In Poincaré (1902a, 92–109).

1904. "L'état actuel et l'avenir de la physique mathématique." *Bulletin des sciences mathématiques* 28: 302–24. English trans. "The Present Crisis of Mathematical Physics" in Poincaré (1913, 297–320).

1905. "Sur la dynamique de l'électron." *Académie des sciences* (Paris). *Comptes rendus* 140: 1504–8.

1906. "Sur la dynamique de l'électron." *Circolo Matematico di Palermo. Rendiconti* 21: 129–75.

1913. *The Foundations of Science*. New York: Science Press.

Pooley, Oliver. 2006. "Points, Particles, and Structural Realism." In Rickles et al. (2006, 83–120).

Poppel, Stephen M. 1977. *Zionism in Germany, 1897–1933: The Shaping of a Jewish Identity*. Philadelphia: Jewish Publication Society.

Pound, Robert V. 2000. "Weighing Photons: I." *Physics in Perspective* 2: 224–68.

2001. "Weighing Photons: II." *Physics in Perspective* 3: 4–51.

Przibram, Karl, ed. 1963. *Briefe zur Wellenmechanik*. Vienna: Springer.

Pyenson, Lewis. 1985. *The Young Einstein: The Advent of Relativity*. Boston: Adam Hilger.

Quine, Willard V. O. 1951. "Two Dogmas of Empiricism." *Philosophical Review* 60: 29–43.

Ramsauer, Carl. 1914a. "Über eine direkte magnetische Methode zur Bestimmung der lichtelektrischen Geschwindigkeitsverteilung." *Annalen der Physik* 45: 961–1002.

1914b. "Über die lichtelektrische Geschwindigkeitsverteilung und ihre Abhängigkeit von der Wellenlänge." *Annalen der Physik* 45: 1121–59.

Rayleigh, Lord. 1943. *The Life of Sir J.J. Thomson, O.M.: Sometime Master of Trinity College, Cambridge*. Cambridge: Cambridge University Press.

Reichenbach, Hans. 1920. *Relativitätstheorie und Erkenntnis Apriori*. Berlin: Springer. English trans. *The Theory of Relativity and A Priori Knowledge*. Berkeley and Los Angeles: University of California Press, 1965.

1922. "La signification de la théorie de la relativité." *Revue Philosophique de la France et de l'Étranger* 94: 5–61. English trans. with omissions by M. Reichenbach, "The Present State of the Discussion of Relativity" in *Hans Reichenbach: Selected Writings*, Vol. 2, eds. M. Reichenbach and R. Cohen, Dordrecht: Reidel, 3–47.

1924. *Axiomatik der relativistischen Raum-Zeit-Lehre*. Braunschweig: Vieweg.

1928. *Philosophie der Raum-Zeit-Lehre*. Berlin: De Gruyter. English trans. *The Philosophy of Space and Time*. New York: Dover, 1958.

1938. *Experience and Prediction: An Analysis of the Foundations and the Structure of Knowledge*. Chicago: University of Chicago Press.

1949. "The Philosophical Significance of the Theory of Relativity." In Schlipp (1949, 289–311).

1951. *The Rise of Scientific Philosophy*. Berkeley: University of California Press.

Reichenbächer, Ernst. 1920. "Inwiefern läßt sich die moderne Gravitationstheorie ohne die Relativität begründen?" *Die Naturwissenschaften* 8: 1008–10.

Reichinstein, David. 1932. *Albert Einstein, sein Lebensbild und seine Welt-anschauung*. Prague: Selbstverlag des Verfassers.

Reiser, Anton (pseudo. Kayser, Rudolf). 1930. *Albert Einstein: A Biographical Portrait*. New York: Albert & Charles Boni.

Renn, Jürgen. 1993. "Einstein as a Disciple of Galileo: A Comparative Study of Conceptual Development in Physics." *Science in Context* 6: 311–41.

1997. "Einstein's Controversy with Drude and the Origin of Statistical Mechanics: A New Glimpse from the 'Love Letters.'" *Archive for History of Exact Sciences* 51: 315–54.

ed. 2005a. *Einstein's Annalen Papers: The Complete Collection, 1901–1922*. Weinheim: Wiley-VCH.

ed. 2005b. *Albert Einstein – Ingenieur des Universums/Albert Einstein – Chief Engineer of the Universe.* 3 Vols. Berlin: Wiley-VCH. Vol. 1. *Einsteins Leben und Werk im Kontext/Einstein's Life and Work in Context;* Vol. 2. *Hundert Autoren für Einstein/One Hundred Authors for Einstein;* Vol. 3. *Dokumente eines Lebensweges/Documents of a Life's Pathway.*

2005c. "Einstein's Invention of Brownian Motion." *Annalen der Physik* 14: 23–37. Supplement published in book form as Renn (2005a).

2006. *Auf den Schultern von Riesen und Zwergen: Einsteins unvollendete Revolution.* New York: Wiley.

ed. 2007a. *The Genesis of General Relativity.* 4 vols. New York and Berlin: Springer. Vol. 1. *Einstein's Zurich Notebook: Introduction and Source,* Michel Janssen, John D. Norton, Jürgen Renn, Tilman Sauer, and John Stachel; Vol. 2. *Einstein's Zurich Notebook: Commentary and Essays,* Michel Janssen, John D. Norton, Jürgen Renn, Tilman Sauer, and John Stachel; Vol. 3. *Gravitation in the Twilight of Classical Physics: Between Mechanics, Field Theory and Astronomy,* eds. Jürgen Renn and Matthias Schemmel; Vol. 4. *Gravitation in the Twilight of Classical Physics: The Promise of Mathematics,* eds. Jürgen Renn and Matthias Schemmel.

2007b. "Classical Physics in Disarray." In Renn (2007a, Vol. 1, 21–80).

2007c. "The Third Way to General Relativity." In Renn (2007a, Vol. 3, 21–75).

2007d. "The Summit Almost Scaled: Max Abraham as a Pioneer of a Relativistic Theory of Gravitation." In Renn (2007a, Vol. 3, 305–30).

2008. "Boltzmann and the End of the Mechanistic Worldview." In *Boltzmann's Legacy,* eds. G. Gallavotti, W. L. Reiter, and J. Yngvason. Zurich: European Mathematical Society Publishing House.

Renn, Jürgen, and Sauer, Tilman. 2003. "Eclipses of the Stars: Mandl, Einstein, and the Early History of Gravitational Lensing." In Ashketar et al. (2003, 69–92).

2007. "Pathways out of Classical Physics: Einstein's Double Strategy in Searching for the Gravitational Field Equation." In Renn (2007a, Vol. 1, 113–312).

Renn, Jürgen, and Schulmann, Robert, eds. 1992. *Albert Einstein – Mileva Marić: The Love Letters.* Princeton, NJ: Princeton University Press.

Renn, Jürgen, and Stachel, John. 2007. "Hilbert's Foundation of Physics: From a Theory of Everything to a Constituent of General Relativity." In Renn (2007a, Vol. 4, 857–973).

Renn, Jürgen, Sauer, Tilman, and Stachel, John. 1997. "The Origin of Gravitational Lensing: A Postscript to Einstein's 1936 *Science* Paper." *Science* 275: 184–6.

Rhodes, Richard. 1986. *The Making of the Atomic Bomb.* New York: Simon and Schuster.

Richards, Joan. 1977. "The Evolution of Empiricism: Hermann von Helmholtz and the Foundations of Geometry." *British Journal for the Philosophy of Science* 28: 235–53.

Richardson, Owen W. 1912a. "Some Applications of the Electron Theory of Matter." *Philosophical Magazine* 23: 594–627.

1912b. "The Theory of Photoelectric Action." *Philosophical Magazine* 24: 570–4.

1914. "The Theory of Photoelectric and Photochemical Action." *Philosophical Magazine* 27: 476–88.

Richardson, Owen W., and Compton, Karl T. 1912. "The Photoelectric Effect." *Philosophical Magazine* 24: 575–94.

Richardson, Owen W., and Rogers, F. J. 1915. "The Photoelectric Effect: III." *Philosophical Magazine* 29: 618–23.

Rickert, Heinrich. 1921. *Der Gegenstand der Erkenntnis: Einführung in die Transzendental-Philosophie*. 4th and 5th ed. Tübingen: Verlag J. C. B. Mohr.

Rickles, Dean, and French, Steven. 2006. "Quantum Gravity Meets Structuralism: Interweaving Relations in the Foundations of Physics." In Rickles et al. (2006, 1–39).

Rickles, Dean, French, Steven, and Saatsi, Juha T., eds. 2006. *The Structural Foundations of Quantum Gravity*. Oxford: Oxford University Press.

Riemann, Bernhard. 1867. "Ueber die Hypothesen, welche der Geometrie zu Grunde liegen." *Abhandlungen der Königlichen Gesellschaft der Wissenschaften zu Göttingen. Mathematische Klasse* 13: 133–50. Reprint: Weyl, Hermann, ed. Berlin: Springer, 1919. English trans. "On the Hypotheses which Lie at the Foundations of Geometry" in *A Source Book in Mathematics*, ed. David Smith, Vol. 2. New York: Dover, 1959, 411–25.

Rindler, Wolfgang. 1956. "Visual Horizons in World Models." *Royal Astronomical Society. Monthly Notices* 116: 662–77.

Rigden, John S. 2005. *Einstein 1905: The Standard of Greatness*. Cambridge, MA: Harvard University Press.

Robertson, Howard P. 1928. "Relativistic Cosmology." *Philosophical Magazine* 5: 835–48.

1929. "On the Foundations of Relativistic Cosmology." *National Academy of Science. Proceedings* 15: 822–9.

1933. "Relativistic Cosmology." *Reviews of Modern Physics* 5: 62–90.

1935. "Kinematics and World Structure." *Astrophysical Journal* 82: 284–301.

1936a. "Kinematics and World Structure. II." *Astrophysical Journal* 83: 187–201.

1936b. "Kinematics and World Structure. III." *Astrophysical Journal* 83: 257–71.

Röhle, Stefan. 2002. "Mathematische Probleme in der Einstein – de Sitter Kontroverse." *Max Planck Institute for the History of Science*, preprint 210, Berlin.

Rosenberger, Ferdinand. 1895. *Isaac Newton und seine physikalischen Prinzipien: Ein Hauptstück aus der Entwickelungsgeschichte der modernen Physik*. Leipzig: Barth.

Rowe, David. 1999. "The Göttingen Response to General Relativity and Emmy Noether's Theorems." In Gray (1999, 189–234).

2006. "Einstein's Allies and Enemies: Debating Relativity in Germany, 1916–1920." In Hendricks et al. (2006, 231–80).

2012. "Einstein and Relativity: What Price Fame?" *Science in Context* 25: 197–246.

Rowe, David, and Schulmann, Robert, eds. 2007. *Einstein on Politics: His Private Thoughts and Public Stands on Nationalism, Zionism, War, Peace, and the Bomb*. Princeton, NJ: Princeton University Press.

Royal Astronomical Society. 1930. "Report of the Royal Astronomical Society Meeting in January 1930." *The Observatory* 53: 33–44.

Rutherford, Ernest. 1918. "Silvanus P. Thomson Memorial Lecture." *Journal of the Röntgen Society* 14: 75–86.

Ryckman, Thomas A. 2005. *The Reign of Relativity*. Oxford: Oxford University Press.

Rynasiewicz, Robert. 1992. "Rings, Holes and Substantivalism: On the Program of Leibniz Algebras." *Philosophy of Science* 45: 572–89.

1994. "The Lessons of the Hole Argument." *The British Journal for the Philosophy of Science* 45: 407–36.

1996. "Absolute versus Relational Space-Time: An Outmoded Debate?" *Journal of Philosophy* 93: 279–306.

1999. "Kretschmann's Analysis of Covariance and Relativity Principles." In Goenner et al. (1999, 431–62).

2000. "The Construction of the Special Theory: Some Queries and Considerations." In Howard and Stachel (2000, 159–201).

2005. "The Optics and Electrodynamics of 'On the Electrodynamics of Moving Bodies.'" *Annalen der Physik* 14: 38–57. Supplement published in book form as Renn (2005a).

Rynasiewicz, Robert, and Renn, Jürgen. 2006. "The Turning Point for Einstein's Annus Mirabilis." *Studies in History and Philosophy of Modern Physics* 37: 5–35.

Salzman, George, and Taub, Abraham H. 1954. "Born-Type Rigid Motion in Relativity." *Physical Review* 95: 1659–69.

Samuel, Herbert V. 1945. *Memoirs*. London: Cresset Press.

Sauer, Tilman. 1999. "The Relativity of Discovery: Hilbert's First Note on the Foundations of Physics." *Archive for History of Exact Sciences* 53: 529–75.

2004. "The Challenge of Editing Einstein's Scientific Manuscripts." *Documentary Editing* 26: 145–65.

2005a. "Einstein's Review Paper on General Relativity Theory." In *Landmark Writings in Western Mathematics, 1640–1940*, ed. I. Grattan-Guiness. Amsterdam: Elsevier, 802–22.

2005b. "Einstein Equations and Hilbert Action: What Is Missing on Page 8 of the Proofs for Hilbert's First Communication on the Foundations of Physics?" *Archive for History of Exact Sciences* 59: 577–90. Reprinted in Renn (2007a, Vol. 4, 975–88).

2006. "Field Equations in Teleparallel Spacetime: Einstein's *Fernparallelismus* Approach toward Unified Field Theory." *Historia Mathematica* 33: 399–439.

2008. "The Einstein-Variçak Correspondence on Relativistic Rigid Rotation." *Proceedings of the Eleventh Marcel Grossmann Meeting on General Relativity*, eds. H. Kleinert and R.T. Jantzen, Singapore: World Scientific, 2008, 2453–55.

Saunders, Simon. 2003. "Indiscernibles, General Covariance, and Other Symmetries: The Case for Non-Reductive Relationalism." In Ashtekar et al. (2003, 151–73).

Sayen, Jamie. 1985. *Einstein in America: The Scientist's Conscience in the Age of Hitler and Hiroshima*. New York: Crown Publishers.

Scheibe, Erhard. 1992. "Albert Einstein: Theorie, Erfahrung, Wirklichkeit." *Heidelberger Jahrbücher* 36: 121–38.

Scheideler, Britta. 2002. "The Scientist as Moral Authority: Albert Einstein between Elitism and Democracy, 1914–1933." *Historical Studies in the Physical and Biological Science* 32: 319–46.

Schilpp, Paul A., ed. 1949. *Albert Einstein: Philosopher-Scientist*. Evanston, IL: Library of Living Philosophers.

Schlick, Moritz. 1910. "Das Wesen der Wahrheit nach der modernen Logik." *Vierteljahrsschrift für wissenschaftliche Philosophie und Soziologie* 34: 386–477.

1915. "Die philosophische Bedeutung des Relativitätsprinzips." *Zeitschrift für Philosophie und philosophische Kritik* 159: 129–75. English trans. "The

Philosophical Significance of the Principle of Relativity" in Mulder and van de Velde-Schlick (1979, 153–89).

1917. *Raum und Zeit in der gegenwärtigen Physik: Zur Einführung in das Verständnis der allgemeinen Relativitätstheorie.* Berlin: Springer. English trans. (from 4th ed. 1922) "Space and Time in Contemporary Physics" in Mulder and van de Velde-Schlick (1979, 207–69).

1918. *Allgemeine Erkenntnislehre.* Berlin: Springer.

1920. "Naturphilosophische Betrachtungen über das Kausalprinzip." *Die Naturwissenschaften* 8: 461–74. English trans. "Philosophical Reflections on the Causal Principle" in Mulder and van de Velde-Schlick (1979, 295–321).

1921. "Kritizistische oder empiristische Deutung der neuen Physik." *Kant-Studien* 26: 96–111.

1925. *Allgemeine Erkenntnislehre.* 2nd ed. of Schlick (1918) Berlin: Springer. English trans. *General Theory of Knowledge.* Chicago and La Salle, IL: Open Court, 1985.

1935. "Sind die Naturgesetze Konventionen?" In *Actes du Congrès International de Philosophie Scientifique, Paris 1935.* Vol. 4. *Induction et Probabilité.* Paris: Hermann 1936, 8–17.

Scholz, Erhard. ed. 2001. *Hermann Weyl's Raum-Zeit-Materie and a General Introduction to His Scientific Work.* Basel: Birkhäuser.

2004. "The Changing Concept of Matter in H. Weyl's Thought, 1918–1930." In Hendricks et al. (2006, 281–306).

Schrödinger, Erwin. 1918. "Über ein Lösungssystem der allgemein kovarianten Gravitationsgleichungen." *Physikalische Zeitschrift* 19: 20–2.

1922. "Dopplerprinzip und Bohrsche Frequenzbedingung." *Physikalische Zeitschrift* 23: 301–3.

1924. "Bohrs neue Strahlungshypothese und der Energiesatz." *Die Naturwissenschaften* 12: 720–4. Reprinted in Schrödinger (1984, 26–30).

1935. "Die gegenwärtige Situation in der Quantenmechanik." *Die Naturwissenschaften* 23: 807–12; 823–8; 844–9.

1956. *Expanding Universes.* Cambridge: Cambridge University Press.

1984. *Beiträge zur Quantentheorie.* Vienna: Verlag der Österreichischen Akademie der Wissenschaften and Braunschweig: Vieweg.

Schücking, Engelbert, and Surowitz, Eugene J. 2007. "Einstein's Apple: His First Principle of Equivalence," http://arxiv.org/abs/gr-qc/0703149.

Schulmann, Robert. 1993. "Einstein at the Patent Office: Exile, Salvation or Tactical Retreat?" *Science in Context* 6: 17–24.

2000. "Albert Einstein." In *Les Juifs et le XXe siecle: Dictionnaire critique,* ed. Elie Barnavi and Saul Friedlaender. Paris: Calmann-Levy, 570–8.

ed. 2012. *Seelenverwandte. Der Briefwechsel zwischen Albert Einstein und Heinrich Zangger (1910–1947).* Zurich: Verlag Neue Zürcher Zeitung.

Schwarzschild, Karl. 1913. "*Leçons sur les hypothèses cosmogoniques* by H. Poincaré and *Researches on the Evolution of the Stellar Systems* by T. J. J. See." *Astrophysical Journal* 37: 294–8.

1916. "Über das Gravitationsfeld eines Massenpunktes nach der Einsteinschen Theorie." *Königlich Preußische Akademie der Wissenschaften* (Berlin). *Sitzungsberichte:* 189–196.

Schweber, Silvan S. 2008. *Einstein and Oppenheimer: The Meaning of Genius.* Cambridge, MA: Harvard University Press.

Sciama, Dennis. 1957. "Inertia." *Scientific American* 196: 99–109.

Seelig, Carl. 1954. *Albert Einstein: Eine dokumentarische Biographie.* Zurich: Europa Verlag. English translation: *Albert Einstein: A Documentary Biography.* Translated by Mervyn Savill. London: Staples, 1956.

1956. *Helle Zeit – Dunke Zeit: In Memoriam Albert Einstein.* Zürich: Europa Verlag.

1960. *Albert Einstein: Leben und Werk eines Genies unserer Zeit.* Zurich: Europa Verlag.

Seeliger, Hugo von. 1895. "Über das Newton'sche Gravitationsgesetz." *Astronomische Nachrichten* 137: 129–36.

Selety, Franz. 1922. "Beiträge zum kosmologischen Problem." *Annalen der Physik* 72: 281–334.

Serber, Robert, and Crease, Robert P. 1998. *Peace and War: A Life on the Frontiers of Science.* New York: Columbia University Press.

Shankland, Robert S. 1963/1973. "Conversations with Einstein." *American Journal of Physics* 31: 47–57 (1963); 41: 895–901 (1973).

ed. 1973. *Scientific Papers of Arthur Holly Compton: X-Ray and Other Studies.* Chicago: University of Chicago Press.

Shapley, Harlow. 1921. "The Scale of the Universe." *Bulletin of the National Research Council* 2: 171–93.

Slater, John C. 1924. "Radiation and Atoms." *Nature* 113: 307–8.

1975. *Solid-State and Molecular Theory: A Scientific Biography.* New York: Wiley.

Smeenk, Christopher, and Martin, Christopher. 2007. "Mie's Theories of Matter and Gravitation." In Renn (2007a, Vol. 4, 623–32).

Smith, Robert W. 1982. *The Expanding Universe: Astronomy's "Great Debate" 1900–1931.* Cambridge: Cambridge University Press.

Smoluchowski, Marian von. 1906. "Zur kinetischen Theorie der Brownschen Molekularbewegung und der Suspensionen." *Annalen der Physik* 21: 756–80.

Sommerfeld, Arnold. 1909. "Über die Verteilung der Intensität bei der Emission von Röntgenstrahlung." *Physikalische Zeitschrift* 10: 969–70.

1911a. "Das Plancksche Wirkungsquantum und seine allgemeine Bedeutung für die Molekularphysik." *Deutsche Physikalische Gesellschaft. Berichte* 9: 1074–93.

1911b. "Application de la théorie de l'élément d'action aux phénomènes moléculaires non-périodiques." In Langevin and de Broglie (1912, 313–72; "Discussion," 373–92).

1968. *Gesammelte Schriften.* 4 vols. Braunschweig: Vieweg.

2004. *Wissenschaftlicher Briefwechsel*, ed. Michael Eckert and Karl Märker. Diepholz: GNT Verlag.

Specker, Hans E., ed. 1979. *Einstein und Ulm: Festakt und Ausstellung zum 100. Geburtstag von Albert Einstein.* Ulm: W. Kohlhammer.

Stachel, John. 1982. "Einstein and Michelson: the Context of Discovery and the Context of Justification," *Astronomische Nachrichten* 303: 47–53. Reprinted in Stachel (2002a, 177–90).

1986a. "What a Physicist Can Learn from the Discovery of General Relativity." In *Proceedings of the Fourth Marcel Grossmann Meeting on General Relativity*, ed. Remo Ruffini. Amsterdam: North-Holland, 1857–62.

1986b. "Einstein and the Quantum: Fifty Years of Struggle." In *From Quarks to Quasars: Philosophical Problems of Modern Physics*, ed. Robert G. Colodny.

Pittsburgh: University of Pittsburgh Press, 349–85. Reprinted in Stachel (2002a, 367–403).

1986c. "Eddington and Einstein." In *The Prism of Science*, ed. Edna Ullmann-Margalit. Dordrecht: Reidel, 225–50. Reprinted in Stachel (2002a, 453–75).

1987. "Einstein and Ether Drift Experiments." *Physics Today* 40: 45–7. Reprinted in Stachel (2002a, 171–6).

1989a. "The Rigidly Rotating Disk as the 'Missing Link' in the History of General Relativity." In Howard and Stachel (1989, 48–62). Reprinted in Stachel (2002a, 245–61).

1989b. "Einstein's Search for General Covariance, 1912–1915." In Howard and Stachel (1989, 63–100). Reprinted in Stachel (2002a, 301–39).

1993a. "The Meaning of General Covariance: The Hole Story." In *Philosophical Problems of the Internal and External World: Essays on the Philosophy of Adolf Grünbaum*, eds. John Earman, Allen I. Janis, Gerald J. Massey, and Nicholas Rescher. Konstanz: Universitätsverlag; Pittsburgh: University of Pittsburgh Press, 129–60.

1993b. "The Other Einstein: Einstein Contra Field Theory." *Science in Context* 6: 275–90. Reprinted in Stachel (2002a, 141–54).

1994. "Lanczos's Early Contributions to Relativity and His Relationship with Einstein." In *Proceedings of the Cornelius Lanczos International Centenary Conference*, eds. J. David Brown, Moody T. Chu, Donald C. Ellison, and Robert J. Plemmons. Philadelphia: SIAM, 201–21. Reprinted in Stachel (2002a, 499–519).

1996. "Albert Einstein and Mileva Marić. A Collaboration That Failed to Develop." In *Creative Couples in the Sciences*, eds. Helena M. Pycior, Nancy C. Slack, and Pnina G. Abir-Am. New Brunswick, NJ: Rutgers University Press, 207–19.

2002a. *Einstein from "B" to "Z."* Einstein Studies 9. Boston: Birkhäuser.

2002b. "The First Two Acts." In Stachel (2002a, 261–92). Reprinted in Renn (2007a, Vol. 1, 81–111).

2002c. "'The Relations between Things' versus 'The Things between Relations': The Deeper Meaning of the Hole Argument." In Malament (2002, 231–66).

ed. 2005. *Einstein's Miraculous Year: Five Papers That Changed the Face of Physics*. Princeton, NJ: Princeton University Press.

2007. "The Story of Newstein or Is Gravity Just Another Pretty Force?" In Renn (2007a, Vol. 4, 1039–976).

2013. "The Hole Argument." *Living Reviews in Relativity*, http://www.livingreviews.org.

Stadler, Friedrich. 1982. *Vom Positivismus zur "Wissenschaftlichen Weltauffassung."* Vienna: Löcker.

1997. *Studien zum Wiener Kreis: Ursprung, Entwicklung, und Wirkung des logischen Empirismus im Kontext*. Frankfurt am Main: Suhrkamp.

Staley, Richard. 2008. *Einstein's Generation: The Origins of the Relativity Revolution*. Chicago: University of Chicago Press.

Stark, Johannes. 1909. "Zur experimentellen Entscheidung zwischen Ätherwellen- und Lichtquantenhypothese. I. Röntgenstrahlung." *Physikalische Zeitschrift* 10: 902–13.

1910a. "Zur experimentellen Entscheidung zwischen der Lichtquantenhypothese und der Ätherimpulstheorie." *Physikalische Zeitschrift* 11: 24–31.

1910b. "Zur experimentellen Entscheidung zwischen Lichtquantenhypothese und Ätherwellentheorie II. Sichtbares und ultraviolettes Spektrum." *Physikalische Zeitschrift* 11: 179–93.

Stein, Howard. 1977. "Some Philosophical Prehistory of General Relativity." In *Foundations of Space-Time Theories*, eds. Clark Glymour, John Earman, and John Stachel. Minneapolis: University of Minnesota Press, 3–49.

Stöltzner, Michael. 2002. "Franz Serafin Exner's Indeterminist Theory of Culture." *Physics in Perspective* 4: 247–319.

2003. *Vienna Indeterminism: Causality, Realism, and the Two Strands of Boltzmann's Legacy (1896–1936)*. PhD diss. University of Bielefeld.

Stone, A. Douglas. 2013. *Einstein and the Quantum: The Quest of the Valiant Swabian*. Princeton, NJ: Princeton University Press.

Straus, David F. 1835–6. *Das Leben Jesu*. 2 vols. Tübingen: C. F. Osiander.

Study, Eduard. 1914. *Die realistische Weltansicht und die Lehre vom Raume*. Braunschweig: Vieweg.

Stuewer, Roger. 1971. "Hertz's Discovery of the Photoelectric Effect." *Proceedings of the XIIIth International Congress in the History of Science, Section VI*. Moscow, 35–43.

1975. *The Compton Effect: Turning Point in Physics*. Canton, MA: Science History Publications.

1999. *History as Myth and Muse*. Amsterdam: University of Amsterdam Press.

2000. "The Compton Effect: Transition to Quantum Mechanics." *Annalen der Physik* 9: 975–89. German trans. in Müller-Krumbhaar and Wagner (2001, 538–40).

Swenson, Loyd S. 1972. *The Ethereal Aether: A History of the Michelson-Morley-Miller Aether-Drift Experiments, 1880–1930*. Austin: University of Texas Press.

Synge, John L. 1960. *Relativity: The General Theory*. Amsterdam: North-Holland.

Thirring, Hans. 1918. "Über die Wirkung rotierender ferner Massen in der Einsteinschen Gravitationstheorie." *Physikalische Zeitschrift* 19: 33–9.

Thomson, George P. 1964. *J. J. Thomson and the Cavendish Laboratory in His Day*. London: Nelson.

Thomson, Joseph J. 1908. "On the Ionization of Gases by Ultra-Violet Light and on the Evidence as to the Structure of Light Afforded by Its Electrical Effects." *Cambridge Philosophical Society Proceedings* 14: 417–24.

1910a. "On the Theory of Radiation." *Philosophical Magazine* 20: 238–47.

1910b. "Letter to the Editors." *Philosophical Magazine* 20: 544.

1913. "On the Structure of the Atom." *Philosophical Magazine* 26: 792–9.

Tolman, Richard C. 1929a. "On the Possible Line Elements for the Universe." *National Academy of Science. Proceedings* 15: 297–304.

1929b. "On the Astronomical Implications of the de Sitter Line Element for the Universe." *Astrophysical Journal* 69: 245–74.

1934. *Relativity, Thermodynamics, and Cosmology*. Oxford: Oxford University Press. Reprint: New York: Dover, 1987.

Torretti, Roberto. 1978. *Philosophy of Geometry from Riemann to Poincaré*. Dordrecht: Reidel.

1983. *Relativity and Geometry*. Oxford: Pergamon Press.

2000. "Spacetime Models of the World." *Studies in History and Philosophy of Modern Physics* 31: 171–86.

2008. "Objectivity: A Kantian Perspective." In Massimi (2008, 81–94).

Trautman, Andrzej. 1962. "Conservation Laws in General Relativity." In *Gravitation: An Introduction to Current Research*, ed. Louis Witten. New York: Wiley, 169–98.

1965. "Foundations and Current Problems of General Relativity." In *Lectures on General Relativity*, eds. A. Trautman, F. A. E. Pirani, and H. Bondi. Englewood Cliffs, NJ: Prentice Hall, 1–248.

Trbuhović-Gjurić, Desanka. 1993. *Im Schatten Albert Einsteins: Das tragische Leben der Mileva Einstein-Marić*. 5th rev. ed. Berne: Paul Haupt.

Trumpler, Robert J. 1930. "Absorption of Light in the Galactic System." *Astronomical Society of the Pacific. Publications* 42: 214–27.

Uffink, Jos. 2006. "Insuperable Difficulties: Einstein's Statistical Road to Molecular Physics." *Studies in History and Philosophy of Modern Physics* 37: 36–70.

van der Waerden, Bartel L., ed. 1967. *Sources of Quantum Mechanics*. Amsterdam: North-Holland.

van Dongen, Jeroen. 2002a. "Einstein and the Kaluza-Klein Particle." *Studies in History and Philosophy of Modern Physics* 33: 185–210.

2002b. *Einstein's Unification: General Relativity and the Quest for Mathematical Naturalness*. PhD diss. University of Amsterdam.

2004. "Einstein's Methodology, Semivectors and the Unification of Electrons and Protons." *Archive for History of Exact Sciences* 58: 219–54.

2007a. "Emil Rupp, Albert Einstein and the Canal Ray Experiments on Wave-Particle Duality: Scientific Fraud and Theoretical Bias." *Historical Studies in the Physical and Biological Sciences: Supplement* 37: 73–120.

2007b. "The Interpretation of the Einstein-Rupp Experiments and Their Influence on the History of Quantum Mechanics." *Historical Studies in the Physical and Biological Sciences: Supplement* 37: 121–31.

2009. "On the Role of the Michelson-Morley Experiment: Einstein in Chicago." *Archive for History of Exact Sciences* 63: 655–63.

2010. *Einstein's Unification*. Cambridge: Cambridge University Press.

2012. "Mistaken Identity and Mirror Images: Albert Einstein and Carl Einstein, Leiden and Berlin, Relativity and Revolution." *Physics in Perspective* 14: 126–77.

Violle, Jules. 1892. *Lehrbuch der Physik*. Part 1, Vol. 1, *Mechanik und Mechanik der festen Körper*. Berlin: Springer.

1893. *Lehrbuch der Physik*. Part 1, Vol. 2, *Mechanik der flüssigen und gasförmigen Körper*. Berlin: Springer.

Vizgin, Vladimir P. 1994. *Unified Field Theories in the First Third of the 20th Century*. Boston: Birkhäuser.

Walker, A. G. 1935. "On the Formal Comparison of Milne's Kinematical System with the Systems of General Relativity." *Royal Astronomical Society. Monthly Notices* 95: 263–9.

1936. "On Milne's Theory of World Structure." *Proceedings of the London Mathematical Society*, Series 2, 42: 90–127.

Walter, Scott. 1999. "Minkowski, Mathematicians, and the Mathematical Theory of Relativity." In Goenner et al. (1999, 45–86).

Wazeck, Milena. 2009. *Einsteins Gegner: Die öffentliche Kontroverse um die Relativitätstheorie in den 1920er Jahren*. Frankfurt and New York: Campus. English translation: *Einstein's Opponents: The Public Controversy about the Theory of Relativity in the 1920s*. Cambridge and New York: Cambridge University Press, 2013.

2013. "Marginalization Processes in Science: The Controversy about the Theory of Relativity in the 1920s." *Social Studies of Science* 43: 163–90.

Weinberg, Steven. 2005. "Einstein's Mistakes." *Physics Today* 58: 31–5.

Weyl, Hermann. 1918a. "Gravitation und Elektrizität." *Königlich Preußische Akademie der Wissenschaften* (Berlin). *Sitzungsberichte*: 465–80.

1918b. *Raum–Zeit–Materie: Vorlesungen über allgemeine Relativitätstheorie.* Berlin: Springer.

1922. *Space–Time–Matter.* London: Methuen. English trans. of 4th ed. of Weyl (1918b).

1924. "Massenträgheit und Kosmos: Ein Dialog." *Die Naturwissenschaften* 12: 197–204.

1927. *Philosophie der Mathematik und Naturwissenschaft.* München and Berlin: Oldenbourg. Trans. in rev. and augmented English ed. as *Philosophy of Mathematics and Natural Science.* Princeton, NJ: Princeton University Press, 1949.

Wheaton, Bruce. 1983. *The Tiger and the Shark: Empirical Roots of Wave-Particle Dualism.* Cambridge: Cambridge University Press.

Whittaker, Edmund T. 1951–3. *A History of the Theories of Aether and Electricity.* London and New York: T. Nelson.

Wien, Wilhelm. 1898. "Über die Fragen, welche die translatorische Bewegung des Lichtäthers betreffen." *Annalen der Physik und Chemie* 65.3 (Beilage): i–xviii.

Wigner, Eugene. 1980. "Thirty Years of Knowing Einstein." In Woolf (1980, 461–8).

Wilson, Mark. 1993. "There's a Hole and a Bucket, Dear Leibniz." In *Philosophy of Science*, eds. Peter A. French, Theodore E. Uehling, and Howard K. Wettstein. Notre Dame, IN: University of Notre Dame Press, 202–41.

Winteler-Einstein, Maja. 1987. "Albert Einstein: Beitrag für sein Lebensbild." In CPAE 1, xlviii–lxvi.

Winternitz, Josef. 1923. *Relativitätstheorie und Erkenntnislehre.* Leipzig and Berlin: Teubner.

Woolf, Harry, ed. 1980. *Some Strangeness in the Proportion: A Centennial Symposium to Celebrate the Achievements of Albert Einstein.* Reading, MA: Addison-Wesley.

Yavelov, Boris. 2002. "Einstein's Zurich Colloquium." In Balashov and Vizgin (2002, 261–96).

Yourgrau, Palle. 2005. *A World without Time: The Forgotten Legacy of Gödel and Einstein.* New York: Basic Books.

Zhai, Zhenming. 1990. "The Problem of Protocol Statements and Schlick's Concept of 'Konstatierungen.'" *PSA 1990: Proceedings of the 1990 Biennial Meeting of the Philosophy of Science Association*, eds. Linda Wessels, Arthur Fine, and Micky Forbes. East Lansing, MI: Philosophy of Science Association, 15–23.

Zuelzer, Wolf. 1982. *The Nicolai Case.* Detroit: Wayne State University Press.

Index